# Lecture Notes in Computer Science 1338

Edited by G. Goos, J. Hartmanis and J. van Leeuwen

Advisory Board: W. Brauer    D. Gries    J. Stoer

Springer
Berlin
Heidelberg
New York
Barcelona
Budapest
Hong Kong
London
Milan
Paris
Santa Clara
Singapore
Tokyo

František Plášil   Keith G. Jeffery  (Eds.)

# SOFSEM'97: Theory and Practice of Informatics

24th Seminar on Current Trends
in Theory and Practice of Informatics
Milovy, Czech Republic, November 22-29, 1997
Proceedings

 Springer

Series Editors

Gerhard Goos, Karlsruhe University, Germany

Juris Hartmanis, Cornell University, NY, USA

Jan van Leeuwen, Utrecht University, The Netherlands

Volume Editors

František Plášil
Charles University, Department of Software Engineering
Malostranské nám. 25, 118 00 Prague, Czech Republic
E-mail: plasil@nenya.ms.mff.cuni.cz

Keith G. Jeffery
CLRC Rutherford Appleton Laboratory
Chilton, Didcot, OX11 0QX Oxfordshire, UK
E-mail: kgj@inf.rl.ac.uk

Cataloging-in-Publication data applied for

**Die Deutsche Bibliothek - CIP-Einheitsaufnahme**

**Theory and practice of informatics ; proceedings** / SOFSEM '97,
24th Seminar on Current Trends in Theory and Practice of
Informatics, Milovy, Czech Republic, November 22 - 29, 1997 /
František Plášil ; Keith G. Jeffery (ed.). - Berlin ; Heidelberg ; New
York ; Barcelona ; Budapest ; Hong Kong ; London ; Milan ; Paris ;
Santa Clara ; Singapore ; Tokyo : Springer, 1997
  (Lecture notes in computer science ; Vol. 1338)
  ISBN 3-540-63774-5


CR Subject Classification (1991): D, F, H.1-3, H.5, I.2-3, G.2

ISSN 0302-9743
ISBN 3-540-63774-5 Springer-Verlag Berlin Heidelberg New York

Typesetting: Camera-ready by author
SPIN 10647901    06/3142 – 5 4 3 2 1 0    Printed on acid-free paper

# Foreword

The SOFSEM (SOFtware SEMinar) is now being held for the 24th time. Having been transformed during the years from a local event to a fully international conference, the current SOFSEM is special in being a mix of a winter school, a conference, and an advanced workshop, each striving for multidisciplinarity in computer science. This aim is reflected in the technical program consisting of a relatively large number of invited talks, refereed papers (known as contributed papers), and refereed poster contributions. In addition, time and space for flash communications, industrial presentations, and exhibitions are provided. The program of SOFSEM usually starts with an opening talk delivered by a distinguished member of the computer science community; this year we are honored that Jan van Leeuwen has agreed to present his keynote address *Challenges in Large-Scale Distributed Systems*.

This volume constitutes the proceedings of SOFSEM'97 held in Milovy, Czech Republic, November 22–29, 1997. This year, 63 papers were submitted from 11 European countries. The selection of the 24 best papers was carried out during a one-day Program Committee (PC) meeting held in Brno. The referees had previously judged all papers according to their contribution to the state of the art, technical soundness, clarity of presentation, and adequacies of the length and bibliography. The Endowment Board (EB) supported by the Advisory Board recommended 22 invited talks focused on the following key topic areas: *Distributed and Parallel Systems, Software Engineering and Methodology, Databases and Information Systems*, and *Fundamentals*.

SOFSEM is the result of considerable effort by a number of people. It is my pleasure to record my thanks to the Advisory Board for their support, to the Endowment Board for their general guidance and enormous effort in finding excellent speakers for the invited talks, and to the Organizing Committee (OC) for making SOFSEM'97 happen. It has been an honor for me to work with the members of the Program Committee and other referees who devoted a lot of effort to reviewing the submitted papers.

My special thanks go to Keith Jeffery, the PC vice-chair, for his extremely cooperative and generous overall support and for sharing with me all his experience from serving as the PC Chair in 1996. Also, I would like to record a special credit to Mirek Bartošek for technical preparation of the proceedings. On a personal note, I am particularly grateful to Jiří Sochor, the PC secretary, who has done a tremendous job in keeping track of all the hundreds PC agenda related issues, among which, his perfect preparation of the PC meeting deserves to be especially recognized. Also, I would like to acknowledge the excellent cooperation I received from Jiří Wiedermann as EB Chair, and Jan Staudek as OC Chair.

I am very grateful to the editors of the LNCS series for their continuing trust in SOFSEM and to Springer-Verlag for publishing the proceedings. In addition, I appreciate highly the financial support of our sponsors which assisted with the invited speakers and advanced students.

Ultimately, the success of SOFSEM will be determined not only by the speakers and the committee members as the core of conference, but generally by all the attendees, who will contribute to the spirit of the conference. So I am very grateful to all of my professional colleagues who have done, and will do, their best to make SOFSEM a high-quality event. Finally, I would like to express my appreciation for having the opportunity to serve as the Program Committee Chair of SOFSEM'97 and welcome you to this event.

September 1997                                František Plášil
                                 SOFSEM'97 Program Committee Chair

# SOFSEM '97 :≡

## Advisory Board

Dines Bjørner — Technical University of Denmark, Lyngby, Denmark
Peter van Emde Boas — University of Amsterdam, The Netherlands
Manfred Broy — Technical University Munich, Germany
Michal Chytil — Anima Praha, s.r.o., Prague, Czech Republic
Georg Gottlob — Vienna University of Technology, Austria
Keith G. Jeffery — CLRC Rutherford Appleton Laboratory, Oxon, UK
Maria Zemánková — NSF, Washington DC, US

## Endowment Board

Jiří Wiedermann, *chair* — Academy of Sciences of the Czech Republic, Prague, CZ
Bronislav Rovan, *vice-chair* — Comenius University, Bratislava, SK
Keith G. Jeffery — CLRC Rutherford Appleton Laboratory, Oxon, UK
Jan Pavelka — DCIT Prague, CZ
František Plášil — Charles University, Prague, CZ
Igor Prívara — Institute of Informatics and Statistics, Bratislava, SK
Jan Staudek — Masaryk University, Brno, CZ

## Program Committee

František Plášil, *chair* — Charles University, Prague, CZ
Keith G. Jeffery, *vice-chair* — CLRC RAL, Oxon, UK
Jiří Sochor, *secretary* — Masaryk University, Brno, CZ
Patrizia Asirelli — I.E.I.–E.N.R., Pisa, IT
Robert G. Babb II — University of Denver, US
Michel Banatre — IRISA/INRIA, Rennes, FR
Guy Bernard — INT, Evry, FR
Viliam Geffert — UPJŠ Košice, SK
Jane Grimson — Trinity College, Dublin, IR
Eduard Gröller — Vienna University of Technology, Vienna, AT
Petr Jančar — Ostrava University, CZ
Martin Kersten — CWI, Amsterdam, NL
Petr Kroha — TU Chemnitz, DE
Antonín Kučera — Charles University, Prague, CZ
Bořivoj Melichar — Czech Technical University, Prague, CZ
Hanspeter Mössenböck — University of Linz, AT
Lenka Motyčková — Masaryk University, Brno, CZ
Mogen Nielsen — Aarhus University, DK
Václav Rajlich — Wayne State University, Detroit, US

| | |
|---|---|
| Peter Ružička | Comenius University, Bratislava, SK |
| Anton Scheber | SOFTEC, Bratislava, SK |
| Santosh Shrivastava | University of Newcastle, UK |
| Hava Siegelmann | Technion, Haifa, IL |
| Jiří Šíma | Academy of Sciences of the Czech Republic, Prague, CZ |
| Arne Sølvberg | The Norwegian University of Sci. & Techn., Trondheim, NO |
| Gerard Tel | Utrecht University, NL |
| Volker Tschammer | GMD Fokus, Berlin, DE |
| Krzysztof Zielinski | University of Mining & Metallurgy, Krakow, PL |

## List of Referees

| | |
|---|---|
| Patrizia Asirelli | Jaroslav Nešetřil |
| Robert G. Babb II | Mogen Nielsen |
| Guy Bernard | Lukáš Petrlík |
| Václav Dvořák | František Plášil |
| Viliam Geffert | Václav Rajlich |
| Jane Grimson | Karel Richta |
| Eduard Gröller | Peter Ružička |
| Václav Hlaváč | Anton Scheber |
| Jan Hlavička | Santosh Shrivastava |
| Petr Jančar | Hava Siegelmann |
| Keith G. Jeffery | Jiří Šíma |
| Martin Kersten | Jiří Sochor |
| Petr Kroha | Arne Sølvberg |
| Antonín Kučera | Gerard Tel |
| Bořivoj Melichar | Volker Tschammer |
| Francois Meunier | Jiří Wiedermann |
| Hanspeter Mössenböck | Krzysztof Zielinski |
| Lenka Motyčková | |

# SOFSEM '97

## Organized by

Czech Society for Computer Science
Slovak Society for Computer Science
Czech ACM Chapter
Czech Research Consortium for Informatics and Mathematics

## In cooperation with

Faculty of Informatics, Masaryk University, Brno
Institute of Computer Science, Masaryk University, Brno
Institute of Computer Science, Academy of Sciences of the Czech Republic, Prague
Department of Software Engineering, Charles University, Prague
Department of Computer Science, Comenius University, Bratislava
CLRC Rutherford Appleton Laboratory, Oxon, UK

## Sponsored by

ApS Brno s.r.o.
Digital Equipment s.r.o.
European Research Consortium for Informatics and Mathematics
Help Service s.r.o.
Hewlett Packard s.r.o.
IBM Czech Republic s.r.o.
Oracle Czech s.r.o.

## Organizing Committee

| | | |
|---|---|---|
| Jan Staudek, *chair* | Petr Hanáček | Petr Přikryl |
| Miroslav Bartošek, *vice-chair* | Zdeněk Malčík | Petr Sojka |
| Zdena Walletzká, *secretary* | Tomáš Pitner | Tomáš Staudek |

# Contents

## Invited Papers

# Contributed Papers

# Computer Chess: Algorithms and Heuristics for a Deep Look into the Future *

Rainer Feldmann

University of Paderborn, Germany

**Abstract.** In this paper we will describe some of the basic techniques that allow computers to play chess like human grandmasters. In the first part we will give an overview about the sequential algorithms used. In the second part we will describe the parallelization that has been developed by us. The resulting parallel search algorithm has been used successfully in the chess program ZUGZWANG even on massively parallel hardware. In 1992 ZUGZWANG became Vize World Champion at the Computer Chess Championships in Madrid, Spain, running on 1024 processors. Moreover, the parallelization proves to be flexible enough to be applied successfully to the new ZUGZWANG program, although the new program uses a different sequential search algorithm and runs on a completely different hardware.

## 1 Introduction

The game of chess is one of the most fascinating two-person zero-sum games with complete information. Besides of being one of the oldest games of this kind it is still played by millions of people all over the world. Moreover, during the last years, chess had been put to a worldwide attention due to the interest of the people in superstars like International Grandmaster Kasparov and due to the fact that computers now start to challenge the worlds best chess players [41].

The playing strength of the top human chess players is due to their strategic planning ability, lookahead, intuition and creativity. The playing strength of the computers, however, is mainly based on their speed. It will be interesting to watch the contest between humans and computers in the future.

On the other side, strategic games like chess are considered to be excellent test environments for planning devices [39]. It is great evidence that algorithms, that enable a computer to play chess successfully, can be of use in other domains where strategic planning is required, e.g. in expert systems, motion planning in dynamic environments, or business planning [61].

The modern history of computer chess started with the work of Shannon [56] and Turing [58]. In 1974 a first Computer Chess World Championship was held in Stockholm seeing a win of the soviet program KAISSA. Software as well as hardware improvements significantly increased the playing strength of the

---

* This work was partly supported by the Leibniz award fund from the DFG under grant Mo 476/99 and by the DFG research project "Selektive Suchverfahren" under grant Mo 285/12-2

computers. Most of the references given at the end of the paper are offsprings of this research in computer chess. Levy and Newborn [40, p 5] mention that the playing strength increased from 1640 ELO points [17] in 1967 to more than 2500 points nowadays. For a detailed description of the history of computer chess we refer to [40].

In this paper we will give an overview about the most important lookahead techniques used in modern chess programs. In the first part we will describe some sequential search algorithms, that enable computers to choose their move based upon a deep look into the future. We will describe the standard enhancements used to speed up the search. Then we will present two selective search algorithms allowing even further lookahead along promising lines of play at the risk of overlooking chances or threats on other lines. The first one is the Null Move Search Algorithm [8, 29, 16] that is widely used in state-of-the-art chess programs today. The second one, the so called Fail High Reduction Algorithm, is an alternative for programs using a sophisticated static evaluation function [25]. We developed it quite recently and showed its superiority to the standard algorithm in a sequence of tests. It is now used in ZUGZWANG and at least one commercially available chess program.

In the second part we will describe a parallel approach to the search algorithm allowing an even deeper lookahead. The resulting distributed algorithm has been used successfully in the chess program ZUGZWANG even on massively parallel hardware [20, 21, 22, 23, 24]. In 1992 ZUGZWANG became Vize World Champion at the Computer Chess Championships in Madrid, Spain, running on 1024 T805 processors. Moreover, the parallelization proves to be flexible enough to be applied successfully to the new ZUGZWANG program, although the new program uses a different sequential search algorithm and runs on a completely different hardware. This new program recently played successfully at the 12th AEGON Man vs. Machine tournament in Den Haag, The Netherland, running on 40 M604 processors. Our parallelization is the first one that has been applied efficiently to massively parallel hardware.

## 1.1 The Problem

Informally spoken, given a chess position $S$ we are interested in the best move for the side to move in $S$. Zermelo [66] formalized the notion of a theoretical win, loss or draw in two-person zero-sum games. From this the following recursive minmax function $\mathcal{F}$ can be deviated: For a terminal position $t$

$$\mathcal{F}(t) := \begin{cases} 1, & \text{if } t \text{ is won for the side to move in } t \\ 0, & \text{if } t \text{ is a draw} \\ -1, & \text{if } t \text{ is lost for the side to move in } t \end{cases}$$

The value of a nonterminal position $v$ with children $v_1, \ldots v_w$ is then given as

$$\mathcal{F}(v) := \max\{-\mathcal{F}(v_1), \ldots, -\mathcal{F}(v_w)\}.$$

If the game does not contain infinitely long sequences of moves, $\mathcal{F}(v)$ can be computed for any position $v$. This is done by building a search tree with root

$v$ and the leaves being the terminal positions reachable from $v$. $\mathcal{F}$ is applied to the leaves and then backed up to the root. A best move in $v$ is then obtained by choosing an edge to one of the children $v_i$ with $\mathcal{F}(v_i) = \mathcal{F}(v)$.

Unfortunately, due to time constraints, in strategic games it is not possible to generate the complete search tree, i.e. the tree cannot be extended to the terminal positions. Therefore, instead of $\mathcal{F}$, a static evaluation function $f$ is computed mapping positions to numbers. $f(v)$ is a heuristic estimation of the chances to win for the side to move in $v$. $f$ is then used as an approximation of $\mathcal{F}$. A simple recursive algorithm, cutting the search tree at depth $d$ and applying $f$ at the leaves is used to compute the minmax value $\mathcal{F}_{f,d}$ of the root. Then a move is chosen as if $\mathcal{F}$ was computed. We can now define the problem somewhat more precisely as

Given a position $v$, a search depth $d$ and a static evaluation function $f$. Compute $\mathcal{F}_{f,d}(v)$.

Although it can be shown in theoretical game tree models that searching deeper may result in an even worse decision at the root of the tree [3, 52, 57], this has barely been observed in practice. For the game of chess an increase of 100-250 ELO points has been observed by searching one ply deeper [59, 60, 48]. Therefore it is essential to search as deep as possible while paying attention to the time constraints posed to the program by the chess clocks.

## 2 Sequential Game Tree Search

The recursive algorithm to compute the minmax value of the root visits every leaf of the search tree. But often one can do better as is shown in the figure below.

Suppose, some game tree search algorithm already computed the value $-5$ for the leftmost child of the root. Then the algorithm starts its search below the second child. After it sees the value 5 a cutoff can be performed pruning the rest of the subtree, since the second child can never improve on the value of the first child.

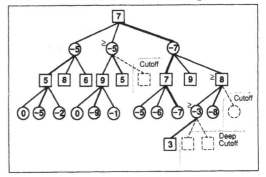

The situation is similar below the node marked $-3$ : If the third child of the root will improve on the result found below the first one, this cannot happen by choosing the move to $-3$ at the node marked 8, since the result at the root would be $\leq 3$. A deep cutoff occurs. The rightmost child of the node marked 8 can be pruned by a cutoff because of the values 7 and $-8$ already computed. Note that instead of an exact minmax value only lower bounds are computed for the nodes where a cutoff occurs. An upper bound is computed for a node if for all its children a lower bound has been computed.

## 2.1 $\alpha\beta$-Search and $\alpha\beta$-Enhancements

The $\alpha\beta$-algorithm shown to the right uses the cut-off technique to speed up the search. Knuth and Moore [36] analyzed the $\alpha\beta$-algorithm started with alphaBeta($root, -\infty, \infty$) and showed that in a uniform tree of width $w$ and depth $d$ it visits $\lceil w^{d/2} \rceil + \lfloor w^{d/2} \rfloor$-1 leaves in the best case but all $w^d$ leaves in the worst case.

```
int alphaBeta(node v, int α, β) {
    int val;
    generate all successors v₁, ..., vw of v;
    if v is a leaf return(f(v));
    val := -∞;
    for i := 1 to w do {
        val := max(val,-alphaBeta(vi, -β, -α));
        if val ≥ β return(val); /* cutoff */
        α := max(α, val);
    }
    return(val);
}
```

Moreover, they showed that the best case is achieved, if the best move is considered first at any inner node of the tree. From then on, research in the field of game tree search algorithms concentrated on the question on how to improve on the move ordering to get full use of the efficiency of the $\alpha\beta$-algorithm. In the following we will describe some of the enhancements.

**The Transposition Table** The $\alpha\beta$-Algorithm is a tree searching algorithm. However, due to transpositions of moves, the same position may be reached at different nodes in the tree. In this case the same subtree would have to be searched twice. Since in general it is not possible to store the whole search tree in the main memory, a hash table is used to store the results already known. An entry of the hash table consists of the components: ($lock, d, x, flag, move$). The lock typically consists of 64 bits. It can be computed efficiently in the same way as the hash function used to access the hash table [67], and in practice suffices to identify the chess position uniquely [64]. The component $d$ gives the search depth, $x$ is the value computed. $flag$ indicates whether $x$ is an upper bound, a lower bound or an exact value, and $move$ is the best move obtained.

When a subtree below a chess position $v$ has been evaluated with search window $[\alpha, \beta]$ and a result $x$ has been computed, an entry is written to the hash table. $x$ is an upper bound, iff $x \leq \alpha$, a lower bound iff $x \geq \beta$ and an exact value iff $\alpha < x < \beta$.

Before the search starts at a chess position $v$, a hash table lookup is done for $v$. If there is an entry ($lock, d, x, flag, move$) for $v$ and $d$ is large enough, the search may be speeded up by immediately returning $x$, if $x$ is an exact value, increasing $\alpha$ if $x$ is a lower bound, or decreasing $\beta$ if $x$ is an upper bound. If $d$ is not large enough the move from the hash table is tried first in the search, since there is good hope that the move found as a best one in a shallow search remains the best one in the deeper search.

For a detailed description of the use of transposition tables see [44]. Collision resolution strategies have been investigated in [13, 10].

**Other Enhancements** Several other $\alpha\beta$-enhancements are widely used:

The *iterative deepening* technique is common in any chess program. Instead of searching the root of the tree directly to a search depth $d$, it is searched successively to search depths $1, 2, \ldots, d$. Many of the results of the shallow searches are stored in the transposition table and help to improve the move ordering for the deeper searches. Moreover, in real time applications like chess, the iterative search guarantees, that there is a best move available even after completion of the 1-ply search. Thus, the process may be stopped at any time and the best move from the last iteration may be delivered as the move to be played.

The *aspiration search* starts the $\alpha\beta$-search at the root by the function call alphaBeta($root, \alpha_0, \beta_0$) for some $-\infty < \alpha_0 < \beta_0 < \infty$. If the search terminates with a result $\alpha_0 < x < \beta_0$, $x$ is correctly determined and some effort may have been saved by the use of the narrow aspiration window $[\alpha_0, \beta_0]$. Otherwise, if $x \leq \alpha_0$ or $x \geq \beta_0$, a re-search with the full window $[-\infty, x]$, or $[x, \infty]$ resp., has to be performed. $\alpha_0$ and $\beta_0$ are usually estimated using the result of the preceding iteration.

Many heuristics to improve the move ordering have been developed. The *killer heuristic* stores a number of moves that led to cutoffs in subtrees already searched [1]. Moves that produce a cutoff in level $d$ are stored in a list for level $d$. These lists usually have only two entries. A benefit is stored for each move and new moves overwrite the one with the smaller benefit. Moves from these killer lists are tried early in the search, if they are found to be legal moves.

The *history heuristic* [54] increments a bonus in a table for moves, that become best or lead to cutoffs in subtrees previously searched. These bonuses are then used to sort the moves at a chess position, if no other information is available.

Bad moves are often refuted by the same good move, no matter where they are applied in the tree. The *countermove heuristic* stores such refuting moves and provides them for early use in the search [62].

The total move ordering then is as follows: First the move from the transposition table is searched. Afterwards positive captures, i.e. captures that look at least equal at first glance, are tried in decreasing order of their potential benefit. If a countermove is available it is tried after the positive captures. Then, the moves from the killer list are tried in decreasing order of their benefit. Afterwards all the remaining moves are searched in decreasing order of their history values.

## 2.2 The Negascout Algorithm

Using the above described heuristics the move ordering usually is very good. Therefore, the first move considered in the search is very often found to be the best move, or at least suffices to produce a cutoff. This fact is used by the Negascout algorithm of Reinefeld [53] shown in figure 1.

The main idea of the algorithm is that it computes the first successor of a position just as the $\alpha\beta$-algorithm would do, but then tries to show that the other successors are inferior to the first one. This is done by searching them with an

```
int Negascout(node v, int α, β, d) {
    int x, low, val, high;
    if d > 0 generate all successors v₁, ..., v_w of v;
    if v is a leaf return(f(v));                          /* static evaluation */
    low := α; high := β; val := −∞;
    for i := 1 to w do {
        x := −Negascout(vᵢ, −high, −low, d − 1);         /* i > 1 : null window */
        if x > low and x < β and i > 1
            x := −Negascout(vᵢ, −β, −x, d − 1);          /* re-search */
        val := max(val, x);
        if val ≥ β return(val);                          /* cutoff */
        low := max(low, x); high := low + 1;
    }
    return(val);
}
```

**Fig. 1.** The Negascout algorithm

artificial search window $[-high, -low]$ of width zero, i.e. $high - low = 1$. Only if this search shows that the corresponding node is superior to the best one found so far, a re-search is done with the full window $[-\beta, -x]$. The Negascout algorithm has the same best case behaviour as the $\alpha\beta$-algorithm, i.e. in a uniform tree of width $w$ and depth $d$ it searches exactly $\lceil w^{d/2} \rceil + \lfloor w^{d/2} \rfloor - 1$ leaves. In the worst case all $w^d$ leaves have to be visited, many of them even twice. In practice the Negascout algorithm appears to be faster than the pure $\alpha\beta$-algorithm [53], and the average running time is very close to the best case running time due to the use of the $\alpha\beta$-enhancements . But the depth of the search is still limited by the exponential running time of the algorithm.

Therefore, many heuristics have been developed to guide the lookahead of the game tree search algorithms into those branches of the tree, which are relevant for the decision at the root, in order to search them deeper than the irrelevant lines. Many of these heuristics are domain dependent, like e.g. check evasion extensions. An overview and an empirical comparison is given in [6]. There has also been some research on developing domain independent heuristics to guide the search in game trees [11, 47, 5, 8, 16, 15, 42, 25]. The effects of some combinations are studied in [9]. The selective search heuristic most widely used today is described in the next section.

## 2.3 Null Move Search

In the late eighties Beal [8] again looked at his idea of the Null Move Heuristic. It is based upon the observation that in many games and game like applications for the side to move there is almost always a better alternative than doing nothing

```
const maxGain = 1.5;      /* pawn units */
const nmReduce = 2;       /* depth reduction for null move searches */

int NmNegascout(node v, int α, β, d, bool nullmove) {
    int x, y, low, val, high;
    if d > 0 generate all successors v₁, ..., v_w of v;
    if v is a leaf return(f(v));                        /* static evaluation */

    /* null move search */
    if (nullmove and d > 1 and x > α−maxGain and not in check) {
        v' := switchSide(v);
        y := −NmNegascout(v',−β,−α,d − nmReduce − 1,false);
        if y ≥ β return(y);                             /* null move cutoff */
        α := max(α,y);
    }

    /* regular Negascout algorithm */
    low := α; high := β; val := −∞;
    for i := 1 to w do {
        x := −NmNegascout(vᵢ, −high, −low, d − 1, true);
        if x > low and x < β and i > 1
            x := −NmNegascout(vᵢ, −β, −x, d − 1, true);   /* re-search at vᵢ */
        val := max(val, x);
        if val ≥ β return(val);                         /* cutoff */
        low := max(low, x); high := low + 1;
    };
    return(val);
}
```

**Fig. 2.** The Negascout Algorithm with Null Move Searches

(the null move). We call this the Null Move Observation (NMO). Beal proposed to use cutoffs found by shallow searches after a null move and thus avoid to compute the cutoffs in deep searches after real moves. A first implementation is presented in [29]. Donninger [16] proposed a null move search algorithm similar to the one shown in figure 2.

The main drawback of the Null Move Search is that it fails in zugzwang positions. In these positions the NMO is not valid, causing chess programs to fail heavily. Therefore, in endgames, chess programs usually switch off the Null Move Search or try to verify the NMO by some extra searches. Another drawback of the Null Move Search is that extra searches are carried out in order to establish the cutoffs by the null move. If the search after a null move fails to produce a cutoff, extra effort has been spent for nothing. It has been observed by many programmers that the Null Move Search improves the computers play in tacti-

```
int FHRNegascout(node v, int α, β, d) {
    int x, low, val, high, eval, δ, t;
    eval := f(v);                                    /* static evaluation */
    δ := d;                                          /* save search depth */
    if (eval ≥ β and α = β − 1) δ := δ − 1;
    if δ > 0 generate all successors v₁, ..., vw of v;
    if v is a leaf return(eval);
    low := α; high := β; val := −∞;
    for i := 1 to w do {
        x := −FHRNegascout(vᵢ, −high, −low, δ − 1);  /* i > 1 : null window */
        if x > low and x < β and i > 1
            x := −FHRNegascout(vᵢ, −β, −x, d − 1);   /* re-search: no FHR */
        val := max(val, x);
        if val ≥ β return(val);                      /* cutoff */
        low := max(low, x); high := low + 1;
    };
    return(val);
};
```

**Fig. 3.** The FHRNegascout Algorithm

cal positions but sometimes performs poorly in positional ones. These effects, however, are hard to quantify and therefore no publications investigating the behavior of the algorithm are available. Nevertheless, the Null Move Search is used widely by professional as well as amateur chess programmers.

## 2.4 Fail High Reductions

The algorithm presented in this section tries to overcome the drawbacks of the Null Move Search algorithm. Again, it is based upon the NMO. It uses the static evaluation function $f$ to compute a value even for inner nodes of the tree. If this static value indicates, that the side to move is already better than $β$, a cutoff is expected and the remaining search depth is reduced by one. These depth reductions, however, are done only in the searches with an artificial zero width window.

A slightly simplified form of the algorithm is given in figure 3. For details we refer to [25].

The FHR-algorithm is used in ZUGZWANG since 1995, and at least one commercially available chess program [34]. In [25] we showed in three different tests that the FHR-algorithm is superior to the regular $αβ$-algorithm with statistical significance. All three tests indicate that the improvement in playing strength is about 120 ELO points. Recently, Ulf Lorenz implemented both the FHR-algorithm and the Null Move Search in his chess program CONNY. The tests performed with these implementations indicate that the playing strengths of the FHR-algorithm and the Null Move Search are equal [43].

## 3   Parallel Game Tree Search

In this part of the paper we will sketch the concept of our parallelization of the game tree search. A detailed description of an implementation on a network of 1024 T805 processors can be found in [23].

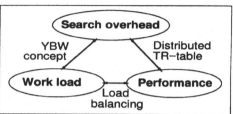

The only hope to get scalable parallelism in a game tree search is to split the game tree and search the resulting subtrees in parallel. For many years it has been an open problem how to parallelize the game tree search and a lot of research has been done in this field [7, 27, 2, 28, 44, 32, 49, 63, 26, 31, 46, 50, 51, 18, 33, 35, 55, 30, 37, 19, 21, 22, 23, 4, 38, 65].

Three difficulties have to be dealt with when searching subtrees in parallel. They are shown in the figure above as a triangle.

- **Processor work load:** Since the game tree search routines use cutoffs during the search, the size of a subtree is not known in advance. Thus, a dynamic load balancing mechanism is required.
- **Search overhead:** Many subtrees are not visited by the sequential search algorithm. The information required to prune a subtree has been found by searching the subtrees to the left. In a parallel system, the information about the left subtrees may not be available in time or may be distributed over the processors. In both cases the parallel algorithm will miss cutoffs and will search more nodes than the sequential would have done. A method to reduce this search overhead is required.
- **Performance:** For the dynamic load balancing and the information sharing, the processors have to communicate. In a message passing system like the ones used, this is done by exchanging messages. The sending and receiving of messages costs CPU cycles, and therefore processors work at a decreased performance in a parallel environment.

The edges between the nodes in the triangle above indicate the tradeoffs between the three losses:

- Increasing the number of messages for the dynamic load balancing may improve the average work load but at the costs of a decrease of the performance. Reducing the numbers of messages for the load balancing will result in a decrease of the average work load.
- Increasing the number of messages for information sharing may improve the efficiency of the parallel search but again at the costs of a decrease of the performance. On the other hand, the search overhead will increase if important information is not available.

— Applying parallelism rashly will result in a perfectly balanced work load, but most of the processors will search subtrees that are not visited by the sequential algorithm. A careful use of parallelism is necessary reducing the average work load of the processors.

## 3.1 Parallel Search in ZUGZWANG

In this section we will describe the main concept of the parallel algorithm used in ZUGZWANG since 1990. The concept is designed for the use on massively parallel distributed systems, i.e. systems without shared memory. The communication is realized by message passing. In the years from 1990 - 1994 we ran ZUGZWANG on a system based upon T800/805 Transputers, with up to 1024 processors. Today ZUGZWANG is running on an M604 based system with 40 processors. The adaptation of the parallelization to the new hardware or software required only minor changes in the implementation, showing the flexibility of our approach.

**The Algorithm** Our parallelization of the game tree search is based upon a decomposition of the tree to be searched, and on a parallel evaluation of the resulting subproblems. The sequential algorithm is a depth first search algorithm. It visits the nodes of the tree from left to right. To do so, a current variation is stored in a stack. All the nodes to the left of this variation have been visited, or, visiting these nodes has been shown to be superfluous. The nodes to the right of this variation have not yet been visited. The idea of the parallelization of such a tree search is now to make available as much of the right brothers of the current variation for parallel evaluation as possible. By this, several processors may start a tree search on disjoint subtrees of the whole game tree.

For a formal definition of the algorithm we refer to [23]. Informally the algorithm can be described as follows:

— Initially all processors but the master are idle. The master gets the root of the search tree, and starts a sequential search generating a current variation.
— Idle processors send a request for work to any other processor. When a processor gets a request for work it looks for free subproblems to the right of its current variation. If one is found it is sent to the requesting processor. Otherwise the request is forwarded to another processor.
— After an idle processor has sent its request for work it waits for a new subproblem.
— After a processor has solved a subproblem it sends a return message to its master, containing the result computed for the subproblem. The master, upon receiving the return message, updates the bounds on its current variation.
— Whenever the bounds of the current variation at a processor are changed, either by computation or by a message, the processor immediately informs all slaves using the corresponding bounds about the update by sending a new search window. The new window may be empty causing the slaves to cancel their subproblems.

The efficiency of such an algorithm depends heavily on efficient solutions for the following problems:

1. *Establishing a master-slave relationship*: Every processor searching a subproblem generates subproblems for itself, which may be searched in parallel by other processors. Thus, it is a potential master. An idle processor is a potential slave. In order to make an idle processor busy, it has to receive a subproblem from one of the working processors for parallel evaluation.

2. *Distribution of information in the distributed algorithm:* The sequential algorithm allows many cutoffs. The number of cutoffs, and thus the efficiency of the sequential search, is heavily influenced by the quality of the first successors of the inner nodes and by the knowledge about the current bounds. The sequential search algorithm uses the information computed during the search of the left parts of the tree to improve the ordering of the successors of inner nodes, as well as to tighten the bounds which allow the cutoffs. In our concept for a parallel game tree search this successive improvement of the information is obtained by communication between the processors.

3. *Restricting the use of parallelism to avoid search overhead:* The game trees occurring in practice are very similar to the minimal game tree, i.e. many cutoffs occurring in the minimal game tree will also occur in the game tree that has to be searched. Even if the available information about cutoff bounds etc. is distributed optimally, in some situations it may not make any sense to use parallelism. The parallel algorithm will visit only nodes, which the sequential one does not search. In situations like this, one will prefer an idle processor to remain idle for a short time and then to start working on a subproblem which is useful to evaluate, rather than to start working on a subproblem immediately, which is irrelevant with high probability.

**The Load Balancing Algorithm** For the load balancing we have chosen an adaptive algorithm:

- Local balancing: A processor chooses a neighbor in the processor network as a target for its first request for work.
- Medium range balancing: A processor that gets a request for work from its neighbor, but is not able to send a subproblem, forwards the request to one of its remaining neighbors, if the request has not been forwarded too far. In this case the request is sent back to the origin.
- Global balancing: Any return message is interpreted as a request and is handled in the same way.

The local and medium range load balancing is cheap in terms of communication costs, since messages are sent only to neighbors in the network, but not sufficient to balance large networks. The global balancing has three effects: As the search proceeds, the request are spread all over the network. Second, slaves tend to stay at the same master and therefore local move ordering heuristics tend to work better. Last but not least it saves a request, since the return message has to be sent anyway.

**The Distributed Transposition Table** Marsland and Popowich [45] showed that local transposition tables do not work well in a parallel system, even if the number of processors is small, since processors can access only the information computed by themselves. Therefore we have implemented the transposition table as a large distributed table, and organize the access to the table by routing messages. Although this is quite communication intensive, it has the advantage that the parallel system has access to a much larger hash table than the sequential one, improving its search behaviour.

In an $N$-processor system, the value $h(v)$ of the hash function for a chess position $v$ is interpreted as the concatenation of the lock of $v$, a processor number $p(v) := h(v) \bmod N$ and a local index $i(v) := \lfloor \frac{h(v)}{N} \rfloor$. Therefore, a processor that wants to access the table for a position $v$ sends a read request to processor $p(v)$ containing the local index $i(v)$ and the lock. A processor, upon getting a read request for a local index $i$, compares the lock of the corresponding entry with the lock of the message. If both are equal the entry from the table is sent back.

A critical delay may occur when a processor wants to read a remote table entry for a position $v$. Sometimes a valid entry is found, and the move contained in this entry has to be searched first below $v$. But this means, that the processor has to wait until the answer arrives. In large networks these waiting times may last longer than the search below $v$. Therefore, after sending a read request the processor continues the search below $v$ immediately. If an answer from the remote table arrives, the entry is checked and the search is restarted if the first move is not the one proposed by the transposition table entry.

In the current version a processor requests the global transposition table for a node $v$ only if $v$ is the root or if an entry has been found for the father of $v$. This keeps the number of messages necessary to access the table reasonably small. The result for a node $v$ is written into the global table only if the remaining search depth below $v$ is larger than 1.


**The Young Brothers Wait Concept** In this section we describe how to restrict parallelism to reduce the amount of work done in irrelevant subtrees, but get enough parallelism out of the application to achieve a good processor work load. The problem is that the sequential version determines the shape of the tree on the basis of the search windows. These are narrowed as the search proceeds. Therefore the shape of the tree is not known in advance. We know on the other side, that the tree, that is searched by the sequential version, usually is very close to the minimal game tree and the shape of the minimal game tree can be determined exactly. This has been done by Knuth and Moore [36] by recursively assigning types to the nodes of the tree as follows:

- The root gets type 1.
- The first child of a type-1 node gets type 1. The other children get type 2.
- The first child of a type-2 node gets type 3.
- All children of a type-3 node get type 2.

By this a type is assigned to exactly the nodes of the minimal tree. If only the minimal game tree is searched a cutoff happens exactly at the nodes of type 2. Hsu [30] implicitly extended the assignment of types to the whole tree by adding the rule

— The right children of a type-2 node get type 2.

The new assignment is shown in the figure below. We use these extended node types to restrict the use of parallelism in the distributed search by assuming that a cutoff will happen at type-2 nodes. As a result the use of parallelism is delayed at type-1 nodes, even further delayed at type-2 nodes but enabled at type-3 nodes immediately after they have been generated. This results in the so called *Young Brothers Wait Concept*:

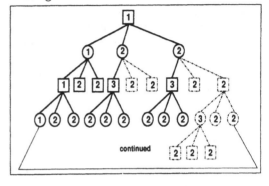

— The parallel search below right children of a type-1 node is allowed only if the search below the leftmost child is complete.
— The parallel search below right children of a type-2 node is allowed only if the searches below all promising children are complete. Here, promising children are the leftmost one, those reached by a move proposed by the transposition table, the killer heuristic, or any positive capturing move.
— The parallel search below a type-3 node may be started immediately after its generation.

Parallelism is used carefully at type-2 nodes. Since we expect a cutoff to happen at these nodes, we evaluate the promising moves sequentially, one by one. Only if these promising moves fails to produce a cutoff the remaining children may be searched in parallel. By this the parallel search algorithm finds a lot of cutoffs found by the sequential one without searching irrelevant subtrees. At type-1 nodes the use of parallelism is delayed until the first child is completely evaluated, since the result from this search is used as a bound for the search below the right children. At type-3 nodes we do not expect a cutoff and therefore allow the parallel search of the children as soon as possible. The Young Brothers Wait Concept is a heuristic to guide the use of parallelism. In case of a mistake at a type-2 node, the search is sequentialized below this node and thus the processor work load may be decreased. In case of a failure at a type-3 node, processors search irrelevant subtrees in parallel increasing the search overhead. However, in [23] we present statistical data showing that this heuristic works very well.

**Results** In [23] we describe experiments showing the efficiency of the proposed approach for a parallel search of the Negascout algorithm using up to 1024 T805

Transputers. The Transputers were connected as a DeBruijn network ($\leq$ 256 processors) or as a 2-dimensional grid ($>$ 256 processors). Here, we give only a small summary:

| Proc. | 1 | 8 | 16 | 32 | 64 | 128 | 256 | 512 | 1024 |
|---|---|---|---|---|---|---|---|---|---|
| depth | 5 | 6 | 7 | 7 | 7 | 8 | 8 | 9 | 9 |
| time | 163 | 125 | 342 | 193 | 113 | 320 | 193 | 384 | 264 |
| SPE | 1.00 | 6.13 | 12.33 | 21.83 | 37.34 | 86.03 | 142.32 | 237.03 | 344.49 |
| SO % | 0.00 | 15.81 | 14.04 | 19.79 | 25.10 | 10.05 | 17.30 | 39.82 | 55.55 |
| LOAD % | 100.00 | 90.10 | 89.36 | 85.48 | 77.86 | 81.98 | 74.13 | 74.56 | 62.67 |
| PERF % | 100.00 | 98.56 | 98.68 | 95.61 | 93.78 | 90.22 | 87.91 | 86.96 | 83.61 |

**Fig. 4.** Results of ZUGZWANG on the Transputer networks

Figure 4 contains the speedup measurements on the basis of a set of 24 chess positions [12]. The speedup (**SPE**) is defined as the running time of the sequential algorithm divided by the running time of the parallel algorithm, both with the same depth parameter. The search overhead (**SO**), processor work load (**LOAD**) and performance (**PERF**) are given in percent compared to the sequential version with the same depth parameter.

For each number of processors we have chosen the search depth such that the average running time is close to the running time under tournament conditions (150 − 300 seconds). The single processor version needs 163 seconds on the average for a 5-ply search while the 1024 processor is able to complete a 9-ply search at 264 seconds on the average. For the tests on the two grid networks we had to reduce the number of accesses to the distributed transposition table since otherwise the performance would have been too bad. As a consequence the search overhead increased.

The distributed search algorithm proposed in this paper was the heart of the chess program ZUGZWANG at the Computer Chess Championships in Madrid, Spain, 1992. ZUGZWANG ran on 1024 T805 processors and finished second.

In the meantime ZUGZWANG uses the FHR-algorithm and runs on a 40 processor machine based upon M604 processors. Since the FHR-algorithm uses the search windows to determine the shape of the tree, the sequential and the parallel search algorithm may not compute the same result, even if the search parameters are the same. Therefore it is very hard to measure the speedup. Different metrics have to be developed to determine the efficiency of the parallel algorithm. This work is still in progress. Here we present only preliminary data, showing that the approach can be used to parallelize the search in very irregular trees, as generated by the Fail High Reductions algorithm. In the data of figure 5 we substituted the term "speedup" with "acceleration".

The loss of performance of about 25% is mainly due to the large number of messages for the distributed transposition table and to the worse ratio of com-

| Proc. | 1 | 4 | 8 | 16 | 32 | 40 | 40 |
|---|---|---|---|---|---|---|---|
| depth | 9 | 10 | 10 | 11 | 11 | 11 | 12 |
| time | 255 | 360 | 215 | 408 | 205 | 161 | 475 |
| ACC | **1.00** | **2.49** | **4.16** | **7.02** | **13.96** | **17.71** | **22.58** |
| SO % | 0.00 | 8.14 | 19.90 | 43.94 | 22.28 | 16.46 | 0.95 |
| LOAD % | 100.00 | 82.23 | 79.19 | 79.38 | 68.99 | 67.94 | 73.48 |
| PERF % | 100.00 | 81.70 | 78.73 | 79.52 | 77.35 | 75.88 | 77.57 |

**Fig. 5.** Results of ZUGZWANG on the M604 based systems

munication speed to computational speed of the M604 based machine compared to the T805 based machines.

At the 12th Aegon Man vs. Machine Tournament in Den Haag, The Netherlands 1997, ZUGZWANG ran on a 44 M604 processor machine and achieved a tournament performance rating of more than 2600 ELO points showing that ZUGZWANG belongs to the worlds top chess programs.

# 4 Conclusions

First, we presented some sequential algorithms used in modern chess programs. We showed how heuristics like the transposition table are used to speed up the sequential search. An even deeper lookahead is obtained by using either the Null Move Search or the Fail High Reductions algorithm. In the second part of the paper we presented our approach to a distributed game tree search algorithm. This approach has been used successfully on massively parallel systems like a 1024-processor Transputer network. Basically the same approach is used for the parallelization of the Fail High Reductions algorithm. Although the trees searched by this algorithm are different in structure, the distributed search algorithm again shows promising results.

Nevertheless, more locality has to be exploited in the distributed algorithm, especially in using the transposition table in a parallel environment. New methods to save a large portion of the store and read messages without increasing the search overhead have to be developed. Another open problem is a problem of testing: How can we compare the sequential and the parallel program if for fixed search parameters they do not deliver the same results ?

**Acknowledgements** I would like to thank my supervisor Burkhard Monien, who enthusiastically supported the computer chess projects at the University of Paderborn since the early beginning. Many of the results presented here have been joined work with Peter Mysliwietz, who left the University of Paderborn in autumn 1996.

# References

1. *S.G. Akl, M. Newborn* **The principal continuation and the killer heuristic** ACM Annual Conference, pp 466-473, 1977

2. *S.G. Akl, D.T. Barnard, R.J Doran* **Design, Analysis and Implementation of a Parallel Tree Search Algorithm** IEEE Transactions on Pattern Analysis and Machine Intelligence, 14(2), pp 192-203, 1982

3. *I. Althöfer* **On Pathology in Game Tree and Other Recursion Tree Models** Habilitation thesis, University of Bielefeld, Germany, 1991

4. *I. Althöfer* **A Parallel Game Tree Search Algorithm with a Linear Speedup** Journal of Algorithms, 15(2), pp 175-198, 1993

5. *T.S. Anantharaman, M. Campbell, F.H. Hsu* **Singular Extensions: Adding Selectivity to Brute-Force Searching** ICCA Journal, 11(4), pp 135–143, 1988.

6. *T.S. Anantharaman* **Extension Heuristics** ICCA Journal, 14(2), pp 47–63, 1991.

7. *G. Baudet* **The Design and Analysis of Algorithms for Asynchronous Multiprocessors** Phd thesis, Carnegie-Mellon University, Pittsburgh, USA, 1978

8. *D.F. Beal* **Experiments with the Null Move** Advances in Computer Chess V, D.F. Beal (ed.), pp 65–79, 1989.

9. *D.F. Beal, M.C. Smith* **Quantification of Search-Extension Benefits** ICCA Journal, 18(4), pp 205–218, 1995.

10. *D.F. Beal, M.C. Smith* **Multiple Probes of Transposition Tables** ICCA Journal, 19(4), pp 227–233, 1995.

11. *H.J. Berliner* **The B\* Tree Search Algorithm: A Best-First Proof Procedure** Artificial Intelligence, 12, pp 23-40, 1979

12. *I. Bratko, D. Kopec* **A Test for Comparison of Human and Computer Performance in Chess** Advances in Computer Chess III, M.R.B. Clarke (ed.), Pergamon Press, pp 31-56, 1982

13. *D.M. Breuker, J.W.H.M. Uiterwijk, H.J. van den Herik* **Replacement Schemes for Transposition Tables** ICCA Journal, 17(4), pp 183–193, 1994.

14. *D.M. Breuker, J.W.H.M. Uiterwijk, H.J. van den Herik* **Replacement Schemes and Two-Level Tables** ICCA Journal, 19(3), pp 175–180, 1996

15. *M. Buro* **ProbCut: An Effective Selective Extension of the $\alpha\beta$ - Algorithm** ICCA Journal, 18(2), pp 71–76, 1995

16. *C. Donninger* **Null Move and Deep Search** ICCA Journal, 16(3), pp 137–143, 1993.

17. *A.E. Elo* **The Rating of Chessplayers, Past and Present** Arco Publishing, New York, 1978

18. *R. Feldmann, B. Monien, P. Mysliwietz, O. Vornberger* **Distributed Game-Tree Search** ICCA Journal, 12(2), pp 65-73, 1989

19. *R. Feldmann, B. Monien, P. Mysliwietz, O. Vornberger* **Distributed Game Tree Search** Parallel Algorithms for Machine Intelligence and Vision, V. Kumar, L.N. Kanal, P.S. Gopalakrishnan (eds.), Springer, pp 66-101, 1990

20. *R. Feldmann, B. Monien, P. Mysliwietz* **A Fully Distributed Chess Program** Advances in Computer Chess VI, D.F. Beal (ed.), pp 1-27, 1991

21. *R. Feldmann, P. Mysliwietz, B. Monien* **Experiments with a Fully Distributed Chess Program** Heuristic Programming in Artificial Intelligence 3, J. van den Herik, V. Allis (eds.), pp 72-87, 1992

22. *R. Feldmann, P. Mysliwietz, B. Monien* **Distributed Game Tree Search on a Massively Parallel System** in: Data structures and efficient algorithms: Final

report on the DFG special joint initiative, Springer, Lecture Notes on Computer Science 594, B. Monien, T. Ottmann (eds.), pp 270-288, 1991

23. *R. Feldmann* **Game Tree Search on Massively Parallel Systems** Doctoral thesis, University of Paderborn, Germany, 1993

24. *R. Feldmann, P. Mysliwietz, B. Monien* **Studying Overheads in Massively Parallel MIN/MAX-Tree Evaluation** Proceedings of SPAA'94, pp. 94-103, 1994

25. *R. Feldmann* **Fail High Reductions** Advances in Computer Chess VIII, H.J. van den Herik, J.W.H.M. Uiterwijk (eds.), University of Maastrich, The Netherlands, pp 111-127, 1997

26. *C. Ferguson, R.E. Korf* **Distributed Tree Search and its Application to Alpha-Beta Pruning** Proceedings AAAI-88, 7th National Conference on Artificial Intelligence, 2, pp 128-132, 1988

27. *R.A. Finkel, J.P. Fishburn* **Parallel Alpha-Beta Search on Arachne** IEEE International Conference on Parallel Processing, pp 235-243, 1980

28. *R.A. Finkel, J.P. Fishburn* **Parallelism in Alpha-Beta Search** Artificial Intelligence, 19, pp 89-106, 1982

29. *G. Goetsch, M.S. Campbell* **Experiments with the Null-Move Heuristic** Computers, Chess, and Cognition, T.A. Marsland and J. Schaeffer (eds.), Springer, pp 159-168, 1990

30. *F.H. Hsu* **Large Scale Parallelization of Alpha-Beta Search: An Algorithmic Architectural Study with Computer Chess** Phd. thesis, Carnegie Mellon University, Pittsburgh, USA, 1990

31. *M. M. Huntbach, F. W. Burton* **Alpha - Beta Search on Virtual Tree Machines** Information Sciences, 44, pp 3-17, 1988

32. *R.M. Hyatt* **Parallel Chess on the Cray X-MP/48** ICCA Journal, 8(2), pp 90-99, 1985

33. *R.M. Hyatt, B.W. Suter, H.L. Nelson* **A parallel alpha/beta tree searching algorithm** Parallel Computing, No. 10, pp 299-308, 1989

34. *G. Isenberg* (author of ISICHESS), personal communication, February 1997

35. *R. M. Karp, Y. Zhang* **On Parallel Evaluation of Game Trees** Proceedings of SPAA'89, pp 409-420, 1989

36. *D.E. Knuth, R.W. Moore* **An Analysis of Alpha - Beta Pruning** Artificial Intelligence, 6, pp 293-326, 1975

37. *H.-J. Kraas* **Zur Parallelisierung des SSS\*-Algorithmus** Doctoral thesis, University of Braunschweig, Germany, 1990

38. *B.C. Kuszmaul* **The Startech Massively-Parallel Chess Program** ICCA Journal, 18(1), pp 3-19, 1995

39. *R. Levinson, F-h. Hsu, J Schaeffer, T.A. Marsland, D.E. Wilkins* **The Role of Chess in Artificial Intelligence Research** Proc. of the 12th IJCAI, Morgan Kaufman Publishers, pp 557-562, 1991

40. *D. Levy, M. Newborn* **How Computers Play Chess** Computer Science Press, 1991

41. *D. Levy* **Crystal Balls: The Meta-Science of Prediction in Computer Chess** ICCA Journal, 20(2), pp 71-78, 1997

42. *U. Lorenz, V. Rottmann, R. Feldmann, P. Mysliwietz* **Controlled Conspiracy-Number Search** ICCA Journal, 18(3), pp 135-147, 1997

43. *U. Lorenz* (author of ULYSSES, CHEIRON, CONNY), personal communication, 1997

44. *T.A. Marsland, M.S. Campbell* **Parallel Search of Strongly Ordered Game Trees** Computing Surveys, 14(4), pp 533-551, 1982

45. *T.A. Marsland, F. Popowich* **Parallel Game Tree Search** IEEE Transactions on Pattern Analysis and Machine Intelligence, 7(4), pp 442-452, 1985
46. *T.A. Marsland, M. Olafsson, J. Schaeffer* **Multiprocessor Tree-Search Experiments** Advances in Computer Chess IV, D.F. Beal (ed.), Pergamon Press, pp 37-51, 1986
47. *D.A. McAllester* **A New Procedure for Growing Min-Max Trees** Artificial Intelligence, 35(3), pp 287-310, 1988
48. *P. Mysliwietz* **Konstruktion und Optimierung von Bewertungsfunktionen beim Schach** Doctoral thesis, University of Paderborn, Germany, 1994
49. *M. Newborn* **A Parallel Search Chess Program** ACM Annual Conference 1985, pp 272-277, 1985
50. *M. Newborn* **Unsynchronized Iterative Deepening Parallel Alpha-Beta Search** IEEE Transactions on Pattern Analysis and Machine Intelligence, 10(5), pp 687-694, 1988
51. *S.W. Otto, E.W. Felten* **Chess on a Hypercube** The Third Conference on Hypercube Concurrent Computers and Applications, 2, pp 1329-1341, 1988
52. *J. Pearl* **Heuristics: Intelligent Search Strategies for Computer Problem Solving** Addison-Wesley Publishing Company, 1984
53. *A. Reinefeld* **Spielbaum - Suchverfahren** Springer, 1989
54. *J. Schaeffer* **The History Heuristic and Alpha-Beta Search Enhancements in Practice** IEEE Transactions on Pattern Analysis and Machine Intelligence, 11(11), pp 1203-1212, 1989
55. *J. Schaeffer* **Distributed Game-Tree Searching** Journal of Parallel and Distributed Computing, 6(2), pp 90-114, 1989
56. *C.E. Shannon* **Programming a Computer for Playing Chess** Philosophical Magazine 41, pp 256-275, 1950
57. *G. Schrüfer* **Minimax-Suchen Kosten, Qualität und Algorithmen** Doctoral thesis, University of Braunschweig, Germany, 1988
58. *A.M. Turing* **Digital Computers Applied to Games** in B.V. Bowden: Faster than Thought: A Symposium on Digital Computing Machines, Pitman, pp 286-310, 1953
59. *K. Thompson* **Computer Chess Strength** Advances in Computer Chess III, M.R.B. Clarke (ed.), Pergamon Press, pp 55-56, 1982
60. *A. Szabo, B. Szabo* **The technology curve revised** ICCA Journal 11(1), 1988
61. *J. v. Neumann, O. Morgenstern* **Theory of Games and Economic Behavior** Princeton University Press, Princeton, USA, 1944
62. *J.W.H.M. Uiterwijk* **The Countermove Heuristic** ICCA Journal 15(1), pp 8-15, 1992
63. *O. Vornberger, B. Monien* **Parallel Alpha-Beta versus Parallel SSS\*** Proceedings IFIP Conference on Distributed Processing, North Holland, pp 613-625, 1987
64. *T. Warnock, B. Wendroff* **Search Tables in Computer Chess** ICCA Journal, 11(1), pp 10-13, 1988
65. *J.-C. Weill* **The ABDADA Distributed Minimax-Search Algorithm** ICCA Journal, 19(1), pp 3-16, 1996
66. *E. Zermelo* **Über eine Anwendung der Mengenlehre auf die Theorie des Schachspiels** 5. Int. Mathematikerkongreß, Cambridge, 2, pp 510-504, 1912
67. *A.L. Zobrist* **A New Hashing Method with Applications for Game Playing** TR-88, University of Wisconsin, Computer Science Department
Reprint in ICCA Journal 13(2), pp 69-73, 1990

# Algorithms for Triangulated Terrains*

Marc van Kreveld

Dept. of Computer Science
Utrecht University
The Netherlands
marc@cs.ruu.nl

**Abstract.** Digital elevation models can represent many types of geographic data. One of the common digital elevation models is the triangulated irregular network (also called TIN, or polyhedral terrain, or triangulated terrain). We discuss ways to represent a TIN in a data structure, and give some of the basic algorithms that work on TINs. These include retrieving contour lines, computing perspective views, and constructing TINs from other digital elevation data. We also give a recent method to compress and decompress a TIN for storage and transmission purposes.

## 1 Introduction

Geographic Information Systems are large software packages that store and operate on geographic data. They are large because they usually include a full database system, and set of functions to operate on spatial data. It is the spatial (or geometric) component that distinguishes geographic information systems (or GIS for short) from standard databases.

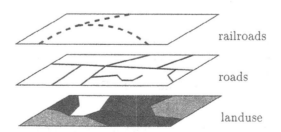

**Fig. 1.** Layers of geographic data in a GIS.

Geographic data comes in many forms. Borders of countries and provences, locations of roads and hospitals, and pollution of the lakes and rivers are types

---

* Research is partially supported by the ESPRIT IV LTR Project No. 21957 (CGAL).

of man-made geographic data. Natural geographic data includes elevation above sea level, annual percipitation, soil type, and much more. GIS store the different types of geographic data in different *map layers*, so there is a map layer with the major roads, one with the rivers, one with the current land use, and one with the elevation above sea level. GIS typically store from ten up to a few hundred map layers.

It is rather useful to have types of data in different layers. A GIS user may wish to see a map on the screen with only cities and railroads, because this particular user plans to travel by train somewhere. Or a physical geographer may wish to see the *overlay* of soil type and amount wind erosion, to study how these two data sets are related.

For any specified location on the Earth or on a map, one can say that some value is associated to it in a particular theme. For instance, at $53°15'$ latitude and $6°$ longitude the particular land use is "agricultural", the elevation is 1 meter above sea level, and the annual percipitation is 790 mm. So the value can be a name, or a number, or something else. In the first case the data is called *nominal*, in the second case it is called *ratio*. (Traditionally, four scales of measurement were used: nominal, ordinal, interval, and ratio [27]. Geographic data can also be a direction or a vector, like wind.)

This paper deals with data on the ratio scale, which can be seen as a function from a 2-dimensional region to the reals. The domain can be referenced by geographic coordinates, for example, but we'll do as if we have a function from the $xy$-plane into the third, $z$-dimension. Elevation above sea level is the most obvious type of data that is modelled by such a function.

One of the problems when storing and computing on elevation data is that the amount of data can be enormous. Currently available for the US is elevation data for points at a regular spacing of 30 meters, which means a few billion points. In the future data sets of considerably smaller spacing will be collected, leading to even larger data sets. A consequence is that only *efficient* algorithms can be used to process the data. This paper surveys the common models to store elevation data, in particular the triangulated irregular network. We then discuss a couple of the basic algorithms that operate on this model, like determining contour lines, visualization by perspective views, and conversion from other elevation models. Then we concentrate on a recent result on efficient compression and decompression for the triangulated irregular network, which is important for background storage and for network transmission.

This paper presents several algorithms for terrains. To analyse and express the efficiency of these algorithms we'll use big-Oh notation. For instance, for a triangulated irregular network determined by a set of $n$ data points, the contour lines of some given elevation can be determined in $O(n)$, or linear time. It is important that the algorithms have running times like $O(n)$ or $O(n \log n)$, because quadratic time usually is too slow in practice for the amount of data involved.

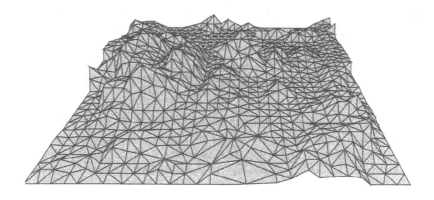

**Fig. 2.** Perspective view of a triangulated irregular network.

## 2 Digital elevation models

In the computer the true geographic elevation (function) has to be approximated by some finite representation of it. This is called a *digital elevation model*. There are three common digital elevation models (or DEMs for short). They are the regular square grid, the contour lines, and the triangulated irregular network.

The regular square grid is a 2-dimensional array where each entry stores an elevation. An entry represents some region on the Earth of, say, 10 by 10 meters, and the stored value is the elevation of the center point of the region.

**Fig. 3.** Contour lines of a terrain.

A contour map consists of a collection of separate contour lines that each have some elevation. Each contour line can be stored as a sequence of control points through which the contour line is assumed to pass. A contour line is a closed curve, or it may have its endpoints on the boundary of the region for which the elevation function is defined. It can be represented by a polygon or polygonal line, or a spline curve.

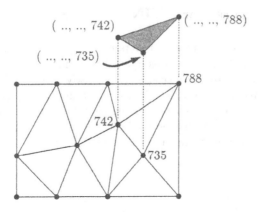

**Fig. 4.** Triangulation with elevation values at vertices.

The triangulated irregular network (or TIN, or polyhedral terrain, or triangulated terrain) is a third way to represent elevation. A triangulation on some finite set $S$ of points is a planar subdivision into triangles that is maximal, and such that only the points of $S$ appear as vertices of the subdivision. When used as an elevation model, the vertices of the triangulation store an elevation value. The elevations on the edges and inside the triangles of the subdivision are obtained by linear interpolation. So if some point $q = (\bar{x}, \bar{y})$ lies inside a triangle with vertices $v_i = (x_i, y_i)$, $v_j = (x_j, y_j)$, and $v_k = (x_k, y_k)$ with elevations $z_i$, $z_j$, and $z_k$, then we consider the unique plane that passes through the three points $(x_i, y_i, z_i)$, $(x_j, y_j, z_j)$, and $(x_k, y_k, z_k)$. Then we determine the value $\bar{z}$ of $q$ such that the point $(\bar{x}, \bar{y}, \bar{z})$ lies on the plane, which gives the interpolated value.

Note that there are many different triangulations possible of a given set of points. All must have the same number of edges and triangles, because triangulations are maximal planar subdivisions. By Euler's relation for planar graphs, the number of edges and triangles is linear in the number of points, the vertices that determine the subdivision. Different triangulations of a point set lead to different elevations at points on edges and inside triangles. For interpolation purposes, it seems natural to choose one that has small, well-shaped triangles. The standard choice is the so-called *Delaunay triangulation* that will be discussed later.

The contour line model isn't used as a way to store elevation data permanently. However, one of the ways to obtain elevation data is by digitizing the contour lines on contour maps, so one may have to deal with the contour model nevertheless. Often, the contour model is converted to the grid or TIN model before further processing.

One of the advantages of the TIN over the grid is that it is adaptive: more data points can be used in regions where there is much elevation change, and fewer points in regions where the elevation hardly changes. One of the disadvantages of the TIN when compared to the grid is that the algorithms usually are somewhat more complex.

# 3 Data structures for a TIN

This section gives two different ways to represent TINs. One is edge based and the other is triangle based. In both cases it will be possible to navigate on the TIN, going from one triangle to an adjacent one efficiently, or finding all triangles that are incident to a particular vertex. The two structures are simplified versions of data structures that can store arbitrary planar subdivsions such as the doubly-connected edge list, winged edge, or quad edge structure, commonly used in GIS, graphics, and computational geometry [6, 8, 14, 32].

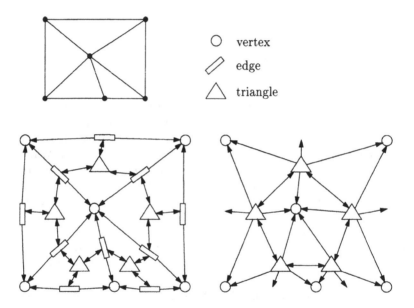

**Fig. 5.** Left, the edge-based and right, the triangle-based structures of the triangulation shown at the top.

In the triangle-based structure for a TIN, any triangle and vertex is represented by an object or record. A triangle object has six references, three to the adjacent triangles, and three to the incident vertices. Vertices are stored by objects that only have the $x$-, $y$-, and $z$-coordinates. Edges are not stored explicitly, but they can be determined from the structure if necessary. The CGAL-library of geometric primitives and algorithms provides this structure [2].

In the edge-based structure for a TIN, any edge is an object that has a dual purpose. It connects two vertices and it separates two triangles. In the structure any edge object has references to two vertex objects (of the vertices it connects) and to two triangle objects (of the triangles it separates; there may only be one). The triangle objects have references to the three edge objects that bound the triangle. Vertices again only store the $x$-, $y$-, and $z$-coordinates. There are no references between vertex objects and triangle objects.

## 4  Visualization and traversal of a TIN

The most common way to show elevation on maps is by contour lines at regular intervals. The map can be enhanced by *hill shading*, a technique where an imaginary light source is placed in 3-dimensional space, and parts of the terrain that don't receive much light are shaded. Another way to visualize a terrain is by a perspective view. The algorithms required for visualization are standard graph algorithms on the TIN structure in both cases.

### 4.1  Contour maps

To determine all contour lines of, say, 1000 meters, on a TIN representing a terrain, observe that any triangle contains at most one line segment that is part of the contour lines of 1000 meters. In fact, the contour lines of 1000 meters are nothing else than the cross-section of the terrain as a 3-dimensional surface, and the horizontal plane $z = 1000$. So, to determine the contour lines on a TIN it suffices to examine every triangle once and see if it contributes to the contour lines. Similarly, to compute hill shading for a TIN, one needs to determine the slope of each triangle and its aspect (the compass direction to which the triangle is facing, in the $xy$-projection). The slope and aspect determine how much a triangle is shaded, given a position of the light source. As for contour lines, it suffices to examine every triangle of the TIN once to compute hill shading for the whole terrain.

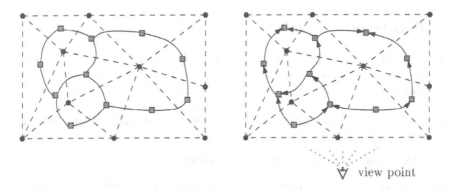

view point

**Fig. 6.** Left, dual graph of the TIN, with grey nodes and solid arcs. Right, dual directed acyclic graph for agiven view point.

A traversal that visits every triangle once is like a depth-first search in a graph dual to the TIN. This graph has a node for every triangle, and two nodes are connected by an arc if the corresponding triangles share an edge in the triangulation. Both the triangle-based and the edge-based TIN structures implicitly store this graph, and depth-first search through all the triangles is easy if a mark bit is available in every object, to see if it has been visited before. So for a TIN with $n$ vertices and, hence, $O(n)$ edges and triangles, one can compute all contour lines of a given elevation in $O(n)$ time by depth-first search. Similarly, one can compute hill shading in linear time.

## 4.2 Perspective views

Another way to visualize a TIN is by a perspective view of the terrain. We can produce such a view using the Painter's Algorithm, where all triangles are drawn from back to front, so that the ones more to the front erase the ones more to the back. What is the appropriate drawing order for the triangles of a TIN? Given the view point and a TIN, every two triangles that share an edge must be drawn in a specific order (unless the line supporting the edge happens to pass through the view point). This necessary condition on the drawing order happens to be sufficient as well: if for all edges of a TIN, the triangle 'behind' the edge is drawn before the triangle 'in front of' the edge, then the drawing order is correct. So the drawing order is a partial order as well, and it can be obtained by a topological sort of a directed graph. Again, this directed graph is implicitly present in either of the two structures for storing a TIN. It is the same dual graph as we used for depth-first search, but now the arcs have a direction. The direction of an arc can be determined by checking the coordinates of the endpoints of the dual edge of that arc, and the coordinates of the view point. So a TIN with $n$ vertices can be drawn in perspective view in $O(n)$ time using the Painter's Algorithm.

## 5 Construction of a TIN

We'll now study algorithms for constructing a TIN from elevation data. First we consider the case that a set of data points with elevations is given, and the problem is to construct a triangulation on that set. Then we assume that the input is a large grid of regularly spaced data points, and the problem is to produce a TIN that approximates the grid to within a specified maximum error.

## 5.1 Delaunay triangulation on a point set

The most popular triangulation of a set of points without doubt is the Delaunay triangulation. For a set $P$ of $n$ points in the plane, three points $p, q, r \in P$ are the vertices of a triangle in the Delaunay triangulation of $P$ if the circle through $p, q, r$ doesn't contain any other points of $P$. If $P$ doesn't contain four points that are co-circular, then the definition just given really defines a unique triangulation. It has the property that, among all triangulations of $P$, it is the one that maximizes the minimum angle of the triangles in the triangulation (in other words, no triangulation has a larger smallest angle). This property implies that

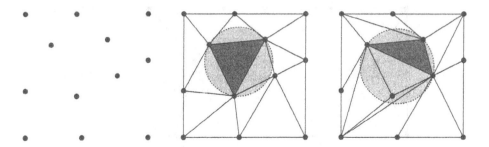

**Fig. 7.** Left, a set of points in the plane. Middle, the Delaunay triangulation of the points, where all triangles have the empty circle property as shown for one triangle. Right, a triangulation of the same points with a triangle that doesn't have the empty circle property.

the triangles generally will be well-shaped, which is important for the interpolation function it defines. More information on the Delaunay triangulation is in the book of Okabe, Boots, and Sugihara [19] and in textbooks on computational geometry [6, 20, 21].

Several algorithms are known that construct the Delaunay triangulation of a set on $n$ points in $O(n \log n)$ time, and this is optimal. We sketch one that is simple and requires $O(n \log n)$ time expected, based on randomized incremental construction. The expectation in the running time is only dependent on the random choices made by the algorithm and is valid for any set of points, independent of the distribution.

In the randomized incremental construction algorithm for the Delaunay triangulation, all points of $P$ are added one by one, and after each insertion, the Delaunay triangulation is restored to incorporate the new point. So we in fact compute a sequence $T_1, \ldots, T_n$ of triangulations, where triangulation $T_i$ contains $i$ points of $P$. To compute $T_{i+1}$, we choose one of the remaining $n - i$ points at random and insert it. So if $p_1, \ldots, p_n$ is a random permutation of $P$, we can simply insert the points in this order, and $T_i$ is the Delaunay triangulation of $p_1, \ldots, p_i$.

One insertion of a point requires locating the triangle of the triangulation that contains the point, and then the actual insertion. We'll skip the location part and assume that the new point $p_{i+1}$ falls inside some triangle $t$ of triangulation $T_i$. For certain, the Delaunay triangulation $T_{i+1}$ will contain the three edges between $p_{i+1}$ and the vertices of $t$. It is not certain, however, that the three triangles of which $p_{i+1}$ is now a vertex really satisfy the Delaunay property: the circle through $p_{i+1}$ and two of the vertices of $t$ may contain other vertices of $T_i$. If this is the case, we'll *flip*: we destroy the edge between the two vertices of $t$ and we create a new edge from $p_{i+1}$ to repair the triangulation.

The correctness of such a flip depends on two facts that can be shown from the Delaunay property of $T_i$. Firstly, all edges that appear in $T_{i+1}$ but not in $T_i$ must have $p_{i+1}$ as one of the endpoints. Secondly, if any triangle $t'$ incident to

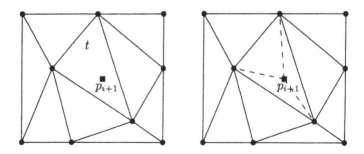

**Fig. 8.** New point $p_{i+1}$ in triangle $t$ of $T_i$, and three of the new edges.

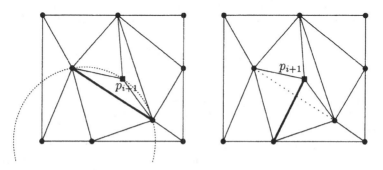

**Fig. 9.** Flipping an edge because the empty circle property is violated.

$p_{i+1}$ doesn't have the empty circle property, then at least the vertex opposite of the edge opposite of $p_{i+1}$ in $t'$ must be in this circle. The flip will connect $p_{i+1}$ to this vertex.

Any flip destroys one triangle incident to $p_{i+1}$ and another triangle, and creates two triangles incident to $p_{i+1}$. Any edge opposite to $p_{i+1}$ of a triangle incident to $p_{i+1}$ must be tested to see if a flip is necessary. If all such edges are tested and none have to be flipped to guarantee the empty circle property, then we have computed $T_{i+1}$, the Delaunay triangulation of $p_1, \ldots, p_{i+1}$.

After locating the triangle of $T_i$ in which $p_{i+1}$ lies, all the flipping and testing requires time linear in the number of new edges created, which is the degree of point $p_{i+1}$ in the triangulation $T_{i+1}$.

One can show that the expected time for the flipping is constant, even though it is linear in the worst case. Consider the triangulation $T_{i+1}$. Since $p_1, \ldots, p_{i+1}$ is a random permutation of the set $\{p_1, \ldots, p_{i+1}\}$ of $i+1$ points, each one of those $i+1$ points is equally likely to have been inserted as the last one. The sums of the degrees of all points in $T_{i+1}$ is $O(i+1)$, since $T_{i+1}$ is a planar triangulation of $i+1$ points. So the average degree of a point is constant, which shows that the expected time for inserting $p_{i+1}$ after locating the triangle $t$ is also constant. This proof idea is called *backwards analysis* [6, 24]. A more complete description

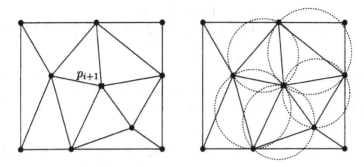

**Fig. 10.** One more flip is needed to obtain $T_{i+1}$. Right, the empty circles of the triangles incident to the new point.

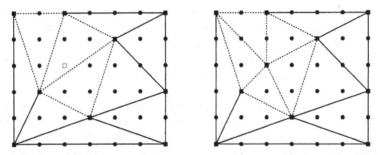

**Fig. 11.** The right TIN shows the situation if the square grid point on the left is the one with maximum error.

of randomized incremental construction of the Delaunay triangulation and its analysis can be found in [4, 6, 13].

## 5.2 Delaunay triangulation to approximate an elevation grid

We proceed with the construction of a TIN from a large grid of elevation data. The algorithm we'll describe selects a subset of the grid points, such that the Delaunay triangulation of this subset is a TIN that approximates the elevation at all grid points to within a prespecified error $\epsilon$. In other words, at all grid points, the absolute difference of its $z$-coordinate and the interpolated $z$-coordinate of the TIN is at most $\epsilon$. The approach is to start with a coarse TIN with only a few vertices, and keep adding more points from the grid to the TIN to obtain a TIN with smaller error. The algorithm was described before by Heller [16], Fjällström [9], and Heckbert and Garland [15].

1. Let $P$ be the set of midpoints of grid cells, with their elevation value. Take the four corner points and remove them from $P$, and put them in a set $S$ under construction.

2. Compute the Delaunay triangulation $DT(S)$ of $S$.

3. Determine for all points in $P$ in which triangle of $DT(S)$ they fall. For points on edges we can choose either one. Store with each triangle of $DT(S)$ a list of the points of $P$ that lie in it.

4. If all points of $P$ are approximated with error at most $\epsilon$ by the current TIN then the TIN is accepted and the algorithm stops. Otherwise, take the point with maximum approximation error, remove it from $P$ and add it to $S$. Continue at step 2.

If we assume a simple and slow implementation of the algorithm, we observe that at most $n$ times a Delaunay triangulation is computed. For each one, the points in $P$ are distibuted among the triangles of $DT(S)$. This requires for the whole algorithm $\Theta(n^3)$ tests of the type point in triangle, if a linear number of points is added to $S$.

A much faster implementation has a worst case performance of $O(n^2 \log n)$ time, and in typical situations even better: typically $O(n \log n)$ time. The algorithm resembles incremental construction of the Delaunay triangulation to some extent. But the TIN construction algorithm must also distribute the points of $P$ and find the one with maximum approximation error. We'll show that these steps can be done efficiently.

Assume that $p \in P$ has been determined as the point with maximum error, bigger than $\epsilon$, and $p$ must be removed from $P$ and added to $S$. Then we locate the triangle $t$ of $DT(S)$ that contains $p$, and we find the vertices that will become neighbors of $p$ in $DT(S \cup \{p\})$. This update step of the Delaunay triangulation is the same as in the incremental construction algorithm. To distribute the points of $P \setminus \{p\}$ over the triangles of $DT(S \cup \{p\})$, we know that only the triangles of which $p$ is a vertex in $DT(S \cup \{p\})$ have changed. So for all triangles of $DT(S)$ that don't exist in $DT(S \cup \{p\})$, we collect the associated lists of points. These points are distibuted among the new triangles and stored in new lists.

The problem that remains is locating the point with maximum error. It is solved as follows. For each triangle of the TIN we determine the point of $P$ inside it with maximum error. These points are stored in the nodes of a balanced binary search tree $\mathcal{T}$ sorted on error. This allows us to locate the point $p$ with maximum error efficiently; it is in the rightmost leaf of $\mathcal{T}$. Before $p$ is moved from $P$ to $S$, the Delaunay triangulation must be changed accordingly. To find the triangle in $DT(S)$ that contains $p$ we'll use a pointer from the node in $\mathcal{T}$ to the triangle record in the TIN structure; such pointers are shown as dashed lines with arrows in Figure 12. The triangle records are shown as grey triangles. After updating the TIN to be $DT(S \cup \{p\})$ we move $p$ from $P$ to $S$.

Then we reorganize the lists that were stored with the triangles. When $p$ was added to the Delaunay triangulation, some triangles were destroyed. The point of $P$ inside each one that had maximum error is deleted from $\mathcal{T}$. The lists of points of the destroyed triangles contain $p$ and the points that must be distributed among the new triangles, and stored in new lists. For each of the new lists we must find the point that has the maximum error in the corresponding triangle, and store it in $\mathcal{T}$.

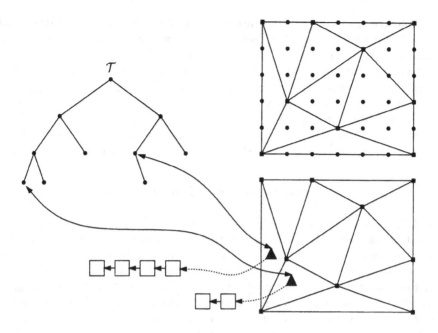

**Fig. 12.** The situation for a TIN with vertices shown as small squares (top right), and the corresponding structure with a few of the pointers between triangle objects, list elements, and tree nodes.

If $k$ is the number of neighbors of $p$ in DT($S \cup \{p\}$), then $k - 2$ triangles were destroyed and $k$ new ones were made. Let $m$ be the number of points in the triangles incident to $p$ in DT($S\cup\{p\}$). Then the iteration that added $p$ as a vertex of the TIN requires $O(k + \log n)$ time for updating the Delaunay triangulation, $O(km)$ time to redistribute the $m$ points over the $k$ triangles, and $O(k \log n)$ time to update the balanced binary tree $\mathcal{T}$. In the worst case, $m$ and $k$ are both linear in $n$, giving an worst case performance of $O(n^3)$. But redistribution of the points can also be done in $O(k + m \log m)$ time by sorting the $m$ points by angle around $p$. Since all new triangles in the TIN are incident to $p$, we can distribute the $m$ points over the $k$ triangles by using the sorted order. The modification improves the worst case running time to $O(n^2 \log n)$.

One can expect that $k$ is usually constant, and after a couple of iterations of the algorithm, $m$ will probably be much smaller than $n$. The more iterations, the smaller $m$ tends to be. One can expect that the algorithm behaves more like the best case than like the worst case, for typical inputs. In the best case, $k$ will be constant, and every list of points stored with a triangle reduces in length considerably each time it is involved in a redistribution. On the average, a new vertex has degree about six, which means that four triangles are destroyed. One can hope that the points of $P$ in these four triangles are distributed more or less evenly over the six new triangles, implying that each of the six new triangles, on the average, only has 2/3 of the points of $P$ when compared to the average of the four triangles that were destroyed. So later iterations in the algorithm tend to go faster and faster, and $m$ decreases from linear in $n$ to a constant. Or the algorithm may stop sooner because the error criterion is met. If $k$ is assumed to be a constant, we needn't use the modification to distribute the points, but simply spend $O(km) = O(m)$ time. Using an amortized analysis technique, one can show that the whole algorithm will take $O(n \log n)$ time under the (best case) assumptions given.

Emperical tests of the running time on real world input has shown that the typical running time seems to be closer to linear than to quadratic [9, 15, 16].

# 6 Compression and decompression of a TIN

In this section we explain a recent result for the compression and decompression of a TIN, assuming that the Delaunay triangulation is used for the data points. As we mentioned in the introduction, terrain data usually is huge in size, which means that a lot of storage is needed for the permanent storage of terrains, and a lot of bandwidth is needed for transmission over a network. We'll give a simple and efficient algorithm for compression and for decompression. The idea applies to some other geometric structures as well, like Voronoi diagrams, convex hulls, and vertical decompositions. It was introduced by Jack Snoeyink and the author of this paper [25, 26].

A structure like the Delaunay triangulation can be compressed by omitting all structural information (egdes and triangles), leaving only the vertices. This is true because the Delaunay triangulation is uniquely determined by its vertices. So to compress a Delaunay triangulation we need only store the vertices by the $x$-, $y$-, and $z$-coordinates. However, if we would do this, it'll take $O(n \log n)$ time to reconstruct the TIN, since it takes this much time to construct the Delaunay triangulation of $n$ points from scratch. We'll show that if the vertices are stored in a particular order, then the reconstruction takes only $O(n)$ time, and it is simpler too. The algorithm to compute this particular order takes $O(n)$ time as well. Interestingly, the sorted order on $x$-coordinate doesn't work to (re)construct the Delaunay triangulation of a point set; the $\Omega(n \log n)$ lower bound remains valid [7].

Our first algorithm produces a permutation of the data; our second takes this permutation and reconstructs the Delaunay triangulation.

## 6.1 Compression

Let $P$ be a set of $n$ points in the plane, of which the Delaunay triangulation is given in some structure. For simplicity we'll assume that the point set lies in some rectangle of which the four vertices are also data points of the set, and that there are no other four co-circular points in $P$. To construct the permutation of $P$, we deconstruct its Delaunay triangulation in phases, by deleting groups of points in a specific way. More precisely (but still as an outline of the algorithm), in one phase we find a subset of the vertices that are an independent set in the

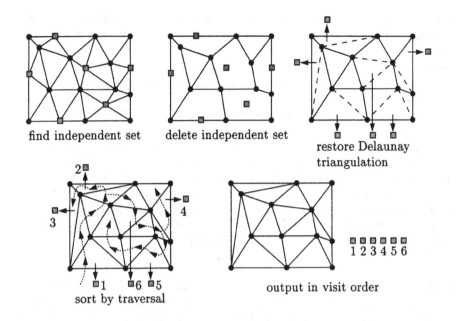

find independent set    delete independent set

restore Delaunay
triangulation

sort by traversal

output in visit order

1 2 3 4 5 6

**Fig. 13.** The steps for one phase in the compression algorithm.

Delaunay triangulation, see Figure 13. Then we delete this subset and restore
the Delaunay triangulation for the remaining vertices, while remembering in
which new triangles the deleted points fall. Next the Delaunay triangulation is
traversed, and on the way we collect the subset of points that were removed. So
the order in which the triangles are visited determines the order of the subset.
This concludes one phase, and we repeat the steps by finding a new independent
set in the smaller Delaunay triangulation.

**Find an independent set $S$.** The algorithm starts with the Delaunay triangulation of a point set $P$. There are three conditions on the independent set
we choose in the Delaunay triangulation, for efficiency or for simplicity of the
algorithm. Firstly, we won't choose any of the four corner points in this step.
Secondly, we'd like all vertices in the chosen independent set to have constant
degree in the graph, say, each chosen vertex has at most ten neighbors. Thirdly,
the independent set should be large, in any case at least some constant fraction
of the whole set. By the four-color theorem, any planar graph on $m$ vertices has
an independent set of size $\lceil m/4 \rceil$. However, it takes quadratic time to find it
[23]. Much simpler algorithms are known to compute an independent set of size
$m/5$ [3]. However, the algorithms don't give only vertices with constant degree.

The following simple algorithm determines an independent set satisfying
these conditions. The idea was introduced by Kirkpatrick [17], and used for
efficient planar point location. Initially, all vertices are unmarked and not cho-

sen. Traverse the vertices of the Delaunay triangulation in any order and choose a vertex if it has degree at most 10, it is not a corner, and it is unmarked. After choosing a vertex we mark the at most 10 neighbors, and then we continue the traversal. This simple algorithm runs in linear time.

One can show that at least 5/121 of all non-corner vertices are chosen. For convenience of analysis we forget about not choosing corner vertices. Let $m$ be the total number of vertices in the Delaunay triangulation. The total number of neighbors of all vertices is at less than $6m$ by Euler's relation. Therefore, there are less than $6m/11$ vertices with degree $> 10$, and more than $5m/11$ vertices with degree $\leq 10$. So, an algorithm that chooses one vertex with degree at most 10 and throws away the neighbors chooses at least 1 of 11 vertices of degree $\leq 10$, and in total 1/11 times $5m/11$ vertices are chosen in the independent set $S$. The analysis given can be refined to get a larger constant than 5/121. A more clever linear time algorithm can guarantee choosing roughly 4/21 of the vertices, and in practice gives a fraction close to 1/3 [26].

**Delete the points of $S$.** After computing the subset $S$ of independent vertices we must delete each one, and restore the Delaunay triangulation. One can show that the only edges that were in $\mathrm{DT}(P)$ but not in $\mathrm{DT}(P \setminus S)$ are the ones incident to the vertices of $S$. We start by removing all these edges. Since $S$ is an independent set and each vertex in $S$ had degree $\leq 10$, the graph we obtain has polygonal faces with at most 10 vertices. The faces can be triangulated separately, and in linear time we determine the Delaunay triangulation of $P \setminus S$. Then we determine for each of the deleted points of $S$ in which new triangle of $\mathrm{DT}(P \setminus S)$ it occurs. This is easy, because a deleted point occurs in one of the at most 7 new triangles formed to triangulate the face it was in. So the total deletion time is linear.

**Traverse to sort the deleted points.** The set of points $S$ that has just been deleted will be placed in some order next. This order is determined by any traversal algorithm of the triangles of the Delaunay triangulation, like the depth-first search algorithm. When we visit a triangle that contains a deleted point, we collect it and add it to the sequence of points already collected.

**End a phase and continue.** We add an end-of-phase marker to the collected sequence of points. Then we decide whether or not to restart by finding an independent set of the current, smaller Delaunay triangulation. We restart if more than five vertices are left. If only the four corner vertices and one more vertex remain, we simply store them in any order, and the algorithm terminates. For the final sequence of all $n$ points, we start with the final five points, then an end-of-phase marker, and then the points chosen and deleted in the last phase, in the order in which they were collected. Then another end-of-phase marker, followed by the points of the second-last phase, and so on.

The steps given above for one phase take $O(m)$ time for a Delaunay triangulation with $m$ vertices. But it is also true that all phases together take $O(n)$ time for an initial Delaunay triangulation with $n$ vertices. Since the problem that remains after a phase has at most $116n/121$ vertices, the running time $T(n)$ satisfies the recurrence $T(n) \leq T(116n/121) + O(n)$ if $n > 5$, and $T(n) = O(1)$ if $n \leq 5$. This recurrence solves to $T(n) = O(n)$ time.

Some final remarks about the compression algorithm. To make the algorithm work as discribed we didn't choose the four corner vertices in the independent set. These wouldn't fall in a triangle of the new, smaller Delaunay triangulation when they are deleted. Therefore we leave the four corners in until the end. Secondly, the $O(\log n)$ markers that were placed in the sequence are used in the decompression algorithm. There are a few ways to avoid them altogether, and only use the order of the points [25, 26].

## 6.2 Decompression

Decompression is done when a computer gets the data from background storage, or receives the data over a network. The data arrives in a sequence, and the algorithm can start to reconstruct the Delaunay triangulation as soon as the first points arrive. According to the compression algorithm, these are the four corner points and one additional point. The following points are inserted in phases. All steps of the algorithm basically are the reverse of some step of compression, and therefore we only sketch it briefly.

The first five points are used to initialize the reconstruction, so we start by computing their Delaunay triangulation. Then one phase, until the next end-of-phase marker, can be inserted. In any phase, the next sequence of points between two end-of-phase markers is inserted into the current triangulation as follows. Traverse the current Delaunay triangulation using *the same traversal algorithm used to collect the points*, but the traversal now serves to locate the points and store them with the triangles containing them. During the traversal we need only test if the next triangle contains the first point of the sequence. We know that the first point in the sequence must be the first point that falls in a triangle, and all following points are in triangles visited later in the traversal. After all points between two end-of-phase markers are located in the triangles, we insert them in the Delaunay triangulation. This is the flipping algorithm described before in this paper. The addition to the Delaunay triangulation ends a phase, and the next sequence of points between end-of-phase markers can be added.

Just like the compression algorithm, decompression takes $O(n)$ time for the Delaunay triangulation of $n$ points. For a comparison, the randomized incremental construction algorithm for the Delaunay triangulation takes $O(\log n)$ time expected to locate the triangle that contains the next point to be added. Then it spends $O(k)$ time for flipping if the next point has degree $k$. And one can agree that $k$ is constant in the expected case. The reconstruction algorithm of this section does the point location for one new point in $O(1)$ time in the amortized sense, that is, the location of all $n$ points in $O(n)$ triangles takes at most

$O(n)$ time together. Then $O(1)$ is used for adding a new point to the Delaunay triangulation, since we have by construction that the new point has constant degree.

# 7 Conclusions and further reading

This paper surveyed a couple of geometric algorithms that can be used when working with digital elevation data. These algorithms were developed in the research areas of computational geometry and GIS. Both areas also have strong connections with computer graphics.

Two simple algorithms for visualization were presented. More efficient methods are known to find contour lines of a terrain, by using preprocessing [10, 28]. De Berg has written a more extended survey on TIN visualization, including the use of levels of detail of terrains in visualization [5]. More generally, visualization in GIS is treated in a book edited by MacEachren and Taylor [18], see also [1, 22, 30].

Algorithms for the construction of TINs from digital elevation in another form has been studied extensively. This can be the triangulation between contour lines, grid to TIN conversion as in this paper, or producing a TIN from point data, with or without an interpolation method. Surveys on digital elevation models contain many references to such methods [29, 31].

Compression of digital elevation data hasn't been studied so much yet. For gridded data, Franklin gives a number of experimental results showing how well standard image compression techniques work for elevation data [11, 12].

More algorithms that operate on terrains and can be used in GIS have been described in a survey of the author of this paper [29].

# References

1. B.P. Buttenfield and W.A. Mackaness. Visualization. In D.J. Maguire, M.F. Goodchild, and D.W. Rhind, editors, *Geographical Information Systems – Principles and Applications*, volume 1, pages 427–443. Longman Scientific & Technical, 1991.
2. CGAL home page. `http://www.cs.ruu.nl/CGAL/`.
3. N. Chiba, T. Nishizeki, and N. Saito. A linear 5-coloring algorithm of planar graphs. *J. of Algorithms*, 2:317–327, 1981.
4. K. L. Clarkson and P. W. Shor. Applications of random sampling in computational geometry, II. *Discrete Comput. Geom.*, 4:387–421, 1989.
5. M. de Berg. Visualization of TINs. In M. van Kreveld, J. Nievergelt, T. Roos, and P. Widmayer, editors, *Algorithmic Foundations of GIS*, Lecture Notes in Comp. Science. Springer-Verlag, 1997. to appear.
6. M. de Berg, M. van Kreveld, M. Overmars, and O. Schwarzkopf. *Computational Geometry – Algorithms and Applications*. Springer-Verlag, Berlin, 1997.
7. H. Djidjev and A. Lingas. On computing Voronoi diagrams for sorted point sets. *Internat. J. Comput. Geom. Appl.*, 5:327–337, 1995.
8. D. P. Dobkin. Computational geometry and computer graphics. *Proc. IEEE*, 80(9):1400–1411, September 1992.

9. P.-O. Fjällström. Polyhedral approximation of bivariate functions. In *Proc. 3rd Canad. Conf. Comput. Geom.*, pages 187–190, 1991.

10. L. De Floriani, D. Mirra, and E. Puppo. Extracting contour lines from a hierarchical surface model. In *Eurographics'93*, volume 12, pages 249–260, 1993.

11. Wm Randolph Franklin. Compressing elevation data. In *Advances in Spatial Databases (SSD'95)*, number 951 in Lecture Notes in Computer Science, pages 385–404, Berlin, 1995. Springer-Verlag.

12. Wm Randolph Franklin and A. Said. Lossy compression elevation data. In *Proc. 7th Int. Symp. on Spatial Data Handling*, pages 8B.29–8B.41, 1996.

13. L. J. Guibas, D. E. Knuth, and M. Sharir. Randomized incremental construction of Delaunay and Voronoi diagrams. *Algorithmica*, 7:381–413, 1992.

14. L. J. Guibas and J. Stolfi. Primitives for the manipulation of general subdivisions and the computation of Voronoi diagrams. *ACM Trans. Graph.*, 4:74–123, 1985.

15. P. S. Heckbert and M. Garland. Fast polygonal approximation of terrains and height fields. Report CMU-CS-95-181, Carnegie Mellon University, 1995.

16. M. Heller. Triangulation algorithms for adaptive terrain modeling. In *Proc. 4th Int. Symp. on Spatial Data Handling*, pages 163–174, 1990.

17. D. G. Kirkpatrick. Optimal search in planar subdivisions. *SIAM J. Comput.*, 12:28–35, 1983.

18. A.M. MacEachren and D.R.F. Taylor, editors. *Visualization in Modern Cartography*. Elsevier Science Inc., New York, 1994.

19. Atsuyuki Okabe, Barry Boots, and Kokichi Sugihara. *Spatial Tessellations: Concepts and Applications of Voronoi Diagrams*. John Wiley & Sons, Chichester, UK, 1992.

20. J. O'Rourke. *Computational Geometry in C*. Cambridge Univ. Press, NY, 1994.

21. F. P. Preparata and M. I. Shamos. *Computational Geometry: An Introduction*. Springer-Verlag, New York, NY, 1985.

22. Relief: depicting a surface on a map.
http://acorn.educ.nottingham.ac.uk/ShellCent/maps/relief.html.

23. N. Robertson, D.P. Sanders, P. Seymour, and R. Thomas. Efficiently four-coloring planar graphs. In *Proc. 28th ACM Symp. Theor. Comp.*, pages 571–575, 1996.

24. R. Seidel. Backwards analysis of randomized geometric algorithms. In J. Pach, editor, *New Trends in Discrete and Computational Geometry*, volume 10 of *Algorithms and Combinatorics*, pages 37–68. Springer-Verlag, 1993.

25. J. Snoeyink and M. van Kreveld. Good orders for incremental (re)construction. In *Proc. 13th ACM Symp. Computational Geometry*, pages 400–402, 1997.

26. J. Snoeyink and M. van Kreveld. Linear time reconstruction of the Delaunay triangulation with applications. In *Proc. 7th Europ. Symp. Algorithms*, Lecture Notes in Comp. Science. Springer-Verlag, 1997.

27. S.S. Stevens. On the theory of scales of measurement. *Science*, 103:677–680, 1946.

28. M. van Kreveld. Efficient methods for isoline extraction from a TIN. *Int. J. of GIS*, 10:523–540, 1996.

29. M. van Kreveld. Digital elevation models and TIN algorithms. In M. van Kreveld, J. Nievergelt, T. Roos, and P. Widmayer, editors, *Algorithmic Foundations of GIS*, Lecture Notes in Comp. Science. Springer-Verlag, 1997. to appear.

30. Visualization techniques for landscape evaluation, literature review.
http://bamboo.mluri.sari.ac.uk/ jo/litrev/chapters.html.

31. R. Weibel and M. Heller. Digital terrain modelling. In D. J. Maguire, M. F. Goodchild, and D. W. Rhind, editors, *Geographical Information Systems – Principles and Applications*, pages 269–297. Longman, London, 1991.

32. M.F. Worboys. *GIS: A Computing Perspective*. Taylor & Francis, London, 1995.

# On the Distributed Realization of Parallel Algorithms

Klaus-Jörn Lange*

Fakultät für Informatik, Universität Tübingen

**Abstract.** This paper discusses some aspects of implementing parallel algorithms on distributed computer systems like a LAN–connected set of workstations. The notions of parallel and distributed computing are represented by their interrelation. The possibility of distributed simulations of parallel models is discussed. Finally, the complexity theoretical consequences will be addressed.

## 1  Introduction

The efficient and correct use of distributed computing systems is one of the most important and challenging topics of the present computer science. The abundance both of distributed problems and of distributed models can be roughly distinguished into two main branches: in the problem area of mastering distributed systems and in that of making efficient use of them. The latter case includes the task of speeding up computations by using parallelism. In this context, only few and rather easy synchronization problems like race analysis or dead-lock avoidance occur. But the process of speeding up cannot totally avoid to take these matters into account since every physical realization of a parallel model necessarily exhibits concurrent phenomena. The topic of this paper is to discuss certain aspects of the distributed realization of parallel algorithms on rather weak "parallel" systems like a set of Ethernet connected workstations. This research has been carried out in the projects KLARA and KOMET at the Technische Universität München and the Universität Tübingen.

The paper is organized as follows. We will first relate the notions of distributed and parallel computing and compare their properties. Then, the possibilities and limitations of bridging the gap between parallel and distributed models are discussed. Finally, some remarks concerning parallel complexity theory are given.

## 2  Parallel and Distributed Computing

This section considers distributed and parallel computing in their interrelation. First, it deals with the comparison of distributed vs parallel problems, models, and programming. Then, it considers the usefulness of parallel models as intermediate steps between parallel problems and distributed models.

* Supported by the DFG, Project La618/3-2

## 2.1 Parallel vs. Distributed

The aim of this subsection is to contrast and to compare the notions of parallel and distributed computing. Things will be (over)simplified in order to exhibit the characteristic distinctions of these two notions. A related and very interesting discussion referring to aspects of concurrency has recently been given by Panangaden [12].

*Parallel problems*, which ask for the possibility of decreasing running time by using parallelism, are a sub-area of *distributed problems* which also include questions of concurrency, nondeterminism, or liveness. While central distributed problems, like for instance correctness, are rather simple and uninteresting in the parallel case, issues like efficiency, the main concern of parallel computing, are up to now nearly irrelevant in the distributing setting. Efficiency is an objective and very crucial measure of success; there is no sense in any parallel solution, as correct or beautiful it might be, which gives no speed-up. While *concurrent problems* ask for the taming of distributed systems, parallel problems simply want to exploit them.

In the following, we use the term *distributed* problems in the general sense as the task of managing systems involving several processing units including coordination of the components of distributed systems. In contrast, we speak about *parallel* problems if we are solely interested in speeding up the computation. We remark here that every "physical" or realized system exhibits concurrent behavior. *Reality is not parallel, but inherently distributed.*

The confrontation of distributed vs. parallel also pertains to models and algorithms. *Distributed algorithms* are designed for distributed models, i.e., for systems composed of several more or less closely coupled processors exchanging messages. There is broad variety of distributed models. Throughout this paper we will consider rather weak distributed systems.

Also within parallel computing we find a large variety of models ranging from PRAMs with global memory to systolic nets with distributed memory. Which model should one choose for one's purposes? It depends, since there is sort of trade-off between ease of software development and hardware availability.

- The stronger the model, the easier to design and express parallel algorithms. In addition, it might be easier to give proofs of correctness.
- The weaker the model, the higher the chance to find an existing parallel system coming close to the chosen model.

Throughout the paper the following notation will be used in connection with parallel algorithms. The number of steps performed by a parallel program is called the parallel running time. Often we denote or bound it by $T(n)$ if $n$ is the size of the input. When there is a related sequential algorithm we denote or bound its running time by $t(n)$. The *speed-up* is then the quotient $t(n)/T(n)$. The number of processors used by the parallel algorithm is denoted by $P(n)$. An algorithm is called *optimal* if the number of processors is linear in the speed-up or, equivalently, if the time-processor-product of the parallel algorithm is linear in the sequential time.

There are essential differences in programming parallel and distributed systems. A *distributed program* for a distributed system consists in a set of sequential programs which communicate with each other by the help of mechanisms like semaphores or message passing. On the other hand, a *parallel program* consists in a sequence of steps which are executed in a synchronized way by all processing units. These differences are indicated in the following diagram:

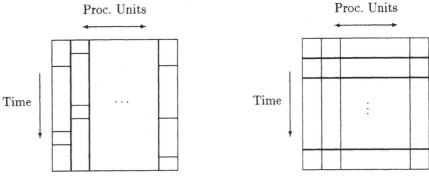

**Distributed Computing**        **Parallel Computing**

Let us consider an example. The exchange of two register values $a$ and $b$ could be done in parallel by performing $a := b$ on one processor and $b := a$ at the same time step on another one. Running the same pieces of program in an asynchronous way leads to unforeseeable results. To put it in a more picturesque scene, consider acrobats throwing and catching dishes "in parallel". Without a perfect synchrony, a pile of broken fragments would result. Alternatively, a protocol of exchange messages like "Are you ready?" – "Yes. Please throw now!" – "Did you get it?" – "Yes, I got it." ... would preserve the dishes – or not $\cdots$ "Did you get it?" – "No. Did you throw it?" – "CLINK!".

## 2.2 Distributed Solutions of Parallel Problems

Now let us discuss the process of solving parallel speed-up problems on available parallel computing systems. Since these necessarily show distributed behavior, typical problems of concurrency theory have to be solved during this process despite the fact that the original question of speeding up the running time needed to compute some function is free of these distributed aspects.

In the following we will assume the available hardware to be rather weak for instance a bunch of workstations coupled by an Ethernet. An appropriate formal model for this situation would be a set of *nodes* or active elements partially connected by a set of directed edges. There is no global clock; instead each node has its own program counter. The nodes communicate by sending and receiving messages along the edges. Typically, communications are very slow compared to local operations.

Any use of a machine or model like that involves the solution of intricate distributed problems. Let us symbolize this long way to go when solving distributed problems by constructing algorithms on distributed models by the following diagram:

The use of distributed systems in order to speed up computations is a subarea of the realm of all distributed problems. The nature of this kind of distributed problems exhibits fewer concurrent aspects and the process of finding a solution on the distributed system should be easier than the general case. In some sense it is a detour to treat speed-up questions simply like arbitrary distributed problems. This might be indicated by the following diagram:

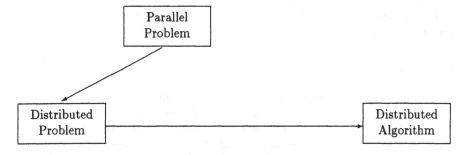

The restricted amount of distribution in parallel problems should offer easier solutions. Further on, efficiency which is most important when trying to speed up running time is usually a rather unimportant and not formalized aspect of distributed problems. These are the reasons to use parallel models which serve as a platform to formalize and express parallel algorithms. But these models leave us with the task of efficiently transforming the parallel algorithm onto a distributed system. This situation is depicted in the following diagram:

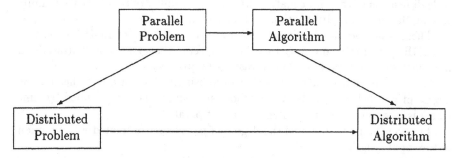

Thus the problem of speeding up the running time needed to solve some computational problem is now divided into two steps: first, construct an algorithm on a parallel model and, second, simulate the parallel model on the distributed system. The first step of coming from a problem to a solution is much easier in the parallel setting than in the distributed one. Since there are no concurrent problems involved the whole process is similar to the sequential case. But the more powerful the chosen parallel model is the easier is the step of finding a parallel solution to a speed-up problem. This is the reason for the abundance of algorithms designed for powerful parallel models like PRAMs or hypercubes.It is significant for these models that in general there are no efficient transformations known leading down to realistic distributed models.

It is the aim of this paper to discuss this "missing" second step under contradicting aspects like efficiency, correctness, ease of handling, or independence of the underlying distributed system.

# 3  Model Discussion

In this section parallel and distributed models are discussed and their main differences brought out.

## 3.1  Constants and Functions

First let us consider the degree of parallelism, i.e. the number of processing units. Off course, every real, existing machine has a bounded number of processors. Even if this number may vary in time it surely doesn't grow with the input size. Hence this number, at first sight, seemingly has to be modeled by a constant. But in order to get a both general and robust (i.e. model independent) theory constants are to be avoided.

On the parallel side of the gap the degree of parallelism will be treated as a resource measured as a function in the input length. This might be compared with the opposition of finite vs infinite memory. Although every existing computing system is finite, the appropriate models are infinite like Turing automata or register machines. In some sense even the simplest algorithm like a matrix multiplication can be regarded as a uniform family of algorithms: For each possible dimension of the input matrix there is one member of the algorithm family solving the matrix multiplication for matrices of this size.

There is no specific algorithmic theory for memory size of 16MB in contrast to 32MB and in the same way there is no specific theory for parallelization on machines with 16 processors in contrast to 32 processors.

It should be added that the danger of abusing this feature is higher for the degree of parallelism than for the memory size, since space is bounded by time which is in general not true for the degree of parallelism.

Thus on the parallel model the degree of parallelism is treated as a function of the input length.

On the other hand, on the distributed side, the number of processing units is modeled by a constant. This makes the first crucial difference between the parallel and the distributed side of the gap. The other differences of parallel and distributed models, described in the following subsection, point to fundamental difficulties in the process of simulating a parallel model on a distributed one.

But this is not the case for the number of processors, since it is no problem to simulate a larger number of processors by a smaller one without increasing the time-processor-product. In the contrary, the high degree of parallelism in many parallel algorithms could be regarded as a resource which can be exploited in order to get efficient implementations on distributed systems.

## 3.2 Local and Global Properties

There is an abundance of platforms for parallel programs reaching from systolic arrays to PRAMs. The common feature of all these models is that they assume *global time*: all participating processors are assumed to perform their steps in a synchronized way; no processor start its $i + 1$st step before all processors finished the execution of their $i$th step. Global time is the basis for *global space*: all processors share a common address space and the access time to the memory is independent of the accessed address.

In the following, when speaking of the parallel model being the platform for some parallel algorithms, we refer to a model like a PRAM with global time and global space using an input dependent number of processors and programmed in the parallel way.

The world of distributed models is an unintelligible jungle, as well. Probably the weakest, but also most common form of an existing distributed computing system is a set of workstations connected by some local area network. Inspired by this the distributed model will have *local time*, i.e. each processing unit will have its own clock or program counter which is not adjusted by any "master clock". Further on, the model has *local space*: each unit has its own local memory and cannot access directly the memory of other unit but has to simulate that by message passing over the network. In addition, the number of units is bounded and independent of the size of the actual input to be solved. On the other hand, this number is not regarded as fixed since it might change by technical faults or successful applications for grants every day. Finally, the distributed model is used by a distributed program.

The gap to be bridged between the parallel algorithm and the distributed algorithm, depicted in the last diagram, can now be given in more detail:

**Parallel Model**　　　　　　　　　　　**Distributed Model**

| Global Time | Local Time |
| Global Space | Local Space |
| $p(n)$ Units | $O(1)$ Units |
| Parallel Program | Distributed Program |

## 4　On the Distributed Realization of Parallel Models

A parallel model like a PRAM with all its properties cannot be realized in a scalable way for several physical reasons [18,19]. But even with a bounded number of processing elements this seems to be at least expensive if it is to be done in a general way. One possibility is to use the fact that the degree of parallelism of a parallel program is much larger then the actual number of processing elements. This *slackness* leads to situation that every processor has to serve a large number of the parallel jobs. Using a powerful interconnection network it is possible to hide the cost for latency of the distributed machine behind the computation time for the many jobs [13,17]. This scheduling approach has the advantage that independently of the parallel program to be executed the running time of the distributed system will with sufficiently high probability be proportional to the parallel time-processor-product. On the other hand, it should be observed that this solution uses randomization and is not deterministic. Further on, its interconnection network is logarithmic and hence this approach is not scalable. In addition, this interconnection construction is intricate and cannot be simulated by a simple network like an Ethernet, not even in the case of a small number of participating processing elements. Remarkable effort in building a PRAM has been made by W Paul and his group [1]. Their activities even include the implementation of a PRAM-specific parallel programming language [6].

A totally different way of bridging this gap is to use the fact that in most existing interconnection networks it is cheaper to send one large message than to send many small ones. This leads to the exploitation of *locality*. The disadvantage of this approach is that it is not as general as the scheduling approach. Not every parallel algorithm provides enough locality and probably not every parallel problem allows for a parallel algorithm with enough locality. This gives the (complexity) theoretical issue to say, exactly which problems are efficiently solvable using locality, and the practical issue to say, how the efficiently solvable problems can be efficiently solved. In the following, we will not consider the strategy of *caching* to try to find implicit locality by working with large neighborhoods of single addresses. Instead we will follow the approach to handle locality explicitly in the parallel algorithm.

## 4.1 Using Locality

In this subsection we consider some aspects of using locality and propose strategies to use it[1]. They are presented a bit more detailed and with some few examples in [5].

Let us shortly review the gap between the parallel Model and the distributed one. We have a parallel algorithm, i.e. a sequence of parallel steps, working with a large number of processors which communicate via a shared memory at no extra cost. We want to execute or simulate this program on a distributed machine consisting of a smaller number of processing units which communicate by exchanging messages via a network. The time needed to exchange a message through the network is much larger than that to access local data.

---

[1] Some of these theoretical results and their implementations can be found under http://www-fs.informatik.uni-tuebingen.de/forschung/komet.html.

Since the degree of parallelism is much larger than the actual number of processors, each node of the distributed system will simulate many of the parallel processes. Those of them which are laid on on the same physical node share an ideal "PRAM atmosphere": there are no synchronization problems and no differences between local and remote communication; they share global time and global space. Problems are caused by those processors which want to interact but are laid on different nodes. One way to cope with these problems is simply to consider only very restricted parallel algorithms.

- Admit only regular algorithms which allow to block many small communications into few large ones.
- Admit only simple algorithms for which the transfer from global time to local time can be done without too much overhead

Off course, this is not possible for all algorithms and is not clear which problems possess such algorithms. Nevertheless, there are simple and regular algorithms for many relevant problems like sorting or matrix multiplication.

Thus, for adequate algorithms we are faced with two major steps: the *distribution of space* and the *distribution of time*. It is a central dogma of this contribution that these two steps have to be performed in this order: first go from global space to local space and afterwards go from global time to local time. The first step asks for the solution of parallel problems and cares for efficiency. The distribution of time poses distributed problems. Its main objective will be the solution of concurrency problems. Hence we propose to work with an intermediate model with local memory but global time. The situation can be depicted as follows:

**Parallel Model**     **Border Model**     **Distributed Model**

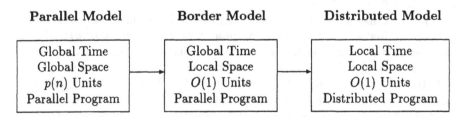

| Global Time | Global Time | Local Time |
| Global Space | Local Space | Local Space |
| $p(n)$ Units | $O(1)$ Units | $O(1)$ Units |
| Parallel Program | Parallel Program | Distributed Program |

The Border model marks the frontier between parallel and concurrency problems. It is used by a *designer* (of space) for parallel algorithms, who cares for efficiency, knows properties of his algorithm, but doesn't know too much about distributed problems or tools offered by concurrency theory. The border model is implemented by an *expert* (of time) dealing with distributed problems. The expert gets formalized, explicit information about synchronization aspects of the parallel algorithm. This information is given in terms of declarations in the parallel program by the designer. The expert implements the parallel model on a distributed one and guarantees the correctness of his simulation as long as the conditions declared by the designer are fulfilled.

**Distribution of Space: Virtual Topologies** In this first step the gap from a purely parallel model to the weaker border model has to be bridged. Every parallel algorithm exhibits in its access to the global memory some pattern which we will call *virtual topology*. This doesn't pertain solely to the arrangement of the processors. For instance, both a typical 2D-grid algorithm like a red-black pebble game and Warshall's algorithm would name their processors by two indices. But while the grid-algorithm uses communications only between neighbors, Warshall's algorithm exhibits a clique communication on each column and on each row. Thus, the virtual topology is given by naming scheme of processing units together with a collection of *multicasts*. A multicast expresses the parallel movement of data in global memory possibly including concurrent read and write features. It consists in one or more sets of names of processing units acting as senders, one or more sets of units acting as receivers and information regarding which data to be send. Typical examples would be *row-multicast*, the sending of entry in a matrix to all element in the same row, or *par-row-multicast*, the execution of row-multicast for all rows in parallel.

The algorithm designer now writes his program by explicitly declaring the virtual topology of his algorithm. This is done by choosing an adequate virtual topology including its multicasts from a predefined set of topology types. The set of available topologies might differ between different implementations of the border model and depends in the topology of the distributed system to be used. This *real topology* might range from clique realized by a slow bus to a quick hypercube network or a two dimensional grid.

The virtual topologies available within an implementation of the border model are not implemented by the algorithm designer. The field of embedding and simulating topologies into other ones is too large to expect every algorithm designer to know all about it. He is only to use the virtual topologies for expressing his parallel algorithms.

The implementation of multicasts, i.e.the translation into communications on the distributed system follows the properties of the real topology. If the real system only allows sending messages from one sender along a connection to one receiver the simulation will transform every multicast statement into a collection of point-to-point communications. If, for example, the real topology provides broadcasting this would be used for the simulation of concurrent reads.

The efficiency of the simulation of a virtual topology off course strongly depends in the real topology and will be different on different implementations of the border model. These efficiencies are represented by *border parameters* which are values expressing the time consumption of the simulations of multicast statements in the actual implementation of the border model. These values can be accessed by the designer or even by his parallel program in order to select the appropriate available virtual topology. Assume, for instance, that for some subtask there is a very good parallel algorithm using a hypercube topology and a slower one using a simple grid. If now the border parameters of the hypercube topology are bad since the actual real topology is inadequate to simulate a hypercube, it might be better to switch to the grid algorithm to solve the subtask.

This decision could even be made at runtime by the parallel program reading the border parameters. Thus the text of the parallel program would be independent of the implementation of the border model. In this way we also regard the number of processing units to be a border parameter.

Thus it is the task of the algorithm designer to select depending in the actual border parameters an appropriate algorithm and to decompose it into as many pieces as there are processing units by dividing the parallel processors appropriately onto the processing units of the border model. This could be connected with some local modifications of the original parallel algorithm in order to increase the possibility to combine many small accesses to memory into one large data movement. Sometimes it can be useful to repeat computations and thus to increase the time processor product but at the same time to decrease the number of communications [5].

**Distribution of Time: Synchronization Types** The second step has to link the border model to the distributed system. Their main difference is the use of global and local time. Even if the program on the border model simply exchanges data via directed channels, this difference is crucial and in general it is not possible to execute directly this parallel program on the distributed machine without further synchronization. This is demonstrated by the following example. Assume the real topology to consist in two units $P_1$ and $P_2$ which are connected by two directed channels. Let now the parallel task be the exchange of the value of two cells of global memory $G_1$ and $G_2$. Than this could be done by the single line OROW-Program (owner read owner write, see [15]) where $P_i$ executes $G_i := G_{3-i}$. Here $P_1$ is the read-owner of $G_2$ and the write-owner of $G_1$ while $P_2$ is the read-owner of $G_1$ and the write-owner of $G_2$. If this program would be executed without global time with high probability the content of one of the two cells would be lost.

A conceptionally simple method to provide global time is *barrier synchronization* where a node can only start a new step if all other nodes have performed their corresponding steps, which requires an enormous overhead of synchronization messages on a distributed machine. In addition, this method forces some processors to wait, even if they actually do not need any results from other processors at this point. Waiting for input values should be the only acceptable delay. This can be realized generally but inefficiently if each node stores each computed value of each step in a local list and sends such a value to another node whenever it receives a corresponding request. The transformation from global to local time should yield more efficient distributed programs in the case of a more restricted communication structures of the corresponding parallel program.

The idea to achieve this is to let the designer to give explicitly information relevant for synchronization of his program. This (s)he does by making *synchronization declarations*. These pertain either to elements of global memory or to single communications in terms of multicast statements. By assigning a *synchronization type* to a (global) variable certain conditions are guaranteed to be fulfilled by the parallel program. Thus a synchronization type is a condition which when fulfilled by a variable or by a communication makes it possible to implement this variable or this communication with a certain expense of synchronization overhead. In general one might say that the weaker the condition of a certain type the more complex and expensive will be the protocol used in implementing the type. This information is then used in the transformation from global time to local time with the aim of minimizing synchronization overhead. It should be observed that the designer isn't confronted with any distributed Problem. He works and thinks in a non distributed world with global time where matters like liveness or correctness are comparable to the sequential case.

Assume, for example, the communications accessing a cell of global memory are *2-cyclic*, that is if a processor $P$ writes into this cell $a$, then each processor $Q$ reading $a$ has to write something into another cell $b$ which has later to be read by $P$ followed later by $P$ writing into $a$ before $Q$ may again read $a$. Then this cell could be declared as being of synchronization type *cycle* (together with the information of the cycle length, in this case 2). Then the access to this cell simply can be simulated by FIFO-queues of length 2 (See [5]). Examples of 2-cyclic algorithms are convolution or red-black-algorithms. Other examples of synchronization types are *write–determined* (each processor when reading from global memory knows the time the data it is going to read has been written [11]) and *zippered* (between any two writings into a cell at different global times there is a reading from that cell and vice versa).

If a global variable doesn't fulfill the conditions of a certain synchronization type in all multicasts accessing this cell it is nevertheless possible to give guarantees for some of them by declaring those which fulfill certain restrictions and thus need less synchronization overhead.

The task of the expert who bridges the gap between border model and distributed system is to implement virtual topologies in an efficient and correct way. To preserve efficiency he tries to find or construct the best possible embedding into the real topology and to use the given synchronization information in an optimal way. His transformations have to result in correct executions on the distributed system as long as the synchronization properties stated by the designer are fulfilled. Thus correctness proofs and the resulting distributed questions occur only for every implementation of a virtual topology and not for every parallel program using this implementation.

## 5 Complexity Theory

This section deals with complexity theoretical aspects of parallel and distributed computing. First, a short overview of current parallel complexity in terms of classes like $NC$ and $P$ is given. This is followed by a discussion of the disadvantages of this approach. Finally, we discuss some possibilities to avoid these problems.

The reader of this section is assumed to be familiar with the basic facts of sequential and parallel complexity theory as they are contained for instance in [2,3].

## 5.1  Parallel Complexity Theory

The original aim of complexity theory is to classify computational problems according to the amount of time or of other resources needed for their solution. The problems of proving matching upper and lower bounds remain in general unsolved and led to the comparison of complexities and to notions like hardness and completeness.

In the the parallel case it is not useful to consider just the time needed for a solution. It seems necessary to bound simultaneously the degree of parallelism. Observe, that complexity classes defined by simultaneous resource bounds are more difficult to handle than the usual ones. For instance, they lack sharp hierarchy theorems.

Using the complexity theoretical tools which yielded the successful notion of $NP$-completeness the class $NC$ of all problems possessing efficient parallel algorithms has been introduced. While the $P$-completeness of a problem stands for its inherent sequential nature, membership in $NC$ is regarded as an indication that the running time of a problem can be decreased dramatically using parallelism. The class $NC$ is extremely robust against modification of the model. It is defined by polylogarithmically time or depth bounded PRAMs, circuits, or alternating machines. This uniform picture changes if logarithmic instead of polylogarithmic time and depth bounds come into consideration. This results in three levels. The most powerful one, *unbounded fan-in parallelism*, is represented by CREW and CRCW PRAMS, circuits of unbounded fan-in, or depth bounded alternation and contains as a typical class $AC^1$. The second, weaker level is characterized by space bounded *sequential computations* and is represented by space bounded sequential models with or without recursion. A typical representative is the class $DSPACE(\log n)$. The lowest level, *bounded fan-in parallelism*, is characterized by circuits of bounded fan-in or time bounded alternation and leads to the class $NC^1$.

## 5.2  Problems of NC-Theory

As explained above, the use of reducibilities led to the opposition of $P$-completeness regarded as a sign of being inherently sequential vs membership in the class $NC$ of all problems which have "efficient parallel algorithms". But this classification has the clear drawback that it looks for the reduction of polynomial running time down to polylogarithmic one, i.e. for exponential speed-up, without caring for the time-processor-product and optimality. Vitter and Simmons were able to construct a $P$-complete problem which has an optimal parallel algorithm with polynomial speed-up [20]. This phenomenon is caused by the use of reducibilities which allow polynomial padding when defining the class $NC$. In order to avoid this it seems necessary to restrict the growth of the reducibilities by linear or nearly linear functions. There are some attempts to define classes capturing polynomial speed-up [20,7]. There is even a "hardest sequential problem" in $P$, which has polynomial speed-up if and only if this is true for every member of

$P$ [14]. But these notions lack the existence of an adequate notion of reducibility and didn't gain the relevance of a notion like $NP$-completeness.

It should be remarked here that the many approaches to define more realistic models as for instance the LogP model [4] are adequate in order to get good estimations of running times, but they don't seem to be appropriate for structural classification of being inherently sequential in contrast to being efficiently parallelizable.

### 5.3 Classification by Criteria

A way out of this dilemma could be to classify not by criteria that are based on reducibilities like hardness and completeness but instead to find other properties of problems or algorithms which may serve as sufficient or necessary criteria for the existence or nonexistence of efficient parallel solutions.

A striking example is the notion of obliviousness or *data–independence* [8]. It was inspired by the observation that most of the algorithms which showed an acceptable speed-up on a net of workstations exhibited a static pattern of access to data. More formal, assume that we work with a CRCW-PRAM algorithm. Then we can consider the read- and the write-access of this algorithm to the global memory formalized as a graph. If these structures are independent of the actual input data but only depend in the size of the input we speak about data-independent read and data-independent write. If both read and write are data-dependent we have no restriction at all and get the full power of unbounded fan-in parallelism. For instance, in logarithmic time we can solve exactly the problems in $AC^1$. If, however, we restrict the write to be data-independent and let the read data-dependent we get space bounded sequential classes! In logarithmic time we now get the class $DSPACE(\log n)$. And if we work with static communication patterns where both the read and the write is data-independent, we end up with bounded fan-in parallelism. In logarithmic time we get exactly the class $NC^1$. This might be interpreted in the way that parallel models are related to unbounded fan-in parallelism while distributed models are related to bounded fan-in parallelism. Since space bounded sequential computations seem to be more powerful than bounded fan-in parallelism (unless $DSPACE(\log n)$ and $NC^1$ coincide) it is no surprise that a property like data-independence cannot be treated by traditional complexity theory: Their reducibilities are originally based on sequential devices which cannot preserve data-(in)dependence.

In this connection the notion of a problem being *recursively divisible* introduced by R. Niedermeier is very interesting [10,9]. He calls a problem $A$ divisible if there is an algorithm solving $A$ in time $T(n)$ using $P(n) = n^\epsilon$ processors on inputs of size $n$ which consists in $O(1)$ alternating layers of simple and regular communications followed by layers consisting in $P(n)$ independent applications of some algorithm solving a problem $B$ on inputs of size $n/P(n)$ as indicated in the following picture. (Here simple and regular is formalized as being $NC^0$-computable.) Obviously, an algorithm like that can be easily implemented on a distributed system with acceptable efficiency.

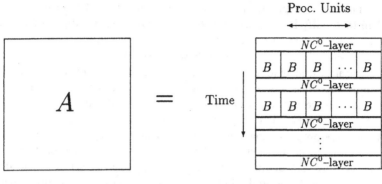

**Algorithm for** $A$

If now $B$ is identical to $A$ we can repeat this process of decomposition. In this case $A$ is called *recursively divisible* [10,9]. Many important computational problems like sorting or matrix multiplication are recursively divisible. All known examples of recursively divisible problems are members of $NC^1$, a class defined by bounded fan-in parallelism. This supports the impression that distributed models are closely related to bounded fan-in parallelism.

## 6   Discussion and open Questions

One aim of this paper was to relate and link rather separated areas. Much of this contribution is unfinished and speculative. Only few constructions have been implemented until now. One reason for this are the many open question of these areas.

A very important task to settle is to handle recursion properly. On the theoretical side a proper definition of benign recursion is missing. Recursion is necessarily dynamical and not static. Hence we have to deal with data-dependence. While there are well-known examples of malign cases of data-dependence like pointer jumping there also many cases of dynamic data-dependent algorithms which exhibit a very local communication pattern like many approximation algorithms. Neither there is a theoretical treatment of this dynamic case nor there seems to exist an adequate language to express and formalize this phenomenon. (An interesting approach to express (non)locality was given by Sabot [16]).

Another weakness of the concept of virtual topologies is that often algorithms exhibit virtual topologies which only come close to a more regular one offered in an implementation of the border model. For these cases something like an editor for topologies would be helpful which could be used by the algorithm designer to tailer an existing virtual topology to something more appropriate. The difficulties to keep efficiency and correctness of the implementation of the modified virtual topology are incalculable. One further problem would be that there would be no adequate border parameters for the modified topology.

One task of the designer is to embed his parallel algorithm using $n$ processors where $n$ depends in the input size into a border program using a much smaller number of $k$ processing units. Once he did this, it is in many cases no problem to modify his construction to be executed with a different number $k'$ of units as long as the topology stays similar. It should be possible to do this computer aided. Questions are here, how to express and formalize this process on the syntactical level and, second, can those problems which are adequate for this (semi-)automatic lay out, be characterized in complexity theoretical terms. These question are also interesting for the case of a weakly dynamic real topology where the set of available nodes may change day by day.

While these problems concerned the distribution of space there are also many open questions in connection with the concept of synchronization types. First, there is the question for more reasonable synchronization types which are general enough to be useful and which are restrictive enough to be implementable without much synchronization overhead. Unclear is the general relation between the different synchronization types and their structural behavior. Is it possible to characterize synchronization types in terms of complexity theory? Are there connections to concepts like completeness or to the other criteria mentioned in subsection 5.3?

A specific feature sometimes found in parallel algorithms is that they are robust against certain differences in local time. Consider for example a monotone algorithms like Warshall's algorithm. The participating memory cells $a_{ij}$ carry either the value 0 (there is no path from $i$ to $j$) or 1 (there is a path from $i$ to $j$)) and never a 1 is overwritten by a 0. That means that the algorithm could tolerate results which are too young, i.e. come out of the future (wrt global time). What has to be avoided is to receive data which is too old. It should be possible to exploit this behavior and to introduce a synchronization type named, may be, *best before*.

## Acknowledgment

I like to thank Henning Fernau, expert for rewriting and formal languages, for rewriting the introduction.

## References

1. F. Abolhassan, J. Keller, and W. J. Paul. On the cost-effectiveness and realization of the theoretical pram model. SFB Report 09/1991, Universität des Saarlandes, Saarbrücken, 1991. revised and extended version of SFB Report 21/1990.
2. J. Balcázar, J. Díaz, and J. Gabarró. *Structural Complexity Theory I.* Springer, 1988.
3. J. Balcázar, J. Díaz, and J. Gabarró. *Structural Complexity Theory II.* Springer, 1990.
4. D. Culler, R. Karp, A. Sahay, K. E. Schauser, E. Santos, R. Subramonian, and T. von Eicken. LogP: Towards a realistic model of parallel computation. In *Proc. 4th ACM SIGPLAN Symposium on Principles and Practice of Parallel Programming*, pages 1–12, 1993.

5. D. Gomm, M. Heckner, K.-J. Lange, and G. Riedle. On the design of parallel programs for machines with distributed memory. In *Distributed Memory Computing, Proc. 2nd European Conference, EDMCC2*, volume 487 of *LNCS*, pages 381–391. Springer, 1991.

6. T. Hagerup, A. Schmitt, and H. Seidl. FORK: A high-level language for PRAMs. Technical Report 22/90, Universität des Saarlandes, Fachbereich 14, Im Stadtwald, 6600 Saarbrücken, 12 1990.

7. C. P. Kruskal, L. Rudolph, and M. Snir. A complexity theory of efficient parallel algorithms. *Theoret. Comput. Sci.*, 71:95–132, 1990.

8. K.-J. Lange and R. Niedermeier. Data-independences of parallel random access machines. In *Proc. of 13th Conference on Foundations of Software Technology and Theoretical Computer Science*, number 761 in LNCS, pages 104–113. Springer, 1993. Accepted for publication by JCSS.

9. R. Niedermeier. Recursively divisible problems. In *Proc. of the 7th ISAAC*, number 1178 in LNCS, pages 183–192. Springer, 1996.

10. R. Niedermeier. *Towards Realistic and Simple Models of Parallel Computation*. PhD thesis, Universität Tübingen, 1996.

11. N. Nishimura. Restricted CRCW PRAMs. *Theoret. Comput. Sci.*, 123:415–526, 1994.

12. P. Panangaden. Does concurrency theory have anything to say about parallel programming? *EATCS Bull.*, 58:140–147, 1996.

13. A. G.. Ranade. How to emulate shared memory. *J. Comp. System Sci.*, 42:307–326, 1991.

14. K. Reinhardt. Strict sequential P-completeness. In *Proc. of 14th STACS*, number 1200 in LNCS, pages 329–338. Springer, 1997.

15. P. Rossmanith. The owner concept for PRAMs. In *Proc. of the 8th STACS*, number 480 in LNCS, pages 172–183. Springer, 1991.

16. Gary Wayne Sabot. *The Paralation Model*. MIT Press Cambridge Massachusetts., 1988.

17. Leslie G. Valiant. A bridging model for parallel computation. *Communications of the ACM*, 33:103–111, 8 1990.

18. Paul M.B. Vitányi. Nonsequential computation and laws of nature. In *VLSI Algorithms and Architectures*, number 227 in LNCS, pages 108–120. Springer, 1986.

19. Paul M.B. Vitányi. Locality, communication, and interconnect length in multicomputers. *SIAM J. Comp.*, 17:659–672, 1988.

20. J. S. Vitter and R.A. Simons. New classes for parallel complexity: A study of unification and other complete problems for P. *IEEE Trans. on Computers*, 35:403–418, 1986.

# The Fundamental Problem of Database Design

J.A. MAKOWSKY* and E.V. RAVVE **

Department of Computer Science
Technion - Israel Institute of Technology
Haifa, Israel
e-mail: {cselena,janos}@cs.technion.ac.il

**Abstract.** We introduce a new point of view into database schemes by applying systematically an old logical technique: *translation schemes*, and their induced formula and structure transformations. This allows us to re-examine the notion of dependency preserving decomposition and its generalization *refinement*. We have previously demonstrated the usefulness of this approach by recasting the theory of vertical and horizontal decompositions in our terminology.

The most important aspect of this approach, however, lies in laying the groundwork for a formulation of the *Fundamental Problem of Database Design*, namely to exhibit desirable differences between translation equivalent presentations of data and to examine refinements of data presentations in a systematic way. The emphasis in this paper is not on results. The main line of thought is an exploration of the use of an old logical tool in addressing the Fundamental Problem.

## 1 Introduction

Some ten years ago a famous and influential American database researcher declared that dependency and design theory for relational databases was *passé*. The classification of various useful dependency classes had reached a satisfactory state [Var87, FV86]. Furthermore, the decision problem for the consequence relation of dependencies had been solved for all but one case, the embedded multivalued dependencies, which was finally solved by C. Herrmann, [Her95]. Work on normal forms of database schemes seemed to reach an end and it was time to proceed in different directions. Nevertheless, work in design theory continued to some extent, mostly in Europe, as witnessed in [PDGV89, MR92, Tha91], but the database community at large showed diminished enthusiasm for this kind of work, [AHV94]. May be this was due to the abundance of undecidability results in dependency theory, or to the lack of decisive progress in design theory, or just a lull in fashionability.

In this lecture we make a modest attempt to revive interest in design theory. We hope to do this with the help of the systematic use of *translation schemes* as the major logical tool to carry out research in design theory.

---
\* Partially supported by a Grant of the German-Israeli Foundation and by the Fund for Promotion of Research of the Technion–Israeli Institute of Technology
\*\* Partially supported by a Grant of the German-Israeli Foundation

## 1.1 ER-Design

Entity Relation ship design theory of databases comprises the following steps:

- Identify a set of entities;
- Identify, for each entity, a fixed set of attributes;
- Identify a set of functional dependencies between those attributes;
- Check whether the entities are in some desirable normal form, say BCNF;
- If not, decompose into this normal form, if possible. (BCNF, if possible, else into 3NF);
- Choose set of relationships (and identify the resulting inclusion dependencies);
- Identify weak entities (and the resulting inclusion dependencies);
- Introduce aggregations.

The steps are iterated, choices are modified, and ultimately some satisfactory design emerges, which can be often specified with functional (FD's) and inclusion dependencies (ID's). But equally often other classes of dependences are necessary. For an excellent sound exposition, cf. [MR92, PDGV89, Tha93] and for an encyclopedic view of dependency theory, cf. [Tha91, Tha94].

If we design the database scheme directly in the relational model, we usually first identify a set attributes and functional dependencies, and only then proceed to the identification of relation schemes (or entities and relationships). Other approaches also start with a first design in Object Data Language (ODL), and then translate into the relational model, but the basic steps are very similar, cf. [MR92, UW97].

## 1.2 Stepwise modification during the design process

In the process of modification, horizontal and vertical decompositions are used. Ideally, these decompositions should be information and dependency preserving. The requirement of preservation of information leads to join dependencies (JD's). The requirement of dependency preservation very quickly becomes complicated even in the case of FD's. In particular, not every database scheme specified by FD's can be decomposed even into BCNF while preserving the dependencies.

In this lecture we introduce a new notion of *information and dependency preserving refinements* of database schemes which is meaningful for arbitrary first order dependencies. The basic underlying notion is a *first order translation scheme*. It is based on the classical syntactic notion of interpretability from logic made explicit by M. Rabin in [Rab65]. More recently, translation schemes have been used in descriptive complexity theory under the name of first order reductions [Dah83] and [Imm87], and in descriptive graph theory [Cou94]. A systematic survey of their use to analyze complexities of logical theories, cf. [CH90].

## 1.3 Outline

The paper is organized as follows: In section 2 we introduce in detail the notion of a translation scheme $\Phi$ and its derived maps $\Phi^*$ on database instances and $\Phi^\#$ on formulae expressing queries or dependencies. The *fundamental property* of translation schemes shows that these maps are dual to each other. This section is fairly abstract allowing arbitrary first order formulae or relational algebra expressions as dependencies. Although of purely logical character, it is written in the notation of databases.

In section 3 we present the general theory of information and dependency preserving transformations. We give a general definition of *information preserving transformation* of database schemes in terms of the existence of a left inverse $\Psi^*$ of a transformation given by a weak reduction $\Phi^*$. We also give a general definition of dependency preserving transformations in terms of the map $\Psi^\#$. As a result of our detailed theoretical analysis we propose a definition of *dependency preserving translation refinements* of database schemes as the key notion of design theory.

In section 4 we formulate precisely what we call the *Fundamental Problem of Database Design*. For this purpose we evaluate first the merits and failures of traditional design theory. Its main tenets consist of normal forms to achieve data independence and avoid update anomalies; attaining of normal forms using dependency preserving decompositions and minimization of interrelational dependencies.

We formulate these issues for arbitrary dependency classes and give a general formulation of what we consider the *Fundamental Conceptual Problem of Database Design*. It consists in identifying properties which distinguish translation–equivalent presentations of data represented by different database schemes, or, when this is not meaningful, to study the properties of the partial pre-order induced by translation–refinement. This leads us to the notion of *separability of a database scheme modulo a set $\Sigma$ of interrelational dependencies*. Clearly, $\Sigma$ should be as simple as possible. We show how the classical normal form theory of 3NF and BCNF fits in this framework. Our view of BCNF, taken from [MR96a] is new in its generality, although it may have been known as folklore before.

In the last section we draw our conclusions and sketch further research in progress. The use of translation schemes can be extended to a full fledged design theory for Entity–Relationship design or, equivalently, for database schemes in ER-normal form, cf. [MR92]. It can also be used to deal with views and view updates, as views are special cases of translation schemes. These issues will be discussed in forthcoming papers.

## Acknowledgments

The material presented in this paper is an outcome of teaching database theory over the years. We would like to thank O. Shmueli, our co-teacher, for many valuable arguments, and our students for their questions and their commitment in projects. An expanded version of this paper will appear as [MR97]. A preliminary version did appear as [MR96b]. We would like to thank B. Thalheim and C. Beeri for inviting the first author to present these results first as an invited speaker at ER'96. We also presented this material as lecture 6 of our joint course *Translations, Interpretations and Reductions* given at ESSLLI'97 in August 1997 at Aix-en-Provence, France. Without these encouragements this study would not have taken its current shape. Further results along these lines will appear in Ms. Ravve's Ph.D. thesis.

# 2  Translation Schemes

In this section we introduce the general framework for syntactically defined translation schemes in terms of databases. The definition is valid for a wide class of logics or query languages, including Datalog or Second Order Logic, but in this paper we restrict ourselves to Relational Calculus in the form of First Order Logic ($FOL$), but occasionally use Relational Algebra expressions when they are more convenient to read. The reader should be able to pass freely between these formalisms.

Intuitively, a translation scheme is a syntactic description of a database transformation. It defines the new database scheme in terms of the old one. The new relations are defined by queries over the old relations. Typically such queries are given as projections of joins or by selecting tuples from a table which satisfy certain conditions. The degree of generality for which we give the definitions is motivated by the need to go beyond this simplistic choice of constructs. Some of our theoretical observations will show that this is unavoidable. Translation schemes are used to define views and restructurings. In the life cycle of a database they are also used to define new generations. In the design phase of a database they are used to define the passage between two design stages, where previous specifications are somehow to be preserved.

## 2.1  The basic definition

We use the following notations:

- $\bar{A}$ is a finite set of attributes;
- $R$ is a relation name;
- $\mathbf{R}$ is a database scheme (i.e. a non empty set of relation names);
- $I(\mathbf{R})(A(\mathbf{R}), B(\mathbf{R}))$ is an instance of $\mathbf{R}$.
- $\Sigma$ is a set of boolean queries (first order unless specified otherwise). We refer to them also as dependencies.

**Definition 1 (Translation schemes $\Phi$).**
Let $\mathbf{R}$ and $\mathbf{S}$ be two database schemes. Let $\mathbf{S} = (S_1, \ldots, S_m)$ and let $\rho(S_i)$ be the arity of $S_i$. Let $\Phi = \langle \phi, \phi_1, \ldots, \phi_m \rangle$ be first order formulae over $\mathbf{R}$. $\Phi$ is $k-$ *feasible for* $\mathbf{S}$ *over* $\mathbf{R}$ if $\phi$ has exactly $k$ distinct free first order variables and each $\phi_i$ has $k\rho(S_i)$ distinct free first order variables. Such a $\Phi = \langle \phi, \phi_1, \ldots, \phi_m \rangle$ is

also called a $k$–$\mathbf{R}$–$\mathbf{S}$–*translation scheme* or, in short, a *translation scheme*, if the parameters are clear in the context. If $k = 1$ we speak of *scalar* or *non–vectorized* translation schemes.

The formulae $\phi, \phi_1, \ldots, \phi_m$ can be thought of as queries. $\phi$ describes the new domain, and the $\phi_i$'s describe the new relations. Vectorization creates one attribute out of a finite sequence of attributes. We shall discuss concrete examples after we have introduced the induced transformation of database instances.

*Remark.* In the logical tradition one also allows in translation schemes to redefine equality. For our applications in databases this is not needed.

## 2.2  Transforming database instances

A (partial) function $\Phi^*$ from $\mathbf{R}$ instances to $\mathbf{S}$ instances can be directly associated with a translation scheme $\Phi$.

**Definition 2 (Induced map $\Phi^*$).**
Let $I(\mathbf{R})$ be a $\mathbf{R}$ instance and $\Phi$ be $k$–feasible for $\mathbf{S}$ over $\mathbf{R}$. The instance $I(\mathbf{S})_\Phi$ is defined as follows:

(i) The universe of $I(\mathbf{S})_\Phi$ is the set $I(\mathbf{S})_\Phi = \{\bar{a} \in I(\mathbf{R})^k : I(\mathbf{R}) \models \phi(\bar{a})\}$.

(ii) The interpretation of $S_i$ in $I(\mathbf{S})_\Phi$ is the set

$$I(\mathbf{S})_\Phi(S_i) = \{\bar{a} \in I(\mathbf{S})_\Phi{}^{\rho(S_i)} : I(\mathbf{R}) \models (\phi_i(\bar{a}) \wedge \phi)\}.$$

Note that $I(\mathbf{S})_\Phi$ is a $\mathbf{S}$ instance of cardinality at most $\mid \mathbf{R} \mid^k$.

(iii) The partial function $\Phi^* : I(\mathbf{R}) \to I(\mathbf{S})$ is defined by $\Phi^*(I(\mathbf{R})) = I(\mathbf{S})_\Phi$. Note that $\Phi^*(I(\mathbf{R}))$ is defined iff $I(\mathbf{R}) \models \exists \bar{x}\phi$.

$\Phi^*$ maps $\mathbf{R}$ instances into $\mathbf{S}$ instances, by computing the answers to the queries $\phi_1, \ldots, \phi_m$ over the domain of $\mathbf{R}$ specified by $\phi$.

*Example 1.*

(i) In database applications translation schemes typically consist of queries which are simple projections or natural joins. This is the case for vertical decomposition via projections and its reconstruction via joins. The classical decomposition theory can be recast via translation schemes.

(ii) Other examples are horizontal decompositions via selections and its reconstruction via unions.

(iii) Let $\Phi$ be given by defining $S(x, y, z)$ by $(R(x, y) \wedge y = z)$. The instances of $S$ look like $R$ but with the $y$-column doubled (replaced by a vector). The translation scheme $\Psi$ which looks at $S$ as a binary relation of $x$'s and tuples $(y, z)$ would be a vectorized translation scheme, where $\Psi^*$ happens to be the inverse of of $\Phi^*$. Such cases of vectorized translation schemes play an important role in our new approach to Boyce Codd Normal Form.

## 2.3  Transforming queries and dependencies

Next we want to describe the way formulae (query expressions) are transformed when we transform databases by $\Phi^*$. For this a function $\Phi^\#$ from first order formulae over **S** to first order formulae over **R** can be directly associated with a translation scheme $\Phi$.

**Definition 3 (Induced map $\Phi^\#$).**
Let $\theta$ be a **S**–formula and $\Phi$ be $k$–feasible for **S** over **R**. The formula $\theta_\Phi$ is defined inductively as follows:

(i) For $S_i \in \mathbf{S}$ and $\theta = S_i(x_1, \ldots, x_l)$ let $x_{j,h}$ be new variables with $j \le l$ and $h \le k$ and denote by $\bar{x}_j = \langle x_{j,1}, \ldots, x_{j,k} \rangle$. We make $\theta_\Phi = \phi_i(\bar{x}_1, \ldots, \bar{x}_l)$.

(ii) For the boolean connectives, the translation distributes, i.e. if $\theta = (\theta_1 \vee \theta_2)$ then $\theta_\Phi = (\theta_{1\Phi} \vee \theta_{2\Phi})$ and if $\theta = \neg\theta_1$ then $\theta_\Phi = \neg\theta_{1\Phi}$, and similarly for $\wedge$.

(iii) For the existential quantifier, we use relativization, i.e. if $\theta = \exists y \theta_1$, let $\bar{y} = \langle y_1, \ldots, y_k \rangle$ be new variables. We make

$$\theta_\Phi = \exists\bar{y}(\phi(\bar{y}) \wedge (\theta_1)_\Phi).$$

(iv) The function $\Phi^\# : FOL$ over $\mathbf{S} \to FOL$ over $\mathbf{R}$ is defined by $\Phi^\#(\theta) = \theta_\Phi$.

(v) For a set of **S**–formulae $\Sigma$ we define

$$\Phi^\#(\Sigma) = \{\theta_\Phi : \theta \in \Sigma \text{ or } \theta = \forall\bar{y}(S_i \leftrightarrow S_i)\}$$

This is to avoid problems with $\Sigma$ containing only quantifierfree formulae, as $\Phi^\#(\theta)$ need not be a tautology even if $\theta$ is.

For both induced maps we have simple monotonicity properties:

**Proposition 4.** *Let* $\Phi = \langle \phi, \phi_1, \ldots, \phi_m \rangle$ *be a* $k$–$\mathbf{R}$–$\mathbf{S}$–*translation scheme,* $K_1, K_2$ *be sets of* **R**–*instances and* $\Sigma_1, \Sigma_2$ *be sets of* **S**–*sentences.*

*(i) If* $K_1 \subseteq K_2$ *then* $\Phi^*(K_1) \subseteq \Phi^*(K_2)$.
*(ii) If* $\Sigma_1 \models \Sigma_2$ *then* $\Phi^\#(\Sigma_1) \subseteq \Phi^\#(\Sigma_2)$.

The following fundamental theorem is easily verified, cf. [EFT94]. Its origins go back at least to the early years of modern logic, cf. [HB70, page 277 ff].

**Theorem 5.**
*Let* $\Phi = \langle \phi, \phi_1, \ldots, \phi_m \rangle$ *be a* $k$–$\mathbf{R}$–$\mathbf{S}$–*translation scheme,* $I(\mathbf{R})$ *be a* **R**–*instance and* $\theta$ *be a* $FOL$–*formula over* **S**. *Then* $I(\mathbf{R}) \models \Phi^\#(\theta)$ *iff* $\Phi^*(I(\mathbf{R})) \models \theta$.

# 3 Information and Dependency Preserving Refinements

In this section we introduce our general approach to information and dependency preserving refinements. We can define information preserving translation schemes $\Phi$ as those for which there is a left inverse for $\Phi^*$ which is derived from a translation scheme. In other words, we can reconstruct the original database instance. Formally we define this as follows:

**Definition 6.** Let $\langle \mathbf{R}, \Sigma^\mathbf{R} \rangle$ be database schemes. Let $\Phi$ be a $\mathbf{R} - \mathbf{S}$-translation scheme and $\Psi$ be a $\mathbf{S} - \mathbf{R}$-translation scheme. $\Phi$ is *information preserving on* $Inst(\Sigma^\mathbf{R})$ with left inverse $\Psi$ if for every relation $R_i \in Inst(\Sigma^\mathbf{R})$ we have

$$R_i \cong \Psi^*(\Phi^*(R_i)))$$

**Assumption 1.** $\Phi^*$ *is a total function and* $\Psi^*$ *is the left inverse of* $\Phi^*$.

Under these assumptions we can give an axiomatization of $\Phi^*(Inst(\Sigma^\mathbf{R}))$.

**Definition 7.** Let $DEP(\mathbf{R})$ be a class of dependencies over $\mathbf{R}$ and denote by $TAUT(\mathbf{R})$ the set of tautologies over $\mathbf{R}$.

(i) $(TAUT^\mathbf{R}_{DEP})^\Phi = \{\theta \in DEP(\mathbf{S}) : \Phi^\#(\theta) \in TAUT(\mathbf{R})$ is a tautology $\}$
(ii) $(\Sigma^\mathbf{R}_{DEP})^\Phi = \{\theta \in DEP(\mathbf{S}) : \Sigma^\mathbf{R} \models \Phi^\#(\theta)\}$
(iii) If $DEP$ consist of all $FOL(\mathbf{R})$-dependencies, we omit it, $(\Sigma^\mathbf{R})^\Phi = (\Sigma^\mathbf{R}_{FOL})^\Phi$.

Note that $(\Sigma^\mathbf{R}_{DEP})^\Phi$ is infinite as defined, provided $DEP$ is infinite.

**Proposition 8.** *Under the assumption above*

$$\Phi^*(Inst(\Sigma^\mathbf{R})) = Inst(\Psi^\#(\Sigma^\mathbf{R})) \cap Inst((TAUT^\mathbf{R})^\Phi)$$

From a database point of view decompositions are made to allow more legal instances than the original database scheme, in other words, we are interested in some database scheme $\langle \mathbf{S}, \Sigma^\mathbf{S} \rangle$ where $\Phi^*(Inst(\Sigma^\mathbf{R})) \subseteq Inst(\Sigma^\mathbf{S})$. Given a transformation of a database scheme $\langle \mathbf{R}, \Sigma^\mathbf{R} \rangle$ into $\langle \mathbf{S}, \Sigma^\mathbf{S} \rangle$ by $\Phi^*$ with left inverse $\Psi^*$, what can we reasonably request about $\Sigma^\mathbf{S}$ ? We definitely want that $\Sigma^\mathbf{R} \models \Phi^\#(\Sigma^\mathbf{S})$ or in terms of $Inst(\mathbf{S})$: $\Phi^*(Inst(\Sigma^\mathbf{R})) \subseteq Inst(\Sigma^\mathbf{S})$.

**Question 9.** *Our only well defined candidate for $\Sigma^\mathbf{S}$ is $(\Sigma^\mathbf{R})^\Phi$. What is the relationship of $Inst((\Sigma^\mathbf{R})^\Phi)$ with $Inst(\Sigma^\mathbf{S})$ ?*

We are given

**Assumptions 2.** $\langle \mathbf{R}, \Sigma^\mathbf{R} \rangle$ *and* $\langle \mathbf{S}, \Sigma^\mathbf{S} \rangle$ *are database schemes.*

(i) $\Phi$ *is a* $\mathbf{R} - \mathbf{S}$-*translation scheme, which is information preserving for* $\Sigma^\mathbf{R}$ *with left inverse* $\Psi^*$ *for* $\Phi^*$.
(ii) $\Phi^*(Inst(\Sigma^\mathbf{R})) \subseteq Inst(\Sigma^\mathbf{S})$.

Informally, dependency preservation says that $\Sigma^{\mathbf{R}}$ somehow follows from $\Sigma^{\mathbf{S}}$. However, put this way, the statement makes no sense as the formulae involved speak about different database schemes.

To make the notion of preservation of dependencies precise we have three options:

(**A**) For **R**-instances: $\Phi^{\#}(\Sigma^{\mathbf{S}}) \models \Sigma^{\mathbf{R}}$
(**B**) For **S**-instances: $\Sigma^{\mathbf{S}} \models \Psi^{\#}(\Sigma^{\mathbf{R}})$
(**C**) For **S**-instances: $(\Sigma^{\mathbf{R}})^{\Phi} \models \Psi^{\#}(\Sigma^{\mathbf{R}})$

To make these three options meaning full we have to take in account two members of assumption 2. To avoid overspecification of the transformed database scheme we may sometimes assume $(\Sigma^{\mathbf{R}})^{\Phi} \models \Sigma^{\mathbf{S}}$.

The relationship between these options is given by the following

**Proposition 10.** *Under assumption 2 we have:*

*(i) Under only the assumption above we have (B) $\rightarrow$ (A).*
*(ii) If $\Phi^{*}$ is also a left inverse of $\Psi^{*}$, we have also (A) $\rightarrow$ (B).*
*(iii) If $(\Sigma^{\mathbf{R}})^{\Phi} \models \Sigma^{\mathbf{S}}$ then (B) $\rightarrow$ (C).*
*(iv) If $\Psi$ is also a weak reduction from $Inst(\Sigma^{\mathbf{S}})$ to $Inst((\Sigma^{\mathbf{R}})^{\Phi})$ then we have also (C) $\rightarrow$ (B).*

We see that (A) is the notion mostly used in the literature. This is justified only if $\Phi^{*}$ is also the left inverse of $\Psi^{*}$ on $\Sigma^{\mathbf{S}}$.

In the light of the above analysis we now propose the following definition:

Let $DEP$ be a class of first order dependencies and $TRANS$ be a class of first order translation schemes.

**Definition 11 (Translation refinement and equivalence).**
Given $\Sigma^{\mathbf{R}} \cup \Sigma^{\mathbf{S}} \subseteq DEP$ and $\langle \mathbf{R}, \Sigma^{\mathbf{R}} \rangle$ and $\langle \mathbf{S}, \Sigma^{\mathbf{S}} \rangle$ two database schemes.

(i) We say that $\langle \mathbf{S}, \Sigma^{\mathbf{S}} \rangle$ is a *translation–refinement of* $\langle \mathbf{R}, \Sigma^{\mathbf{R}} \rangle$ if there is $\Phi \in TRANS$ which is both information and $(B)$-dependency preserving.
(ii) We say that $\langle \mathbf{R}, \Sigma^{\mathbf{R}} \rangle$ and $\langle \mathbf{S}, \Sigma^{\mathbf{S}} \rangle$ are *translation–equivalent* under transformations from $TRANS$ if one can be converted into the other by $\Phi^{*}, \Psi^{*} \in TRANS$ which are information and $(B)$-dependency preserving.

If $DEP$ and $TRANS$ are omitted we allow arbitrary first order dependencies and translation schemes.

A natural choice for $\Sigma^{\mathbf{S}}$ is given by $\Psi^{\#}(\Sigma^{\mathbf{R}}) \cup \Sigma_0$ with $\Sigma_0 \subseteq (TAUT^{\mathbf{R}})^{\Phi}$.

**Proposition 12.** *Under the assumptions 2 we have that for every choice of $\Sigma_0 \subseteq (TAUT^{\mathbf{R}})^{\Phi}$ if $\Sigma^{\mathbf{S}}$ is equivalent to $\Psi^{\#}(\Sigma^{\mathbf{R}}) \cup \Sigma_0$ then $\langle \mathbf{S}, \Sigma^{\mathbf{S}} \rangle$ is a translation–refinement of $\langle \mathbf{R}, \Sigma^{\mathbf{R}} \rangle$.*

*Proof.* Immediate from the definition of translation refinement.

The converse of proposition would assert that every refinement is axiomatizable by some choice of $\Sigma_0$. But we only know that $\Psi^\#(\Sigma^\mathbf{R}) \cup (TAUT^\mathbf{R})^\Phi \models \Sigma^\mathbf{S}$ and $\Sigma^\mathbf{S} \models \Psi^\#(\Sigma^\mathbf{R}) \cup \Sigma_1$ with $\Sigma_1 = \{\theta \in (TAUT^\mathbf{R})^\Phi : \Sigma^\mathbf{S} \models \theta\}$, which does not seem to determine $\Sigma_0$.

**Problem 13.** *Under what assumptions is the converse of proposition 12 true ?*

Let $DEP$ be a class of dependencies. Proposition 12 gives some clue on how to specify refinements. In many cases $\Psi^\#(\Sigma^\mathbf{R})$ would be a good choice provided it were in $DEP$. If not, we still can look for $\Sigma_0 \in (TAUT^\mathbf{R})^\Phi$ such that $\Psi^\#(\Sigma^\mathbf{R}) \cup \Sigma_0$ is equivalent to some set of dependencies in $DEP$.

In contrast, for $\Sigma^\mathbf{S} = (\Sigma^\mathbf{R})^\Phi$ the situation is more complicated:

**Theorem 14.** *Under the assumptions 2 restricted to $DEP$ the following are equivalent:*

*(i)* $\langle \mathbf{S}, (\Sigma^\mathbf{R})^\Phi \rangle$ *is a* translation–refinement *of* $\langle \mathbf{R}, \Sigma^\mathbf{R} \rangle$.
*(ii)* $(\Sigma^\mathbf{R})^\Phi \models \Psi^\#(\Sigma^\mathbf{R})$.
*(iii)* $Inst((\Sigma^\mathbf{R})^\Phi = \Phi^*(Inst(\Sigma^\mathbf{R})))$.

*Proof.* Combine the definition of translation refinements with proposition 10.

In the case of $DEP = FD$, $\Psi^\#(\Sigma^\mathbf{R})$ is not equivalent to a set of FD's, whereas $(\Sigma^\mathbf{R})^\Phi$ is equivalent to a set of FD's.

The problem of the existence of dependency preserving refinements is intimately related to the choice of restrictions imposed on $DEP$ and $TRANS$.

**Problem 15.** *For what choices of $DEP$ and $TRANS$ is there a natural choice for $\Sigma^\mathbf{S} \subseteq DEP$ ?*

# 4 The Fundamental Problem of Database Design

## 4.1 Why Normal Forms ?

Given a collection of attributes and a collection of data dependencies the database design problem consists in finding a database scheme with certain desirable additional properties. Some of these properties addressed in the literature are

- **Normal forms:** 3NF, Boyce–Codd NF, 4NF, 5NF, Domain Key NF are normal forms for individual relation schemes, ER-Normal Form is a normal form for the database scheme, [Ull82, Mai83, MR92, Fag81]. All these normal forms are attempts to avoid insertion and deletion anomalies and to provide maximum data independence. They are related to the predominant role played by key dependencies. Given a set of dependencies $\Sigma$ let $\Sigma_{Key}$ be the set of FD's which follow from $\Sigma$ and are keys of some relation. Then BCNF, 4NF and 5NF are defined by requiring that $\Sigma_{Key} \models \Sigma$. Domain Key NF requires that $\Sigma$ is implied by $\Sigma_{Key}$ together with the domain specifications. In the Entity–Relationship design approach the entities and relationships are in these normal forms.

- **Preservation of dependencies:** When decomposing or normalizing a database scheme $\langle \mathbf{R}, \Sigma^{\mathbf{R}} \rangle$ into a new schema $\langle \mathbf{S}, \Sigma^{\mathbf{S}} \rangle$ we want that the decomposition is information preserving. To have more data independence we may allow new instances of $\langle \mathbf{S}, \Sigma^{\mathbf{S}} \rangle$ which do not come from instances of $\langle \mathbf{R}, \Sigma^{\mathbf{R}} \rangle$. All we want is that $\Sigma^{\mathbf{R}}$ is somehow implied by $\Sigma^{\mathbf{S}}$. For this purpose we have defined our notion of dependency preserving refinement.
- **Keep interrelational dependencies simple:** In ER-design the only interrelational dependencies are inclusion dependencies between relationship tables and entity tables, if restricted to the projection on they keys of the entities, in other words they are keybased. In general, it is desirable to avoid interrelational dependencies or, to keep them as simple as possible. Another way to keep them simple is to assure various forms of acyclicity. This approach is both inherent in ER-Normal form and in the definition of acyclic database schemes via the existence of jointrees. When dealing with EID's or more general sets of dependencies, this is not trivial to formulate precisely.

Partial successes in such design endeavors were achieved and hopes were entertained for the development of computer aided tools for database design. However, serious theoretical problems for the development of such tools exist. On the one hand, we have algorithmic unsolvability of some of the underlying concepts, on the other hand, we have to narrow our design goals when we try to balance algorithmic solvability with engineering requirements.

## 4.2 Generalizing normal forms

For a fixed dependency class $DEP$ and a database scheme $\langle \mathbf{R}, \Sigma^{\mathbf{R}} \rangle$ with $\mathbf{R} = \{\mathbf{R_1}, \ldots \mathbf{R_n}\}$ and $\Sigma^{\mathbf{R}} \subseteq DEP(\mathbf{R})$ we would like to find criteria for transparent design.

It may be useful to single out two subsets of $DEP$:

$RDEP_i$: the dependencies with which we specify the single relation scheme $R_i$ and

$IDEP$: the inter-relational dependencies with which we specify referential integrities.

Traditionally one has $RDEP$ consist of

**FD** all functional dependencies;
**FD(key)** all functional dependencies which define keys;
**MVD** the multivalued dependencies; or
**JD** the join dependencies.

For $IDEP$ we have often the empty set or $IDEP = IND$, the set of inclusion dependencies.

We put

$$\Sigma_i = \{\sigma \in RDEP_i : \Sigma^{\mathbf{R}} \models \sigma\}$$

and

$$\Delta = \{\sigma \in IDEP : \Sigma^{\mathbf{R}} \models \sigma\}$$

The classical normal forms can be expressed now as follows:

**BCNF:** $DEP = FD$, $RDEP = FD(key)$, $IDEP = \emptyset$ and $\bigcup_i \Sigma_i \models \Sigma^{\mathbf{R}}$;
**4NF:** $DEP = MVD$, $RDEP = FD(key)$, $IDEP = \emptyset$ and $\bigcup_i \Sigma_i \models \Sigma^{\mathbf{R}}$;
**JDNF:** (Project-Join NF) $DEP = JD$, $RDEP = FD(key)$, $IDEP = \emptyset$ and
$\bigcup_i \Sigma_i \models \Sigma^{\mathbf{R}}$.

**Problem 16.** *Find a similar formulation for* $3NF$.

If $DEP = IDEP \cup \bigcup_i RDEP_i$ then trivially

$$\Delta \cup \bigcup_i \Sigma_i \models \Sigma^{\mathbf{R}}.$$

For design in ER-diagrams the following seems interesting:

**Problem 17.** *(i) Let* $DEP = EID$, $RDEP = FD$ *and* $IDEP = IND$. *Characterize those* $\Sigma^{\mathbf{R}} \subseteq EID$ *for which*

$$\Delta \cup \bigcup_i \Sigma_i \models \Sigma^{\mathbf{R}}$$

*(ii) Similarily for* $DEP = EID$, $RDEP = FD$ *and* $IDEP = aIND$, *where* $aIND$ *are the* acyclic inclusion dependencies.

### 4.3 The consequence problem for dependencies

The *consequence problem* is of the form

$$\Sigma_1 \cup \Sigma_2 \models \theta$$

or, in the case we restrict our attention to finite databases,

$$\Sigma_1 \cup \Sigma_2 \models_{fin} \theta$$

where $\Sigma_i \subseteq DEP_i$ and $\theta \in DEP_0$ for various dependency classes $DEP_j$. The set $FID$ consisting of universal Horn formulas is called the set of *full implicational dependencies*. It does not contain the inclusion dependencies, unless the right hand side of the inclusion dependency is an atomic formula. *acyclicIND* denotes the subset of the inclusion dependencies consisting of of acyclic systems of inclusion dependencies, cf. [AHV94].

Decidability of the consequence problem
(Mitchell 1983, Chandra-Vardi 1985; Sciore 1986)

| $DEP_1$ | $DEP_2$ | $DEP_0$ | $\models$ | $\models_{fin}$ |
|---|---|---|---|---|
| $IND$ | $FD$ | $FD$ | no | no |
| $IND$ | $FD$ | $IND$ | no | no |
| $acyclicIND$ | $FID$ | $FID$ | yes | yes |

Even if we restrict our dependencies to subsets $\Sigma$ of FD's and IND's, the set $\Sigma_{FD}$ of FD's which are consequences of $\Sigma$ is not, in general, recursive, [AHV94]. But for the definition of all the normal forms the set $\Sigma_{FD}$ has to be computed. Worse, even if we restructure our database scheme with a translation scheme $\Phi$ which consists of projections and joins only, the possible choices of new dependencies which preserve the old ones, always involves the set $\Psi^{\#}(\Sigma)$, as we argued in section 3.

In this last section we want to **disregard** algorithmic unsolvability and concentrate on the conceptual aspects.

We shall outline how to use first order translation schemes to make the design problem a bit more precise in the presence of more complicated dependencies.

## 4.4  Formulating the Fundamental Problem of Database Design

For a fixed dependency class $DEP$ and a database scheme $\langle \mathbf{R}, \Sigma^{\mathbf{R}} \rangle$ with $\mathbf{R} = \{R_1, \ldots R_n\}$ and $\Sigma^{\mathbf{R}} \subseteq DEP(\mathbf{R})$ we would like to find criteria for transparent design.

Our first requirement is data independence.

**Definition 18 Data independence:.** We say that $\mathbf{R}$ *is independent over* $\Sigma^{\mathbf{R}}$ if there is no first order query $\theta \in FOL(\mathbf{R} - \{R_i\})$ such that $\Sigma^{\mathbf{R}} \models (\theta \leftrightarrow R_i)$.

We require here that no relation of $\mathbf{R}$ is first order definable over the others in the presence of $\Sigma^{\mathbf{R}}$. The choice of first order queries is natural but not unique. If we consider two relations where one is the transitive closure of the other, then they are independent as defined here but interdefinable using DATALOG queries. Database schemes specified by FD's only are always independent. This case also shows that this notion of independence is not enough to allow independent data updates. On the other hand, with $IND$'s alone we can get dependent relations, e.g. $R_0 = \pi_X R_1$.

The following is straight forward:

**Proposition 19.** *Let* $\theta \in FOL(\mathbf{R} - \{R_i\})$ *be a first order query which defines* $R_i$. *Let*

$$\Phi = \langle R_1, \ldots, R_{i-1}, \theta, R_{i+1}, \ldots, R_n \rangle$$

*be a translation scheme. Then* $\langle \mathbf{R} - \{R_i\}, \Phi^{\sharp}(\Sigma^{\mathbf{R}}) \rangle$ *is translation-equivalent to* $\langle \mathbf{R}, \Sigma^{\mathbf{R}} \rangle$.

Iterating this proposition we can always achieve independence, but $\Phi^{\sharp}(\Sigma^{\mathbf{R}})$ may be outside of $DEP$, depending on the complexity of $\theta$.

**Problem 20.** *Find criteria for independent relations. In particular, are database schemes specified by $FD$'s and acyclic$IND$'s always independent ?*

Our next requirement concerns $\Sigma^{\mathbf{R}}$. In the degenerate case $\Sigma^{\mathbf{R}}$ does not express any interaction between the relations.

**Definition 21.** Let $\langle \mathbf{R}, \Sigma^{\mathbf{R}} \rangle$ be a database scheme. We say that $\Sigma^{\mathbf{R}}$ is *separable over* $\mathbf{R}$ if there exist for each $i$ a $\Sigma_i \subset DEP(R_i)$ such that $\bigcup_i \Sigma_i$ is equivalent to $\Sigma^{\mathbf{R}}$.

Separability expresses the fact that no interrelational dependencies are needed. If a database scheme is separable, there are obvious candidate for the $\Sigma_i$'s. The following is immediate.

**Proposition 22.** *For* $\langle \mathbf{R}, \Sigma^{\mathbf{R}} \rangle$ *we define* $\Sigma_i = \{\phi \in DEP(R_i) : \Sigma^{\mathbf{R}} \models \phi\}$. $\Sigma^{\mathbf{R}}$ *is separable over* $\mathbf{R}$ *iff* $\bigcup_i \Sigma_i$ *is equivalent to* $\Sigma^{\mathbf{R}}$.

Note that it is undecidable to check whether a given $\Sigma^{\mathbf{R}}$ is separable over $\mathbf{R}$ already for $DEP = FD \cup IND$, as the corresponding consequence problem is undecidable, cf. [AHV94].

Separability generalizes the notion of independence.

**Proposition 23.** *If* $\Sigma^{\mathbf{R}}$ *is separable over* $\mathbf{R}$ *then* $\mathbf{R}$ *is independent over* $\Sigma^{\mathbf{R}}$. *But the converse does not hold.*

*Proof.* Let $\theta$ define $R_0$ over $\mathbf{R} - \{R_0\}$ and $\Sigma^{\mathbf{R}}$. In other words $\Sigma^{\mathbf{R}} \models \theta \leftrightarrow R_0$ which contradicts separability.

Conversely, two relations with one inclusion dependency give an independent set of relations which are not separable.

A database scheme with only functional dependencies is always trivially separable with $\Sigma_i$ the FD's for $R_i$. If a database scheme is not separable, we might ask whether this fact is due to poor design, i.e. whether there is a translation refinement (or a translation equivalent database scheme) which is separable ?

**Definition 24.** Let $\langle \mathbf{R}, \Sigma^{\mathbf{R}} \rangle$ be a database scheme. $\langle \mathbf{R}, \Sigma^{\mathbf{R}} \rangle$ *splits* if there exists a database scheme $\langle \mathbf{S}, \Sigma^{\mathbf{S}} \rangle$ such that

(i) $\mathbf{S}$ contains at least two relation symbols;
(ii) $\Sigma^{\mathbf{S}} \subseteq DEP(\mathbf{S})$;
(iii) $\langle \mathbf{S}, \Sigma^{\mathbf{S}} \rangle$ is a translation–refinement of $\langle \mathbf{R}, \Sigma^{\mathbf{R}} \rangle$;
(iv) $\mathbf{S}$ is independent over $\Sigma^{\mathbf{S}}$ and
(v) $\Sigma^{\mathbf{S}}$ is separable over $\mathbf{S}$.

If $\langle \mathbf{R}, \Sigma^{\mathbf{R}} \rangle$ splits we call the database scheme $\langle \mathbf{S}, \Sigma^{\mathbf{S}} \rangle$ which witnesses this fact a *splitting* of $\langle \mathbf{R}, \Sigma^{\mathbf{R}} \rangle$.

The classical synthesis algorithm into 3NF is a splitting. More generally, dependency preserving decomposition into BCNF, as defined traditionally, requires a splitting.

**Problem 25.** *Find deeper (model theoretic) criteria for separability and splitting.*

If a database schemes is specified by FD's only, it always splits. But if the database scheme originates from an ER-scheme, the presence of inclusion dependencies will prevent, in general, splitting. The *universal instance* of [Ull82] is not necessarily a solution to the splitting problem, as the resulting translation scheme is not dependency preserving.

Splitting is, in general, too strong a requirement. What we really want is that the interaction between the relations be as simple to express as possible. To capture this notion we specify a class $IDEP$ of interrelational dependencies. A good candidate for $IDEP$ are the inclusion dependencies $IND$. We do not require that $IDEP \subseteq DEP$ as we could take for $DEP = FID$ and $IDEP = IND$. If $DEP \subseteq IDEP$ all the notions trivialize.

**Definition 26.** Let $\langle \mathbf{R}, \Sigma^{\mathbf{R}} \rangle$ be a database scheme.

(i) We say that $\Sigma^{\mathbf{R}}$ is *separable over* $\mathbf{R}$ *modulo* $IDEP$ if there exist $\Sigma_i \subset DEP(R_i)$ and $D \subseteq IDEP$ such that $\bigcup_i \Sigma_i \cup D$ is equivalent to $\Sigma^{\mathbf{R}}$.

(ii) $\langle \mathbf{R}, \Sigma^{\mathbf{R}} \rangle$ *splits modulo* $IDEP$ if there exists a database scheme $\langle \mathbf{S}, \Sigma^{\mathbf{S}} \rangle$ such that

(ii.a) $\mathbf{S}$ contains at least two relation symbols;

(ii.b) $\Sigma^{\mathbf{S}} \subseteq DEP(\mathbf{S})$;

(ii.c) $\langle \mathbf{S}, \Sigma^{\mathbf{S}} \rangle$ is a translation–refinement of $\langle \mathbf{R}, \Sigma^{\mathbf{R}} \rangle$;

(ii.d) $\mathbf{S}$ is independent over $\Sigma^{\mathbf{S}}$ and

(ii.e) $\Sigma^{\mathbf{S}}$ is separable modulo $IDEP$ over $\mathbf{S}$.

Clearly, every database scheme derived from an ER-scheme splits modulo $IND$. In this case we can choose the $\Sigma_i \subseteq FD$ and such that each $\langle S_i, \Sigma_i \rangle$ is in BCNF, cf. [MR92]. In other words, modulo $IND$ every database scheme $\langle \mathbf{R}, \Sigma^{\mathbf{R}} \rangle$ with $\Sigma^{\mathbf{R}} \subseteq FD$ splits into BCNF.

**Problem 27.** *Characterize the $\Sigma^{\mathbf{R}}$ for which a given database scheme $\langle \mathbf{R}, \Sigma^{\mathbf{R}} \rangle$ splits modulo $IND$ into BCNF.*

It is not clear whether BCNF is the best choice. However, a general approach to normal forms should take into account splitting modulo $IND$.

> **The Fundamental Problem of Database Design Theory,** consists in the systematic study of non–splitting database schemes.

## 4.5 Normal form theorems revised

In [MR97] and [Rav98] two improved Normal Form Theorems are proved. They show that dependency preserving BCNF and even 4NF can always be achieved if we allow a wider range of translation-refinements. Furthermore, the query independence of the relations can be preserved and the inter-relaional dependencies can be restricted to inclusion dependencies.

**Theorem 28 (Boyce-Codd Normal Form).** *Let* $\langle \mathbf{R}, \Sigma^{\mathbf{R}} \rangle$ *be a relation scheme with* $\Sigma^{\mathbf{R}} \subseteq FD \cup IND$. *There is database scheme* $\langle \mathbf{S}, \Sigma^{\mathbf{S}} \rangle$ *with* $\Sigma^{\mathbf{S}} \subseteq FD \cup IND$ *which is a translation-refinement with weak reductions* $\Phi$ *and (left inverse)* $\Psi$ *such that:*

*(i) Both* $\Phi$ *and* $\Psi$ *are compositions of projections and joins, but* $\Psi$ *uses vectorization.*

*(ii)* $\Phi$ *is dependency preserving in the sense of (B-DEP).*

*(iii) For* $F^{\mathbf{S}} = \{f \in FD : \Sigma^{\mathbf{S}} \models f\}$, $\langle \mathbf{S}, F^{\mathbf{S}} \rangle$ *is in Boyce-Codd Normal Form.*

*(iv) Hence,* $\Sigma^{\mathbf{S}}$ *is separable over* $\mathbf{S}$ *modulo* $IND$.

*(v) Furthermore, if* $\Sigma^{\mathbf{R}}$ *is independent over* $\mathbf{R}$ *then* $\Sigma^{\mathbf{S}}$ *is independent over* $\mathbf{S}$.

**Theorem 29 (Fourth Normal Form).** *Let* $\langle \mathbf{R}, \Sigma^{\mathbf{R}} \rangle$ *be a relation scheme with* $\Sigma^{\mathbf{R}} \subseteq FD \cup MVD$. *There is database scheme* $\langle \mathbf{S}, \Sigma^{\mathbf{S}} \rangle$ *with* $\Sigma^{\mathbf{S}} \subseteq FD \cup IND$ *which is a translation-refinement with weak reductions* $\Phi$ *and (left inverse)* $\Psi$ *such that:*

*(i) Both* $\Phi$ *and* $\Psi$ *are compositions of projections and joins, but* $\Psi$ *uses vectorization.*

*(ii)* $\Phi$ *is dependency preserving in the sense of (B-DEP).*

*(iii) For* $\Theta^{\mathbf{S}} = \{\theta \in FD \cup MVD : \Sigma^{\mathbf{S}} \models \theta\}$, $\langle \mathbf{S}, \Theta^{\mathbf{S}} \rangle$ *is in Fourth Normal Form.*

*(iv) Hence,* $\Sigma^{\mathbf{S}}$ *is separable over* $\mathbf{S}$ *modulo* $IND$.

*(v) Furthermore, if* $\Sigma^{\mathbf{R}}$ *is independent over* $\mathbf{R}$ *then* $\Sigma^{\mathbf{S}}$ *is independent over* $\mathbf{S}$.

## 5   Conclusions and Further Research

We have introduced the use of translation schemes into database design theory. We have shown how they capture disparate notions such as information preservation and dependency preservation in a uniform way. We have shown how they relate to normal form theory and have stated what we think to be the Fundamental Problem of Database Design.

As the material presented grew slowly while teaching database theory, its foundational and didactic merits should not be underestimated. Over the years our students of the advanced database theory course confirmed our view that traditional database design lacks coherence and that this approach makes many issues accessible to deeper understanding.

Our approach via dependency preserving translation–refinements can be extended to a full fledged design theory for Entity–Relationship design or, equivalently, for database schemes in ER-normal form, cf. [MR92]. It is also the appropriate framework to compare transformations of ER-schemes and to address the Fundamental Problem of ER–Database Design.

Translation schemes can also be used to deal with views and view updates, as views are special cases of translation schemes. The theory of *complementary views* from [BS81] can be rephrased elegantly in this framework. It is connected with the notion of translation schemes invariant under a relation and implicit definability, [Kol90]. Order invariant translation schemes play an important role in descriptive complexity theory, [Daw93] and [Mak94]. The theory of *independent* complementary views of [KU84] exhibits some severe limitations on the applicability of [BS81]. In spite of these limitations it seems worthwhile to explore the connection between independent views and transformation invariant for certain relations further.

The latter two applications are currently being developed by the authors and their students.

# References

[AHV94] S. Abiteboul, R. Hull, and V. Vianu. *Foundations of Database.* Addison Wesley, 1994.

[BS81] F. Bancilhon and N. Spyratos. Update semantics of relational views. *ACM Transactions on Database Systems,* 6(4):557-575, 1981.

[CH90] K.J. Compton and C.W. Henson. A uniform method for proving lower bounds on the computational complexity of logical theories. *Annals of Pure and Applied Logic,* 48:1-79, 1990.

[Cou94] B. Courcelle. Monadic second order graph transductions: A survey. *Theoretical Computer Science,* 126:53-75, 1994.

[Dah83] E. Dahlhaus. Reductions to NP-complete problems by interpretations. In E. Börger et. al., editor, *Logic and Machines: Decision Problems and Complexity,* volume 171, pages 357-365. Springer Verlag, 1983.

[Daw93] A. Dawar. *Feasible Computation Through Model Theory.* PhD thesis, Department of Computer Science, University of Maryland, 1993.

[EFT94] H.D. Ebbinghaus, J. Flum, and W. Thomas. *Mathematical Logic, 2nd edition.* Undergraduate Texts in Mathematics. Springer-Verlag, 1994.

[Fag81] R. Fagin. A normal form for relational databases that is based on domains and keys. *ACM Transactions on Database Systems,* 6(3):387-415, 1981.

[FV86] R. Fagin and M. Vardi. The theory of data dependencies. In M. Anshel and W. Gewirtz, editors, *Proceedings of Symposia in Applied Mathematics,* volume 34 of *American Mathematical Society,* pages 19-71. RI, 1986.

[HB70] D. Hilbert and P. Bernays. *Grundlagen der Mathematik, I,* volume 40 of *Die Grundleheren der mathematischen Wissenschaften in Einzeldarstellungn.* Springer Verlag, Heidelberg, 2nd edition, 1970.

[Her95] C. Herrmann. On the ubdecidability of implication between embedded multivalued database dependencies. *Information and Computation,* 122:221-235, 1995.

[Imm87] N. Immerman. Languages that capture complexity classes. *SIAM Journal on Computing,* 16(4):760-778, Aug 1987.

[Kol90] P.G. Kolaitis. Implicit definability on finite structures and unambiguous computations. In *LiCS'90,* pages 168-180. IEEE, 1990.

[KU84] A. Keller and J.D. Ullman. On complementary and independent mappings on databases. *Proceedings of ACM SIGMOD Annual Meeting on the Management of Data,* 14(2):145-148, 1984.

[Mai83]     D. Maier. *The Theory of Relational Databases*. Computer Science Press, 1983.

[Mak94]     J.A. Makowsky. Capturing complexity classes with Lindström quantifiers. In *MFCS'94*, volume 841 of *Lecture Notes in Computer Science*, pages 68–71. Springer Verlag, 1994.

[MR92]      H. Mannila and K.J. Räihä. *The Design of Relational Databases*. Addison-Wesley, 1992.

[MR96a]     J.A. Makowsky and E. Ravve. Dependency preserving refinment of database schemes. Technical Report, April 1996, Department of Computer Science, Technion–Israel Institute of Technology, Haifa, Israel, 1996.

[MR96b]     J.A. Makowsky and E. Ravve. Translation schemes and the fundamental problem of database design. In *Conceptual Modeling - ER'96*, volume 1157 of *Lecture Notes in Computer Science*, pages 5–26. Springer Verlag, 1996.

[MR97]      J.A. Makowsky and E. Ravve. Dependency preserving refinements and the fundamental problem of database design. In *Data and Knowledge Engineering*, to appear 1997.

[PDGV89]    J. Paredaens, P. De Bra, M. Gyssens, and D. Van Gucht, editors. *The Structure of the Relational Database Model*, volume 17 of *EATCS Monographs on Theeoretical Computer Science*. Springer Verlag, Heidelberg, 1989.

[Rab65]     M.A. Rabin. A simple method for undecidability proofs and some applications. In Y. Bar Hillel, editor, *Logic, Methodology and Philosophy of Science II*, Studies in Logic, pages 58–68. North Holland, 1965.

[Rav98]     E. Ravve. *Ph.D. Thesis*, Department of Computer Science, Technion - Israel Institute of Technology, to be completed in 1998.

[Tha91]     B. Thalheim. *Dependencies in Relational Databses*, volume 126 of *Teubner–Texte zur Mathematik*. B.G. Teubner Verlagsgesellschaft, Leipzig, 1991.

[Tha93]     B. Thalheim. Foundation of entity-relationship modeling. *Annals of Mathematics and Artificial Intelligence*, 7:197–256, 1993.

[Tha94]     B. Thalheim. A survey on database constraints. Reine Informatik I-8/1994, Fakultät für Mathematik, Naturwissenschaften und Informatik, 1994.

[Ull82]     J.D. Ullman. *Principles of Database Systems*. Principles of Computer Science Series. Computer Science Press, 2nd edition, 1982.

[UW97]      J. Ullman and J. Widom. *A First Course in Databases System*. Prentice-Hall, 1997.

[Var87]     M. Vardi. Fundamentals of dependency theory. In *Trends in Theoretical Computer Science*, pages 171–224. Computer Science Press, 1987.

# Solving and Approximating Combinatorial Optimization Problems (Towards MAX CUT and TSP)

Jaroslav Nešetřil[1] and Daniel Turzík[2] *

[1] Charles University
Department of Applied Mathematics
Malostránské nám. 25
118 00 Praha 1, Czech Republic

[2] Institute of Chemistry and Technology
Department of Mathematics
Technická 5
166 28 Praha 6, Czech Republic

**Abstract.** We present a brief outline of recent development of combinatorial optimization. We concentrate on relaxation methods, on polynomial approximate results and on mutual relationship of various combinatorial optimization problems. We believe that this complex web of results and methods is typical for the modern combinatorial optimization. This paper is an introduction to our full paper [53].

## 1 Introduction

### 1.1

Combinatorial optimization deals with a special type of optimization problems with the property that set of all (feasible) solutions forms a finite set. 15 years ago, on the eve of the Khachiyan's ellipsoid method, we wrote a survey [52] for SOFSEM 1980. The present paper cannot be an update of that paper. The development has been too rapid and too extensive to be able to cover even its highlights. The present paper is also essentially shorter than [52]. But with all the differences the main theme is the same and we want to begin along the same lines.

### 1.2

The basic problem of Combinatorial Optimization can be described as follows:

Given a finite set $S$ of objects and a weight function $w : S \longrightarrow \mathbf{R}$ find

$$\max\{w(s) \; ; \; s \in S\}$$

---

* Partially suported by GAČR 0194 and GAUK 194 grants.

In most cases the set $S$ is highly structured and typically it can be described as a certain collection of subsets of an (underlying) set $X$, i.e. $S \subseteq \mathcal{P}(X)$. Then in turn we may view $S$ as a finite subset of $\mathbf{R}^V$ of $|V|$-dimensional vector space whose coordinates are indexed by elements of $V$. Narrowing our focus still further, the most important case is the following:

There exists a weight function $w : V \longrightarrow \mathbf{R}$ such that the weight $w(s)$ of an object $s = (s_v; v \in V) \in \mathbf{R}^V$ is given by

$$w(s) = \sum_{v \in V} w(s_v) .$$

Thus we assume that the weight is a linear function of its coordinates and we speak about Linear Objective Combinatorial Optimization problem (shortly LOCO problem). All problems considered in this paper are LOCO problems.

In many cases the size of $S$ is huge (exponential) yet the problem can be solved in time bounded by the polynomial function on $|V| = n$.

### 1.3

Combinatorial optimization shares one important aspect with combinatorics and discrete mathematics:

It is not a cohesive theory which is recognized and studied per se but rather a collection (or a ZOO) of results, algorithms and even trics which are best described by a series of examples. The following are some of the best known and traditional examples of LOCO problems. We describe them first informally, LOCO formulation will follow.

### 1.4  Examples of LOCO problems

SP: In a given configuration find a shortest path between two given points.

MST: In a given configuration find a minimum scheme which guarantees mutual communication.

CHP: In a given city find the shotest route of a (hypothetical, tireless) postman to visit each street at least once.

MF: Find a maximum flow through a network.

TSP: In a given region find a shortest tour of a travelling salesman (visiting all the palces).

MM: For every pair $(x, y)$ of SOFSEM '97 participants there is a known measure $e(x, y) \geq 0$ of their empathy. Schedule an accomodation in two bed rooms to maximize the success of this meeting.

MC: Given a set of animosities find a best partition (say by a fence) in 2 classes minimizing the animosities within the classes.

MS: Given a set of senders find maximal set among them which do not interfere.

## 1.5

All these problems can be formulated as LOCO problems exactly and concisely by means of graphs as follows. This reformulation also explains the above acronyms:

SP= Shortest Path
   In an edge-weighted graph $G$ find a shortest path between two vertices.

MST= Minimum Spanning Tree
   Given a graph $G = (V, E)$ together with a weight function $w : E \longrightarrow \mathbf{R}$ find a tree $(V, E')$ for which
$$\sum_{e \in E'} w(e)$$
is minimal.

CHP= Chinese Postman Problem
   Given a weighted graph $G = (V, E)$, $w : E \longrightarrow \mathbf{R}$ find a walk
$$v_0, e_1, v_1, \ldots, e_t, v_t = v_0$$
which contains all edges of $G$.

MF= Maximal Flow in a Network
   (This needs a few formal definitions.) A network is a directed graph $G = (V, E)$ (i.e. $E \subseteq V \times V$) with edge capacities $c : E \longrightarrow \mathbf{R}^+$ and two specified vertices $z$ (the source) and $s$ (the sink). A flow in a network $(G, c, z, s)$ is a function $f : E \longrightarrow \mathbf{R}^+$ which obeys capacities (i.e. $0 \le f(e) \le c(e)$ for every edge $e \in E$) and which preserves conservation law for its inner vertices (i.e. vertices different from $z$ and $s$). By this we mean that
$$\sum_{(x,y) \in E} f(x, y) = \sum_{(y,x) \in E} f(y, x)$$
for every $x \ne z, s$. The size $|f|$ of a flow $f$ is defined by
$$|f| = \sum_{(z,y) \in E} f(z, y) - \sum_{(x,z) \in E} f(x, z) .$$

The problem of maximal flow can be formulated as follows: Given a network $(G, c, z, s)$ find the maximal size of a flow.

TSP= Travelling Salesman Problem
   Given a graph $G = (V, E)$ with edge weights $w$ find an ordering of (all) its vertices $v_0, v_1, \ldots, v_n = v_0$ such that
$$\sum_{i=1}^{n} w(v_{i-1}, v_i)$$
is minimal. If the weight function satisfies the triangle inequality (or if $w$ is given by the euclidean distance) than we speak about metric (or euclidean) TSP.

MM= Maximal Matching

Given a graph $G = (V, E)$ with edge weights $w$ find a set $M$ of disjoint edges such that

$$\sum_{e \in M} w(e)$$

is maximal.

MC= Max Cut = Maximal Cut

Given a graph $G = (V, E)$ with edge weights find a bipartite subgraph $(V, E')$ for which

$$\sum_{e \in E'} w(e)$$

is maximal. This is a formulation of Max Cut problem for positive edge weights.

MS= Maximal Stable Set

Given a graph $G$ find a maximal set $A \subseteq V$ which does not contain any edge (such a set is called a stable set or independent set; its maximal size is denoted by $\alpha(G)$).

## 1.6

These problems are comming from different areas (both theory and applications) and their history is to a great degree disjoint. Perhaps more importantly, the computational complexity of these problems have a very different behaviour. This was studied recently in great detail and one can say that this research is one of the driving forces of the whole theoretical computer science. One of the aims of this paper is to provide a reader with at least few glimpses of this fascinating development.

On the other hand side combinatorial optimization is very close to applied problems and one should say perhaps surprisingly successfull in solving large projects. Here one should mention very successfull design of chips (see [37]; chips designed by University of Bonn were reportedly used in machines in a recent match with Kasparov) and a recent business-success of CPLEX company (visit *http:/www.ilog.com/html/press_cplex.html* ).

## 1.7

But despite of all these differences the above problems form a compact body of results and techniques and one can say that together form a key part of modern Combinatorial Optimization. This fact has double meaning for our presentation:

First, as the above problems reflect most of the development of combinatorial optimization, the subject is vast and we have so restrict. So even in the full version [53] we decided to concentrate on two problems: Travelling Salesman Problem and Max Cut Problem.

Secondly, as the above problems form a closely knight group we can illustrate the main trends of the development by considering only few typical examples.

One of our main goals is to demonstrate mutual relationship and similarities between above problems and techniques devised for their solution. This will be done here. This paper provides then a framework for the more delailed study of particular cases of TSP and Max Cut in the full version of our paper [53].

## 2    A Few Connections and Implications

### 2.1

The Max Cut and TSP were not chosen randomly. Quite to the contrary. They both present traditional problems of combinatorial optimization. It is often said that these problems are important as they are applied and appear in many practical situations. This is of course a logical and safe claim in todays world but in reality problems appear quite rarely in a purity demanded by TSP and Max Cut formulation. Of course there are exceptions and they are recorded e.g. in [1], [7], [63]. But perhaps one shouldn't claim that this is the most important motivation for studiyng such problems. What is perhaps more to the point is that both these problems became sort of a testing ground and cornerstones of modern combinatorial optimization [2], [27], [45], [6], [22], [55], [61], [59], [58]. These problems were studied for a long time and despite all the efforts they still appear hard.

### 2.2

There is a theoretical justification for this hardness. In fact there is a manifold evidence for it and this has been one of the main driving forces of theoretical computer science. The following table gives the basic information about the (worst case) complexity of our problems.

| SP | MST | CHP | MF | TSP | MM | MC | MS |
|----|-----|-----|----|-----|----|----|----|
| P  | P   | P   | P  | H   | P  | H  | H  |

Here P stands for the existence of polynomial time algorithm and H for NP-hardness; as all the corresponding decision problems belong to NP these decision problems actually belong to NP-complete problems.

### 2.3

But while our problems appear to be hard yet because of their formulation simplicity ( and, yes, "application appeal") they attracted considerable attention and from the early days efforts have been made to solve large (or modestly large) problems. Because of the inherent complexity of the problems no direct approach worked and no simple master algorithm has been devised. Rather a complex web of trics, mutual interplays, partial and approximate solutions, and heuristical arguments have been found. To a certain degree this in fact established Combinatorial Optimization as we know it today. One can say that the main goal of this short survey is to provide an interested reader with a description of a few facets of this (what we believe) fascinating development.

## 2.4

Thus we concentrate mainly on the theoretical aspects (with an eye on potential and recorded experiments; as we know the theory is here very close to applied work [7], [32] , [37], [45]). We introduce all the key notions and try to display main interconnections between various problems.

## 2.5

The above complexity table does not exhaust the complicated 50 years or so history of these problems. Many interesting results were obtained on both ends of our spectrum. Perhaps more rapid development was on the polynomial side of the spectrum.

## 2.6

The interesting history of polynomial solutions of MST starts with Boruvka [8] and Jarník [31]. The manifold uprovements of the MST algorithms would fill a rather long paper by itself (and this has been done e.g. in [25],[42] and [51]). Let us just note that all efforts so far (narrowly) missed the main goal: to find a linear deterministic algorthm for MST problem. (The only problem in our list with known linear algorithm is SP Problem).

However the ideas behind the speedup of existing algorithms and their refinements led to isolation important structures (Set Union problem, Fibonacci heaps, verification algorithm), see [66], [34], [40] and thus provided a background for much of the recent development.

## 2.7

Conceptually one of the easiest algorithms is so called a greedy algorithm due to Kruskal [43]. This algorithm can be used as an heuristic for most (if not all) combinatorial problems which have a hereditary set of feasible partial solutions: We start with an an initial ( say trivial) solution and extend this solution step by step so that each step the extension is optimal. This greedy heuristic leads to the notion of matroid and greedoid, see [69], [41], which can be seen as the combinatorial structures supporting the optimality of greedy algorithm. The greedy algorithm is a useful heuristic even in the case when it does not leads to an optimum. Examples of this include also TSP.

## 2.8

Network flow algorithms play a central role ([66] is one of the best references for this). The same is true for the seemingly isolated problem of maximal weighted matching (Problem MM). Mathing theory and matching algorithms present perhaps the best developed field of graph theory and [49] is still best source.

## 2.9

These problems are important as they until recently influence other problems. For example the best heuristic for the metric TSP ( due to Christofides) was based on the MM problem .

It is important that the MM problem can be sometimes regarded as a flow problem. Also Max Cut problem can be viewed as a Maximal Flow problem but we have to accept negative capacities. It is a bit surprising that this extension of flow problems leads to such drastical change of the complexity of the problem. It is worth to note that this is not the case in the MST problem.

On the other hand side even the shortest path problem includes Travelling salesman problem if we accept negative weights.

Continuing with this introductory overview we mention that the polynomiality of (all pairs) shortest path problem (i.e. SP), together with Maximal Matching Problem implies that Chinese Postman problem is polynomial. On the other hand side we may view Chinese Postman Problem as a flow (or circulation) problem of minimal cost.

## 2.10

This does not exhaust all the implications and translations between the above problems. In fact one of the essential features of Combinatorial Optimization (and in a sense of complexity theory too) is an abundance of reductions and translations.

# 3  A Case Analysis

## 3.1

In a certain sense the Maximal Stable Set problem (i.e. MS) is universal as it captures e.g. any subgraph type problem. It is also a prototype of an NP-complete problem and one of the problems which were most thoroughly investigated. The reader may ask what is the connection of this problem to our two problems. Well for example we shall see bellow that the best known method for approximate solutions follow the same line.

## 3.2

Imagine that we want to solve a maximal stable set problem in a (large) graph $G = (V, E)$ with $n$ vertices. After trying several heuristics (such as greedy algorithm) we would like to get something more sophisticated which would give us some performance quarantee (if only in some cases). The standard approach which goes back to the beginning of fifties is to consider geometric form and a linear relaxation of our (combinatorial) problem:
We denote by $\mathbf{R}^V$ the $n$-dimensional vector space with coordinates indexed by elements of $V$. Thus any independent subset $A \subseteq V$ may be thought as vector (i.e. incidence 0-1 vector) of $\mathbf{R}^V$. The convex hull of the set of all independent sets generates a polyhedron, *a stable set polyhedron* of $G$, which we denote by $MS(G)$. Obviously all the vertices of $MS(G)$ have integral coordinates and they correspond to independent sets. Thus it suffices to find the vertex $x$ of $MS(G)$ with largest weight $\sum_{v \in V} x_v$. However this is just a reformulation of MS problem. (It is also an integer LP problem - so no hope to solve it - at least presently). However Linear Programming provided an important methodology which constitutes the key part of polyhedral combinatorics.

## 3.3

We restate the problem by writing a set of linear constraints in the following way:
$$\max \sum_{v \in V} x_v$$

subject to constraints

$$\left. \begin{array}{ll} x_v + x_{v'} \leq 1 & \{v, v'\} \in E \\ x_v \geq 0 & v \in V . \end{array} \right\} \tag{1}$$

Now these inequalities determine a polyhedron $MS_L(G)$ which obviously contains polyhedron $MS(G)$. Not only that all the integral vectors in $MS_L(G)$ belong to $MS(G)$ and thus the linear programming problem (1) present a *linear relaxation* of MS problem.

## 3.4

As one sees easily the polyhedron $MS_L(G)$ properly contains $MS(G)$ and thus an optimal solution to (1) does not necessarily gives a solution of MS. What are the obstacles? E.g. for odd cycle $C_5$ the polyhedron $MS_L(C_5)$ has the vector $(1/2, 1/2, 1/2, 1/2, 1/2)$ as its vertex. However, in any case two important facts are valid:

i.  The optimal solution to problem (1), i.e. optimization over polyhedron $MS_L$ provides an upper bound for the propblem MS.
ii. We can add the "obstacle" inequalities to the system (1) in a hope that we shall locate the position of polyhedron $MS$ within $MS_L$ more accurately.

Thus the new system of inequalities induces again a linear relaxation of the original problem. Gomory [23] and Chvátal [10] showed a systematic way of generating new obstacle inequalities (by taking linear combinations and rounding), such that every (integer) problem may be transformed to a linear programming problem. This method - cutting plane method - proved to be extremaly usefull in many instances. However recently it has been shown [60] that there are problems which need exponentially many cutting planes to be added. Also these problems are related to MS problem.

## 3.5

Returning to our main scheme, for Maximal Stable Set problem three more accurate relaxations were isolated which lead to convex sets $MS_{ODD}(G)$, $MS_{CL}(G)$, $MS_{TH}(G)$ as follows:
$MS_{ODD}(G)$ is the set of all vectors in $\mathbf{R}^V$ satisfying (1) and conditions

$$\sum_{v \in C} x_v \leq \frac{|C| - 1}{2} \, , \ C \text{ an odd cycle in } G. \tag{2}$$

$MS_{CL}(G)$ is the set of all vectors in $\mathbf{R}^V$ satisfying (1) and conditions

$$\sum_{v \in K} x_v \leq 1 \, , \ K \text{ a clique (i.e. a complete subgraph) in } G. \tag{3}$$

Both $MS_{ODD}(G)$ and $MS_{CL}(G)$ are polyhedra in $\mathbf{R}^V$.

## 3.6   $MS_{TH}(G)$

This is based on "Lovász theta function" first defined in the seminal paper by Lovász [46], see [48] for the following formulation:
$MS_{TH}(G)$ is defined as the set of all vectors $d = (d_v \, ; \, v \in V) \in \mathbf{R}^V = \mathbf{R}^n$ for which there exists a positive semidefinite $(n + 1) \times (n + 1)$ matrix $Z = (z_{ij})$ satisfying

$$\left. \begin{array}{ll} z_{00} = 1 & \\ z_{i0} = z_{0i} = z_{ii} = d_i & i = 1, \ldots, n \\ z_{ij} = 0 & \{i, j\} \in E \, . \end{array} \right\} \tag{4}$$

(If $d$ is the incidence vector of an independent set in $G$, then putting $z_{ij} = d_i d_j$, $z_{00} = 1$ and $z_{0i} = z_{i0} = d_i$ we get a matrix $Z$ which satisfies the above properties. As $Z = (1, d) \cdot (1, d)^T$ the matrix $Z$ is also positive semidefinite. This shows that $MS_{TH}(G)$ is a relaxation of the stable set polyhedron $MS(G)$.)

## 3.7

The convex set $MS_{TH}(G)$ fails to be a polytope. However as one can decide polynomially whether a given matrix is positive semidefinite, one can solve a separation problem for $MS_{TH}(G)$ polynomially and thus by one of the main result of [27] one can optimize over $MS_{TH}(G)$. This has important consequences, which we outline now.

**3.8**

All these convex sets lead to important classes of graphs and their important properties:

Graphs with $MS_{ODD}(G) = MS(G)$ are called $t$-perfect graphs (and their study was proposed by Chvátal [11]). >From all the classes this is perhaps the most esoteric class (which has not yet been characterized by other means and, particularly, no P membership algorithm for deciding whether a graph is $t$-perfect is known).

**3.9**

Still the MS problem for $t$-perfect graphs can be solved in polynomial time. That may seem to be surprising on the first glance but according to the general theory [27] it suffices to solve a corresponding separation problem:

Given $y \in \mathbf{Q}^V$ we want to decide whether $y \in MS_{ODD}(G)$ or otherwise to find a hyperplane of type (1) or (2) which is violated by $y$. Checking (1) is easy. Thus let us assume $y_v \geq 0$ and $y_v + y_{v'} \leq 1$ for every edge. We want to check whether (2) holds. This is possible to solve by the following trick:

In our situation define a vector $z = (z_e \; ; \; e \in E) \in \mathbf{R}^E$ by

$$z_e = 1 - y_v - y_{v'} \geq 0$$

for an edge $e = \{v, v'\}$. Then

$$\sum_{e \in C} z_e = |C| - 2 \sum_{v \in C} y_v$$

and thus (2) is equivalent to the validition of the condition

$$\sum_{e \in C} z_e \geq 1$$

for every odd circuit $C$ of $G$. However we can think of $z_e$ as nonnegative edge weights of $E$ and thus it suffices to check whether (with the weights $z_e$) the shortest odd circuit has length $\geq 1$.

**3.10**

Thus the MS problem for $t$-perfect graphs was finally reduced to the shortest odd circuit in an edge weighted (undirected) graph. This problem is possible to solve by another neat trick:

Given a graph $G = (V, E)$ we define a new graph $G' = (V', E')$ by

$$V' = V \times \{0, 1\} \; , \qquad E' = \{\{(v, i), (v', j)\} \; ; \; \{v, v'\} \in E \; , \; i \neq j\} \; .$$

The shortest path from $(v, 0)$ to $(v, 1)$ is necessarily of an odd length and thus we may solve our problem by $|V|$ iterations of the shortest path algorithm.

## 3.11

The shortest odd cycle problem is an interesting variation of SP problem with a long history and important connections. The reader can try to convict himself that to detect and to find shortest odd cycle (in directed) and shortest even cycle (in undirected graphs) are easy problems. However the problem of deciding whether a given directed graph contains an even cycle is much harder and also more important. This problem has been shown to be related to coloring of hypergraphs [65], to problems in statistical physics [36] and to the problem when the permanent of a given 0-1 matrix $A$ (a problem believed to be not easily algoritmicaly solvable [68]) can be reduced to the easy computation of the determinant of a matrix $A'$ which arises from $A$ by changing some 1 to $-1$ (this problem goes back to Polya), see also [39], [19]. Very recently all these problems were solved by surprising characterization theorem and a polynomial algorithm [13]. This algorithm involves among others MM polynomial algorithm. The complexity of the algorithm is $O(n^3)$. Interestingly and unexpectedly the present known complexity of most naturally defined problems seems to be bounded by $n^3$. We do not know any reason for it.

## 3.12

The above hierarchy of convex sets defines furher classes of graphs. It is easy to prove that $MS(G) = MS_L(G)$ iff the graph $G$ is bipartite (i.e. iff $G$ does not contain any odd cycle). However the graphs defined by $MS(G) = MS_{TH}(G)$ coincide with graphs satisfying $MS(G) = MS_{CL}(G)$ and these graphs are called perfect graphs. It follows from the above theory that there is a polynomial time algorithm to solve MS problem for perfect graphs. This is a non-trivial result for which no other proof is known (and which thus relies on the ellipsoid method [27]).

## 3.13

In general, MS problem is difficult and NP-complete even when restricted to very special classes of graphs (such as planar graphs). The most polynomial cases are covered by the above general polyhedral approach (there are some exceptions see e.g. [5]). Starting with [24] it became a promising line of research to look for approximate algorithms. In the case of MS problem this mean the following:

Denote by $\alpha(G)$ the maximal size of an independent set in the graph $G$. Let $A$ be an algorithm which finds a (not necessarily largest) independent set of size $A(G)$ in $G$. We say that $A$ is an $c$-approximation algorithm if

$$\alpha(G) \leq c \cdot A(G) .$$

We also say that MS problem (and quite analogously any CO problem) admits a polynomial approximation scheme iff for every $c > 1$ there exists a polynomial algorithm $A$ which is $c$-optimal. It is wellknown that there are NP-complete problems which admit a polynomial approximation scheme (such as bin packing problem [17]). However for many problems the existence of a polynomial approximation scheme remained until recently an open and fundamental problem.

## 3.14

For the maximal stable set problem it had been observed in [20] that an existence of $c > 1$ polynomial approximation algorithm yeilds the existence of a polynomial approximation scheme. However the existence of such approximation scheme was disproved in a dramatic way in [18], [4] and finally [3] where it has been shown that for an $\varepsilon > 0$ the maximal independent set in a graph $G = (V, E)$ cannot be approximated within a factor $|V|^\varepsilon$.

These results were the first deep results about non-approximability of combinatorial optimization problems and they extend (using [56]) to most interesting combinatorial problems.

## 3.15

Both for the Max Cut problem and TSP this development was especially interesting. This will be discussed in more details in our talks. Let us just state here that the best known $c$-approximation algorithm for the Max Cut problem has ratio $c = 1.1383$ [21] and it is based on a semidefinite relaxation of the Max Cut polytope due to Delorme and Poljak [15]. This is a great progress from an obvious 2-approximation algorithm (see [57]). It has been shown recently [44] that this relaxation is closely related to Lovász theta function. Currently the best non-approximability result is for $c = 1.0624$ [28]. For the TSP problem, like for the Maximal Stable Set problem, there is no polynomial approximation algorithm [3]. However for metric TSP problem there is an easy $3/2$ approximation algorithm [9] which is still the best. Again for matric TSP no polynomial approximation scheme (under $P \neq NP$), exists [3]. To a great surprise the euclidean TSP behaves differently. Recently Arora [2] produced polynomial approximation scheme not only for Euclidean TSP, but also polynomial approximation schemes for other euclidean optimization problems among them euclidean Steiner Tree Problem (see also [35]). These results hold in any fixed dimension $d$ (and the dependence on $d$ in approximation algorithms is double exponential). This dependence on $d$ is necessary as was recently shown in an important paper by Trevisan [67].

## 3.16

The problem to determine the approximation status of a particular optimization problem received recently lot of attention and in several cases the full solution was achieved [28], [38], [67] (such as 3-SAT problem [28]). However for metric TSP and the Max Cut problem the situation is far from beeing solved. This will be discussed in great detail in the full version of this paper [53].

## 3.17 Conclusion

We tried to outline some of the main directions of contemporary Combinatorial Optimization. We concentrate mainly on theoretical questions where the progress has been remarkable. Still we had to omit several important issues such as classification problem (see [16], [29], [30], [38], [64]) or randomized algorithms (see [33], [47], [50]). We also only mentioned few aspects of TSP and Max Cut problems; the present paper should be seen as an introduction to the manysided research related to these problems. However all this theoretical work found its way to large scale computing and indeed some practical problems. Significant and unexpectedly complex (and indeed very complex) problems have been solved. This is documented e.g. in [1], [7], [12], [14], [26], [32], [54], [62], [61]. The paper [53] will describe some of these methods in a greater detail.

# References

1. D.Applegate, R.Bixby, V.Chvátal, W.Cook: Finding cuts in the TSP. DIMACS Tech. Report, 95–105.
2. S.Arora: Polynomial time approximation schemes for Euclidean TSP and other geometric problems. In: Proc. of the 37th IEEE FOCS, 1966.
3. S.Arora, C.Lund, R.Motwani, M.Sudan, M.Szegedy: Proof verification and intractability of approximation problems. 33rd FOCS, 1992, 14–23.
4. S.Arora, S.Safra: Probabilistic checking of proofs: a new characterization of NP. 33rd FOCS, 1992, 2–13.
5. E.Balas, V.Chvátal, J.Nešetřil: On the maximum weight clique problem. J. of Oper. Research (1985).
6. F.Barahona: On cuts and matchings in planar graphs. Mathematical Programming 60(1993), 53–68.
7. R.E.Bland, D.F.Shallcross: Large travelling salesman problems arising from experiments in X-ray crystalography: a preliminary report on computation. Oper.Res.Lett. 8(1989), 125–128.
8. O.Boruvka: O jistém problému minimálním (about a certain minimal problem). Práce mor. přírodověd. spol. v Brně III, 3(1926), 37–58.
9. N.Christofides: Worst-case analysis of a new heuristic for the travelling salesman problem. Technical report, Carnegie-Mellon University, 1976.
10. V.Chvátal: Edmonds polytopes and hierarchy of combinatorial problems. Discrete Math. 4(1973), 305–337.
11. V.Chvátal: On certain polytopes associated with graphs, J.Comb.Th. B, 18(1975), 138–154.
12. H.Crowder, M.W.Padberg: Solving large-scale symmetric travelling salesman problems to optimality. Management Sci 26(1980), 495–509.
13. W.McCuaig, N.Robertson, P.D.Seymour, R.Thomas: Permanents, Pffafian orientations and even directed circuits. (to appear; see also a preliminary version in STOCK (1997)).
14. G.B.Dantzig, D.R.Fulkerson, S.M.Johnson: Solution of large-scale travelling salesman problem, Oper.Res. 2(1954), 393–410.
15. C.Delorme, S.Poljak: Laplacian eigenvalues and the max-cut problem. Mathematical Programming 63(1993), 557–574.

16. T.Feder, M.Vardi: Monotonne monadic SNP and constraint satisfaction. Proceedings of the 25th ACM STOC, ACM, 1993.

17. W.Fernandez de la Vega, G.S.Lueker: Bin packing can be solved within $1 + \varepsilon$ in linear time. Combinatorica 1(1981), 349–355.

18. U.Fiege, S.Goldwasser, L.Lovász, S.Safra, M.Szegedy: Interactive proofs and the hardness of approximating cliques. Journal of the ACM 43(1966), 268–292.

19. A.Galluccio, M.Loebl: Even cycles and H-homomorphismus. (to appear)

20. M.R.Garey, D.S.Johnson: The complexity of near optimal coloring. J. Assoc. Comput. Math. 23(1976), 43–49.

21. M.Goemans, D.Williamson: .878-approximation algorithm for Max-Cut and Max-2-SAT. 25th STOC, 1994, 422–431.

22. M.Goemans, D.Williamson: Improved approximation algorithm for maximum cut and satisfiability problems using semidefinite programming. Journal of the ACM 42(1995), 1115–1145.

23. R.E.Gomory: Outline of an algorithm for integer solutions to linear programs. Bull. Amer. Math. Soc. 64(1958), 275–278.

24. R.L.Graham: Bounds for certain multiprocessing anomalies. Bell System Tech. J. 45(1969), 1563–1581.

25. R.L.Graham, P.Hell: On the history of the minimum spanning tree problem. Annals of the History of Computing 7(1985), 43–57.

26. M.Grötschel, O.Holland: Solution of large-scale symmetric travelling salesman problems. Math.Programing 51(1991),141–202.

27. M.Grötschel, L.Lovász, A.Schrijver: Geometric algorithms and combinatorial optimization. Springer Verlag, 1988.

28. J.Hastad: Some optimal inapproximability results. In: Proceedings 29th STOC (1997), 1–10.

29. P.Hell, J.Nešetřil: On the complexity of H-coloring. J.Comb.Th. B, 48(1990), 92–110.

30. P.Hell, J.Nešetřil, X.Zhu: Duality and polynomial testing of graph homomorphism. Trans.Amer.Math.Soc. 348, 4(1996), 1281–1297.

31. V.Jarník: O jistém problému minimálním (about a certain minimal problem). Práce mor. přírodověd. spol. v Brně IV, 4(1930), 57–63.

32. M.Jünger, G.Reinelt, S.Thienel: Practical problems solving with cutting plane algorithms. In: Combinatorial Optimization (W.Cook, L.Lovász, P.Seymour, ed.), DIMACS Series vol.20, AMS(1995), 111–152.

33. G.Kalai: A subexponential randomized simplex algorithm. Proc. 24th STOCK, 1992, 475–482.

34. D.Karger, P.N.Klein, R.E.Tarjan: A randomized linear-time algorithm to find minimum spanning trees. J.Assoc.Comp.Mach. 42(1995), 321–328.

35. M.Karpinski, A.Zelikovsky: New approximation algorithms for the Steiner tree problems. Technical Report ECCC TR95-030, 1995.

36. P.W.Kasteleyn: Graph theory and crystal physics. In: Graph Theory and Theoretical Physics (F.Harary, ed.), Academic Press, New York, 1967, 43–110.

37. A.Kehrbaum, B.Korte: Calculi. West Deutcher Verlag, 1995.

38. S.Khanna, M.Sudan, D.P.Williamson: A complete classification of the approximability of maximization problems derived from Boolean constraint satisfaction. In: Proceedings 29th STOC (1997), 11–20.

39. V.Klee, R.Ladner, R.Mauber: Sign-solvability revisited. Linear Algebra Appl. 59(1984), 131–158.

40. J.Komlos: Linear verification for spanning trees. Combinatorica 5(1985), 57–65.

41. B.Korte, L.Lovász, R.Schrader: Greedoids. Springer 1990.
42. B.Korte, J.Nešetřil: Vojtěch Jarník's work in combinatorial optimization. KAM Series 95, 315.
43. J.B.Kruskal: On the shortest spanning subtree of a graph and the travelling salesman problem. Proc.Amer.Math.Soc. 7(1856), 48–50.
44. M.Laurent, S.Poljak, F.Rendl: Connections between semidefinite relaxation of the Max Cut and Stable Set Problems. (to appear)
45. E.L.Lawler, J.K.Lenstra, A.H.G.Rinnooy Kan, D.B.Shmoys: The travelling salesman problem. John Wiley, 1985.
46. L.Lovász: On the Shannon capacity of a graph. IEEE Transactions on Information Theory 25(1979), 1–7.
47. L.Lovász: Randomized algorithms in combinatorial optimization. In: Combinatorial optimization (ed. W.Cook, L.Lovász, P.Seymour), DIMACS Series vol. 20, AMS 1995, 153–180.
48. L.Lovász, A.Schrijver: Cones of matrices and set-functions and 0-1 optimization. Siam Journal of Optimization 1(1991), 166–190.
49. L.Lovász, M.Plummer: Matching Theory. North Holland 1986.
50. J.Matoušek, M.Sharir, E.Welzl: A subexponential bound for linear programming, Proc. 8th Ann.Symp. on Comput. Geom. (1992),1–8.
51. J.Nešetřil: A few remarks on the history of MST-problem. Archivum Mathematicum 33(1997), 16–22.
52. J.Nešetřil, S.Poljak: Geomatrical and algebraical connections of combinatorial optimization. SOFSEM '80, Bratislava 1980, 35–78.
53. J.Nešetřil, D.Turzík: Solving and approximating Combinatorial Optimization Problems. (to appear).
54. M.W.Padberg, G.Rinaldi: Facet identification for the symmetric travelling salesman polytope. Math.Programming 47(1990), 219–257.
55. C.H.Papadimitriou: Euclidean TSP is NP-complete. Theoretical Computer Science 4(1977), 237–244.
56. C.H.Papadimitriou, M.Yannakakis: Optimization, approximation and complexity classes. Journal of Computer and System Sciences 43(1991), 425–440.
57. S.Poljak, D.Turzík: A polynomial algorithm for constructing a large bipartite subgraph with an application to satisfiability problem. Canadian Math. J. 24(1982), 519–524.
58. S.Poljak, D.Turzík: Maximum cut and circulant graphs. Discrete Math. 108(1992), 379–392.
59. S.Poljak, Zs.Tuza: Maximum cuts and largest bipartite subgraphs. In: Combinatorial Optimization (W.Cook, L.Lovász, P.Seymour, ed.), DIMACS Series vol.20, AMS(1995), 181–244.
60. P.Pudlák: Lower bounds for resolution and cutting plane proofs and monotonne computations. J. Symbol. Logic (to appear).
61. G.Reinelt: TSPLIB - A travelling salesman problem library. ORSA J.Computing 3(1991), 376–384.
62. G.Reinelt: Fast heuristics for large geometric travelling salesman problems. ORSA J.Comput. 4(1992), 206–217.
63. D.B.Shmoys: Computing near-optimal solution to combimatorial optimization problems. In: Combinatorial Optimization (W.Cook, L.Lovász, P.Seymour, ed.), DIMACS Series vol.20, AMS(1995), 355–398.
64. T.Schaefer: The complexity of satisfiability problems. Proceedings of the 10th ACM STOC, ACM, 1978.

65. P.Seymour: On the two coloring of hypergraphs. Quart.J.Math. Oxford 25(1975), 303–312.

66. R.E.Tarjan: Data structures and network algorithms. CBMS-NSF Regional Conf. Series in Applied Math., SIAM 44, 1983.

67. L.Trevisan: When Hamming meets Euclid: The approximability of geometric TSP and MST. In: Proceedings 29th STOC (1997), 21–29.

68. L.G.Valiant: The complexity of computing the permanent. Theoret.Comp.Sci. 8(1979), 189–201.

69. D.Welsh: Matroid theory. Academic Press 1976.

# The Computational Power of Continuous Time Neural Networks

Pekka Orponen

Department of Mathematics
University of Jyväskylä, Finland*

**Abstract.** We investigate the computational power of continuous-time neural networks with Hopfield-type units. We prove that polynomial-size networks with saturated-linear response functions are at least as powerful as polynomially space-bounded Turing machines.

## 1 Introduction

In a paper published in 1984 [12], John Hopfield introduced a continuous-time version of the neural network model whose discrete-time variant he had discussed in his seminal 1982 paper [11]. The 1984 paper also contains an electronic implementation scheme for the continuous-time networks, and an argument showing that for sufficiently large-gain nonlinearities, these behave similarly to the discrete-time ones, at least when used as associative memories.

The power of Hopfield's discrete-time networks as general-purpose computational devices was analyzed in [21, 22]. In this paper we conduct a similar analysis for networks consisting of Hopfield's continuous-time units; however we are at this stage able to analyze only the general asymmetric networks, and the very interesting subclass of continuous-time networks with symmetric interconnections has to wait for further research. Also, our analysis is restricted to networks with saturated-linear response functions, although computer experiments do indicate that our constructions work equally well for e.g. the standard sigmoid nonlinearities.

Under the above assumptions, we prove that sequences of networks of polynomially increasing size can compute all the functions in the class PSPACE/poly, i.e. the same functions as are computed by polynomially space-bounded nonuniform Turing machines. Such analyses of the computational power of continuous-time processes are at the moment relatively rare in the literature, but we expect their number to grow along with the current increase of interest in analog computation. (The only other analysis that explicitly addresses computational complexity issues seems to be that in [27]. Computability aspects have been studied more often; see, e.g. [2, 5, 6, 19, 20, 24, 25], and the survey [23].)

---

* Address: P. O. Box 35, FIN–40351 Jyväskylä, Finland. E-mail: orponen@math.jyu.fi. Part of this work was done during the author's visit to the Technical University of Graz, Austria.

## 2 A Continuous-Time Neural Network Model

An electrical model of Hopfield's continuous-time neuron is shown in Fig. 1. Here $\sigma$ denotes the characteristic of the nonlinear amplifier, and $\rho_i$ and $C_i$ are its input resistance and capacitance, respectively. In the following analyses we shall consider only the saturated-linear characteristic

$$\sigma(z) = \begin{cases} -1, \text{ for } z < -1, \\ z, \quad \text{for } -1 \le z \le 1, \\ 1, \quad \text{for } z > 1 \ . \end{cases}$$

The input voltage of the amplifier is denoted by $u_i$, and the output voltage by $y_i$. In order to establish inhibitory interconnections between such units, also the inverted output voltages $\bar{y}_i = -y_i$ are needed.

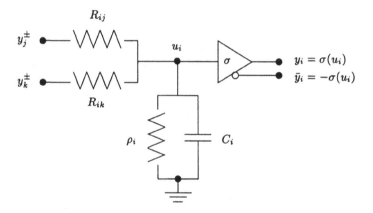

**Fig. 1.** An electrical model of Hopfield's continuous-time neuron.

The unit $i$ indicated in the figure draws input from units $j$ and $k$ via two resistors, whose resistances are denoted by $R_{ij}$ and $R_{ik}$. The voltages $y_j^\pm$ and $y_k^\pm$ are obtained from the appropriate output terminals of units $j$ and $k$, depending on whether the inputs are excitatory or inhibitory.

By Kirchhoff's current law, the circuit equations for a network of $p$ such units can be written as

$$C_i \frac{du_i}{dt} + \frac{u_i}{\rho_i} = \sum_{j=1}^{p} \frac{1}{R_{ij}}(y_j^\pm - u_i), \quad \text{for} \quad i = 1, \ldots, p \ .$$

By choosing the circuit parameters appropriately and normalizing the time constants to 1, we can use such a network to implement any system of first-order nonlinear differential equations of the form

$$\frac{du_i}{dt} = -u_i + \sum_{j=1}^{p} h_{ij}\sigma(u_j), \qquad i = 1,\ldots,p \ . \tag{1}$$

(We essentially choose $R_{ij} = 1/h_{ij}$ and normalize; for details see [12].) This is the formulation of the network we are going to use: i.e., we assume that the state $u_i$ of each unit $i$, with input connections of weight $h_{ij}$ from the other units, develops as described by equation (1).

## 3 Simulating Turing Machines by Networks

We shall consider classes of Boolean functions computed by networks of continuous-time units. The appropriate definitions may be framed in e.g. the following manner. Let $N$ be a network of $p$ units, including a special indicator unit $u_{done}$. Let $\iota : \{0,1\}^n \to \mathcal{R}^p$ and $\rho : \mathcal{R}^p \to \{0,1\}^m$ be two "simple" mappings, used to translate a binary input string of length $n$ to an initial network state, and a final network state into a binary output string of length $m$. Given an input string $x \in \{0,1\}^n$, the network is initialized in state $\iota(x)$; this initial state should in particular satisfy $\sigma(u_{done}) = -1$. The network is then allowed to run until $\sigma(u_{done})$ achieves value 1: the computation is *well-behaved* if during its course the value of $\sigma(u_{done})$ increases monotonically from $-1$ to 1, and when $\sigma(u_{done}) = 1$, the value of $\rho(u)$ stays constant. The result of the computation is then the final stable value of $\rho(u)$.

Assuming the above notion of a well-behaved computation, a network $N$ thus computes a partial mapping

$$f_N : \{0,1\}^n \to \{0,1\}^m \ .$$

The mapping is partial, because we take its value to be undefined for inputs that do not lead to a well-behaved computation.

For simplicity, we consider from now on only networks with a single-bit output (i.e. $m = 1$); the extensions to networks with multiple-bit outputs are straightforward. The *language recognized* by an $n$-bit input, single-bit output network $N$ is defined as

$$L(N) = \{x \in \{0,1\}^n \mid f_N(x) = 1\} \ .$$

A *sequence* of networks $(N_n)_{n\geq 0}$ recognizes a language $A \subseteq \{0,1\}^*$, if each network $N_n$ recognizes the language $A \cap \{0,1\}^n$. A sequence of networks has *polynomial size* if there is a polynomial $q(n)$ such that for each $n$, the network $N_n$ has at most $q(n)$ units.

Let $\langle x,y \rangle$ be some standard pairing function mapping pairs of binary strings to binary strings (see, e.g. [4, p. 7]). A language $A \subseteq \{0,1\}^*$ belongs to the nonuniform complexity class PSPACE/poly ([3], [4, p. 100], [13]), if there are a polynomial space bounded Turing machine $M$, and an "advice" function

$f : N \rightarrow \{0,1\}^*$, where for some polynomial $q$ and all $n \in N$, $|f(n)| \leq q(n)$, and for all $x \in \{0,1\}^*$,

$$x \in A \quad \Leftrightarrow \quad M \text{ accepts } \langle x, f(|x|) \rangle .$$

It was shown in [21] that all languages in PSPACE/poly can be recognized by polynomial-size sequences of discrete Hopfield networks, even symmetric ones. (Recall that in a discrete Hopfield net the units have bipolar, i.e. $\pm 1$ states, and each unit $i$ updates its state from time $t$ to time $t+1$ according to the rule $y_i(t+1) = \text{sgn}(\sum_j h_{ij} y_j(t))$, where $\text{sgn}(z) = 1$ if $z \geq 0$ and $-1$ if $z < 0$.) As the construction without the symmetricity restriction is not too difficult, we shall for completeness outline it here[2].

Let $A \in$ PSPACE/poly via a machine $M$ and advice function $f$. Let the space complexity of $M$ on input $\langle x, f(|x|) \rangle$ be bounded by a polynomial $q(|x|)$. Without loss of generality (see, e.g. [4]) we may assume that $M$ has only one tape, halts on any input $\langle x, f(|x|) \rangle$ in time $c^{q(|x|)}$, for some constant $c$, and indicates its acceptance or rejection of the input by printing a 1 or a 0 on the first square of its tape.

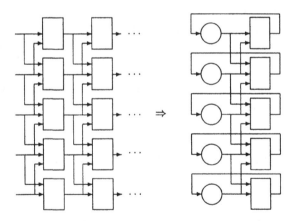

**Fig. 2.** Simulation of a space bounded Turing machine by an asymmetric recurrent net.

Following the standard simulation of Turing machines by combinational circuits [4, pp. 106–112], it is straightforward to construct for each $n$ a feedforward circuit that simulates the behavior of $M$ on inputs of length $n$. (More precisely,

---

[2] In fact, this version of the result, i.e. Turing machine simulation by asymmetric networks with synchronously updated units, was already obtained by Lepley and Miller in the unpublished report [16]. The result was extended to symmetric networks in [21], and to networks with asynchronous updates in [22].

the circuit simulates computations $M(\langle x, f(n)\rangle)$, where $|x| = n$.) This circuit consists of $c^{q(n)}$ "layers" of $O(q(n))$ parallel wires, where the $t$th layer represents the configuration of the machine $M$ at time $t$ (Fig. 2, left). Every two consecutive layers of wires are interconnected by an intermediate layer of $q(n)$ constant-size subcircuits, each implementing the local transition rule of machine $M$ at a single position of the simulated configuration. The input $x$ is entered to the circuit along input wires; the advice string $f(n)$ appears as a constant input on another set of wires; and the output is read from the particular wire at the end of the circuit that corresponds to the first square of the machine tape.

One may now observe that the interconnection patterns between layers are very uniform: all the local transition subcircuits are similar, with a structure that depends only on the structure of $M$, and their number depends only on the length of $x$. Hence we may replace the exponentially many consecutive layers in the feedforward circuit by a single transformation layer that feeds back on itself (Fig. 2, right). The size of the recurrent network thus obtained is then only $O(q(n))$. When initialized with input $x$ loaded onto the appropriate input units, and advice string $f(n)$ mapped to the appropriate initially active units, the network will converge in $O(c^{q(n)})$ update steps, at which point the output can be read off the unit corresponding to the first square of the machine tape. It is also easy to arrange for a separate unit $u_{done}$ whose value flips from $-1$ to $1$ when the simulated machine $M$ enters a halting state.

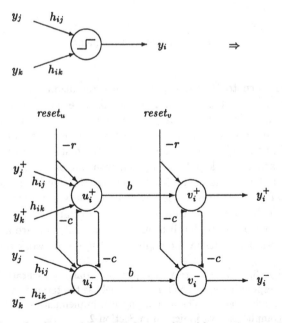

**Fig. 3.** Simulation of a discrete-time unit by two bistable continuous-time unit pairs.

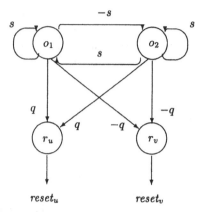

**Fig. 4.** Deriving reset pulses from an oscillating unit pair.

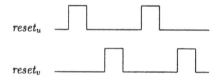

**Fig. 5.** Reset pulse sequence.

Let us then turn to the continuous-time simulation. We shall simply take the discrete network of Fig. 2 (or any other discrete network, for that matter), and replace it unit by unit with a computationally equivalent continuous-time system. Each discrete-time unit $i$ is replaced by two *bistable pairs* of continuous-time units as indicated in Fig. 3[3]. The units in the bistable pairs are periodically reset by alternating "clock pulses" derived from an *oscillating pair* of units as shown in Fig. 4. The sequencing of the reset signals is schematically depicted in Fig. 5. As regards the values of the various parameters, we shall just note here that they should be chosen so that $r \gg c \gg b > 0$, and $s > 1$; we shall discuss appropriate choices in more detail later. For simplicity, we are always assuming that the connection weights in the simulated discrete network are integers.

---

[3] Note that Figs. 3, 4, and 6 are really just pictorial representations of the equations (1). Correspondingly, we shall in the sequel use the terms "unit" and "variable" completely interchangeably. The actual physical implementation of these "units" is rather more complicated, as indicated in Section 2.

The computational idea underlying this construction is the following. The pair of units labeled $v_i^{\pm}$ in Fig. 3 represents the state of the simulated discrete unit $i$ at each discrete time $t$ in a redundant manner, so that $y_i^{+} = \sigma(v_i^{+}) = 1$ if $y_i(t) = 1$, and $y_i^{-} = \sigma(v_i^{-}) = 1$ if $y_i(t) = -1$. As long as no reset signals arrive from the clock subnetwork this representation is stable because of the inhibitory connections of weight $-c \ll -b$ between the units $v_i^{+}$ and $v_i^{-}$.

Assume then that a $reset_u$ signal from the clock subnetwork drives the states of all of the $u$ units in the network close to the values $u^{\pm} = -r$. When the signal falls off, the $u$ units start to compete for activation, based on the inputs they receive from the $v$ units (which have not been reset, and remain stable). It can be seen that the net input to unit $u_i^{+}$ (resp. $u_i^{-}$) is positive if and only if in the discrete network $y_i(t + 1) = 1$ (resp. $-1$). Thus, as a result of this biased competition, the $u$ units will converge to values that represent the states of the discrete units at time $t + 1$: $\sigma(u_i^{+}) = 1$ if $y_i(t + 1) = 1$, and $\sigma(u_i^{-}) = 1$ if $y_i(t + 1) = -1$.

When the $u$ pairs have stabilized, a $reset_v$ signal from the clock similarly drives all the $v$ units close to $-r$. When this signal falls off, the values of the $u$ units are simply copied into the $v$ units by the competitive mechanism, biased by connections of weight $b$ from the $u$ units to the $v$ units. After the $v$ pairs have stabilized, the network is ready for another step of the simulation. (Thus, we could actually double the speed of the simulation by having the $u$ and $v$ units represent the discrete units at even and odd times $t$, and computing new values into the $v$ units instead of just copying the $u$ values.)

We shall analyze the simulation in a general way in Section 4, but let us first look at an example. Fig. 6 shows the continuous-time implementation of a single discrete-time unit with a self-connection of weight $-1$, i.e. a discrete oscillator. The relevant parameter values are indicated in the figure. Note that each of the $u$ and $v$ units has an internal bias of weight $-8$: this has the effect of rescaling the $reset$ signals arriving at them from range $[-1, 1]$ with weight $-8$ to range $[0, 1]$ with weight $-16$. Also the $r$ units, used for deriving the $reset$ signals from the oscillating $o_1/o_2$ pair, have internal biases of weight $-8$, in order to drive their outputs to $-1$ when they receive zero input (i.e., when the inputs arriving from the $o_1$ and $o_2$ units cancel each other).

Figures 7–14, obtained from a MATLAB [18] numerical integration of the corresponding differential equations, illustrate the behavior of the system. Consider for instance Fig. 7, which shows the time development of the internal states of the two oscillating units $o_1$ and $o_2$. (The solid line represents $o_1$, the dashed line $o_2$.) The differential equations governing the corresponding variables are:

$$\dot{o}_1 = -o_1 + s\sigma(o_1) + s\sigma(o_2) ,$$
$$\dot{o}_2 = -o_2 + s\sigma(o_2) - s\sigma(o_1) , \tag{2}$$

where we have for brevity adopted the Newtonian dot notation for the time derivatives, and dropped explicit references to the time variable. Here $\sigma$ is the saturated-linear response function, and $s = 16$. The initial conditions are $o_1(0) = -31$, $o_2(0) = 1$. A state space plot of $o_1$ vs. $o_2$ appears in Fig. 8. (To show

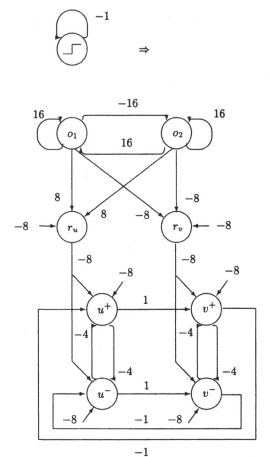

**Fig. 6.** Continuous-time simulation of a discrete oscillator.

the emergence of the limit cycle more clearly, we have here chosen the initial conditions $o_1(0) = -1$, $o_2(0) = 1$.) Fig. 9 shows the reset pulses derived from the oscillator, i.e. the values $reset_u = \sigma(r_u)$ and $reset_v = \sigma(r_v)$, where $r_u$ and $r_v$ are governed by the equations

$$
\begin{aligned}
\dot{r}_u &= -r_u + q\sigma(o_1) + q\sigma(o_2) - q \ , \\
\dot{r}_v &= -r_v - q\sigma(o_1) - q\sigma(o_2) - q \ ,
\end{aligned}
\tag{3}
$$

for $q = 8$. (The solid line corresponds to $reset_u$, the dashed one to $reset_v$.)

Computationally, the most interesting graphs are those in Figs. 10 and 11: these show the development of the states of the $u$ and $v$ units, starting from the initial conditions $u^+(0) = v^+(0) = 3$, $u^-(0) = v^-(0) = -3$. (The solid lines in both figures represent the "+"-units, the dashed lines the "−"-units.) The

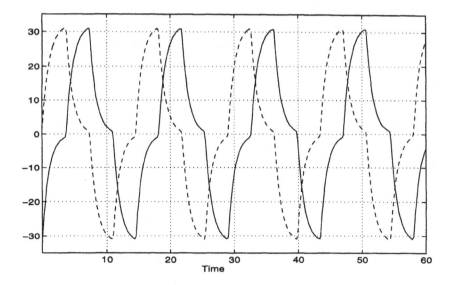

**Fig. 7.** Internal states of the units $o_1$ and $o_2$.

corresponding output signals, i.e. the values $\sigma(u^\pm)$ and $\sigma(v^\pm)$ are shown in Figs. 12 and 13. For completeness, let us write down also the equations for the $u$ and $v$ units, as inferred from the diagram in Fig. 6:

$$
\begin{aligned}
\dot{u}^+ &= -u^+ + h\sigma(v^+) - c\sigma(u^-) - r \cdot reset_u - r, \\
\dot{u}^- &= -u^- + h\sigma(v^-) - c\sigma(u^+) - r \cdot reset_u - r, \\
\dot{v}^+ &= -v^+ + b\sigma(u^+) - c\sigma(v^-) - r \cdot reset_v - r, \\
\dot{v}^- &= -v^- + b\sigma(u^-) - c\sigma(v^+) - r \cdot reset_v - r,
\end{aligned}
\tag{4}
$$

where $h = -1$, $b = 1$, $c = 4$, and $r = 8$.

One can observe in Fig. 10 first the resetting of the $u^+$ and $u^-$ units in response to the $reset_u$ signal at about time $t = 4$, and then the emergence of the competition between the two units as the reset signal is switched off, at about time $t = 8$. The competition is eventually won by the $u^-$ unit, because it receives a weighted input signal of strength 1 from unit $v^-$, whereas unit $u^+$ receives a weighted input signal of strength $-1$ from unit $v^+$. (This initial part of the computation is shown in more detail in Fig. 14.) In Fig. 11 one may then observe how the new states of the $u$ units are copied into the $v$ units, after these have been cleared by the $reset_v$ signal, at about time $t = 15$. The cycle starts again, but from inverted initial conditions, at about time $t = 19$ with the next $reset_u$ signal.

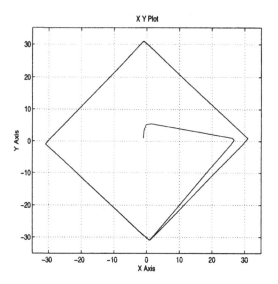

**Fig. 8.** State space plot of $o_1$ vs. $o_2$.

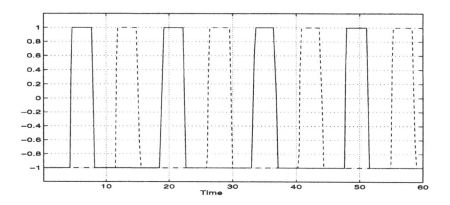

**Fig. 9.** Reset pulses $reset_u$ and $reset_v$ derived from the oscillating pair.

## 4 General Analysis

Let us then look more generally into the behavior of the equations governing the various components of the simulation, starting with the oscillating pair of units $o_1$ and $o_2$. From equation (2) one can see that this system has a fixed point at origin — in fact, some amount of tedious algebra, considering separately the cases $o_1$ 7 0, $o_2$ 7 0, shows that this is the *only* fixed point. The Jacobian matrix

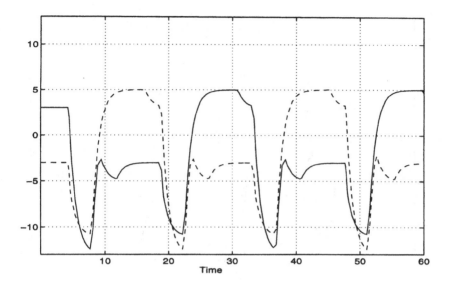

**Fig. 10.** Internal states of the units $u^+$ and $u^-$.

of the system at origin is

$$J = \begin{pmatrix} -1+s & s \\ -s & -1+s \end{pmatrix},$$

with eigenvalues $(s-1) \pm si$, so the fixed point at origin is repelling if and only if $s > 1$. It is easy to see that if the initial conditions satisfy $-2s \leq o_1, o_2 \leq 2s$, then the state of the system stays in this region; thus, by the the Poincaré–Bendixson theorem [10] this region contains a limit cycle of the system.

While determining the exact location and period of the cyclic trajectory, as a function of $s$, is difficult, a tedious iterative solution of the equations (2), made feasible by the computer algebra system Maple [9], shows that for large $s$, the trajectory passes close to the points $\pm(2s-1, 1)$, $\pm(1, -2s+1)$, and its period grows as $4 \ln 2s + O(s^{-1})$. In particular, then, the oscillation of $o_1$ and $o_2$ may be made arbitrarily slow by increasing the parameter $s$ — although the period does grow only logarithmically in $s$.

In the analysis of the discrete network simulation we shall proceed piecewise, by considering the behavior of the continuous-time system separately within each linear region of the response function $\sigma$. The large amounts of calculation required by this brute-force approach have again been performed with the help of the Maple system. Even so, we shall make two simplifying assumptions. First, we only consider a single update step of a single pair of $u$ units, where the units move, as a response to their net inputs, from an initial state of $\sigma(u^+) = 1$, $\sigma(u^-) = -1$ to the state $\sigma(u^+) = -1$, $\sigma(u^-) = 1$ (as in the example considered above). This simplification is justified, because (i) all the $u$ units change

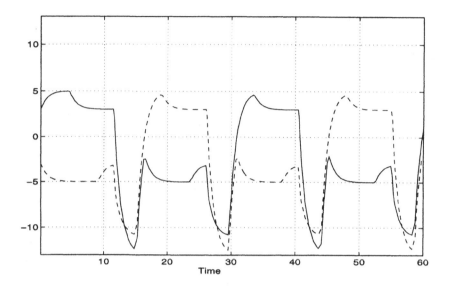

**Fig. 11.** Internal states of the units $v^+$ and $v^-$.

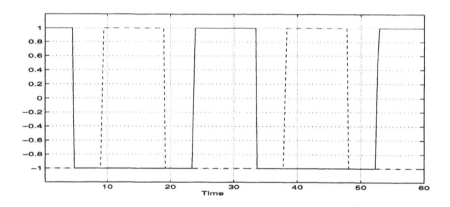

**Fig. 12.** Output signals of the units $u^+$ and $u^-$.

state synchronously, so looking at one pair suffices; (*ii*) the opposite move from $\sigma(u^+) = -1$, $\sigma(u^-) = 1$ to $\sigma(u^+) = 1$, $\sigma(u^-) = -1$ is symmetric, and thus does not need to be considered; (*iii*) any move where the states of the units stay unchanged is more robust than the one considered, because then the unit with output 1 stays continually ahead in the competition: it both has a larger initial value, and receives excitatory, as opposed to inhibitory, input from other units; and finally (*iv*) the $v$ units are similar to $u$ units with a single excitatory input

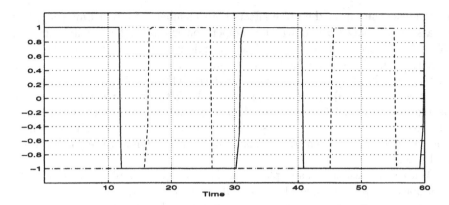

**Fig. 13.** Output signals of the units $v^+$ and $v^-$.

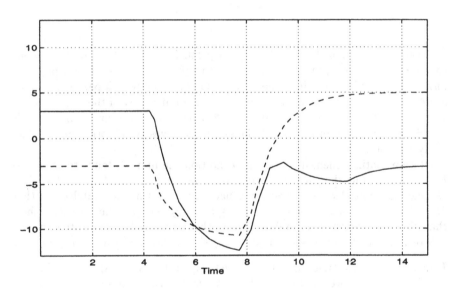

**Fig. 14.** Expanded view of the internal states of the units $u^+$ and $u^-$.

connection, so they need not be considered separately.

The second assumption we make is that the *reset* pulses are sharp, i.e. each pulse switches from $-1$ to 1 and back at precisely defined moments, instead of making a continuous transition. This simplification can be justified by observing that (*i*) slow rise times don't actually matter, as long as the pulse stays high for sufficiently long to effect the intended reset; and (*ii*) although the critical competition between $+/-$ unit pairs is initiated during the falling phase of the reset pulse, the tail of the pulse affects both members of a pair uniformly, and so a slow fall only slows down the competition without affecting its outcome. What *is* important, however, is that the quiescent period between the *reset* signals is sufficiently long for the $+/-$ unit pairs to converge sufficiently close to their new limiting values. A Maple analysis of the oscillator and reset signal equations (2) and (3) shows that the rise and fall times of the reset pulses are asymptotic to $2/q + O(q^{-2})$, and the high and quiescent times are asymptotic to $\ln 2s + O(s^{-1} + q^{-1})$. Thus, the rise and fall times can be made arbitrarily short and the high and quiescent times arbitrarily long by adjusting the parameters $s$ and $q$ appropriately.

With these assumptions in place, we shall now proceed to consider the unit equations (1). As discussed above, initially $\sigma(u^+) = 1$, $\sigma(u^-) = -1$, and unit $u^+$ receives from the rest of the network some total input of $-h < 0$, whereas $u^-$ receives a net input of $h > 0$. Assuming that both of the reset signals are initially off, the equations governing the pair of units are:

$$\dot{u}^+ = -u^+ - h + c \ ,$$
$$\dot{u}^- = -u^- + h - c \ .$$

One can thus see that if the system is not perturbed, and $c \gg h$, unit $u^+$ tends to value $-h + c \gg 0$, and unit $u^-$ tends to value $h - c \ll 0$. To simplify the formulas, let us fix a specific value for the competition strength parameter $c$, say $c = 4w$, where $w \geq 1$ is the maximum total net input to any of the $u$ units. Also, we might just as well assume that $h = 1$, because this is the minimum possible bias for $u^-$ over $u^+$: for any $h > 1$ the system behaves even more robustly than analyzed below.

Recapitulating, then, in the initial condition the unit $u^+$ tends to value $c - h = 4w - 1$, and the unit $u^-$ tends to value $-c + h = -4w + 1$. We shall assume that initially the system has had sufficient time to stabilize so that $4w - 2 \leq u^+(0) \leq 4w - 1$ and $-4w + 1 \leq u^-(0) \leq -4w + 2$; and we shall show that after a sequence of six update phases, the corresponding opposite situation will be reached, so that $-4w - 1 \leq u^+(T_6) \leq -4w$ and $4w \leq u^-(T_6) \leq 4w + 1$.

**Phase 1:** At some time $T_0 \geq 0$ the *reset$_u$* signal turns on. Again, to simplify the equations, we shall fix a definite value $r = 2c = 8w$ for the strength of the reset connection. Thus, at time $T_0$ the equations governing the unit pair change to (cf. equation (4)):

$$\dot{u}^+ = -u^+ - h + c - 2r = -u^+ - 1 - 12w \ ,$$
$$\dot{u}^- = -u^- + h - c - 2r = -u^- + 1 - 20w \ .$$

Solving this system of linear first-order o.d.e.'s (with the help of the Maple system) with the initial conditions $4w - 2 \leq u^+(T_0) \leq 4w - 1$, $-4w + 1 \leq u^-(T_0) \leq -4w + 2$ shows that the value of $u^+$ reaches 1 at some time $T_1 = T_0 + t_1$, where $\ln \frac{16w-1}{12w-2} \leq t_1 \leq \ln \frac{16w}{12w-2}$. At this time variable $u^-$ has a value bounded by $-8w + 3 \leq u^-(T_1) \leq -\frac{128w^2 - 68w - 1}{16w - 1}$. Approximately, then, $u^+$ decreases from about $4w$ to 1 in roughly time $t_1 = \ln 4w$, and in this time $u^-$ decreases from about $-4w$ to between $-8w + 3$ and $-8w + 5$. Because we wish to show that in the eventual competition between the units, $u^-$ will win and obtain a positive value, and $u^+$ will obtain a negative value, we are mainly interested in lower bounds on $u^-$ and upper bounds on $u^+$: to keep from cluttering the presentation we shall in the sequel list only these bounds.

**Phase 2:** At time $T_1$ the system equations change again, because the value of $u^+$ enters the region where the response function is $\sigma(u^+) = u^+$. The new equations are

$$\dot{u}^+ = -u^+ - 1 - 12w \ ,$$
$$\dot{u}^- = -u^- + 1 - 4wu^+ - 16w \ .$$

Solving this system again with the help of Maple, one obtains for the time when $u^+$ reaches $-1$ the value $T_2 = T_1 + t_2$, where $t_2 = \ln \frac{6w+1}{6w} \approx \frac{1}{6w}$. At this time $u^-(T_2) \geq -8w + 1$.

**Phase 3:** At time $T_2$ the system equations change to

$$\dot{u}^+ = -u^+ - 1 - 12w \ ,$$
$$\dot{u}^- = -u^- + 1 - 12w \ ,$$

so that as long as the $reset_u$ signal stays on, the unit states approach exponentially the values $u^+ = -12w - 1$, $u^- = -12w + 1$. We shall assume that the clock subnetwork oscillates so slowly that the reset signal stays on for at least an additional time of $t_3 = \ln 2r = \ln 16w$. One obtains then at time $T_3 = T_2 + t_2$ the bounds $u^+(T_3) \leq -12w$, $u^- \geq -12w + 1$.

**Phase 4:** At time $T_3$ the $reset_u$ signal switches off, and the system equations change to

$$\dot{u}^+ = -u^+ - 1 + 4w \ ,$$
$$\dot{u}^- = -u^- + 1 + 4w \ .$$

Thus the values of both $u^+$ and $u^-$ start to increase, and one can see that $u^-$ reaches $-1$ sooner than $u^+$. (It both has a larger initial value, and rises more steeply than $u^+$.) More precisely, $u^-(T_4) = -1$ at some time $T_4 = T_3 + t_4$, where $t_4 \leq \ln(4 - \frac{2}{w+1})$, and at this time $u^+(T_4) \leq -\frac{5}{2}$.

**Phase 5:** It is now already clear that unit $u^-$ will win the competition, but let us nevertheless follow the system for a few more phases. At time $T_4$ the system equations become

$$\dot{u}^+ = -u^+ - 1 - 4wu^- \ ,$$
$$\dot{u}^- = -u^- + 1 + 4w \ ,$$

and they have this form until either $u^-$ reaches 1 or $u^+$ reaches $-1$. Actually, one can verify that the latter never happens, and when $u^-$ reaches 1, at time $T_5 = T_4 + t_5$, where $t_5 = \ln(w + \frac{1}{2w})$, the value of $u^+$ is still less than $-2$.

**Phase 6:** At time $T_5$ the system equations change to

$$\dot{u}^+ = -u^+ - 1 - 4w \ ,$$
$$\dot{u}^- = -u^- + 1 + 4w \ ,$$

so that in the limit $u^-$ converges to the value $4w + 1$, and $u^+$ converges to $-4w - 1$. Assuming again that the system has at least time $t_6 = \ln 2r = \ln 16w$ to stabilize before the $reset_v$ signal turns on, the values at time $T_6 = T_5 + t_6$ satisfy $u^-(T_6) \geq 4w$, $u^+(T_6) \leq -4w$, as desired. (In fact, in time $\ln 16w$ both variables will converge to within $\frac{1}{4}$ of their asymptotic values.)

Adding up all the transition times one obtains for the total time $T_6 - T_0$ the bound

$$T_6 \leq \ln 4w + \frac{1}{6w} + \ln 16w + \ln(4 - \frac{2}{w+1}) + \ln(w + \frac{1}{2w}) + \ln 16w$$
$$\approx 4 \ln 8w \ .$$

To guarantee correct behavior, the clock network should oscillate with a period of at least twice this bound (recall that in our somewhat inefficient simulation technique we must update both the $u$ and $v$ units during a single clock cycle). Thus, we should choose for the parameter $s$ a value such that $4 \ln 2s \geq 8 \ln 8w$, i.e. $s \geq 32w^2$. However, this is assuming the reset pulses are perfectly sharp: theoretically one should choose for $s$ a somewhat larger value, to compensate for the nonzero rise and fall times. On the other hand, in the example simulation in Section 3, we used successfully the value $s = 16$ instead of a larger value $s \geq 32$, as would be suggested by these calculations.

## 5    Conclusion and Open Problems

We have shown that continuous-time neural networks based on Hopfield-type units with saturated-linear amplifiers are universal computational devices, in the sense that (sequences of) networks with polynomially many units are capable of simulating polynomially space-bounded Turing machines. Some of the many open questions suggested by this work are the following.

Our present simulation construction is somewhat unsatisfying because of its heavy reliance on the clock pulses provided by the oscillator subnetwork. It would be most interesting to develop computation and analysis techniques for nonoscillating networks. One especially interesting case where at least infinite undamped oscillations are precluded are networks with symmetric interconnections. In the discrete-time case also symmetric networks are capable of efficient Turing machine simulation [21], but the continuous-time case seems to be rather more complicated. Apparently not even simple convergence time bounds for symmetric continuous-time networks are known. (For discrete symmetric networks such bounds were first obtained in [8], and the convergence behavior of such networks is by now well understood. For discrete-time networks with continuous unit states, asymptotic convergence has also been established [7, 15, 17], but no time bounds are known.)

A technical issue concerns more general response functions. Our computer simulations indicate that networks with, e.g., *tanh* nonlinearities have qualitatively the same behavior as networks using the saturated-linear response function. However the piecewise-linear analysis of network behavior used above obviously doesn't extend to this case. In the discrete-time, continuous-state setting sigmoidal response functions were analyzed in [14], and shown to be at least as powerful as saturated-linear ones.

Finally, we have only obtained a *lower* bound on the computational power of continuous-time networks, and one may also inquire about the corresponding *upper* bounds. For comparison, note that in the discrete-time case Siegelmann and Sontag [26] have shown that any Turing machine can be simulated uniformly for all input lengths by a *single* network with saturated-linear response functions and arbitrary-precision real number states. Also, several simulations of arbitrary Turing machines by continuous-time and finite-dimensional, but non-network-like systems are known (e.g. [2, 5, 19, 20]).

# References

1. Anderson, J. A., Rosenfeld, E. (eds.) *Neurocomputing: Foundations of Research.* The MIT Press, Cambridge, MA, 1988.

2. Asarin, E., Maler, O. On some relations between dynamical systems and transition systems. *Proc. 21st Internat. Colloq. on Automata, Languages, and Programming (Jerusalem, Israel, July 1994),* 59–72. Springer-Verlag, Berlin, 1994.

3. Balcázar, J. L., Díaz, J., Gabarró, J. On characterizations of the class PSPACE/poly. *Theoret. Comput. Sci. 52* (1987), 251–267.

4. Balcázar, J. L., Díaz, J., Gabarró, J. *Structural Complexity I.* Springer-Verlag, Berlin, 1988 (2nd Ed. 1995).

5. Branicky, M. Analog computation with continuous ODEs. *Proc. Workshop on Physics and Computation 1994 (Dallas, Texas, Nov. 1994),* 265–274. IEEE Computer Society Press, Los Alamitos, CA, 1994.

6. Branicky, M. Universal computation and other capabilities of hybrid and continuous dynamical systems. *Theoret. Comput. Sci. 138* (1995), 67–100.

7. Fogelman, F., Mejia, C., Goles, E., Martínez, S. Energy functions in neural networks with continuous local functions. *Complex Systems 3* (1989), 269–293.

8. Fogelman, F., Goles, E., Weisbuch, G. Transient length in sequential iterations of threshold functions. *Discr. Appl. Math. 6* (1983), 95–98.

9. Char, B. W, Geddes, K. O., Gonnet, G. H., Leong, B. L., Monagan, M. B., Watt, S. M. *First Leaves: A Tutorial Introduction to Maple V.* Springer-Verlag, New York, NY, 1992.

10. Hirsch. M. W., Smale, S. *Differential Equations, Dynamical Systems, and Linear Algebra.* Academic Press, San Diego, CA, 1974.

11. Hopfield, J. J. Neural networks and physical systems with emergent collective computational abilities. *Proc. Nat. Acad. Sci. USA 79* (1982), 2554–2558. Reprinted in [1], pp. 460–464.

12. Hopfield, J. J. Neurons with graded response have collective computational properties like those of two-state neurons. *Proc. Nat. Acad. Sci. USA 81* (1984), 3088–3092. Reprinted in [1], pp. 579–583.

13. Karp, R. M., Lipton, R. J. Turing machines that take advice. *L'Enseignement Mathématique 28* (1982), 191–209.

14. Kilian, J., Siegelmann, H. T. The dynamic universality of sigmoidal neural networks. *Information and Computation 128* (1996), 48–56.

15. Koiran, P. Dynamics of discrete time, continuous state Hopfield networks. *Neural Computation 6* (1994), 459–468.

16. Lepley, M., Miller, G. Computational power for networks of threshold devices in an asynchronous environment. Unpublished manuscript, Dept. of Mathematics, Massachusetts Inst. of Technology, 1983.

17. Marcus, C. M., Westervelt, R. M. Dynamics of iterated-map neural networks. *Phys. Rev. A 40* (1989), 501–504.

18. *MATLAB Reference Guide.* MathWorks Inc., Natick, MA, 1992.

19. Moore, C. Unpredictability and undecidability in physical systems. *Phys. Rev. Lett. 64 (1990),* 2354–2357.

20. Moore, C. Generalized shifts: unpredictability and undecidability in dynamical systems. *Nonlinearity 4* (1991), 199–230.

21. Orponen, P. The computational power of discrete Hopfield nets with hidden units. *Neural Computation 8* (1996), 403–415.

22. Orponen, P. Computing with truly asynchronous threshold logic networks. *Theoret. Comput. Sci. 174* (1997), 97–121.

23. Orponen, P. A survey of continuous-time computation theory. *Advances in Algorithms, Languages, and Complexity* (eds. D.-Z. Du, K.-I Ko), 209–224. Kluwer Academic Publishers, Dordrecht, 1997

24. Pour-El, M. B., Richards, I. *Computability in Analysis and Physics.* Springer-Verlag, Berlin, 1989.

25. Rubel, L. A. Digital simulation of analog computation and Church's Thesis. *J. Symb. Logic 54* (1989), 1011–1017.

26. Siegelmann, H. T., Sontag, E. D. On the computational power of neural nets. *J. Comput. System Sciences 50* (1995), 132–150.

27. Vergis, A., Steiglitz, K., Dickinson, B. The complexity of analog computation. *Math. and Computers in Simulation 28* (1986), 91–113.

# A Foundation for Computable Analysis

Klaus Weihrauch

FernUniversität Hagen, D – 58084 Hagen
e-mail: Klaus.Weihrauch@Fernuni-Hagen.de

**Abstract.** While for countable sets there is a single well established computability theory (ordinary recursion theory), Computable Analysis is still underdeveloped. Several mutually non–equivalent theories have been proposed for it, none of which, however, has been accepted by the majority of mathematicians or computer scientists. In this contribution one of these theories, TTE (Type 2 Theorie of Effectivity), is presented, which at least in the author's opinion has important advantages over the others. TTE intends to characterize and study exactly those functions, operators etc. known from Analysis, which can be realized correctly by digital computers. The paper gives a short introduction to basic concepts of TTE and shows its general applicability by some selected examples.
First, Turing computability is generalized from finite to infinite sequences of symbols. Assuming that digital computers can handle (w.l.o.g.) only sequences of symbols, infinite sequences of symbols are used as names for "infinite objects" such as real numbers, open sets, compact sets or continuous functions. Naming systems are called representations. Since only very few representations are of interest in applications, a very fundamental principle for defining effective representations for $T_0$–spaces with countable bases is introduced. The concepts are applied to real numbers, compact sets, continuous functions and measures. The problem of zero-finding is considered. Computational complexity is discussed. We conclude with some remarks on other models for Computable Analysis. The paper is a shortened and revised version of [Wei97].

## Contents

# 1 Introduction

Classical computability theory considers the natural numbers $\omega$ or the words $\Sigma^*$ over some finite alphabet $\Sigma$ as basic sets. Many definitions of computability have been proposed and justified by intuition, most of which turned out to be equivalent to computability by Turing machines. This lead to a claim generally referred to as Church's Thesis, which can be formulated as follows: A partial function $f :\subseteq \Sigma^* \longrightarrow \Sigma^*$ is computable in the intuitive sense or by a physical device such as a digital computer, if and only if it can be computed by a Turing machine. Although Church's Thesis cannot be proved it is generally accepted, and today "computable" is the usual abbreviation of "Turing computable". Church's Thesis does not apply to functions on the real numbers. Several theories for studying aspects of effectivity in Analysis have been developed in the past (Intuitionistic Logic [TD88], Bishop [Bis67], Grzegorczyk [Grz55], Ceitin [Cei59], Hauck [Hau83], Traub et al. [TWW88], Pour–El/Richards [PER88], Ko [Ko91], Blum/Shub/Smale [BSS89], [WS81], [KW85], ...), and several generalizations of Church's Thesis have been suggested. Although each of these generalizations is based on easily comprehensible intuition, they are not equivalent. Moreover, the relations between the theories and computability definitions are not yet fully explored. None of the theories has been accepted by the majority of mathematicians or computer scientists. This somewhat confusing situation may explain but does not excuse the fact that computer scientists have neglected Computable Analysis in research and almost disregarded it in teaching.

In this article we present one of the existing approaches to Computable Analysis, henceforth called "Type 2 Theory of Effectivity" (TTE for short). TTE is based on a definition of computable real functions given by the Polish logician A. Grzegorczyk in 1955 [Grz55], where computable operators transform fast converging sequences of rational numbers representing real numbers. TTE extends this original approach in several directions:

1. Also continuous operators are considered where "continuity" can be interpreted as a very fundamental kind of "constructivity".

2. By using representations effectivity can be studied not only on the real numbers but on many other spaces used in Analysis.

3. A general concept of "admissible" (effective) representations is derived from basic principles.

4. Computational complexity ("bit–complexity") is included as a part of TTE.

Since intuition has not lead to a single concept of computability for real functions, TTE intends to characterize and study those real functions, which can be computed by physical devices. The main difficulty stems from the fact that the set of real numbers is not countable and that therefore real numbers cannot be coded by natural numbers or finite words but must be considered as infinite objects (e.g. infinite decimal fractions). By laws of Physics, however, any device can store only finitely many bits of information at any moment, and any input or output channel can transfer only finitely many bits in a finite amount of time. TTE solves this by considering finite objects as approximations of infinite ones and by representing infinite objects by infinite sequences of finite objects.

In recursion theory computability is defined explicitly (w.l.o.g.) on $\Sigma^*$. Computability is transferred to other countable sets $M$ by notations $\nu :\subseteq \Sigma^* \longrightarrow M$, where words serve as names. TTE considers additionally the set $\Sigma^\omega$ of infinite sequences as standard infinite objects, defines computability (and other effectivity concepts) on $\Sigma^\omega$ explicitly and transfers them to sets $M$ with maximally continuum cardinality by representations $\delta :\subseteq \Sigma^\omega \longrightarrow M$, where infinite sequences of symbols serve as names (example: infinite decimal fractions as names for real numbers).

In Section 2 we introduce computable functions from tuples of finite or infinite sequences to finite or infinite sequences $f :\subseteq Y_1 \times \ldots \times Y_k \longrightarrow Y_0$ ($Y_0 \ldots Y_k \in \{\Sigma^*, \Sigma^\omega\}$) by means of Type 2 machines, which are Turing machines with one–way finite or infinite input and output tapes. We observe that for such a computable function, every finite portion of the result depends only on a finite portion of the input. This finiteness property can be expressed as continuity w.r.t. the Cantor topology on $\Sigma^\omega$. We give examples of computable functions and prove that multiplication of infinite decimal fractions is not even continuous, hence not computable. Section 3 introduces the vocabulary for transferring effectivity concepts from $\Sigma^\omega$ to represented sets. Although every representation $\delta :\subseteq \Sigma^\omega \longrightarrow M$ of a set $M$ induces some type of effectivity on $M$, most of the countless representations are not interesting, only "effective" ones are important. Effectivity is usually based on some structure on $M$. A type of structure we consider is a topological $T_0$–space with a notation of a subbase [Eng89]. For each such "information structure" we introduce a standard representation with very interesting natural topological and computational properties, which justify to call them "effective" or "admissible". As a fundamental result in TTE, for admissible representations every computable function is continuous. Only admissible representations will be used in the following. They provide a very natural and sufficiently general concept to introduce computability in almost all branches of Analysis. In Section 5 we investigate computability on the real line, based on the Cauchy representation which is admissible for this topological space. We show that the representation by infinite decimal fractions is not admissible, we give examples for computable functions, and we show that no (non–trivial) set of real numbers is recursive. In Section 6 we apply the general concept to the set of compact subsets of $\mathbb{R}$. Several admissible representations, which express different types of knowledge, are introduced and compared. Compactness is an instructive example for a classical concept which branches into several parts in TTE. In Section 7 we consider zero–finding for continuous real functions Computational Complexity is considered in Section 8. For functions on $\Sigma^\omega$, complexity and input lookahead (input information) are measured as functions of the output precision. For transferring complexity and input lookahead, distinguished admissible representations are needed. For the real line we introduce the "modified binary representation" (which uses the digits 0, 1 and -1) [Wie80]. The resulting complexity concept ("bit complexity") has been used by Ko [Ko91] and others.

The first presentation of the concepts of TTE is [KW84]. More complete versions are [Wei85] and [KW85] as well as Part 3 of [Wei87]. A formally simpler access to TTE is "A simple introduction to computable analysis" [Wei95], more details can be found in [Wei94]. Some further publications are added without reference from the text. This paper is a revised and shortened version of a contribution to the Bulletin of the EATCS 57, October 1995 and of [Wei97].

We shall denote the set $\{0, 1, 2, \ldots\}$ of natural numbers by $\omega$. A partial function $f$ from $A$ to $B$ is denoted by $f :\subseteq A \longrightarrow B$. Usually, $\Sigma$ will be a sufficiently large finite alphabet.

## 2 Computability and Continuity on Infinite Sequences of Symbols

For defining computability on finite and infinite sequences of symbols we introduce *Type 2 machines*.

A Type 2 machine is a multi–tape Turing machine $M$ with $k$ ($k \in \omega$) one–way input tapes and a single one–way output tape together with a type specification $(Y_1, \ldots, Y_k, Y_0)$, where $Y_0, \ldots, Y_k \in \{\Sigma^*, \Sigma^\omega\}$ and $Y_i = \Sigma^*(\Sigma^\omega)$ means that Tape $i$ is provided for finite (infinite) inscriptions. Let $f_M :\subseteq Y_1 \times \ldots \times Y_k \longrightarrow Y_0$ be the function computed by a Type 2 machine $M$.

Notice, that the machine $M$ must compute forever, if $f_M(y_1, \ldots, y_k) = y_0 \in \Sigma^\omega$. The above semantics considers infinite input and output tapes which do not exist physically and infinite computations which cannot be completed in reality. For a computation of a Type 2 machine any finite prefix of the output can be obtained in finitely many computation steps from finite prefixes of the inputs. Finite initial parts of computations concerning only finite prefixes of inputs and outputs, however, can be realized by digital computers. There are several other computability concepts which are equivalent (via appropriate encodings) to Type 2 computability (enumeration operators, partial recursive operators, partial recursive functionals, ..., see [Rog67]). In the following we shall say "computable" instead of "computable by a Type 2 machine". We illustrate the concepts by some examples.

1. Let $A \subseteq \Sigma^*$ be recursively enumerable. Define $f :\subseteq \Sigma^\omega \longrightarrow \Sigma^*$ by $f(p) := (0$ if $p \in x\Sigma^\omega$ for some $x \in A$, *div* otherwise). Then $f$ is computable (easy proof).

2. There is a computable function $f :\subseteq \Sigma^\omega \longrightarrow \Sigma^\omega$ which divides infinite decimal fractions by 3. (Use school method beginning from the left.)

3. The function $f :\subseteq \Sigma^\omega \longrightarrow \Sigma^*$ defined by $f(p) = (0$ if $p = 0^\omega$, 1 otherwise) is not computable. The proof is similar to that of 4.

4. No computable function multiplies infinite decimal fractions by 3. For a proof assume that some Type 2 machine $M$ multiplies infinite decimal franctions by 3. Consider the input $p = 0.3^\omega$. Then $f_M(p) = 0.9^\omega$ or $f_M(p) = 1.0^\omega$. Consider the case $f_M(p) = 0.9^\omega$. After some number $t$ of computation steps $M$ has written "0." on the output tape. During this computation $M$ has read at most the first $t$

symbols from the input tape. Now, consider the input $q := 0.3^t 40^\omega$. Also for this input $M$ must write "0." during the first $t$ steps. But for correct multiplication the output must begin with "1.". (The case $f_M(p) = 1.0^\omega$ is treated accordingly.)

The above proof of non–computability is very typical for TTE. It uses only the following basic finiteness property for computable functions on $\Sigma^\omega$ or $\Sigma^*$: If $f(z) = x$, then every finite portion of $x$ depends only on (is already completely defined by) a finite portion of $z$. This finiteness property can be expressed by continuity w.r.t. two topologies [Eng89] which we consider as fixed in the following:

– the discrete topology on $\Sigma^*$ : $\tau = \{A \mid A \subseteq \Sigma^*\}$,
– the Cantor topology on $\Sigma^\omega$ : $\tau_C := \{A\Sigma^\omega \mid A \subseteq \Sigma^*\}$. (The set $\{w\Sigma^\omega \mid w \in \Sigma^*\}$ is a basis of $\tau_C$.)

As a simple but fundamental result we obtain that every computable function $f :\subseteq Y_1 \times \ldots \times Y_k \longrightarrow Y_0$ is continuous.

Already the conventions for Type 2 machines that inputs must be read symbol by symbol and the outputs must be written one–way symbol by symbol suffice to prove the theorem. Since the finiteness property, or more generally speaking continuity, can be interpreted as a very fundamental kind of effectivity or constructivity, it is of separate interest in TTE. We extend well known definitions from recursion theory. Recursive and recursively enumerable subsets of $Y_1 \times \ldots \times Y_k$ can be defined straightforwardly.

The standard numbering $\varphi : \omega \longrightarrow P^{(1)}$ of the partial recursive number functions is an important object in recursion theory [Rog67]. It is determined uniquely except for equivalence by the smn– and the utm–theorem. Correspondingly, for any $a, b \in \{*, \omega\}$ there is a notation $\xi^{ab} : \Sigma^* \longrightarrow P^{ab}$ of the set of computable functions $f :\subseteq \Sigma^a \longrightarrow \Sigma^b$ with similar properties. Moreover, for continuous functions from $\Sigma^\omega$ to $\Sigma^*$ or $\Sigma^\omega$ there are admissible representations [Wei85, Wei87, Wei94]. Let $F^{\omega*}$ be the set of all continuous functions $f :\subseteq \Sigma^\omega \to \Sigma^*$ with open domain, and let $F^{\omega\omega}$ be the set of all continuous functions $f :\subseteq \Sigma^\omega \to \Sigma^\omega$ with $G_\delta$ - domain. Consider $a \in \{*, \omega\}$. There is a total function $\eta^{\omega a}$ from $\Sigma^\omega$ onto $F^{\omega a}$ with the following properties:

- Every continuous function $f :\subseteq \Sigma^\omega \to \Sigma^a$ has an extension in $F^{\omega a}$;
- The function $u :\subseteq \Sigma^\omega \times \Sigma^\omega \to \Sigma^a$ defined by $u(p, q) := \eta^{\omega a}(p)(q)$ is computable (the utm-theorem);
- For any computable function $g :\subseteq \Sigma^\omega \times \Sigma^\omega \to \Sigma^a$ there is a computable function $s : \Sigma^\omega \to \Sigma^\omega$ such that $g(p, q) = \eta^{\omega a}(s(p))(q)$ (the computable smn-theorem).

# 3 Notations and Representations, Induced Effectivity

In TTE it is assumed that computers can only read and write finite or infinite sequences of symbols. Computability is transferred to other sets by notations and representations where finite or infinite sequences are used as names. We introduce the vocabulary now.

A notation of a set $M$ is a surjective function $\nu :\subseteq \Sigma^* \longrightarrow M$, a representation of a set $M$ is a surjective function $\delta :\subseteq \Sigma^\omega \longrightarrow M$. A naming system is a notation or a representation. Examples for notations are the binary notation $\nu_{bin} :\subseteq \Sigma^* \longrightarrow \omega$ of the natural numbers and the notation $\nu_Q :\subseteq \Sigma^* \longrightarrow \mathbb{Q}$ of the rational numbers by signed binary fractions (e.g. $\nu_{bin}(101) = 5$, $\nu_Q(-10/-111) = 2/7$). In the following we shall abbreviate $\nu_Q(w)$ by $\bar{w}$. Well known is the decimal representation $\delta_{dec} :\subseteq \Sigma^\omega \longrightarrow \mathbb{R}$ of the real numbers by infinite decimal fractions. Naming systems can be compared by reducibility, where we consider continuous and computable translations.

For functions $\gamma :\subseteq Y \longrightarrow M$ and $\gamma' :\subseteq Y' \longrightarrow M'$ with $Y, Y' \in \{\Sigma^*, \Sigma^\omega\}$ we call $\gamma$ *reducible to* $\gamma'$, $\gamma \leq \gamma'$, iff $(\forall y \in dom(\gamma))$ $\gamma(y) = \gamma' f(y)$ for some computable function $f :\subseteq Y \longrightarrow Y'$. We call $\gamma$ and $\gamma'$ *equivalent*, $\gamma \equiv \gamma'$, iff $\gamma \leq \gamma'$ and $\gamma' \leq \gamma$.

*Topological reducibility* $\leq_t$ and *topological equivalence* $\equiv_t$ are defined accordingly by substituting "continuous" for "computable".

We have, e.g., $\nu_{bin} \leq \nu_Q \leq \delta_{dec}$, but $\delta_{dec}|^\mathbb{Q} \not\leq_t \nu_Q$. Next, we define how naming systems transfer effectivity concepts from $\Sigma^*$ and $\Sigma^\omega$ to the named sets.

Let $\gamma_i :\subseteq Y_i \longrightarrow M_i$ $(i = 1, \ldots, k)$ be naming systems.

1. $x \in M_1$ is called $\gamma_1$–computable, iff there is a computable element $y \in Y_1$ with $\gamma_1(y) = x$.
2. $X \subseteq M_1 \times \ldots \times M_k$ is called $(\gamma_1, \ldots, \gamma_k)$–*open* ($-r.e.$, $-recursive$), iff $\{(y_1, \ldots, y_k) \in Y_1 \times \ldots \times Y_k \mid (\gamma_1(y_1), \ldots, \gamma_k(y_k)) \in X\}$ is open ($r.e.$, *recursive*) in $dom(\gamma_1) \times \ldots \times dom(\gamma_k)$.

For any naming system $\gamma :\subseteq Y \longrightarrow M$, the set $\tau_\gamma := \{X \subseteq M \mid X$ is $\gamma$–open $\}$ is called the *final topology* of $\gamma$.

For a $\gamma$–open set $X$, a property $\gamma(p) \in X$ is guaranteed already by a finite prefix of $p$, for a $\gamma$–r.e. set additionally there is a "proof–system" for showing this. Examples will be given below. We define relativized computability and continuity of functions and relations.

For $i = 0, \ldots, k$ let $\gamma_i :\subseteq Y_i \longrightarrow M_i$ be naming systems.

1. A relation $Q \subseteq M_1 \times \ldots \times M_k \times M_0$ is called $(\gamma_1, \ldots, \gamma_k, \gamma_0)$–*computable* ($-continuous$), iff there is some computable (continuous) function $f :\subseteq Y_1 \times \ldots \times Y_k \longrightarrow Y_0$ with $(\gamma_1(y_1), \ldots, \gamma_k(y_k), \gamma_0 f(y_1, \ldots, y_k)) \in Q$ if $\exists x.(\gamma_1(y_1), \ldots, \gamma_k(y_k), x) \in Q$.
2. A function $F :\subseteq M_1 \times \ldots \times M_k \longrightarrow M_0$ is called $(\gamma_1, \ldots, \gamma_k, \gamma_0)$–*computable* ($-continuous$), iff there is some computable (continuous) function $f :\subseteq Y_1 \times \ldots \times Y_k \longrightarrow Y_0$ with $F(\gamma_1(y_1), \ldots, \gamma_k(y_k)) = \gamma_0 f(y_1, \ldots, y_k)$ whenever $F(\gamma_1(y_1), \ldots, \gamma_k(y_k))$ exists.

As an example, the real function $x \mapsto x/3$ is $(\delta_{dec}, \delta_{dec})$–computable, but $x \mapsto x \cdot 3$ is not even $(\delta_{dec}, \delta_{dec})$–continuous (see Section 2). Computability of the relation $Q$ is an effective version of the mere existence statement $(\forall x_1, \ldots, x_k)(\exists x_0) Q(x_1, \ldots, x_k, x_0)$. A function $f :\subseteq M_1 \times \ldots \times M_k \longrightarrow M_0$ is called a choice function for $Q$, iff $Q(x_1, \ldots, x_k, f(x_1, \ldots, x_k))$ for all $(x_1, \ldots, x_k) \in dom(Q)$. A relation may be computable without having a computable choice function.

## 4 Effective Representations

Simple considerations show that two naming systems of a set $M$ are equivalent (topologically equivalent), iff they induce the same computational (topological) properties on $M$.

Which among the numerous representations of the set $\mathbb{R}$ of real numbers induce the "right" kind of computability? This question has no answer for the mere set $\mathbb{R}$ but a definite answer for any *information structure* on $\mathbb{R}$.

1. An *information structure* on a set $M$ is a triple $(M, \sigma, \nu)$, where $\nu :\subseteq \Sigma^* \longrightarrow \sigma$ is a notation of a set $\sigma \subseteq 2^M$ of subsets of $M$, which identifies points (i.e. $\{Q \in \sigma \mid x \in Q\} = \{Q \in \sigma \mid y \in Q\} \Longrightarrow x = y$ for all $x, y \in M$).

2. The standard representation $\delta_\nu :\subseteq \Sigma^\omega \longrightarrow M$ for an information structure $(M, \sigma, \nu)$ is defined by: $\delta_\nu(p) = x :\iff \{w \mid x \in \nu(w)\} = \{w \mid \#w\# \text{ is a subword of } p\}$

We assume tacidly that $dom(\nu) \subseteq (\Sigma \backslash \{\mathfrak{c}, \#\})^*$. The elements $Q$ of $\sigma$ are called *atomic properties*. A standard name $p \in \Sigma^\omega$ of $x \in M$ is a list of all those atomic properties $Q \in \sigma$ which hold for $x$. Since $\sigma$ identifies points, $x$ is uniquely defined by $p$. There are many important examples for information structures $(M, \sigma, \nu)$ :

$\quad$ (1) $M = \Sigma^\omega$, $p \in \nu(w) \iff w$ is a prefix of $p$;
$\quad$ (2) $M = 2^\omega$, $A \in \nu(w) \iff \nu_{bin}(w) \in A$;
$\quad$ (3) $M = \mathbb{R}$, $x \in \nu(w) \iff \bar{w} < x$ (remember $\bar{w} = \nu_Q(w)$);
$\quad$ (4) $M = \mathbb{R}$, $x \in \nu(w) \iff x < \bar{w}$;
$\quad$ (5) $M = \mathbb{R}$, $x \in \nu(w) \iff (w = u\mathfrak{c}v, \bar{u} < x < \bar{v})$;
$\quad$ (6) $M = \mathbb{P}$, $f \in \nu(w) \iff (w = 0^i\mathfrak{c}0^j, f(i) = j)$;
$\quad$ (7) $M = \tau_\mathbb{R}$, $U \in \nu(w) \iff (w = u\mathfrak{c}v, [\bar{u}; \bar{v}] \subseteq U)$;
$\quad$ (8) $M = C(\mathbb{R})$, $f \in \nu(w) \iff (w = u\mathfrak{c}v\mathfrak{c}y\mathfrak{c}z, f[\bar{u}; \bar{v}] \subseteq (\bar{y}; \bar{z}))$

(where $\mathbb{P} = \{f \mid f :\subseteq \omega \longrightarrow \omega\}$ and $C(\mathbb{R}) = \{f : \mathbb{R} \longrightarrow \mathbb{R} \mid f \text{ continuous}\}$). The standard representations of the real numbers for the information structures in (3), (4) and (5) are denoted by $\rho_<, \rho_>$ and $\rho$, respectively.

An information structure defines a $T_0$–topology $\tau_\sigma$ on $M$, which has $\sigma$ as a subbase [Eng89]. On the other hand, for any $T_0$–space $(M, \tau)$ with notation $\nu :\subseteq \Sigma^* \longrightarrow \sigma$ of a subbase, $(M, \sigma, \nu)$ is an information structure. The user, who wants to operate effectively on a set $M$, has to specify the set $\sigma$ of atomic properties and the notation $\nu$ of $\sigma$. By $\sigma$ a concept of "approximation" is introduced on $M$ (mathematically by the topology $\tau_\sigma$), and the notation $\nu$ expresses how atomic properties can be handled concretely, i.e. $\nu$ fixes computability on $\sigma$ and $M$. The case of real numbers in the above example shows that for a set $M$ different information structures may be of interest. Of course, only the most relevant ones can be investigated in more details. The choice of the notation $\nu$ of $\sigma$ should be justified, but usually there is some "standard notation" which is generally accepted as "effective".

The standard representation $\delta_\nu$ is continuous and open, and, as an outstanding property it is $t$–complete in the class of continuous representations:

For any representation $\delta$ of $M$: $\delta \leq_t \delta_\nu \iff \delta$ is continuous.

There is a striking formal similarity of this theorem with a formulation of the utm–theorem and the smn–theorem for the "admissible Goedel numberings" of the partial recursive functions [Rog67]. The representations which are $t$–equivalent to the standard representation $\delta_\nu$ are called *admissible*. The admissible representations for an information structure are defined already by their final topology. They are the "topologically effective" representations. Most important is the following "main theorem" which holds for admissible representations:

For admissible representations $\delta_i : \subseteq \Sigma^\omega \longrightarrow M_i$ with final topologies $\tau_i$ ($i = 0, \ldots, k$), a function $f : \subseteq M_1 \times \ldots \times M_k \longrightarrow M_0$ is continuous, iff it is $(\delta_1, \ldots, \delta_k, \delta_0)$–continuous.

Consequently, for admissible representations $(\delta_1, \ldots, \delta_k, \delta_0)$–computable functions are continuous. Roughly speaking, in a natural setting only continuous functions can be computable.

For a separable metric space with a notation $\alpha$ of a countable dense subset the *Cauchy representation* $\delta_\alpha$ is admissible, where $\delta_\alpha$ is defined by:

$\delta_\alpha(p) = x : \iff p = u_0 \sharp u_1 \sharp \ldots$ ($u_i \in dom(\alpha)$) with
$(\forall k)(\forall i > k) d(\alpha(u_i), \alpha(u_k)) < 2^{-k}$ and $x = \lim_{i \to \infty} \alpha(u_i)$. Notice, that only "fast" converging sequences of elements of the dense set $A$ are used as names.

# 5  Computability on the Real Line

The main subject in Analysis is the real line, i.e. the topological space $(\mathbb{R}, \tau_\mathbb{R})$, where $\tau_\mathbb{R}$ is the set of open subsets of $\mathbb{R}$. We consider the information structure $(\mathbb{R}, \sigma, \nu)$ from Example 4.2(5) with $x \in \nu(w) \iff (w = u \natural v$ and $\bar{u} < x < \bar{v})$ as basis of our effectivity theory on the real line. The standard representation $\delta_\nu$ (Def. 4.1) is equivalent to the Cauchy representation $\rho_C := \delta_{\nu_Q}$, where fast converging Cauchy sequences of rational numbers serve as names.

Like $\delta_\nu$, the Cauchy representation is admissible with final topology $\tau_\mathbb{R}$. The requirement of admissibility with final topology $\tau_\mathbb{R}$ excludes many other representations. Since $x \mapsto 3x$ is not $(\delta_{dec}, \delta_{dec})$–continuous (see Example 2.2.(4)), $\delta_{dec}$ cannot be admissible with final topology $\tau_\mathbb{R}$ by Theorem 4.4. Since $\tau_\mathbb{R}$ is the final topology of $\delta_{dec}$, it cannot be admissible at all by the main theorem for admissible representations. Furthermore, there is no injective and no total admissible representation of $\mathbb{R}$ with final topology $\tau_\mathbb{R}$. The "naive" Cauchy representation $\delta_{naive} : \subseteq \Sigma^\omega \longrightarrow \mathbb{R}$, where $\delta_{naive}(p) = x : \iff p = u_0 \sharp u_1 \sharp \ldots$ and $(\bar{u}_0, \bar{u}_1, \ldots)$ is a Cauchy sequence with limit $x$, has $\{\emptyset, \mathbb{R}\}$ as its final topology and is not admissible. If $\delta_{naive}(p) = x$, then no finite prefix of $p$ contains any information about $x$.

We shall now study effectivity induced on $\mathbb{R}$ by the Cauchy representation $\rho_C$. By Definition 3.3(1), a real number $x$ is $\rho_C$–computable, iff $x = \rho_C(p)$ for some computable sequence $p \in \Sigma^{\omega}$. Every rational number is computable. $\sqrt{2}, e, \pi$ and many other important real numbers are computable. Every "fast" converging $(\nu_{bin}, \rho_C)$-computable Cauchy sequence has a computable limit. However, for any set $A \subseteq \omega$, $x_A := \Sigma\{2^{-i} \mid i \in A\}$ is computable, if and only if $A$ is recursive. Let $A$ be r.e., but not recursive. There is some injective computable function $f : \omega \longrightarrow \omega$ with $A = range(f)$. Define $a_n := \Sigma\{2^{-f(i)} \mid i \leq n\}$. Then $(a_n)_{n \in \omega}$ is an increasing $(\nu_{bin}, \nu_Q)$-computable sequence converging to the non–computable number $x_A$. Since the function $f$ must enumerate many numbers $k \in A$ very late the sequence $(a_n)_{n \in \omega}$ converges very slowly. However, the number $x_A$ is $\rho_<$-computable.

A simple topological consideration shows that $\emptyset$ and $\mathbb{R}^n$ are the only subsets of $\mathbb{R}^n$ which are recursive w.r.t. $\rho_C$. $\{x \in \mathbb{R} \mid x \neq 0\}$, and $\{(x, y) \in \mathbb{R}^2 \mid x < y\}$ are examples of sets which are r.e. but not recursive. Moreover, there is no representation $\delta \subseteq \Sigma^{\omega} \longrightarrow \mathbb{R}$ of the real numbers at all, such that $\{(x, y) \in \mathbb{R}^2 \mid x < y\}$ becomes $(\delta, \delta)$-recursive [Wei95]. The relation $x < y$ is "easily definable" but "absolutely" non–recursive. (Notice that some computability models on $\mathbb{R}$ have decidability of $x < y$ as a basic assumption [BSS89]).

By our definitions, a function $f :\subseteq \mathbb{R} \longrightarrow \mathbb{R}$ is $(\rho_C, \rho_C)$-computable, iff some Type 2 machine transforms any Cauchy sequence of rational numbers converging fast to $x \in dom(f)$ to some Cauchy sequence of rational numbers converging fast to $f(x)$. Notice that by main theorem for admissible representations every real function computable w.r.t. $\rho_C$ is continuous. Many of the commonly used functions from "classical" analysis are computable. In the following, for the real numbers "computable" abbreviates "computable w.r.t. $\rho_C$ ".

1. The real functions $(x, y) \mapsto x + y$, $(x, y) \mapsto x \cdot y$, $(x, y) \mapsto \max(x, y)$, $x \mapsto -x$, $x \mapsto 1/x$ are computable.
2. Let $(a_i)_{i \in \omega}$ be a $(\nu_{bin}, \rho_C)$-computable sequence and let $R_0 > 0$ be the radius of the power series $\Sigma a_i x^i$. Then for any $R < R_0$ the real function $f_R(x) := (\Sigma a_i x^i$ if $|x| < R$, $div$ otherwise) is computable.

For details of the proofs see [Wei95]. By 5.3.(2), $exp$, $sin$, $cos$ and many other functions are computable. Notice also, that the computable functions are closed under composition. Theorem 5.3(2) cannot be generalized to the case $R = R_0$! By identifying the complex plane $\mathbb{C}$ with $\mathbb{R}^2$, computability of complex functions can be reduced to computability of real functions. The complex functions addition, multiplication, inversion, exp, sin, log etc. are computable.

# 6    Compact Subsets of the Real Line

The power set $2^{\mathbb{R}}$ of $\mathbb{R}$ has no representation since its cardinality is too large. Therefore, by means of TTE it is impossible to express computability of functions such as sup $:\subseteq 2^{\mathbb{R}} \longrightarrow \mathbb{R}$. The open, the closed and the compact subsets of $\mathbb{R}$, however, have representations since they have continuum cardinality. As an example we introduce and discuss some admissible representations of the compact subsets of the real line. Compactness is a good example for a classical concept which branches into many parts in TTE. A subset $X \subseteq \mathbb{R}$ is compact, iff it is closed and bounded or, equivalently by the Heine/Borel theorem, iff every open covering has a finite subcovering. We introduce several representations $\delta.. :\subseteq \Sigma^{\omega} \longrightarrow K(\mathbb{R})$ of the set $K(\mathbb{R})$ of compact subsets of $\mathbb{R}$ by information structures $(K(\mathbb{R}), \sigma.., \nu..)$ as follows.

$$\delta_o : X \in \nu_o(w) : \iff w = u\natural v \text{ and } X \cap [\bar{u}; \bar{v}] = \emptyset$$
$$\delta_n : X \in \nu_n(w) : \iff w = u\natural v \text{ and } X \cap (\bar{u}; \bar{v}) \neq \emptyset$$
$$\delta_{ob} : X \in \nu_{ob}(w) : \iff w = y\natural 0^n \text{ and } X \in \nu_o(y) \text{ and } X \subseteq (-n; n)$$
$$\delta_{obn} : X \in \nu_{obn}(w) : \iff w = x\natural z \text{ and } X \in \nu_{ob}(x) \cap \nu_n(z)$$
$$\kappa_w : X \in \nu_w(w) : \iff w = u_1\natural v_1\natural \ldots \natural u_k\natural v_k \text{ and } X \subseteq (\bar{u}_1; \bar{v}_1) \cup \ldots \cup (\bar{u}_k; \bar{v}_k)$$
$$\kappa_s : X \in \nu_s(w) : \iff w = u_1\natural v_1\natural \ldots \natural u_k\natural v_k \text{ and } X \subseteq (\bar{u}_1; \bar{v}_1) \cup \ldots \cup (\bar{u}_k; \bar{v}_k)$$
$$\text{and } X \cap (\bar{u}_i; \bar{v}_i) \neq \emptyset \text{ for } i = 1, \ldots, k$$

Thus: $\delta_o(p) = X$, iff $p$ enumerates the "outside" of $X$ (i.e. $\delta_o$ is equivalent to a restriction of $\delta_{cl}$); $\delta_n(p) = X$, iff $p$ enumerates the open "neighbours" of $X$; $\delta_{ob}(p) = X$, iff $p$ enumerates the outside and additionally bounds of $X$; $\delta_{obn}(p) = X$, iff $p$ enumerates the outside, the bounds and the neighbours of $X$; $\kappa_w(p) = X$, iff $p$ enumerates all finite coverings (with rational intervals) of $X$; $\kappa_s(p) = X$, iff $p$ enumerates all minimal finite coverings of $X$. The above namings $\delta..$ represent different types of knowledge about the named objects. Reducibilities express dependences between these types. For the first four representations we have $\delta_{obn} \leq \delta_{ob} \leq \delta_o$ and $\delta_{obn} \leq \delta_n$, while $\not\leq_t$ holds for all other combinations. The following equivalences are two computable versions of the Heine/Borel theorem [KW87]):

$- \delta_{ob} \equiv \kappa_w; \delta_{obn} \equiv \kappa_s$.

Every non–empty compact set $X \in K(\mathbb{R})$ has a maximum. On the non–empty compact subsets

- $Max$ is $(\delta_n, \rho_<)$-computable and $(\delta_{ob}, \rho_>)$-computable,
- $Max$ is **not** $(\delta_n, \rho_>)$-continuous, $(\delta_0, \rho_>)$-continuous or $(\delta_{ob}, \rho_<)$-continuous.

Notice in particular, that without knowing a concrete upper bound not even a $\rho_>$–name of $Max(X)$ can be determined from a $\delta_0$–name of $X$.

The above representations behave differently w.r.t. union and intersection. While union is computable w.r.t. all of them, intersection is e.g. $(\kappa_w, \kappa_w, \kappa_w)$-computable but not even $(\kappa_s, \kappa_s, \delta_n)$-continuous.

Some more details can be found in [KW87, Wei94, Wei95].

# 7 Continuous Functions and Zero–Finding

In this section we shall consider the set $C[0; 1]$ of continuous functions $f : [0; 1] \longrightarrow \mathbb{R}$. Let $\delta_F$ be the standard representation of $C[0; 1]$ for the information structure $(C[0; 1], \sigma, \nu)$ with

$- f \in \nu(w) : \iff w = u\natural v\natural x\natural y$ and $f[\bar{u}; \bar{v}] \subseteq (\bar{x}; \bar{y})$.

Thus $\delta_F(p) = f$, iff $p$ enumerates all $(a, b, c, d) \in \mathbb{Q}^n$ with $f[a; b] \subseteq (c; d)$. Any atomic information about $f$ is a rectangle bounding the graph of $f$. $\delta_F$ is admissible, where the final topology is the well–known compact–open topology [Eng89]. The following useful theorem emphasizes the effectivity of the representation $\delta_F$. Define $apply : C[0; 1] \times \mathbb{R} \longrightarrow \mathbb{R}$ by $apply(f, x) := f(x)$. Then for any representation $\delta$ of $C[0; 1]$:

- $apply$ is $(\delta, \rho_C, \rho_C)$-continuous $\iff \delta \leq_t \delta_F$,
- $apply$ is $(\delta, \rho_C, \rho_C)$-computable $\iff \delta \leq \delta_F$.

Therefore, $\delta_F$ is complete in the set of all representations of $C[0; 1]$, for which the apply–function is computable. Notice the formal similarity with the smn– and the utm–theorem from recursion theory. The following theorem summarizes some effectivity properties [KW87]:

1. $f : [0; 1] \longrightarrow \mathbb{R}$ is $(\rho_C, \rho_C)$-computable $\iff f$ is $\delta_F$-computable.
2. The composition $(f, g) \mapsto fg$ on $C[0; 1]$ is $(\delta_F, \delta_F, \delta_F)$-computable.
3. The function $H : C[0; 1] \times K(\mathbb{R}) \longrightarrow K(\mathbb{R})$, $H(f, X) := f(X)$, is $(\delta_F, \kappa_s, \kappa_s)$-computable.
4. $Max : C[0; 1] \longrightarrow \mathbb{R}$ with $Max(f) := \max\{f(x) \mid 0 \leq x \leq 1\}$ is $(\delta_F, \rho_C)$-computable.
5. $Int : C[0; 1] \times \mathbb{R} \longrightarrow \mathbb{R}$ with $Int(f, a) := \int_0^a f(x)dx$ is $(\delta_F, \rho_C, \rho_C)$-computable.
6. $Diff :\subseteq C[0; 1] \longrightarrow C[0; 1]$, defined by $Diff(f) = g$ iff $g \in C[0; 1]$ is the derivative of $f$, is not $(\delta_F, \delta_F)$-continuous.

In accordance with experience from Numerical Mathematics, the maximum can be determined, integration is easy but differentiation cannot be performed effectively on $\delta_F$-names. We consider the problem of finding zeroes for functions from $C[0; 1]$. The following can be proved:

1. The function $Z_{\min} :\subseteq C[0; 1] \longrightarrow \mathbb{R}$, where $Z_{\min}(f) = x : \iff x = \min f^{-1}\{0\}$, is $(\delta_F, \rho_<)$-computable but not $(\delta_F, \rho_C)$-continuous.
2. The function $Z_1 :\subseteq C[0; 1] \longrightarrow \mathbb{R}$, where $Z_1(f) = x : \iff x$ is the only zero of $f$, is $(\delta_F, \rho_C)$-computable.
3. The relation $R := \{(f, x) \mid f \in C[0; 1], x \in \mathbb{R}, f(0) < 0 < f(1), f(x) = 0\}$ is not $(\delta_F, \rho_C)$-continuous.
4. The relation $R := \{(f, x) \mid f \in C[0; 1], x \in \mathbb{R}, f(0) < 0 < f(1), f(x) = 0, f^{-1}\{0\}$ contains no proper interval$\}$ is $(\delta_F, \rho_C)$-computable but has no $(\delta_F, \rho_C)$-continuous choice function.

As a positive result, the minimal zeroes can be approximated from below, and a computable function determines the zeroes for all functions which have exactly one zero. By the intermediate value theorem, every function $f \in C[0; 1]$, which changes its sign, has a zero. But there is not even a continuous function, which for any $\delta_F$-name of a function $f$ (which changes its sign) determines some $\rho_C$-name of a zero of $f$. We obtain a weak positive result, if we consider only functions $f$, for which $f^{-1}\{0\}$ contains no proper interval. An instructive example is the special problem of finding a zero of $f_a : \mathbb{R} \longrightarrow \mathbb{R}$ from $a \in [-1; 1]$, where $f_a(x) := x^3 - x + a$ [Wei95].

# 8  Measure and Integration

The concepts introduced so far can be used to define natural computability concepts in measure theory. In this section we sketch some ideas shortly. For details see [Wei96]. We only use some basic concepts from measure theory which can be found in any textbook on this topic, e.g. [Bau 74]. For simplicity, we consider only the real interval $[0; 1]$ as basic space. Let $Int := \{(a, b), [0; a), (b, 1], [0; 1] \mid 0 < a < b < 1, a, b \in \mathbb{Q}\}$ be the set of open subintervals of $[0; 1]$ with rational boundaries. Let $\mathbf{B}$ be the set of Borel subsets of $[0; 1]$, i.e. the smallest $\sigma$–Algebra containing $Int$. Let $\mathbf{M}$ be the set of probability measures on the space $([0; 1], \mathbf{B})$. Every measure $\mu \in \mathbf{M}$ is uniquely determined by its restriction to $Int$.

We define explicitly an information structure $(\mathbf{M}, \sigma, \nu)$ (Section 4) which induces canonically a topology and a computability theory on $\mathbf{M}$. Let $I :\subseteq \Sigma^* \longrightarrow Int$ be some standard notation of the set $Int$ of intervals (with $dom(I) \subseteq (\Sigma \backslash \{\mathfrak{e}, \sharp\})^*$). (Remember $\bar{u} := \nu_Q(u)$, see Section 3.)

**Definition 1.** Define the information structure $(\mathbf{M}, \sigma, \nu)$ by

$$\mu \in \nu(u \mathfrak{e} v) : \Longleftrightarrow \bar{u} < \mu(I_v)$$

for all $u \in dom(\nu_Q)$, $v \in dom(I)$ and $\mu \in \mathbf{M}$.

Let $\tau_m$ be the topology on $\mathbf{M}$ with subbase $\sigma$ and let $\delta_m$ be the standard representation for this information structure (Section 4).

Notice, that $\sigma$ identifies points in $\mathbf{M}$ : $\mu = \mu'$ iff $\bar{u} < \mu(I_v) \Longleftrightarrow \bar{u} < \mu'(I_v)$ for all $u \in dom(\nu_Q)$ and $v \in dom(I)$. Notice also, that only (rational) lower bounds of $\mu(J)$ $(J \in Int)$ are avialable immediately from $\delta_m$–names.

Let us call a definition of a representation "$\mathbb{Q}$-stable", iff the following holds: if we replace the set $\mathbb{Q}$ of rational numbers in the definition by a set like $\mathbb{Q}_{10}$ (the finite decimal fractions) with some standard notation, we obtain an equivalent representation. The representations $\rho_<, \rho_C, \rho_>$ and $\delta_F$ are $\mathbb{Q}$–stable. An easy consideration shows that also the above representation $\delta_m$ of measures is $\mathbb{Q}$–stable. However, if we replace "$\bar{u} < \mu(I_v)$" in the definition of $\delta_m$ by "$\bar{u} > \mu(I_v)$", "$\bar{u} \leq \mu(I_v)$", "$\bar{u} \geq \mu(I_v)$", "$\bar{u} < \mu(I_v) < \bar{w}$", "$\bar{u} \leq \mu(I_v) < \bar{w}$" etc., then we obtain representations which are not $\mathbb{Q}$-stable. Therefore, among all variants, $\delta_m$ is the only natural definition.

The representation $\delta_m$ of the set of measure $\mathbf{M}$ has several interesting effectivity properties.

**Theorem 2.** *The function $(a, \mu, \mu') \longmapsto a\mu + (1-a)\mu'$ is $(\rho_C, \delta_m, \delta_m, \delta_m)$-computable for $0 \leq a \leq 1$.*

Let $\delta_F$ be the standard representation of $C[0;1]$ from Section 7.

**Theorem 3.** *The function $(f,\mu) \mapsto T_f(\mu)$ for continuous $f : [0;1] \longrightarrow [0;1]$ and $\mu \in \mathbf{M}$ is $(\delta_F, \delta_m, \delta_m)$-computable (where $T_f(\mu)(A) := \mu f^{-1}A$ for all $A \in \mathbf{B}$).*

As an application consider hyperbolic iterated function systems $\mathbf{S} = ([0;1], f_1, \ldots, f_N, p_1, \ldots, p_N)$ with contracting functions $f_1, \ldots, f_N : [0;1] \longrightarrow [0;1]$ and probabilities $p_1, \ldots, p_N \in \mathbb{R}_+$, $(p_1 + \ldots + p_N = 1)$. Then

$$(\mu, f_1, \ldots, f_N, p_1, \ldots, p_N) \longmapsto \mu' \quad \text{where} \quad \mu'(A) := \sum_{i=1}^{N} p_i \mu f^{-1}(A)$$

is $(\delta_m, \delta_F, \cdots, \delta_F, \rho_C, \cdots, \rho_C, \delta_m)$-computable. And the operator determining the unique invariant measure of $\mathbf{S}$ is $(\delta_F, \cdots, \delta_F, \rho_C, \cdots, \rho_C, \delta_m)$-computable.

Also integration of continuous functions is computable. We formulate a fully effective version.

**Theorem 4.** *The function $(f,\mu) \mapsto \int f d\mu$ for $f \in C[0;1]$ and $\mu \in \mathbf{M}$ is $(\delta_\alpha, \delta_m, \rho_C)$-computable.*

Notice, that in this case $\rho_C$-names, i.e. lower and upper and not only lower rational bounds can be determined.

By the following theorem the finite portions of information provided by $\delta_m$-names are exactly what is needed for effective integration of continuous functions.

**Theorem 5.** *Let $\delta$ be any representation of the set $\mathbf{M}$ of probability measures such that the function $(f,\mu) \mapsto \int f d\mu$ for $f \in C[0;1]$ and $\mu \in \mathbf{M}$ is $(\delta, \delta_m, \rho_C)$-computable. Then $\delta \leq \delta_m$. The statement holds accordingly for "continuous" instead of "computable".*

This means that our representation $\delta_m$ is *complete* in (is "the poorest representation in") the set of all representations, for which the above integration operator is relatively computable or continuous, respectively, in both arguments. As a consequence, the final topology $\tau_m$ of the representation $\delta_m$ is the well-known *weak topology* on the set of measures. Therefore, our computability theory on measures fits smoothly in classical measure theory.

# 9  Computational Complexity

We introduce computational complexity for Type 2 machines and transfer it by representations to other sets. For an ordinary Turing machine $M$, $Time_M(x)$ is the number of steps which $M$ with input $x$ works until it halts. Such a definition is useless for Type 2 machines with infinite output. We introduce as a further parameter a number $k \in \omega$ and measure the time until $M$ has produced the first $k$ output symbols:

$Time_M(y)(k) :=$ the number of steps which the machine $M$ with input $y$ needs for writing the first $k + 1$ output symbols.

Another important information is the input lookahead , $Ila_M(y)(k)$, which is the number of input symbols which $M$ requires for producing the first $k$ output symbols. The parameter $k$ can be interpreted as precision of the intermediate result.

We transfer these concepts to represented sets as follows. For a representation $\delta :\subseteq \Sigma^\omega \longrightarrow M$ and a set $X \subseteq M$ a function $f :\subseteq M \longrightarrow M$ is computable on $M$ in time $t : \omega \longrightarrow \omega$ and input lookahead $s : \omega \longrightarrow \omega$, iff there is a Turing machine $N$ with (1) $f\delta(p) = \delta f_N(p)$, (2) $Time_N(p)(n) \leq t(n)$, (3) $Ila_N(p)(n) \leq s(n)$ for all $n \in \omega$, whenever $p \in \delta^{-1}(X)$.

Unfortunately, this definition of complexity is useless for many representations, e.g. for our Cauchy representation $\rho_C$ of the real numbers. In fact, every $(\rho_C, \rho_C)$-computable real function is $(\rho_C, \rho_C)$-computable in linear time, since every output $q \in dom(\rho_C)$ can be padded arbitrarily by using rational numbers with large numerators and denominators. Informally, the set of $\rho_C$–names of any real number is too big. We solve this problem for the real numbers by using an admissible representation with uniformly bounded redundancy, the "modified binary representation" $\rho_m$, which uses the binary digits 0, 1 and -1 (which we denote by $\bar{1}$) [Wie80], i.e. $\rho_m(a_n \ldots a_0 \cdot a_{-1}a_{-2} \ldots) = \sum\{a_i \cdot 2^i \mid i \leq n\}$ where $a_i \in \{1, 0, -1\}$. (Like decimal representation, ordinary binary representation induces an unnatural computability theory.) Easy proofs show (1) $\rho_m \equiv \rho_C$ and (2) $\rho_m^{-1}(X)$ is compact if $X$ is compact. By (1) $\rho_C$ and $\rho_m$ induce the same computability theory, and by (2) $max\{Time_N(p)(n)|p \in \rho_m^{-1}(X)\}$ exists for each $n$.

The resulting computational complexity is sometimes called "bit–complexity". It is essentially equivalent to definitions used by Brent [Bre76], Schönhage [Sch90], Ko [Ko91] and others. We mention only a single concrete result by Mueller [Mül87]: If a real analytic function is $(\rho_m, \rho_m)$– computable in time $O(n^k)$, $(k \geq 3)$, then its integral and derivative are computable in time $O(n^{k+2})$ .

Admissible representations $\delta$ such that $\delta^{-1}K$ is compact, if $K$ is compact, exist for all separable metric spaces, i.e. uniform complexity can be defined on compact metric spaces.

## 10    Conclusion

The approach presented here (TTE) connects Abstract Analysis and Turing computability. It provides a simple language for expressing a variety of types of effectivity ranging form continuity (interpreted as "constructivity" or "finite dependence") via computability to computational complexity of points, sets, relations and functions. It is claimed that "Type 2 computability" models "computability by machines" adequately (TTE is a realistic model of computation).

Computability aspects in Analysis have been discussed already from the beginning of Computability Theory. In his famous paper [Tur36], A. Turing defines computable real numbers by computable infinite binary fractions. In a correction [Tur37] he proposes to use computable sequences of nested intervals with rational boundaries instead. As we already mentioned, TTE extends a definition of computable real functions by A. Grzegorczyk [Grz55]. Two years later, he introduced several definitions equivalent to the first one [Grz57]. Grzegorczyk's characterization "$f : [0;1] \rightarrow \mathbb{R}$ is computable, iff $f$ maps every computable sequence of computable numbers to a computable sequence of computable numbers and $f$ has a computable modulus of uniform continuity" has been generalized to computable Banach spaces by Pour–El and Richards [PER88]. TTE includes this generalization. Another characterization by Grzegorczyk (by means of monotone (w.r.t. inclusion) computable functions mapping intervals with rational boundaries to intervals with rational boundaries) is the computational background of Interval Analysis [Moo79, Abe88]. Domain Theory approaches consider computability on spaces the elements of which are real numbers as well as approximations of real numbers (e.g. intervals with rational boundaries) [Abe80, WS81, Kus84, Bla95a, Eda95, DG96]. Domain computability restricted to the real numbers coincides with TTE–computability.

The "Russian" approach to Computable Analysis handles programs of computable real numbers. Let $\psi$ be some standard notation (by means of "Turing – programs") of all computable elements of $\Sigma^\omega$. Then $\rho_C \psi$ is a partial notation of the set of all computable real numbers $\mathbb{R}_C$. A function $f :\subseteq \mathbb{R}_C \rightarrow \mathbb{R}_C$ is Ceitin computable, iff it is $(\rho_C \psi, \rho_C \psi)$–computable. Every TTE–computable (w.r.t. $\rho_C$) function $f :\subseteq \mathbb{R} \rightarrow \mathbb{R}$ with $dom(f) \subseteq \mathbb{R}_C$ is Ceitin computable. The converse requires certain restrictions on $dom(f)$ [Cei59]. A necessary and sufficient condition has been given recently by Hertling [Her96]. Computational complexity based on the bit–model [Bre76, Sch90, Ko91] can be considered as a part of TTE. While in the Russian approach only the computable real numbers are considered as entireties, the "real RAM" model [BM75, BSS89, PS85] considers all real numbers as entities. A real RAM is a random access machine operating on real numbers with assignments "$x := y + z$", "$x := y \cdot z$", ... and branchings "$x < 0$" as basic operations and branchings, respectively. Time complexity is measured by the number of operations as a function of the input dimension. The real RAM model is unrealistic since almost all real RAM–computable functions are not continuous, hence not realizable by physical machines, and almost all TTE–computable functions (which are physically realizable), e.g. $e^x$ and $sin(x)$, are not real RAM–computable.

The computational model used in "Information Based Complexity" (IBC) [TWW88] extends the real RAM model [Nov95]. Also this model cannot be realized mathematically correctly by digital computers. Moreover, the output of one machine (in general) cannot be used as the input for a second machine. A slightly modified real RAM, the "feasible real RAM", however, can be realized physically [BH96]. Here, the precise tests "$x < 0$" are replaced by finite precision tests. A different way to study effectivity in analysis is by using constructive logic (e.g. [TD88]). Bishop's "Constructive analysis" [Bis67, BB85] is the most famous example. Troelstra [Tro92] compares TTE and constructive mathematics. This interesting work has to be continued.

Comutational complexity in Analysis ("bit-complexity") will certainly be a very important and fruitful area of future research in computer science. While much work on polynomial complexity and beyond has been done already by Ko [Ko91], only for very few concrete problems upper and lower complexity bounds have been investigated in detail, e.g. [Bre76, Alt85, Mül87, Sch90, Blä95b, Sch97].

# References

[Abe80] Oliver Aberth. *Computable Analysis*. McGraw-Hill, New York, 1980.

[Abe88] Oliver Aberth. *Precise Numerical Analysis*. Brown Publishers, Dubuque, 1988.

[Alt85] Helmut Alt. Multiplication is the easiest nontrivial arithmetic function. *Theoretical Computer Science*, 36:333–339, 1985.

[BB85] Errett Bishop and Douglas S. Bridges. *Constructive Analysis*, volume 279 of *Grundlehren der mathematischen Wissenschaft*. Springer, Berlin, 1985.

[BH96] Vasco Brattka and Peter Hertling. Feasible real random access machines. In Keith G. Jeffrey, Jaroslav Král, and Miroslav Bartošek, editors, *SOF-SEM'96: Theory and Practice of Informatics*, volume 1175 of *Lecture Notes in Computer Science*, pages 335–342, Berlin, 1996. Springer. 23rd Seminar on Current Trends in Theory and Practice of Informatics, Milovy, Czech Republik, November 23-30, 1996.

[Bis67] Errett Bishop. *Foundations of Constructive Analysis*. McGraw-Hill, New York, 1967.

[Bla95a] Jens Blanck. Domain representability of metric spaces. In Ker-I Ko and Klaus Weihrauch, editors, *Computability and Complexity in Analysis*, volume 190 of *Informatik-Berichte*, pages 1–10. FernUniversität Hagen, September 1995. CCA Workshop, Hagen, August 19-20, 1995.

[Blä95b] Markus Bläser. Uniform computational complexity of the derivatives of $C^\infty$-functions. In Ker-I Ko and Klaus Weihrauch, editors, *Computability and Complexity in Analysis*, volume 190 of *Informatik-Berichte*, pages 99–104. FernUniversität Hagen, September 1995. CCA Workshop, Hagen, August 19-20, 1995.

[BM75] A. Borodin and I. Munro. *The Computational Complexity of Algebraic and Numeric Problems*. Elsevier, New York, 1975.

[Bre76] R.P. Brent. Fast multiple-precision evaluation of elementary functions. *Journal of the Association for Computing Machinery*, 23(2):242–251, 1976.

[BSS89] Lenore Blum, Mike Shub, and Steve Smale. On a theory of computation and complexity over the real numbers: $NP$-completeness, recursive functions and universal machines. *Bulletin of the American Mathematical Society*, 21(1):1–46, July 1989.

[Cei59] G.S. Ceitin. Algorithmic operators in constructive complete separable metric spaces. *Doklady Akad. Nauk*, 128:49–52, 1959. (in Russian).

[DG96] Pietro Di Gianantonio. Real number computation and domain theory. *Information and Computation*, 127:11–25, 1996.

[Eda95] Abbas Edalat. Domain theory and integration. *Theoretical Computer Science*, 151:163–193, 1995.

[Eng89] Ryszard Engelking. *General Topology*, volume 6 of *Sigma series in pure mathematics*. Heldermann, Berlin, 1989.

[Grz55] Andrzej Grzegorczyk. Computable functionals. *Fundamenta Mathematicae*, 42:168–202, 1955.

[Grz57] Andrzej Grzegorczyk. On the definitions of computable real continuous functions. *Fundamenta Mathematicae*, 44:61–71, 1957.

[Hau83] Jürgen Hauck. Konstruktive reelle Funktionale und Operatoren. *Zeitschrift für mathematische Logik und Grundlagen der Mathematik*, 29:213–218, 1983.

[Her96] Peter Hertling. Unstetigkeitsgrade von Funktionen in der effektiven Analysis. Informatik Berichte 208, FernUniversität Hagen, Hagen, November 1996. Dissertation.

[Ko91] Ker-I Ko. *Complexity Theory of Real Functions*. Progress in Theoretical Computer Science. Birkhäuser, Boston, 1991.

[Kus84] Boris Abramovich Kushner. *Lectures on Constructive Mathematical Analysis*, volume 60 of *Translation of Mathematical Monographs*. American Mathematical Society, Providence, 1984.

[KW84] Christoph Kreitz and Klaus Weihrauch. A unified approach to constructive and recursive analysis. In M.M. Richter, E. Börger, W. Oberschelp, B. Schinzel, and W. Thomas, editors, *Computation and Proof Theory*, volume 1104 of *Lecture Notes in Mathematics*, pages 259–278, Berlin, 1984. Springer. Proceedings of the Logic Colloquium, Aachen, July 18-23, 1983, Part II.

[KW85] Cristoph Kreitz and Klaus Weihrauch. Theory of representations. *Theoretical Computer Science*, 38:35–53, 1985.

[KW87] Christoph Kreitz and Klaus Weihrauch. Compactness in constructive analysis revisited. *Annals of Pure and Applied Logic*, 36:29–38, 1987.

[Moo79] Ramon E. Moore. Methods and applications of interval analysis. *SIAM Journal on Computing*, 1979.

[Mül87] Norbert Th. Müller. Uniform computational complexity of Taylor series. In Thomas Ottmann, editor, *Proceedings of the 14th International Colloquium on Automata, Languages, and Programming*, volume 267 of *Lecture Notes in Computer Science*, pages 435–444, Berlin, 1987. Springer.

[Nov95] Erich Novak. The real number model in numerical analysis. *Journal of Complexity*, 11(1):57–73, 1995.

[PER88] Marian B. Pour-El and J. Ian Richards. *Computability in Analysis and Physics*. Perspectives in Mathematical Logic. Springer, Berlin, 1988.

[PS85] Franco P. Preparata and Michael Ian Shamos. *Computational Geometry*. Texts and Monographs in Computer Science. Springer, New York, 1985.

[Rog67] Hartley Rogers. *Theory of Recursive Functions and Effective Computability*. McGraw-Hill, New York, 1967.

[Sch90] A. Schönhage. Numerik analytischer Funktionen und Komplexität. *Jahresbericht der Deutschen Mathematiker-Vereinigung*, 92:1–20, 1990.

[Sch97] Matthias Schröder. Fast online multiplication of real numbers. In Rüdiger Reischuk and Michel Morvan, editors, *STACS 97*, volume 1200 of *Lecture Notes in Computer Science*, pages 81–92, Berlin, 1997. Springer. 14th Annual Symposium on Theoretical Aspects of Computer Science, Lübeck, Germany, February 27 - March 1, 1997.

[TD88] A.S. Troelstra and D. van Dalen. *Constructivism in Mathematics, Volume 1*, volume 121 of *Studies in Logic and the Foundations of Mathematics*. North-Holland, Amsterdam, 1988.

[Tro92] A.S. Troelstra. Comparing the theory of representations and constructive mathematics. In E. Börger, G. Jäger, H. Kleine Büning, and M.M. Richter, editors, *Computer Science Logic*, volume 626 of *Lecture Notes in Computer Science*, pages 382–395, Berlin, 1992. Springer. Proceedings of the 5th Workshop, CSL'91, Berne Switzerland, October 1991.

[Tur36] Alan M. Turing. On computable numbers, with an application to the "Entscheidungsproblem". *Proceedings of the London Mathematical Society*, 42(2):230–265, 1936.

[Tur37] Alan M. Turing. On computable numbers, with an application to the "Entscheidungsproblem". A correction. *Proceedings of the London Mathematical Society*, 43(2):544–546, 1937.

[TWW88] Joseph F. Traub, G.W. Wasilkowski, and H. Woźniakowski. *Information-Based Complexity*. Computer Science and Scientific Computing. Academic Press, New York, 1988.

[Wei85] Klaus Weihrauch. Type 2 recursion theory. *Theoretical Computer Science*, 38:17–33, 1985.

[Wei87] Klaus Weihrauch. *Computability*, volume 9 of *EATCS Monographs on Theoretical Computer Science*. Springer, Berlin, 1987.

[Wei94] Klaus Weihrauch. Effektive Analysis. Correspondence course 1681, FernUniversität Hagen, 1994.

[Wei95] Klaus Weihrauch. A simple introduction to computable analysis. Informatik Berichte 171, FernUniversität Hagen, Hagen, July 1995. 2nd edition.

[Wei96] Klaus Weihrauch. Computability on the probability measures on the Borel sets of the unit interval. In Ker-I Ko, Norbert Müller, and Klaus Weihrauch, editors, *Computability and Complexity in Analysis*, pages 99–112. Universität Trier, 1996. Second CCA Workshop, Trier, August 22-23, 1996.

[Wei97] Klaus Weihrauch. A foundation for computable analysis. In Douglas S. Bridges, Cristian S. Calude, Jeremy Gibbons, Steve Reeves, and Ian H. Witten, editors, *Combinatorics, Complexity, and Logic*, Discrete Mathematics and Theoretical Computer Science, pages 66–89, Singapore, 1997. Springer. Proceedings of DMTCS'96.

[Wie80] E. Wiedmer. Computing with infinite objects. *Theoretical Computer Science*, 10:133–155, 1980.

[WS81] Klaus Weihrauch and Ulrich Schreiber. Embedding metric spaces into cpo's. *Theoretical Computer Science*, 16:5–24, 1981.

# Towards Machines That Can Think

Jiří Wiedermann[*]

Institute of Computer Science
Academy of Sciences of the Czech Republic
Pod vodárenskou věží 2, 182 07 Prague
Czech Republic
e–mail `wieder@uivt.cas.cz`

**Abstract.** Recent progress in cognitive computing suggests that we might approach the point when the algorithmic principles of brain–like computing will be revealed and the study, design and realization of thinking machines will start to be an issue in computer science. For this purpose, we shall present a brief overview of related results from a machine oriented complexity theory.

## 1 Introduction

In 1950, in his seminal work Computing Machinery and Intelligence [20] A. M. Turing wrote: *I believe that in about fifty years' time it will be possible to program computers, with a storage capacity of about $10^9$, to make them play an imitation game[2] so well that an average interrogator will not have more than 70% chance of making the right identification after five minutes of questioning. The original question, 'Can a machine think?' I believe to be too meaningless to deserve discussion. Nevertheless I believe that at the end of century the use of words and general educated opinion will have altered so much that one will be able to speak of machine thinking without expecting to be contradicted".*

While with his estimate of the storage capacity of today's computers Turing was astonishingly right, it appears that his further guess concerning the time when machine intelligence will become a reality was too optimistic. Sure, nowadays hardly anyone in computer science, artificial intelligence and related sciences doubts that in principle machines can think, in some form. At the same time, we do not know of any machine that would have passed Turing's test. Thus, the real issue is, whether at the present time machines can already think, and if yes, then how. I.e., do we have plausible arguments leading to the design of thinking machines?

It is the goal of this paper to present the known facts supporting the above arguments from the viewpoint of computer science.

---

[*] This research was supported by GA ČR Grant No. 201/95/0976 "HypercompleX" and partly by INCO–Copernicus Contract IP961095 ALTEC–KIT

[2] Turing's test for whether a machine can think: after a time limited conversation via a terminal one has to decide whether the dialogue was done with a man or a with a machine [20]

In computer science the main approach to the problem of cognition, thinking, or intelligence in general, is to see these activities as a large scale interactive information processing phenomenon supported by a large scale computational machinery.

In understanding this complex task, nature has confronted us with different roadblocks, which in their variety and entirety, have not been challenged successfully either in computer science, or in any other science. The nature of the above mentioned barriers is diverse. To overcome them one would ideally like to at least see a kind of three level model of cognitive systems similar to the case of any abstract or formal computational system (cf. [29]).

At the highest "machine independent" level it is necessary to have some kind of functional specifications stating at a fairly high level (but as formally as possible, and as completely as possible) what kind of cognitive information-processing tasks will be performed and how they should interact. Informally and partially this is described as "higher brain functions" in other disciplines like cognitive psychology. It seems that intelligence cannot be captured entirely in the language of logic and that some non–logical calculus of interaction is required (see [26] for an interesting discussion).

Then, at the next lower level, one would need a more refined model dealing with only several basic cognitive operations out of which more complex mental operations on the previous level can be built. It is perhaps here where computer science can help in identifying this elementary set of mind operations. Of course, other sciences, like psychology and neurobiology can offer valuable guidance in this search.

Finally, one needs a plausible "machine" model of a brain in which the basic mind operations can be realized efficiently. Here again neurobiology and neuroanatomy, together with machine oriented complexity theory and artificial intelligence, can be of tremendous help.

No satisfactory specifications of any of the above mentioned three levels are known. It appears that a collaboration of all of the above mentioned branches of science are necessary needed in order to cope with the related problems[3]. At the border of all of these disciplines, a new discipline of computer science seems to have emerged that may be appropriately called *cognitive computing* [24].

Roughly speaking, each cognitive computational model, that we shall deal with in the sequel, will concentrate on a different level or on different levels with a various intensity of details, in the above mentioned sketch of a three level architecture of cognitive systems. In all cases there will be at least some loose reference to the remaining levels.

The survey is organized as follows.

In Section 2 we shall introduce a simplistic view of the brain as an interactive

---

[3] In 1974, in his address [3] devoted to mathematical models of brains de Bruijn lists the following disciplines that could contribute to the solution of related problems: biology, psychology, animal psychology, psychiatry, parapsychology, pedagogy, sociology, history, archeology, medicine, pharmacology, physics, chemistry, philosophy, linguistics, engineering, computer science, mathematics

machine, or transducer. This view is common to all approaches included in this overview.

Then, in Section 3 we shall describe a family of the simplest models of a brain, viz. that of *finite neural nets*.

In the next four sections we shall present a model called *neuroidal tabula rasa* by Valiant [22], Goldschlager's model of *memory surface* from [7], then *the brain as a molecular computer* by de Bruijn [4] and we shall end with the recent model of *cogitoids* by the present author [30].

In Section 8, in the light of previously presented models we shall give our final answer to the original question whether thinking machines can be built at the present time.

As a suitable expository reading of results related to brains, machines, mathematics and computer science the books [1], [2], [6], and [14] are to be recommended.

## 2   Brains as Interactive Machines

The proper and the most general formal devices for describing cognitive activities seem to be *interactive machines* (cf. [26], [27]) whose task is to cope effectively with their environment by means of interaction. Incrementally they process infinite streams of input information supplied by the environment and produce similar streams of output information that is interpreted as the behaviour of machines. Since each increment to a previous input to such a machine may present a reaction of its environment to the machine's previous behaviour, such machines really allow for modeling interaction between them and their environment.

It is assumed that the streams of input data are produced by some preprocessors that, in practice, may themselves be a part of the brain. These preprocessors process the data obtained by sensors into an encrypted, possibly simplified, form that is more suitable to be processed by the interactive machine itself. Similar situation also holds for output streams that are sent to postprocessors whose task is to decrypt data and instruct effectors appropriately.

By processing further parts of a potentially infinite input stream, the interactive machine is allowed to change its internal structure — so to speak to re–program itself repeatedly and to store some useful derived data — in order to learn from the past.

An important feature of our interactive model is that in many cases it cannot obtain the input data on its input ports "on demand" — rather, only when the data is available, when it is "offered" by the environment, or when it is spotted by the machine. Thus, the device works asynchronously with its environment and is at least partially input driven. Timing, i.e., data arrival time from individual preprocessors also plays a crucial role: it appears that in more elaborated models of cognition simultaneous or ensuing appearance of data is of vital importance for learning from experience.

By modeling the cognitive activities of the brain by using the interactive machine of the above mentioned kind, one in principle obtains a more powerful computational and descriptional tool than presented by standard Turing machines that only operate over inputs of finite length (cf. [27]).

## 3 Neural Nets

We shall begin our overview on models for cognitive computing with the simplest devices that are inspired by our ideas on how real biological brains work. The respective models are presented by various kinds of neural nets that differ in the realization of their various types of basic computational elements — artificial neurons.

In general, a neuron is a device that is able to compute values of a function $f : \mathbb{S}^n \to \mathbb{T}$ in a single computational step. Individual sorts of neurons differ from each other by the choice of function $f$ (inclusively its domain and range). The elements $\mathbf{x} \in \mathbb{S}^n$ are called inputs while the elements $y \in \mathbb{T}$ are called outputs.

By using the outputs of some neurons in the place of inputs to other neurons we get a so–called *neural net*. The topology of the resulting net can be appropriately depicted with the help of an oriented graph in which neurons are its nodes, incoming edges denote the inputs to the neurons and outcoming edges denote the outputs from the neurons.

Any finite neural net can be seen as an interactive machine in the following way. Inputs that are not connected to the outputs of some neurons can be interpreted as *input ports* while outputs that are not connected to the inputs of some neurons present *output ports* of the net. All neurons in the net work synchronously. At the beginning of a computation the output values of all neurons are set to some prescribed values. Then, at each time step each neuron reads its inputs and computes its output according to its definition. Especially, at each computational step the net reads symbols, appearing at each input port, and sends one symbol to each output port.

Besides its topology the computational power of a neural net depends crucially on the nature of function $f$ computed by the individual neurons. Therefore the computational power of each net must be investigated for each specific instance of $f$ separately.

The simplest case of an artificial neuron is presented by the model by McCulloch and Pitts from 1943 [13]. A so–called *first generation neuron* is seen here as a highly idealized mathematical abstraction of real biological neurons, occurring in brains, that computes the values of the weighted threshold Boolean function.

Let $\mathbf{x} = (x_1, \ldots, x_n)$ and $\mathbf{w} = (w_1, \ldots, w_n)$ be integer vectors and $t$ an integer constant. Then a Boolean function $f[\mathbf{w}, t]$ of $n$ Boolean variables $x_1, \ldots, x_n$ is called a *weighted threshold* function if and only if

$$f[\mathbf{w}, t](\mathbf{x}) = \begin{cases} 1, & \sum_{i=1}^n w_i x_i \geq t \\ 0, & \sum_{i=1}^n w_i x_i < t \end{cases}$$

The vector $\mathbf{w}$ is called the vector of *weights* while $t$ is called a *threshold*. It is said that a neuron *fires* if the sum of its overall stimuli, expressed by the weighted sum $\sum_{i=1}^n w_i x_i$, surpasses the threshold $t$.

In order to intuit the computational power and limits of finite neural nets of a first generation, the simplest case of single input port nets has been studied in the framework of formal languages and complexity theory. In the respective model, the input stream is read bit by bit and with a possible delay a stream of 1s or 0s indicating whether the input sequence, read so far, is accepted, is produced at the output. The respective nets are called *neuromata*. They have been studied intensively in works [18] and [19]. The respective results point to the fact that such devices have the computational power equivalent to that of finite automata — i.e., they recognize regular languages.

*Second generation* neurons (also called *analog neurons*) differ from the first generation by working over domains of real inputs and outputs. The biological motivation (that in practice seems to hold only partially) is that firing neurons produce electrical pulses with an intensity that varies in a certain range. This is modeled by the respective variables that take on continuous real values, and hence the neuron computes the values of some real function. This is achieved by applying a so–called *activation function* $\psi : \mathbb{R} \to \mathbb{R}$ to the weighted sum of inputs to the neuron at hand. The output value is, then, also a real number. Thus, if $y_k$ is the input to $i$ from a neuron that is connected to $i$ via the edge $(k, i) \in E$, carrying the weight $w_{ki}$, then the output $y_i$ of $i$ is defined as

$$y_i = \psi \Big( \sum_{(k,i) \in E} w_{ki} y_k \Big)$$

Depending on the activation function (that can be e.g. a sigmoidal function, or linear saturated function, or piecewise polynomial function) and on the weights (that can either be integers, or rationals, or reals), the computational power of the respective nets has a range rom that of the first generation (regular languages) up to super–Turing capabilities. For an overview see e.g. [15], [16], or [17]. Thus, among the second generation neural nets the most powerful computational devices that are know to us exist. To illustrate the computational efficiency of the respective networks, it can be shown that a finite number of analog neurons can simulate the Turing machine on inputs of arbitrary length (cf. [9] or [17]). The corresponding computation relies heavily on the ability of the underlying network to compute with real numbers with arbitrary precision.

From a biological point of view networks computing with reals are unrealistic — no biological system seems to be able to cope with analog values to an arbitrary precision. The above belief has recently obtained an unexpected support by the theory — in [11] it has been shown that the presence of arbitrarily small amounts of analog noise reduces the power of analog neural nets, and even the power of the the third generation (see in the sequel), to the power of the first generation networks (i.e., essentially to that of finite automata). This excludes the possibility that biological systems could possess the super–Turing computational capabilities.

*Third generation neural nets* are based on the model of *spiking neurons* that seem to better correspond to the computation of biological neural systems that employ pulses in order to transmit information among their computational units (i.e., neurons).

Whereas the timing was trivialized in both preceding generations (by assuming synchronousness in most of the cases), the timing of individual computational steps plays a key role for computations in networks of spiking neurons. In fact, the output of a spiking neuron $v$ consists of the set (cf. [10])

$$F_v = \{t_1, t_2, \ldots | v \text{ fires at time } t_i, 1 \leq i\} \subseteq \mathbb{R}^+ = \{x \in \mathbb{R} : x \geq 0\}$$

Thus, the neurons transmit information by encoding it into gaps among successive firings. For details see e.g. the paper by Maass [10] where the computational power of spiking neuron networks is investigated and compared with that of preceding generations. It is shown that these nets are computationally more powerful than the other neural net models. A concrete, biologically relevant function is exhibited which can be computed by a single spiking neuron, but which requires hundreds of hidden units on a sigmoidal neural net of the second generation. A recent overview on the computational power of a pulse computation can be found in [12].

By devising and investigating increasingly more faithful models of biological neurons we are hopefully approaching the true computational efficiency of real brains. Unfortunately, unless some major computational feature of real neurons has been overlooked, it appears that no matter how faithfully we shall be able to simulate the activity of real neurons this effort is not going to substantially contribute to our quest for discovering the basic high level principles by which the mind works.

To summarize the computational power of neural nets from the point of view of complexity theory, the realistic models among them have equal power as the first generation nets. They differ merely in the efficiency of the respective devices: for the same computational task there may exist networks of a higher generation which are smaller than the networks of the lower generation.

In the framework of cognitive computing, the previous results can be interpreted as ones expressing that neural nets present efficient devices that can realize the basic cognitive task of pattern recognition, or an associative retrieval of them, with complexity characteristics, depending on the size and number of recognized patterns and on the computational power of individual neurons. The patterns are represented as finite segments of otherwise infinite input streams.

## 4 Neural Tabula Rasa

The main feature that seems to be missing in the previous neural net models is the ability to learn. The main obstacle, in this sense, seems to be the fact that these nets, once defined, cannot modify their computational behaviour. This appears to be a condition *sine qua non* for any learning device. This defect is eliminated in so–called *neuroidal nets* that present a programmable type of neural nets, which was first proposed by Valiant in 1988 [21]. The potential to model brain–like computations was further investigated in 1994 in the monograph [22], by the same author, and finally a neuroidal architecture for cognitive computation has been recently proposed in [25].

The main building element of neuroidal networks is *a neuroid* which is a combination of a neural threshold element (of the first generation) with the idea of a finite automaton and additional features that enable a neuroid to change its state, weights, and/or threshold depending on the activities of neighbouring connecting neuroids and current values of previous parameters. The idea that real neurons can store information not only in the synaptic strength (which is modeled by the respective weights) but also as *state* information is biologically plausible and experimentally confirmed [25].

The rules for changing respective parameters are formally described with the help of a *transition function*, similarly as it is the case of finite automata. readily, this enables the programming of individual neuroids (or more often: groups of neurons) to behave differently at different times.

E.g., one can design a single neuroid that, once "seeing" a certain input which contains at least one 1 will fire in the future iff it will be again confronted with a similar input. Such a similar input should have ones in the same places as the original input (and possibly in other places). Initially, the respective neuroid has all its weights set to 1. When confronted with an input consisting of $k \geq 1$ ones, it resets the weights at ports containing zero to 0 and its threshold to $k$. Then it enters into a "final" state in which no further parameter changes are performed any longer.

In this way one can program neuroids to implement various atomic cognitive tasks. Connecting neuroids in a net gives rise to a so–called *neuroidal tabula rasa* (NTR).

The most important task of NTR is its ability to represent various semantic *items* $x_1, \ldots, x_t$ which describe any aspect of the modeled reality. Each item is represented by a group of neuroids that fire under certain conditions on the inputs or internal computations corresponding to the moment when the NTR is somehow "reminded" of the item. The size of the group is called a *replication factor* and it must be chosen appropriately in order to ensure that the following NTR operations will work efficiently with a high probability (see the sequel).

By programming the neuroids accordingly, on its existing knowledge base of represented items the NTR is able to build further items hierarchically by using the following operations in any order:

1. *allocating* new neuroids to represent new items;
2. *memorizing* a new item $x_{t+1} = x_i \wedge x_j (1 \leq i < j \leq t)$ in the sense that $x_{t+1}$ will fire at later times whenever both $x_i$ and $x_j$ fire (but not under other conditions);
3. *associating* $x_i$ with $x_j$ in the sense that $x_i$ will fire whenever $x_j$ fires;
4. *detect correlations* between pairs $x_i, x_j$ that fire simultaneously;
5. do each of the above operations with relational or multi–object expressiveness rather than propositionally.

The respective operations are realized with the help of the so–called *vicinal algorithms*. They are probabilistic algorithms. Their most basic feature is that whenever communication is to be established between two items, not directly connected in the NTR, the algorithms establish the necessary connection via neuroids that are connected to both items. These neuroids are called *the frontier* of both items. Such a frontier, with the replication factor $r$, should exist with a high probability for any two randomly chosen groups of neuroids, with about the same replication factor $r$, representing two different items. This alone imposes a certain condition on the topology of the graph underlying the NTR.

Besides the frontier properties, there is a further attribute that is required of the NTR and can be supplied by randomness. This further property is needed to ensure that the nodes chosen for representation of a new item will be, to a large measure, among those not previously chosen. Valiant calls this the *hashing* property since it corresponds to that notion in computer science. The corresponding theory is related to some extent to the problem of universal hashing.

These two properties jointly are fulfilled e.g. in the class of random graphs on $N$ nodes with the expected degree $\sqrt{N/r}$ [22].

For a detailed description of the respective vicinal algorithms see the monograph [22] where also the discussion of other cognitive tasks (e.g., memorization, correlational learning, relation expression learning, reasoning) can be found. Unfortunately, the resulting model of brain–like computations tends to be quite complex and it appears to be difficult to cope with e.g. higher brain functions more rigorously within this framework.

Further Valiant's elaboration of related issues in cognitive computing and the architecture of the respective systems can be found in [23], [24], and [25].

## 5   Memory Surface Model

Memory surface model by L. Goldschlager [7] appeared untimely in 1984 when the interest in cognitive computing in computer science was not as high as nowadays. From today's perspective this model seems to be of upmost interest since it provides the first coherent view on brain operations from a computational, or algorithmic point of view.

The model can be viewed as a directed graph comprised of millions of points, which will be called *columns*. The graph is to be seen in the three-dimensional Euclidean space and therefore it makes sense to speak about the distance of its nodes and the direction of its edges. Each column is connected to a number of other columns nearby and no long distance connections are required. The connection between coincident columns are in both directions. Moreover, columns may have (directed, one way) connections from some pre–processing and/or to some post–processing components. There is a *weight* $m_d$ assigned to each outcoming edge (indexed by) $d$.

Columns present the basic functional unit of memory surface. All communication among columns and pre- and post–processors will be in the form of *trains of pulses* running along the directed edges of the underlying graph. These pulses have the same shape and amplitude, but their frequency can vary with time.

The computational activity of each column is fully described by the following rules:

- whenever a pulse arrives at the column along some edge it behaves as if it is travelling in a straight line through the column, providing that the edges are more or less evenly spaced around the column
- the number of pulses arriving at the column along all incoming edges will be summed over a short time period. If the column happens to have a connection from the pre–processor, the pulses arriving along that connection will also be added into the overall sum of arriving pulses, but with a higher weighting factor.
- then a pulse train will be produced, whose instantaneous frequency $f$, called the *activity* of the column, is proportional to the value of the sum. This pulse train is further transmitted with the unchanged instantaneous frequency $f$ to the post–processor, if there is such a connection, and with frequency $m_d f$ along each outcoming edge $d$.
- the mechanism, which sums the incoming pulses to the columns, will exhibit short term *habituation*. That is, if the sum is repeatedly large for a long time thus giving rise to a large value of $f$, then the mechanism will tire and begin reducing the value of $f$. Conversely, after a period of time when the sum is repeatedly low, the habituation or tiredness will slowly wear off. The habituation may only last a fraction of a second;
- for any edge $d$, whenever a pulse arrives from that direction, $m_d$ will be incremented slightly, provided that $f$ exceeds a certain value at the time the pulse arrives. The values of $m_d$ for all edges represent the memory stored in that column. The increased values of $m_d$ will persist for a long time, perhaps weeks, months or even (in the case of very large $m_d's$) years, but they too will slowly decay over time.

The function of each column, as described above, can be implemented in a variety of ways and the way of implementation is not essential for further explanation. Nevertheless, for the sake of plausibility, Goldschlager sketches column implementation with the help of a model of some variant of "spiking neurons" (see Section 3) which, however, are not specified in sufficient details that would enable a more rigorous treatment of the proposed implementation.

Define a *pattern* P to be any set of columns, together with their relative activities. Many of these activities will be zero, representing the fact that the corresponding column does not participate in this particular pattern, and many will be non–zero, thus representing an important part of the pattern. In practice, each pattern will consist of many thousands of active columns. Each pattern will represent a concept that the brain can handle. These concepts may be less abstract such as "dog", or more abstract such as "ownership".

The simplest patterns which correspond to the least abstract or most fundamental concepts, which the brain can handle are just those patterns which result from some combination of receptors firing in response to some actual events occurring in the environment of the brain. Symmetrically, those patterns which cause effectors to produce some useful impact on the environment, are also among the simplest patterns.

The basic mechanism that causes various new and old patterns to emerge, with various activities, is that of the simultaneous excitation of the respective columns. Namely, when these columns are simultaneously excited by pulses coming from other active columns, or from pre–processors, to such a degree that they fire, the new patterns become active. Moreover, the outcoming pulses will cause the increase of weights of corresponding edges, memorizing thus the patterns with increased strength. Also, the pulses will arrive at post–processors causing some appropriate action of effectors. Due to their habituation, the patterns that caused the activation of new patterns will slowly decay while the newly activated ones will survive for a while.

In this way the memory surface works as an interactive machine.

The process just described is also responsible for memory surface learning of correlations among often simultaneously excited patterns. These correlations take the form of *associations* established among those patterns via edges with increased weight. New concepts emerge by association of simpler concepts, which are related by contiguity in time or place, cause and effect, and by resemblance.

*Contiguity in place* refers to the fact that when objects are often observed to be physically together, they become associated. The frequent simultaneous observation of the objects will cause the patterns on the memory surface, which represent these objects, to be simultaneously active for a long total duration.

*Contiguity in time* refers to the fact that when events often occur in a sequence over time, those events will become associated. Examples are remembering a song and learning the alphabet. E.g., in the latter case, as soon as the concept (of a symbol) $A$ becomes active, its columns will tend to stimulate each other and thus the pattern will persist on a memory surface for some time. However, due to habituation, the strength of the pattern will decrease with time. By the time the concept $B$ becomes active, it will be simultaneous with the weaker strength·pattern $A$. Therefore an association of $A$ with $B$ will be formed by the standard association mechanism. And so on with the next members of the sequence. This, by the way, explains why sequences, which are learned in one direction, are hard to recall in the reverse direction.

*Cause and effect* is to be understood as a special case of sequence.

When two concepts share some common set of features, they *resemble* one another. Hence, due to the very nature of an associating mechanism, two different concepts with a common set of features will have associative links to this set of features, even if the patterns representing the two concepts were never present simultaneously. Clearly, at the same time, the set of common features of two or more concepts presents an abstraction, or a generalization of the concepts involved.

With the help of the above mentioned mechanisms one can also explain the functioning of higher brain functions.

For instance, how does a brain progress from one thought to another? Here a mechanism, similar to one causing the association of members of some sequence, is in action. At any moment of time, some sets of patterns will be active having various strength on the memory surface. Each of these patterns will tend to excite all the patterns, to which it is associated, with strength proportional to its own strength and to the strength (weights) of the associative links. Meanwhile, the currently active patterns will tend to habituate and their strength will die away.

So the memory surface will exhibit a *flow of activity* from concepts to associated concepts which we perceive as *train of thoughts*.

*Creativity* consists of a lucky, so far unprecedented simultaneous activation of two patterns with many features in common. In this way a new, so-far "unknown", abstract concept will be invoked. The luck may be provided by the external environment, or by a random stimulation in sleep. *Sleep* has two survival phases, namely to help the memory of infrequently used concepts and associations, and to encourage creativity.

The concept of *self* is a collection of various concepts that are all related to activities directly controlled by memory surface.

*Consciousness* is a complex of different concepts that all relate to the idea of awareness.

*Awareness of environment* is simply the activation of the appropriate concepts, of the memory surface which model that aspect of the environment. *Self-awareness* refers to the fact that, in addition to the awareness concept, the concept of self is simultaneously active on the memory surface. Finally, *awareness on one's own train of thought*, often called *introspection*, means forming an association between the concept of self and the other concepts currently active on the memory surface.

*Conscious thoughts* can be regarded as those patterns which are active simultaneously with the concept of self, and whose associations are sufficiently strong to form an association with the concept of self. *Unconscious thoughts* are those which are active when the concept of self does not happen to be active, and those patterns where activity is too weak to form an association with the concept of self.

## 6 Brain as a Molecular Computer

The next model that we are going to present originates from a series of papers by de Bruijn [3], [4], and [5]. After Turing, during the nineteen seventies de Bruijn was apparently one of the first pioneers in mathematics and computer science who paid serious attention to problems related to mathematical brain modeling.

His model is quite different from previous ones: it is not based on the preceding ideas of seeing the brain as a set of relatively simple, neuron–like computing elements connected by a fixed communication network.

The model can be seen as a gigantic spatially distributed parallel computer with tiny processors, equipped with sizable associative memories working at a molecular level, that only communicate with pre- and post–processors in a completely unpredictable way.

Within the framework of interactive machines, the operational view of the model is as follows: the pre–processors are broadcasting a potentially infinite stream of primitive signals $p_1, p_2, \ldots$ during the whole life of the system. On the other hand, they also often broadcast a query in which they are giving a signal they broadcasted in the past and are asking for the signal which immediately followed. The model should be able to record the stream and to answer those queries in a real–time manner. As one could expect, the basic mechanism used for answering the queries is that of *associative retrieval*.

This mechanism is implemented in processors whose role is played by *cells*. Each cell is seen as a bowl containing a large, but fixed number of compounds $A, B, C, \ldots$. Hence, the number of reactions that can ever take place in the bowl is finite. In their simplest form the reactions are of the form $A + B \rightarrow C$. This is not necessarily a faithful description of a chemical reaction; rather, it should describe the fact that adding $A$ and $B$ to the existing mixture in the bowl will produce $C$.

The bowl can then act as a huge parallel computer, with different $A$s and $B$s which represent inputs to some operations, and $C$s the outputs, and with the whole computation performed with the help of reactions of the above mentioned type. This is why De Bruijn calls the respective mixture *a thinking soup*. If one thinks of DNA–like molecules of the length $k$, then the number of possible compounds may be of the order $c^k$ with a constant $c > 1$. This "molecular hardware" is potentially available, but most of it is never used.

Operating with such a thinking soup requires inputs and outputs. Inputs do not necessarily need to consist of chemical ingredients inserted into the bowl at the beginning of its computations. There may be other phenomena, like electric or mechanical signals that initiate the production of particular compounds. Similarly, the outputs might include sensors that transform chemical information into other kinds of signals. So, according to de Bruijn, an input and output should be called an active and passive *smelling*.

It is quite possible that a thinking soup helps biological system to *think*, along with neural networks, but de Bruijn does not seem to be able to provide any plausible explanations of such a phenomenon. Rather, he provides an interesting proposal of how the associative memory can be handled in the thinking soup.

Add an input $p$ to the thinking soup, followed immediately by the signal $q$. These inputs generate compounds $A$ and $B$ in the soup. Let these compounds generate a third compound, $C$: $A + B \rightarrow C$. Assume that there are mechanisms that restrict this reaction to the case where $A$ is followed by $B$ and not vice versa, and that this is the only way how $C$ can appear in the bowl. Further, let the compound $C$ be quite stable, and let it survive even after $A$ and $B$ have vanished completely. Then the whole process $A + B \rightarrow C$ can be seen as an information storage and the respective reaction might be appropriately called a *storage reaction*.

In order to obtain memory retrieval, one should assume that there is a companion, the so–called *retrieval reaction* $A + C \rightarrow B$. It has the effect that if $p$ should ever reappear producing $A$, then $C$ subsequently helps to produce $B$; the result is that $q$ is obtained as the output. Of course one has to assume that $A$ did not already trigger the retrieval reaction at the time of storage reaction: perhaps when $C$ was formed, $A$ was entirely consumed by this reaction, or was too weak to trigger the retrieval reaction.

Every brain cell and, indeed, any cell in principle (think of unicellular organisms) may be able to record many thousands of different associations in this way.

In order to explain how one could compute by using the thinking soup within cells, one has to return to the operational view of the model mentioned above. Assume preprocessors are continuously broadcasting a potentially infinite sequence of primitive signals. In the case that the respective signal has not been memorized yet in the cell, it gets memorized in the cell's thinking soup with the help of a storage reaction. Otherwise, the signal is interpreted as a query whose answer is retrieved with the help of retrieval operations and sent further to postprocessors.

Such an arrangement has a major drawback: the same information gets redundantly stored in each and every cell and therefore sooner or later the capacity of the systems will become exhausted. What is needed, is a mechanism that would store the items with significantly less redundancy in order to make better use of full storage capacity available in all cells. ,

To solve this problem de Bruijn offers an ingenious solution: at every moment not every cell is active and is ready to perform store or retrieve operations. The periods of activity of each cell are randomly chosen. Under such a scenario, signals only get stored into, or queries are only performed over, the set of currently active cells that is called *active window*. Since this is a continuously varying set deBruijn calls it "*roaming random set*". It is clear that by decreasing the size of the active window the storage capacity of the system increases while its retrieval reliability decreases. This is because some previously stored information does not need to reside in the cells of a current roaming random set and therefore a query asking for that particular information cannot be answered.

Here we are confronted with a nice optimization problem: what should be the size of a randomly chosen active window in order to guarantee, with a high probability, that there will be at least one cell from any window that was active in the past? The probabilistic analysis reveals that within the population of $N$ cells the size of active window should be of order $\sqrt{N}$ [4] at any moment. Thus, most of the time, most of the cells can rest. For instance, considering $10^{12}$ cells in a human brain, it is enough that only about $10^6$ from among them are active for the period of about half a second. This still leads to several hours of rest for each cell, on average, with an expected size of about 40 cells being in the intersection of a current active window and any past randomly chosen window. It is clear that with such a redundancy the system achieves a high level of fault tolerance. For more details, see the original paper [4].

To finish the description of de Bruijn's molecular computer two questions must be answered. What is the mechanism that implements broadcasting of signals among pre–processors, cells, and post–processors, and how do the cells determine when to be active and when to take a rest?

For both purposes the same mechanism can be envisaged. Namely, it is biologically plausible that not only are the neurons in the brain connected randomly via their dendrites, but moreover, the places where they "touch" each other can also be randomly set "on" or "off" for signal transmission in both directions. Some of the neurons are connected to pre–processing parts of the brain. From here, primitive signals are broadcasted along the connections that are at this very moment "on", essentially to a randomly chosen set of cells. This collection of cells is the active window of the moment. It is assumed that the topology of the underlying graph is such that the signals can arrive at any cell after only a few steps between the cells.

Besides the enormous storage capacity, de Bruijn sees the main advantage of his model in the fact that within its framework one is able to explain consciousness and unconsciousness.

These two notions are to be understood as particular *modes of interaction* between pre and post–processors and the memory. This interaction takes place in the active window. The information flowing from pre–processors is immediately stored in the cells of the active window where it remains very easily available during the decay period. During that period the pre and post–processors might repeatedly recall the stored memory items — *"thoughts"*, and to *think* about them. This is called *reflection*. During the short decay period they can be recalled, and new thoughts can be formed *about* them and stored. All this has to be done within the decay period for otherwise the information is much harder to get.

In the course of the above mentioned process there may be a considerable traffic between pre–processors and the cells of the roaming random set, and between the latter and the post–processors. But it may very well happen that only a small part of the cells from the active window is involved in the reflection. De Bruijn calls this part the *conscious* part, and all the rest the *unconscious* part. There is no sharp borderline between the two.

For more arguments in favor of the above mentioned model see [3] and [5].

# 7   Cogitoids

Unlike in all of the previous cases, forget about the "machine" model of the brain now. Let us concentrate instead on the intended brain functions at the middle level of the three level architecture of a cognitive information processing system mentioned in Section 1.

Building on Goldschlager's ideas the respective approach was taken in [30]. The idea here was to consider the basic cognitive entities (concepts) and operations above them (such as forming of new concepts and both excitatory and inhibitory associations among them) as the basic properties of the model. They are there and we do not care about how they are realized. Further, this model should provably posses a spontaneous ability of continuous learning. It should present a kind of universal learning machine that detects certain time–, space–, and similarity related patterns in the observed streams of data and learns the appropriate behaviour mostly by rehearsal and by punishment and rewards.

The resulting formal model of a *cogitoid* looks as follows.

The basic entity it handles are the *concepts*. They are modeled by subsets of a (large) finite universe $\mathcal{U} = \{f_1, f_2, \ldots, f_n\}$. The elements $f_i$ of the universe are called *features*. Any object possessing every feature (and possibly some others) of a concept $A$ is called an instance of $A$.

The concepts are denoted by upper case letters: $A, B$, etc. There is a special subset of concepts that is called *affects*. Affects correspond to positive or negative feelings of animals.

If $A, B$ are two concepts and $A \subset B$, then $A$ is called an *abstraction* of $B$ and $B$ is called a *concretization* of $A$. This is because any instance of $A$ is at the same time an instance of $B$.

Especially, for any $A$ and $B$ $A \cup B$ is a concretization of either $A$ or $B$ while $A \cap B$ is their abstraction. Note that the set of all concepts over $\mathcal{U}$ with operations of set union and set intersection form a Boolean algebra with $\emptyset$ being its least element and $\mathcal{U}$ its greatest element.

With the help of the above mentioned two operations of concept union and intersection, new concepts can be formed from existing ones.

In a cogitoid, concepts may be explicitly related via associations. Associations emerge among concepts that occur in series or among similar concepts.[4] Formally, an ordered pair of form $(A, B)$ of concepts is called an *association*, denoted also as $A \to B$. We say that $A$ is associated with $B$. There are two types of associations: *excitatory* and *inhibitory*. Among any pair of concepts there may be both types of associations.

Two concepts $A$ and $B$ resemble each other in the concept $C$ iff $A \cap B = C$ and $C \neq \emptyset$. This knowledge is also represented as an association $A \to B$.

At any time $t$ any concept may be either *present* or *absent* in a cogitoid. If present, then a concept may be either in an *active* or in a *passive* state.

Also, at each time $t$ there are two quantities assigned to each concept: its *strength* and its *quality*. The strength of a present concept is always a non–negative integer while absent concepts have the strength zero. The quality of concepts can be positive, negative, or undefined. Positive affects have always positive quality, while negative affects have always negative quality. The quality of other concepts may be arbitrary and depends on the history of concept formation or on the context in which a concept is invoked (activated).

Similarly, the strength is also assigned to each excitatory or inhibitory association.

Currently passive concepts may be activated either directly from the environment (via external stimuli that activate the corresponding features), or by internal stimuli via associations from other active concepts.

In the latter case, in order to activate, concepts should be sufficiently excited. The concepts get excited via associations. The strength of excitation depends on the strength and type of all associations leading from active concepts to the concept at hand. This concept is excited to the level that is proportional to the sum of strengths of all excitatory associations from currently active concepts

---

[4] This is a slightly different presentation from that given in [30] where a resemblance of concepts was not represented explicitly by associations.

decreased by the sum of strengths of all inhibitory associations from currently active concepts.

The cogitoid is seen as an interactive transducer that reads an infinite sequence of Boolean vectors of form $(x_1, x_2, \ldots, x_n)$. Each such a vector denotes an instance of a concept $I$ that is determined by those features that correspond to ones in the above vector. The concept $I$ represents an object that is "observed" by a cogitoid at its input.

The computation of $\mathbb{C}$ proceeds in rounds. At the end of each round a set of concepts is active. This set presents an output of the cogitoid — its behaviour, its reaction to the previous input. Let $\mathcal{A}_t$ be the set of concepts active at the end of the $t$–th computational round in a cogitoid $\mathbb{C}$.

Each round consists of six phases:

**Phase 1:** *Producing the output and reading the input:* The concepts in $\mathcal{A}_t$ are sent to the output. All concepts in the set $\mathcal{I} = \{X | X \subseteq I\}$ corresponding to all abstractions of $I$ are activated. This models the formation of concepts by their simultaneous appearance.

**Phase 2:** *Activating new concepts by internal stimuli:* First a single new concept $N$ from among all currently passive concepts gets activated. This is done with the help of a *selection mechanism* which inspects the excitation of all currently passive concepts from concepts in $\mathcal{I} \cup \mathcal{A}_t$ and subsequently activates the most excited concept $N$.

Simultaneously with activating $N$, also the set $\mathcal{N}$ of all abstractions of $N$ gets activated.

**Phase 3:** *Assigning quality to concepts.* The quality of affects is constant all the time and it will determine the quality of all other currently active concepts to which active affects are related. The concepts whose quality cannot be determined by the preceding rule, get undefined quality.

**Phase 4:** *Long-term memorization:* The strength of all currently activated concepts is increased by a small amount.

Similarly, the strength of associations between each concept in the set $\mathcal{A}_t$ and each in $\mathcal{N}$ is increased. This models the emergence of associations by cause and effect.

Finally, the associations by resemblance are updated by increasing the strength of associations between each active concept in $\mathcal{I} \cup \mathcal{A}_t$ and each resembling present passive concept.

In the above mentioned process, if the association to be strengthened is between the concepts $A$ and $B$, then if the quality of $A$ was positive or negative or undefined, respectively, then the excitatory or inhibitory association, or both associations, respectively, between $A$ and $B$ are strengthened.

Note that increasing the strength of associations in some cases means that new associations are established (since until that time associations can be seen as those with strength zero).

**Phase 5:** *Gradual forgetting:* If positive, then the strength of all concepts that are not currently active and the strength of all associations among them is decreased by a small amount.

**Phase 6:** *Deactivation:* The concepts in the set $A_t$ are deactivated and the set $N$ becomes the set of all active concepts $A_{t+1}$.

Note that the sequence $\{A_t\}_{t\geq 0}$ models the "train of thoughts" in our cogitoid.

The notion of the above described cogitoid can be formalized with the help of sets and mappings.

In [30] it is shown that for any cogitoids it is possible to perform basic cognitive tasks such as abstraction formation, associative retrieval, causality learning, retrieval by causality, and similarity–based behaviour.

The next domain of behaviour that can be acquired by cogitoids is that of *Pavlovian conditioning.* This is a phenomenon in which an animal can be conditioned (learned) to activate a concept as a response to an apparently unrelated stimulating concept (cf. [22], p. 217).

For instance, one may first "train" a cogitoid to establish a strong association $S \rightarrow R$. Then, we may repeatedly confront such a cogitoid with a further, so far unseen concept $A$, with $A \cap S = \emptyset$, that is presented to it jointly with $S$, as $S \cup A$. After a while we shall observe that $A$ alone will elicit the response $R$. Nevertheless, after a few of such "cheating" from our side the cogitoid will abstain from eliciting $R$ when seeing merely $A$ (in psychology this is called *extinction*). Also more complicated instances of Pavlovian conditioning can be observed in *arbitrary* cogitoids. The only condition is that the cogitoids must be large enough to accommodate all the necessary concepts.

In order to explain Pavlovian conditioning no use of negative operant concepts and related inhibitory associations are necessary.

Cogitoids are also able to realize so–called *operant behaviour.* This is a behaviour acquired, shaped, and maintained by stimuli occurring *after* the responses rather than before. Thus, the invocation of a certain response concept $R$ is confirmed as a "good one" (by invoking the positive operant concept $P$) or "bad one" (the negative operant concept $N$) only after $R$ has been invoked. It is the reward $(P)$, or punishment $(N)$ that act to enhance the likelihood of $R$ being re–invoked under similar circumstances as before.

The real problem here is hidden in the last statement which says that $R$ should be re–invoked (or not re–invoked) only under similar circumstances as before. Thus, inhibition, or excitation of $R$ must not depend on $S$ alone: in some contexts, $R$ should be inhibited, and in others, excited. Such a context is called an *operant context*; it is represented by a concept that appears invariantly as the part of the input of a cogitoid during the circumstances at hand. Thanks to cogitoid learning abilities, this operant context gets tied to the respective operant concept which, later on, causes that all associations emerging from this pair will inherit the quality of the operand concept at hand. Therefore, in the future, these associations will inhibit or excite $R$ as necessary.

It appears that by a similar mechanism that ties a certain operant concept to some temporarily prevailing operant context one can also explain a more complicated case of the so–called *delayed reinforcing* when the reinforcing stimulus — a punishment or a reward — does not necessarily appear immediately after the step that will be reinforced.

All of the latter statements concerning the learning abilities of cogitoids can be formalized and rigorously proven (see the original paper [30]).

In the forthcoming paper [31] it is further shown in a kind of a thought experiment that when a cogitoid is exposed to a similar sequence of inputs as the human brain during its existence specific substructures start to develop in a cogitoid's memory. These structures correspond to episodic memories, frames for frequently performed activities, roles to be played by different objects in different contexts, attentional mechanisms for different contexts, and to habits. The latter represent the executive part of cogitoids memory. Within this framework language acquisition and generation can also be explained. Eventually, an emergence of behaviour, which resembles the behaviour which originates in the human mind, is to be expected.

## 8 Conclusions

Can machines think already at the present? An honest answer would be no, unfortunately, not yet, not in any interesting meaning of this word.

The good news is that we already have some clues of how the design of such machines should be approached, as we have seen in the paper.

The bad news is that even if we knew for sure how such machines should work, the next obstacle to be solved is their size. Namely, the human brain consists of about $10^{12}$ neurons and about $10^{15}$ dendrites. To simulate each neuron separately would require computers with memory capacities of that order. In order to keep pace with the brain one parallel computational step requiring an update of each neuron would call for realizing still a substantially larger number (than $10^{15}$) of operations per second. No parallel computer can do that. Interestingly, the total memory and computational capacity of Internet seems to roughly correspond to the required task, if we forget about the speed... Some models seem to be less computationally demanding than others: it is here where the true computational abilities of real biological neurons can play a significant practical role. If de Bruijn is right in his speculations that the processing and memory capacity of the brain is also hidden at the molecular level, then a long way is to be expected before we arrive at working models competing with human brains.

The next bad message is that when we insist on a human–like machine intelligence machine learning should apparently take the form of education similar to that which we undergo during our lives. Providing a machine with respective inputs can turn out to be an insurmountable technical problem. Along these lines, direct brain–machine connection, or making use of methods similar to EEG biofeedback, can present possible approaches. For machines the most suitable source of knowledge is that in a textual form. The question whether one can build cognitive machines educated purely with the help of textual material is therefore of upmost interest. Such a machine would then be inevitably crippled in its own perceiving experience and therefore could be hardly able to pass the Turing test. Nevertheless it can still be of enormous interest.

On the positive side, it can very well happen that we will discover ways of building and training true intelligent systems realized on computational models that are substantially different from neuron–based models and tailored to the possibilities of contemporary or foreseeable computing technologies. Even better (worse?), such systems might outperform our brains in cognitive abilities which is similar to the case that exists now in many application areas of computer science ...

The most important lesson to be taken from the above described models of cognitive computing is perhaps the following one: although the respective models appear to be formally different, in fact, from the view point of their cognitive abilities they are not. Higher level issues of more elaborated models can be simulated on simpler models. This hopefully points to a certain robustness of the respective class of interactive cognitive machines and supports the original Turing claim that mental processes can be described in some kind of logical model and that discrete state machines are a relevant description of brain operations [8].

# References

1. Arbib, M. A.: Brains, Machines, and Mathematics. Second Edition. Springer Verlag, New York, 1987, 202 p.
2. Arbib, M. A. (Editor): The Handbook of Brain Theory and Neural Networks. The MIT Press, Cambridge — Massachusetts, London, England, 1995, 1118 p.
3. de Bruijn, N.G.: Mathematical Models for the Living Brain. Address to the Royal Netherlands Academy of Sciences and Letters, Section of Science, Amsterdam, 21 December 1974, with a postscript, added March 1975
4. de Bruijn, N.G.: A Model of Information Processing in Human Memory and Consciousness. Nieuw Archief voor Wiskunde, Vierde serie Deel 12 No. 1–2 maart/juli 1994, pp. 35–48
5. de Bruijn, N.G.: Can People Think? Journal of Consciousness Studies, Vol. 3, No. 5/6, 1996, p.425–447
6. Churchland, P.S. — Sejnowski, T.J.: The Computational Brain. The MIT Press, Cambridge — Massachusetts, London, England, 1992, 544 p.
7. Goldschlager, L.G.: A Computational Theory of Higher Brain Function. Technical Report 233, April 1984, Basser Department of Computer Science, The University of Sydney, Australia, ISBN 0 909798 91 5
8. Hodges, A.: Alan Turing and Turing Machine. In: The Universal Turing Machine: A Half–Century Survey, R. Herken (ed.), Springer–Verlag Wien, New York, 1994, pp. 1–13
9. Indyk, P.: Optimal Simulation of Automata by Neural Nets. Proc. of the 12th Annual Symp. on Theoretical Aspects of Computer Science STACS'95, LNCS Vol. 900, pp. 337–348, 1995
10. Maass, W.: Networks of Spiking Neurons: The Third Generation of Neural Network Models. NeuroCOLT Technical Report Series NC–TR–96–045, TU Graz, May 1996, 22 p.
11. Maass, W. — Orponen, P.: On the Effect of Analog Noise in Discrete– Time Analog Computing. Manuscript, 1996
12. Maass, W. — Ruf, B.: On Computation with Pulses. A manuscript, 1997

13. McCulloch, W. S. — Pitts, W. H.: A logical calculus of ideas immanent in nervous activity. Bull. of Math. Biophysics, 5:115, 1943

14. Parberry, Ian: Circuit Complexity and Neural Networks. The MIT Press, Cambridge, Massachusetts, London, England, 1994, 270 p., ISBN 0–262–16148–6

15. Siegelmann, H.T.: Recurrent Neural Networks. In: Computer Science Today — Recent Trends and Developments (J. van Leeuwen, ed.), LNCS Vol. 1000, Springer Verlag, Berlin, 1995, pp. 29–45

16. Siegelmann, H. T. — Sonntag, E.D.: Analog Computation via Neural Networks. *Theoretical Computer Science*, 131, 1994, pp. 331–360

17. Siegelmann, H. T. — Sonntag, E.D.: On Computational Power of Neural Networks. *J. Comput. Syst. Sci.*, Vol. 50, No. 1, 1995, pp. 132–150

18. Šíma. J. — Wiedermann, J.: Neural Language Acceptors. In: Developments in Language Theory, Proc. of the Second International Conference, Magdeburg, June 1995, World Scientific Publishing Co.

19. Šíma, J. — Wiedermann, J.: Theory of Neuromata. ICS Technical Report 15/95, ICS AS CR Prague, submitted for publication

20. Turing, A.M.: Computing Machinery and Intelligence. *Mind*, Vol. **59**, 1950, p. 433–460

21. Valiant, L.: Functionality in Neural Nets. Proc. 7th Nat. Conf. on Art. Intelligence, AAAI, Morgan Kaufmann, San Mateo, CA, 1988, pp. 629–634

22. Valiant, L.G.: Circuits of the Mind. Oxford University Press, New York, Oxford, 1994, 237 p., ISBN 0–19–508936–X

23. Valiant, L.G.: Rationality. In: Proc. 8th Ann. Conference on Computational Learning Theory COLT'95, Santa Cruz, California, ACM Press, 1995, p. 3–14

24. Valiant, L.G.: Cognitive Computation (Extended Abstract). Proc. 38th IEEE Symp. on Fond. of Comp. Sci., IEEE Press, 195, p. 2–3

25. Valiant, L.G.: A Neuroidal Architecture for Cognitive Computation. Harvard University, Cambridge, MA, TR–11–96, November 1996, 29 pp.

26. Wegner, P.: Tutorial Notes: Models and Paradigms of Interaction. See Peter Wegner's Home Page, 1995.

27. Wegner, P.: Why Interaction is More Powerful Than Algorithms. Communication of the ACM, Vol. 40, No. 5, May 1997, p. 80–91

28. Wiedermann, J.: Complexity Issues in Discrete Neurocomputing. Neural Network World, 4, 1994, pp. 99–119

29. Wiedermann, J.: Parallel Machine Models: How They Are and Where They Are Going. In: Proc. 22nd Seminar on Current Trends in Theory and Practice of Informatics SOFSEM'95, LNCS Vol. 1012, Springer Verlag, Berlin, 1995, pp. 1–30

30. Wiedermann, J.: The Cogitoid: A Computational Model of Mind. Technical Report No. V–685, Institute of Computer Science, September 1996, 17 pp.

31. Wiedermann, J.: Towards Algorithmic Explanation of Mind Evolution and Functioning. In preparation, 1997.

# Computational Complexity of Continuous Problems

Henryk Woźniakowski

Columbia University, U.S.A., and University of Warsaw, Poland

**Abstract.** Computational complexity studies the intrinsic difficulty of solving mathematically posed problems. Discrete computational complexity studies discrete problems and often uses the Turing machine model of computation. Continuous computational complexity studies continuous problems and tends to use the real number model.

Continuous computational complexity may be split into two branches. The first deals with problems for which the information is *complete*. Informally, information may be complete for problems which are specified by a finite number of inputs. Examples include matrix multiplication, solving linear systems or systems of polynomial equations. We mention two specific results. The first is for matrix multiplication of two real $n \times n$ matrices. The trivial lower bound on the complexity is of order $n^2$, whereas the best known upper bound is of order $n^{2.376}$ as proven by D. Coppersmith and S. Winograd. The actual complexity of matrix multiplication is still unknown. The second result is for the problem of deciding whether a system of $n$ real polynomials of degree 4 has a real root. This problem is NP-complete over the reals as proven by L. Blum, M. Shub and S. Smale.

The other branch of continuous computational complexity is IBC, *information-based complexity*. Typically, IBC studies infinite-dimensional problems for which the input is an element of an infinite-dimensional space. Examples of such inputs include multivariate functions on the reals. Information is often given as function values at finitely many points. Therefore information is *partial* and the original problem can be solved only *approximately*. The goal of IBC is to compute such an approximation as inexpensively as possible. The error and the cost of approximation can be defined in different settings including the worst case, average case, probabilistic, randomized and mixed settings.

In the second part of the talk we concentrate on multivariate problems. By a multivariate problem we mean an approximation of a linear or nonlinear operator defined on functions of $d$ variables. We wish to compute an $\varepsilon$-approximation with minimal cost. We are particularly interested in large $d$ and/or in large $1/\varepsilon$. Typical examples of such problems are multivariate integration and approximation as well as multivariate integral equations and global optimization.

Many multivariate problems are *intractable* in the worst case deterministic setting, i.e., their complexity grows exponentially with the number $d$ of variables. This is sometimes called the "curse of dimension". This holds for multivariate integration for the Korobov class of functions as proven in our recent paper with Ian Sloan.

The exponential dependence on dimension $d$ is a complexity result and one cannot get around it by designing clever algorithms. To break the curse of dimension of the worst case deterministic setting we have to settle for a weaker assurance. One way is to settle for a *randomized* setting or *average* case setting.

In the randomized setting, it is well known that the classical Monte Carlo algorithm breaks the curse of dimension for multivariate integration. However, there are problems which suffer the curse of dimension also in the randomized setting. An example is provided by multivariate approximation,

In the average case setting, the curse of dimension is broken for multivariate integration independently of what is a probability measure on the class of functions. However, in general, the proof is not constructive. For the Wiener sheet measure, the proof is constructive and we know almost optimal algorithms. For multivariate approximation, the curse of dimension is broken only for some probability measures. For instance, it is broken for the Wiener sheet measure, and it is not broken for the isotropic Wiener measure as proven by G. Wasilkowski.

One of the major problems of IBC is to characterize multivariate problems for which the curse of dimension is broken in the randomized or average case setting. Another problem is to characterize which multivariate problems are tractable or strongly tractable in various settings. That is, for which multivariate problems complexity depends polynomially on $d$ and $1/\varepsilon$ (tractability) or polynomially only on $1/\varepsilon$ (strong tractability).

# Path Layout in ATM Networks

Shmuel Zaks
(email: **zaks**@cs.technion.ac.il)

Department of Computer Science
Technion, Haifa 32000, Israel

**Abstract.** This paper surveys recent results in the area of virtual path layout in ATM networks. We focus on the one-to-all (or broadcast) and the all-to-all problems. We present a model for theoretical studies of these layouts, which amounts to covering the network with simple paths, under various constraints. The constraints are the hop count (the number of paths traversed between the vertices that have to communicate), the load (the number of paths that share an edge or a vertex), and the stretch factor (the total length traversed between pairs of vertices, comparing with the distance between them). The results include recursive constructions, greedy and dynamic programming algorithms, lower bounds proofs, and NP-complete results.

## 1 Introduction

### 1.1 Motivation

The advent of fiber optic media has dramatically changed the classical views on the role and structure of digital communication networks. Specifically, the sharp distinction between telephone networks, cable television networks, and computer networks, has been replaced by a unified approach.

The most prevalent solution for this new network challenge is called *Asynchronous Transfer Mode* (ATM for short), and is thoroughly described in the literature [29, 27, 34]. ATM is based on relatively small fixed-size packets termed *cells*. Each cell is routed independently, based on two small routing fields at the cell header, called *virtual channel index* (VCI) and *virtual path index* (VPI). At each intermediate switch, these fields serve as indices to two routing tables (the VCI serves as an index to one table and the VPI to the other), and the routing is done in accordance to the predetermined information in the appropriate entries.

Routing in ATM is hierarchical in the sense that the VCI of a cell is ignored as long as its VPI is not null. This algorithm effectively creates two types of predetermined simple routes in the network - namely routes which are based on VPIs (called *virtual paths* or VP s) and routes based on VCIs and VPIs (called *virtual channels* or VCs). VCs are used for connecting network users (e.g., a telephone call); VP s are used for simplifying network management — routing of VCs in particular. Thus the route of a VC may be viewed as a concatenation of complete VP s.

As far as the mathematical model is concerned, given a communication network, the VP s form a virtual network on top of the physical one which we term the *virtual path layout* (VPL for short), on the same vertices, but with a different set of edges (typically a superset of the original edges). Each VC is a simple path in this virtual network.

The VP layout must satisfy certain conditions to guarantee important performance aspects of the network (see [1, 26] for technical justification of the model for ATM networks). In particular, there are restrictions on the following parameters:

**The load:** The number of virtual edges or vertices that share any physical edge. This number determines the size of the VP routing tables, since at each incoming port which a VP goes through, a separate entry is allocated for routing cells that belong to the VP (see [12] for a detailed description of the routing mechanism in ATM). As the ATM standard [29] limits the maximum size of VP routing tables to 4096 entries, this resource is critical in networks with a few hundreds of vertices (see [21] for a justification).

**The hop count:** The number of VPs which comprise the path of a VC in the virtual graph. This parameter determines the efficiency of the setup of a VC since the routing tables at the end of each VP must be updated to support the new VC. The importance of a low hop count to the efficiency of the network is very high, especially for data applications [8, 36, 37].

**The stretch factor:** The ratio between the length of the path that a VC takes in the physical graph and the shortest possible path between its endpoints — this parameter controls the efficiency of the utilization of the network. Most result in this paper deal with a stretch factor of one (that is, the connection is done along shortest paths).

In many works (e.g., [3, 4, 26, 10]), a general routing problem is solved using a simpler sub-problem as a building block; In this sub-problem it is required to enable routing between all vertices to a single vertex (rather than between any pair of vertices). This restricted problem for the ATM VP layout problem is termed the *rooted (or one-to-all) VPL* problem [26].

This paper surveys some recent results, for the one-to-all and the all-to-all problems. The results are usually of two types: either bounding the (maximal or average) load $\mathcal{L}$ as a function of the other parameters (usually the number of vertices $N$ and a bound $\mathcal{H}$ on the hop count), or bounding the maximal hop count $\mathcal{H}$ as a function of the other parameters (usually the number of vertices $N$ and a bound $\mathcal{L}$ on the maximal load. For this latter problem we comment that, for the one-to-all problem this is like determining the radius of the graph, formed by the virtual paths, that has to be embedded in the physical network, while for the all-to-all problem this is like determining the diameter of this graph.

## 1.2  Related Works

A few works have tackled the VP layout problem, some using empirical techniques [1, 33], and some using theoretical analysis [26, 10]; However, none of these works has attempted to combinatorially characterize the optimal solution, and achieve a tight upper bound for the problem. In addition, most of these works have considered only one of the relevant performance measures, namely the worst case load measure. Of particular practical interest is the weighted hop count measure, since it determines the expected time for setting up a connection between a pair of users, given the relative frequency of connection requests between network vertices. A similar problem was empirically handled in [24].

The VP layout problem is closely related to graph-embedding problems since in both cases it is required to embed one graph in another graph. However, while in most embedding problems both graphs are given, here we are given only the physical (host) graph, and we can *choose* the embedded graph (in addition to the choice of the embedding itself).

Most of the performance parameters are also different between these cases:

- While the association between the host graph and the embedded graph is made by the *dilation* parameter in embedding problems, here it is made by the *stretch factor*. In other words, in embedding problems it is important to minimize the length of each individual embedded edge, while in this model it is important to minimize the length of paths.
- The *hop count* parameter is closely related to the *distance* in the virtual graph, however, while the distance depends only on one graph, the hop count also depends on the physical graph (unless the factor is unbounded).
- The *edge load* parameter is identical to the *congestion* in embedding problems, and the different terminology is due to the loaded meaning of congestion in the communication literature.

These differences have a significant impact on the techniques and results in this model.

Another problem which is related to ours is that of keeping small routing tables for routing in conventional computer networks. This problem was widely studied [3, 4, 19, 20, 30, 31, 35] and yielded interesting graph decompositions and structures, but it differs from ours in some major aspects which deemed most of these solutions impractical for our purposes. The main difference stems from the fact that in our case there is no flexibility as to the routing scheme itself since it is determined by the ATM standard [29].

A related criterion, minimizing the *number* of VPs to achieve a required maximal hop count, is discussed in [7] mainly for database optimization purposes, and yields very different results. For ATM, however, this criterion is of lesser importance since no such global constraint exists.

The notion of *forwarding index* was studied for communication networks (see, e.g., [9, 28]). The forwarding index in a given network with a given set of paths corresponds to the load in our case; in [9] the authors study load on vertices, while in [28] they study both load on vertices and on edges. However, these studies (that deal with the all-to-all problem) do not consider the notion of hop count.

## 2 The Model

We model the underlying communication network as an undirected graph $G = (V, E)$, where the set $V$ of vertices corresponds to the set of switches, and the set $E$ of edges to the set of physical links between them. In addition, we are given a set $\zeta$ of pairs of distinct nodes in $V$, between which a communication has to be established. (A system $(G = (V, E), \zeta)$ was termed as *connection network* in [14, 15].) We concentrate on two extreme cases:

- The *one-to-all* case: in this case a connection is required from one specified vertex to all others; namely, $\zeta = \{(r, u)|u \in V, u \neq r\}$, and $r$ is the specified vertex (usually termed the *root*).
- The *all-to-all* case: in this case a connection is required between all pairs of vertices; namely, $\zeta = \{(v, u)|v, u \in V, u \neq v\}$.

In the following definitions, the network is $G$, and the set $\zeta$ corresponds to either of the above two cases.

**Definition 1.** A *virtual path layout (VPL for short)* $\Psi$ is a collection of simple paths in $G$, termed *virtual paths* (VP s for short).

**Definition 2.** The *load* $\mathcal{L}(e)$ of an edge $e \in E$ in a VPL $\Psi$ is the number of VP s $\psi \in \Psi$ that include $e$. This notion is also referred to as its *congestion*, or *cutwidth* (as used in other references). The *load* $\mathcal{L}(v)$ of a vertex $v \in V$ in a VPL $\Psi$ is the number of VP s $\psi \in \Psi$ that include $v$.

**Definition 3.** The *maximal edge load* $\mathcal{L}_{max}(\Psi)$ of a VPL $\Psi$ is $\max_{e \in E} \mathcal{L}(e)$. The *maximal vertex load* of a VPL $\Psi$ is $\max_{v \in V} \mathcal{L}(e)$. Unless otherwise specified, the loads referred to in this paper are edge loads.

**Definition 4.** The *average (edge) load* of a VPL $\Psi$ is $\mathcal{L}_{avg}(\Psi) \equiv \frac{1}{|E|} \sum_{e \in E} \mathcal{L}(e)$.

**Definition 5.** The *hop count* $\mathcal{H}(u, v)$ between two vertices $u, v \in V$ in a VPL $\Psi$ is the minimum number of VP s whose concatenation forms a path in $G$ connecting $u$ and $v$. If no such VP s exist, define $\mathcal{H}(u, v) \equiv \infty$.

**Definition 6.** The *maximal hop count* of a VPL $\Psi$ is

$$\mathcal{H}_{max}(\Psi) \equiv \max_{(u,v) \in \zeta} \{\mathcal{H}(u, v)\}.$$

**Definition 7.** Let $\psi = (u, v)$ be a VP. Then the *dilation* of $\psi$, denoted $|\psi|$, is the number of physical links that $\psi$ traverses. Let $\Psi$ be a VPL, then the *total load* of $\Psi$ is $\mathcal{L}_{tot}(\Psi) \equiv \sum_{\psi \in \Psi} |\psi|$.

In both problems, in order to minimize the load, one can use a VPL $\Psi$ which has a VP on each physical link, i.e., $\mathcal{L}_{max}(\Psi) = 1$, however such a layout has a large $(O(N))$ hop count. The other extreme is connecting a direct VP from the root to each other vertex, yielding $\mathcal{H}_{max} = 1$ but a large load $(O(N)$ for the one-to-all problem and $O(N^2)$ for the all-to-all problem). This paper discusses recent results for intermediate cases, which suggest trade-offs between these two extreme cases, for various network topologies.

# 3  Summary of Results

In this section we briefly summarize recent results in this area. All of them use the model presented in Section 2, that was presented in [11, 26]. Some of the results are described in more details in Section 4.

We discuss the problems of one-to-all and all-to-all layouts. As will become apparent from the discussion, these problems are closely related to the radius and the diameter of the graph, that represents the virtual paths, that we are embedding within our physical network (see especially the discussion in Section 4).

Unless otherwise specified, $N$ denotes the number of vertices in a given network, and $\Delta$ denotes the maximum degree of a vertex in it; in addition, $\mathcal{H}$ denoted a given bound for the hop count and $\mathcal{L}$ denotes a given bound on the load.

## 3.1  Chain Networks

1. One-to-all (or: broadcast)
   (a) Constructions with a load bounded by $\mathcal{H} \cdot N^{\frac{1}{\mathcal{H}}}$ are presented in [23].
   (b) Optimal constructions for chain networks with stretch factor of one are presented in [25], and for a general stretch factor in [13]. They are described in Sections 4.1.
       In both of these papers the results obtained are symmetric in the load and the hop count; a simple explanation for these symmetries is given in Section 4.1, following [17].

2. All-to-all
   (a) In [32] the authors define $Hops_N(\mathcal{L})$ as the maximal number of hops connecting any pair of vertices, in a chain (with $N$ vertices, where the load is bounded by $\mathcal{L}$). They prove that $\sqrt{2N} - 5 < Hops_N(2) < \sqrt{2N} + 2$, and that $\frac{1}{2}N^{\frac{1}{\mathcal{L}}} < Hops_N(\mathcal{L}) < \mathcal{L} \cdot N^{\frac{1}{\mathcal{L}}}$, for any $\mathcal{L} \geq 1$.
   (b) In [13] the all-to-all path layouts is studied using a geometric approach. For a detailed discussion, see Section 4.2. Stated in graph-theoretic terms, these layouts are translated into embeddings (or linear arrangements) of the vertices of a graph with $N$ vertices onto the points $1, 2, \cdots, N$ of the $x$-axis. The authors look for a graph with minimum diameter $D_{\mathcal{L}}(N)$, for which such an embedding is possible, given a bound $\mathcal{L}$ on

the cutwidth of the embedding. The results can be summarized as follows: For an $N$-vertex chain and for every $\mathcal{L} \geq 1$, $D_{\mathcal{L}}(N)$ is bounded above by $(\mathcal{L}! \cdot N)^{1/\mathcal{L}} + 2$, and is bounded below by $max\{\frac{1}{2}[(\mathcal{L}! \cdot N)^{1/\mathcal{L}} - \mathcal{L}], \frac{1}{2}[N^{1/\mathcal{L}} - 1], \frac{\log N}{\log(2\mathcal{L}+1)}\}$. The relation of these results to those in [32, 2, 38] are summarized in Section 4.2.

(c) In [38] it is proved that the hop count is $\Theta(\frac{\log N}{\log \mathcal{L}})$, if $\mathcal{L} \geq \log^{1+\epsilon} N$, for any fixed $\epsilon > 0$.

## 3.2 Ring Networks

1. In [2] the authors study *chordal rings* ((a ring network enhanced by adding non-crossing edges (*chords*), *express rings* (chordal rings in which the chords are oriented either clockwise or counterclockwise), and *multi rings* (in which subsidiary rings are appended to edges of a ring and, recursively, to edges of appended subrings. The authors first demonstrate the topological equivalence of these structures. They then show that for every $N$ and $\mathcal{L}$, there exists an $N$-vertex express ring (with load bounded by $\mathcal{L}$) and whose diameter is bounded by $2^{-\frac{1}{\mathcal{L}}} \cdot \mathcal{L} \cdot N^{\frac{1}{\mathcal{L}}} + 1$, and that there is no construction with diameter smaller than $\frac{\mathcal{L}}{2 \cdot e} N^{\frac{1}{\mathcal{L}}}$. The discussion involves geometric considerations that are similar to those later used in [13] to obtain optimal constructions ( see Section 4.2). The relation between the diameter and the hop count is clear; see also [13] for more discussion and results. It is also shown that the insistence that the arcs in an express ring be non-crossing at most doubles the diameter of the augmented ring.

2. In [13] the all-to-all path layouts is studied for augmented ring and augmented path networks (for more details, see Section 4.2).

## 3.3 Tree Networks

1. In [26] the authors describe a greedy algorithm to determine the existence of a one-to-all layout in tree networks, which, for a given bound on the hop count, determines a layout with smallest possible vertex load (for more algorithms of related problems, see also [21, 23]). This approach proved to be quite helpful for such constructions of virtual layouts.

2. In [23] the authors first describe a recursive construction for a path layout for the one-to-all problem with the load bounded by $\mathcal{H} \cdot N^{\frac{1}{\mathcal{H}}}$. A lower bound of $\Omega(\frac{1}{\Delta \cdot 2^{\frac{1}{\mathcal{H}}}} \cdot N^{\frac{1}{\mathcal{H}}})$ is also proved, which proves that for "realistic" networks the load is $\Theta(N^{\frac{1}{\mathcal{H}}})$.

For the all-to-all problem an upper bound of $\frac{\mathcal{H}}{2(2^{\frac{2}{\mathcal{H}}}-1)} N^{\frac{2}{\mathcal{H}}}$ and a lower bound of $\frac{1}{\Delta \cdot (8\mathcal{H})^{\frac{2}{\mathcal{H}}}} N^{\frac{2}{\mathcal{H}}}$ are shown.

## 3.4  Mesh Networks

1. In [23] the authors study constructions for the all-to-all problem on an $a \times b$ mesh. They show a construction that achieves a load of $a^{\frac{2}{h_a}}$, where $h_a = \frac{\mathcal{H}}{\frac{\log x}{\log y}+1}$. For an $\sqrt{N} \times \sqrt{N}$ mesh this implies a bound of $\frac{\mathcal{H}}{4(2^{\frac{4}{\mathcal{H}}}-1)}N^{\frac{2}{\mathcal{H}}}$.

2. In [32] the authors study the all-to-all problem. They define $Hops_{a\times b}(\mathcal{L})$ to be the maximal number of hops connecting any pair of vertices, in an $a \times b$ mesh. They show that $Hops_{k\times n}(\mathcal{L}) = \Theta(n^{\frac{1}{k\mathcal{L}}})$, where $k$ and $\mathcal{L}$ are constant, and that $Hops_{n\times n}(\mathcal{L}) = \Theta(\log n)$.

3. In [38] the authors show that the number of hops for the $\sqrt{N} \times \sqrt{N}$ mesh is $\Theta(\frac{\log N}{\log \mathcal{L}})$ for $\mathcal{L} \geq 2$, for the all-to-all problem.

4. In [5] a lower bound of $\Omega((\frac{N}{\mathcal{H}})^{\frac{1}{\mathcal{H}}})$ is shown for the one-to-all problem, and a construction with an upper bound of $\mathcal{H} \cdot N^{\frac{1}{\mathcal{H}}}$ for the load is shown. For the all-to-all problem they present a construction with load bounded by $2 \cdot (\frac{\mathcal{H}}{3}+1) \cdot N^{\frac{3}{2\mathcal{H}}}$.

## 3.5  General Networks

1. All-to-all
   (a) In [26] the authors describe a recursive construction, that, for a given $k$, yields a layout with a stretch factor bounded by $8k$, and a vertex load bounded by $O(\mathcal{H} \cdot k \cdot \log N \cdot N^{\frac{1}{k}+\frac{2}{\mathcal{H}}})$. The technique uses ideas from [4], which seem to be quite useful in these layout designs.
   (b) In [32] the authors define $Hops_N(\mathcal{L})$ to be the maximal number of hops connecting any pair of vertices (for a network with $N$ vertices and a bound $\mathcal{L}$ for the load). They show that $Hops_N(\mathcal{L}) > \frac{\log N}{\log(\Delta\mathcal{L})} - 1$, for any $\mathcal{L} \geq 1$ (same bound is presented in [38]).
   (c) In [38] the authors construct a virtual path layout with hop count $O(\frac{diam(G)\cdot \log \Delta}{\log \mathcal{L}})$, where $diam(G)$ is the diameter of the network $G$, and $\Delta \geq 3$. In the case of unbounded degree networks with diameter $O(\log N)$ these hop numbers are optimal for any $c \geq \Delta$. For any $\mathcal{L} \geq 1$ and bounded degree network with diameter $O(\log N)$ they show a construction with the hop count of $\Theta(\frac{\log N}{\log \mathcal{L}})$.
   (d) In [23] the authors describe a recursive construction for graphs with bounded treewidth (see [6]). They show a construction with load bounded by $O(\frac{k \cdot \mathcal{H} \cdot N^{\frac{2}{\mathcal{H}}}}{2((1.5)^{\frac{2}{\mathcal{H}}}-1)})$, where $k$ is the bound on the treewidth.

2. One-to-all
   (a) In [23] a recursive construction is presented, with a bound of $\sqrt{\mathcal{H}} N^{1+\frac{1}{\mathcal{H}}}$ for the maximal load.
   (b) Decision Problems
       In [16] the complexity of deciding the existence of layouts of one-to-all virtual paths which have maximum hop count $\mathcal{H}$ and maximum (edge)

load $\mathcal{L}$ , for a stretch factor of one, was studied. It was proved that the problem of determining the existence of such layouts is NP-complete for every given values of $\mathcal{H}$ and $\mathcal{L}$, except for the cases $\mathcal{H} = 2, \mathcal{L} = 1$ and $\mathcal{H} = 1$, any $\mathcal{L}$, for which the authors give polynomial-time layout constructions, based on network flow algorithms.

In [26] it was shown that determining the existence of a one-to-all layout of virtual paths, which has maximum hop count $\mathcal{H}$ and maximum vertex load $\mathcal{L}$ , for an unbounded stretch factor, is NP-complete.

3. In [24] dynamic maintenance of virtual path layout was discussed. The authors describe methods to adjust the layout of these paths to the dynamics of changes in the usage of the networks by its end-users.

4. It is certainly of interest to consider problems which are not the one-to-all or all-to-all. An interesting result, which gives a lower bound on the load, and also uses the parameter of the total number of pairs to be connected, was recently developed in [5]. The lower bound is $\frac{1}{\Delta}(\frac{\Delta}{4\mathcal{H}}(\frac{N_C}{|C|} - \frac{4}{\Delta}))^{\frac{1}{\mathcal{H}}}$, where $C$ is a cut in the network and $N_C$ is the number of pairs between which a communication is required and they are separated by this cut. For the one-to-all problem, this gives a lower bound of $\frac{1}{\Delta}(\frac{N-5}{4\cdot\mathcal{H}})^{\frac{1}{\mathcal{H}}}$.

### 3.6 Miscellaneous

1. In [22] the problem of path layout is considered, where the vertices are of three types: those that can switch only virtual paths, those that can switch only virtual channels, and those that can switch both. Few solutions are presented; among them is a greedy algorithm, which optimizes the network overhead for a request/response and the utilization of bandwidth and routing table resources.

2. In [18] the authors present a measure of maximum vertex load (for more results on this notion, see also [21]), which the total number of virtual paths that end at this vertex. They show that determining the existence of a one-to-all layout for a given network, with a vertex load bounded by $\mathcal{L}$ and a hop count bounded by $\mathcal{H}$, is NP-complete for any $\mathcal{H}$ and $\mathcal{L}$, except for the cases $\mathcal{H} = 1$, any $\mathcal{L}$ and $\mathcal{H} = 2, \mathcal{L} = 1$ (the proof techniques is borrowed from [16]). Specific bounds are given for chain, mesh and torus networks.

3. Other specific graphs are also mentioned in the literature, but the area is far from being fully explored. For example, the Hypercube and the de-Bruijn graphs are studied in [38].

## 4 Detailed Description

In this section we present in some more details the structure of optimal layouts for the one-to-all problem on a chain network, for either stretch factor one or for a general stretch factor, which are optimal for the various measures. We then discuss the use of geometry in deriving optimal bounds for the all-to-all problem in chains, augmented paths and ring networks.

## 4.1 One-to-All Problem on Chain Networks

This simple topology, which is also quite practical, admits a precise treatment, as summarized in this section.

We consider four performance measures, and achieve optimal solutions for each measure:

- Given an upper bound on the maximum hop count, minimize the maximum load ($\mathcal{L}_{max}$),
- Given an upper bound on the maximum load, minimize the maximum hop count ($\mathcal{H}_{max}$),
- Given an upper bound on the maximum hop count, minimize the average load ($\mathcal{L}_{avg}$), and
- Given an upper bound on the maximum load, and vertex weights (representing the frequency of connection requests between the vertex and the root), minimize the average hop count ($\mathcal{H}^w_{avg}$).

All these measures have practical implications in different cases: As the hop count is proportional to the setup time of a new connection, the worst case hop measure represents hard deadlines for this overhead (typical to real time applications), while the average hops measure is useful for general purpose networks. Since the load represents the utilization of the routing tables, maximum load is important in cases where the layout is large and may overflow the limited space of routing entries (4096 per routing table [29]), whereas average load measures are relevant for general purpose networks, in which many independent layouts coexist and try to minimize local bottlenecks at any location in the network.

The following discussion presents results that use shortest path layouts. These results are from [25]. At the end of the section we mention the extensions of these results, that yield exact layouts for the case of a general stretch factor. $N$ denotes the number of vertices in the chain, and the layout is from the leftmost vertex to all others (the extension for the case where the layout is to all vertices on the left and right is trivial; the extensions [13, 17] clearly apply to this case explicitly).

We first establish a canonic form of an VPL, which will simplify the rest of the discussion.

**Lemma 8.** *Given a chain network, for every optimality measure ($\mathcal{L}_{max}$, $\mathcal{L}_{avg}$, $\mathcal{H}_{max}$, or $\mathcal{H}^w_{avg}$) there exists an optimal VPL in which every vertex $i \geq 2$ is the right-most endpoint of a single VP. In other words, this VPL induces a tree rooted at vertex 1 with the VPs corresponding to tree edges.*

**Definition 9.** *Let $l_1 < l_2$. Two VPs denoted $(l_1, r_1)$ and $(l_2, r_2)$ constitute a crossing if $l_1 < l_2 < r_1 < r_2$. A VPL is called crossing-free if no pair of VPs constitute a crossing.*

**Theorem 10.** *For each performance measure ($\mathcal{L}_{max}$, $\mathcal{H}_{max}$, $\mathcal{L}_{avg}$, and $\mathcal{H}^w_{avg}$) there exists an optimal VPL which is crossing-free.*

INDUCEVPL($T$): Induce an VPL according to a tree $T$ with $N$ vertices.

1. Label the vertices of $T$ in depth-first order. Let $\lambda(u)$ be the label of a vertex $u \in T$, $1 \leq \lambda(u) \leq N$.
2. For every edge $(u, v) \in T$ connect a VP between $\lambda(u)$ and $\lambda(v)$.
3. Return $\Psi_T$, the collection of generated VP s.

**Fig. 1.** Procedure INDUCEVPL($T$).

In Lemma 8 we showed that an VPL induces a tree. The next lemma shows that the converse holds too, namely, any tree induces an VPL.

**Lemma 11.** *Let $T$ be an ordered tree. Then procedure INDUCEVPL($T$) (see Figure 1 for the pseudo-code, Figure 2 for an example) induces a crossing-free VPL.*

We now consider optimal VPL layouts for the worst-case (maximal) load and hop count measures. Specifically, if the load is required to be $\mathcal{L}_{max} \leq \mathcal{L}$ we characterize the layout with the minimal worst case hop count, and if the hop count is $\mathcal{H}_{max} \leq \mathcal{H}$ we characterize the layout with the minimal worst-case load. This is done using a new class of trees $\mathcal{T}(\mathcal{L}, \mathcal{H})$ (see next definition). These trees contain all VPL s on a chain that satisfy the above load and hop constraints.

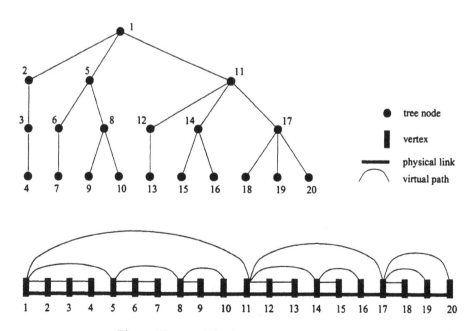

**Fig. 2.** The tree $\mathcal{T}(3,3)$ and its induced VPL.

**Definition 12.** The ordered tree $T(\mathcal{L}, \mathcal{H})$ is defined recursively as follows. The root $r$ has $\mathcal{L}$ children. The $i^{\text{th}}$ child from the left is the root of a $T(i, \mathcal{H} - 1)$ subtree, for $1 \leq i \leq \mathcal{L}$. A tree $T(\mathcal{L}, 0)$ or $T(0, \mathcal{H})$ is a single vertex (see Figure 2 for an example).

**Remarks:** An internal vertex of $T(\mathcal{L}, \mathcal{H})$, which is the $i^{\text{th}}$ child (from the left) of its parent, has $i$ children. The tree $T(1, \mathcal{H})$ is a rooted chain of $\mathcal{H} + 1$ vertices. Note also that $T(\mathcal{L}, \mathcal{H})$ has height $\mathcal{H}$ and maximum degree $\mathcal{L}$.

**Lemma 13.** *The tree* $T(\mathcal{L}, \mathcal{H})$ *contains* $\binom{\mathcal{L} + \mathcal{H}}{\mathcal{H}}$ *vertices.*

**Definition 14.** An ordered tree $T$ is *subsumed* in $T(\mathcal{L}, \mathcal{H})$ if its root is subsumed in the root of $T(\mathcal{L}, \mathcal{H})$ and the subtrees of the root's children in $T$ are (recursively) subsumed in the subtrees of a subset of the children of the root in $T(\mathcal{L}, \mathcal{H})$.

It is easy to see that the VPL of a subsumed tree $T$ has lower load and hop counts than that of the tree $T(\mathcal{L}, \mathcal{H})$ it is subsumed in, since $T$ may be obtained from $T(\mathcal{L}, \mathcal{H})$ by deleting subtrees. The next corollary follows.

**Corollary 15.** *Let $T$ be an ordered tree that is subsumed in $T(\mathcal{L}, \mathcal{H})$, and let $\Psi_T$ be the output $\Psi_T = $ INDUCEVPL$(T)$. Then $\mathcal{L}_{max}(\Psi_T) \leq \mathcal{L}$ and $\mathcal{H}_{max}(\Psi_T) \leq \mathcal{H}$.*

**Lemma 16.** *For every crossing-free VPL $\Psi$, with $\mathcal{L}_{max}(\Psi) \leq \mathcal{L}$ and $\mathcal{H}_{max}(\Psi) \leq \mathcal{H}$ there exists a tree $T$ which is subsumed in $T(\mathcal{L}, \mathcal{H})$ such that $\Psi = $ INDUCEVPL$(T)$.*

**Theorem 17.** *Given $N$ and $\mathcal{L}$, Let $\mathcal{H}$ be such that*

$$\binom{\mathcal{L} + \mathcal{H} - 1}{\mathcal{L}} < N \leq \binom{\mathcal{L} + \mathcal{H}}{\mathcal{L}}.$$

*Then $\mathcal{H}_{opt}(N, \mathcal{L}) = \mathcal{H}$.*

A similar result holds for the maximum load measure:

**Theorem 18.** *Given $N$ and $\mathcal{H}$, let $\mathcal{L}$ be such that*

$$\binom{\mathcal{L} + \mathcal{H} - 1}{\mathcal{H}} < n \leq \binom{\mathcal{L} + \mathcal{H}}{\mathcal{H}}.$$

*Then $\mathcal{L}_{opt}(N, \mathcal{H}) = \mathcal{L}$.*

Given a chain with $N = N(\mathcal{L}, \mathcal{H})$ there exists a unique VPL with $\mathcal{L}_{max}(\Psi) = \mathcal{L}$ and $\mathcal{H}_{max}(\Psi) = \mathcal{H}$, whereas several such VPLs exist for other values of $N$ (i.e., with $N(\mathcal{L}, \mathcal{H} - 1) < N < N(\mathcal{L}, \mathcal{H})$). This symmetry is not a coincidence, as explained in the sequel.

If $N = N(\mathcal{L}, \mathcal{H})$ then $(\mathcal{H}! \cdot N)^{1/\mathcal{H}} - \frac{\mathcal{H}+1}{2} \leq \mathcal{L} \leq (\mathcal{H}! \cdot N)^{1/\mathcal{H}} - 1$. This is an improvement to the upper bound $\mathcal{L} \leq \mathcal{H} \cdot N^{1/\mathcal{H}}$ of [10], since $(\mathcal{H}!)^{1/\mathcal{H}} < \mathcal{H}$.

In [26], a greedy algorithm for finding a VPL for the more general case of tree networks is presented and proven to be optimal with respect to the $\mathcal{H}_{max}$ measure. The algorithm does not give insight into the structure of the obtained VPL, and in particular no upper bound is easily derived from it. When $N = N(\mathcal{L}, \mathcal{H})$ the previous theorem gives a precise characterization of this greedy solution. (The model in [26] differs from ours in that the load is measured on the vertices rather on the edges of the network. Therefore the greedy algorithm should be modified to use the edge-load constraint. We refer in the above to this modified algorithm in the comparison.)

We now turn to discuss the average load. We start with the case where the maximal number of hops is limited to $\mathcal{H}$, and it is required to find the layout with smallest average load. Since a layout $\Psi_{opt}$ that minimizes $\mathcal{L}_{avg}$ also minimizes its total load $\mathcal{L}_{tot}(\Psi_{opt})$, hence the following definition.

**Definition 19.** Let $\mathcal{L}_{tot}(N, \mathcal{H})$ denote the minimal total load of any VPL on $N$ vertices with at most $\mathcal{H}$ hops, namely

$$\mathcal{L}_{tot}(N, \mathcal{H}) \equiv \min_{\Psi}\{\mathcal{L}_{tot}(\Psi) : \mathcal{H}_{max}(\Psi) \le \mathcal{H}\}.$$

The rationale behind our dynamic programming algorithm for finding optimal $\mathcal{L}_{tot}$ ($\mathcal{L}_{avg}$) layouts is the following. Let $\Psi_{opt}$ be the optimal VPL (that achieves $\mathcal{L}_{tot}(\Psi_{opt}) = \mathcal{L}_{tot}(n, \mathcal{H})$). Let $(1, d+1)$ be the longest VP connected to the root. Since by Theorem 10 we can assume that $\Psi_{opt}$ is crossing-free, it follows that no VP of $\Psi_{opt}$ connects a vertex $i \le d$ with a vertex $j > d+1$, thus $\Psi_{opt}$ can be split into two disjoint optimal layouts, one on vertices $1, \ldots, d$ and the other on $d+1, \ldots, N$. However, the second layout (rooted at $d+1$) may use only $\mathcal{H} - 1$ hops since one hop is used to traverse the VP $(1, d+1)$. Thus if $d$ is known, then $\mathcal{L}_{tot}$ satisfies the recurrence

$$\mathcal{L}_{tot}(N, \mathcal{H}) = d + \mathcal{L}_{tot}(d, \mathcal{H}) + \mathcal{L}_{tot}(N - d, \mathcal{H} - 1),$$

so clearly

$$\mathcal{L}_{tot}(N, \mathcal{H}) = \min_{1 \le d \le N-1}\{d + \mathcal{L}_{tot}(d, \mathcal{H}) + \mathcal{L}_{tot}(N - d, \mathcal{H} - 1)\}. \qquad (1)$$

There are two simple "boundary" cases: (i) If $1 \le N \le \mathcal{H} + 1$ then clearly $\mathcal{L}_{tot}(N, \mathcal{H}) = N - 1$, (ii) If $\mathcal{H} = 1$ then we must connect a direct VP to each vertex, so $\mathcal{L}_{tot}(N, 1) = N(N - 1)/2$. The above argument leads to a natural dynamic programming algorithm, with time complexity of $O(N^2 \mathcal{H})$. Moreover, the exact value of $\mathcal{L}_{tot}(N, \mathcal{H})$ can be determined, using the following theorem.

**Theorem 20.** *Let $n$ and $\mathcal{H}$ be given. Let $\mathcal{L}$ be the largest integer such that $N \ge \binom{\mathcal{L} + \mathcal{H}}{\mathcal{L}}$, and let $r = N - \binom{\mathcal{L} + \mathcal{H}}{\mathcal{L}}$. Then*

$$\mathcal{L}_{tot}(N, \mathcal{H}) = \mathcal{H}\binom{\mathcal{L} + \mathcal{H}}{\mathcal{L} - 1} + r(\mathcal{L} + 1).$$

We now turn to the unweighted average hops measure, given a maximum bound $\mathcal{L}$ on the load. This problem can be solved by an algorithm similar to that of the average load.

**Definition 21.** Consider a crossing-free optimal VPL $\Psi_{opt}$ for a chain with $n$ vertices and maximum load $\mathcal{L}$ (which achieves the minimum $\mathcal{H}_{tot}(\Psi)$). Define $\mathcal{H}_{tot}(N, \mathcal{L}) \equiv \mathcal{H}_{tot}(\Psi_{opt})$.

Let $(1, d+1)$ be the longest VP connected to the root. Again, it follows that there exists no VP connecting the vertices $1, \ldots, d$ to the vertices $d+2, \ldots, N$, and thus the layouts in these two segments are disjoint and should both be optimal in $\Psi_{opt}$. In this case however, the layout on the vertices $1, \ldots, d$ should not exceed the load $\mathcal{L} - 1$ (since together with the VP $(1, d)$ the load should not exceed $\mathcal{L}$). By the above discussion, it is evident that

$$\mathcal{H}_{tot}(N, \mathcal{L}) = \min_{1 \leq d \leq N-1} \{\mathcal{H}_{tot}(d, \mathcal{L} - 1) + (N - d) + \mathcal{H}_{tot}(N - d, \mathcal{L})\}. \quad (2)$$

The first and third components of the sum are the values of $\mathcal{H}_{tot}$ in the two separate segments, and the second component is the cost of an additional hop incurred by all vertices in the segment $d+1, \ldots, N$. The boundary cases here are $\mathcal{H}_{tot}(N, 1) = N(N-1)/2$ (if the maximum load is 1 then the only possible VPs are identical to the network edges), and if $N \leq \mathcal{L} + 1$ then $\mathcal{H}_{tot}(N, \mathcal{L}) = N - 1$ (since we can afford to construct direct VPs from all vertices to the root).

A dynamic programming algorithm follows, with time complexity $O(N^2 \mathcal{L})$. Moreover, the exact value of $\mathcal{H}_{tot}(N, \mathcal{L})$ can be determined, using the following theorem.

**Theorem 22.** *Let $N$ and $\mathcal{L}$ be given. Let $\mathcal{H}$ be the maximal such that $N \geq \binom{\mathcal{L}+\mathcal{H}}{\mathcal{H}}$, and let $r = N - \binom{\mathcal{L}+\mathcal{H}}{\mathcal{H}}$. Then*

$$\mathcal{H}_{tot}(N, \mathcal{L}) = \mathcal{L}\binom{\mathcal{L}+\mathcal{H}}{\mathcal{H}-1} + r(\mathcal{H}+1). \quad \square$$

These results are extended for the weighted cases in [25].

Lemma 13 and theorems 20 and 22 suggest a symmetry between $\mathcal{H}$ and $\mathcal{L}$ for the case of chain networks. This symmetry was explored in [13, 17], as briefly explained in the sequel.

The authors show how to transform any VPL with hop count bounded by $\mathcal{H}$ and load bounded by $\mathcal{L}$ into a VPL (its *dual*) with hop count bounded by $\mathcal{L}$ and load bounded by $\mathcal{H}$. In particular, this mapping will transform $T(\mathcal{L}, \mathcal{H})$ into $T(\mathcal{H}, \mathcal{L})$, and vice versa. To show this, they use transformation between any VPL with $x$ edges and binary trees with $x$ vertices (in a binary tree, each internal vertex has a left child and/or a right child). This is done in three steps, as follows.

**Step 1**: Given a planar VPL $\Psi$ we transform it into a binary tree $T = b(\Psi)$,

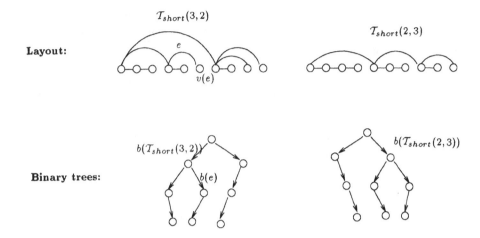

**Fig. 3.** An example of the transformation using binary trees

under which each edge $e$ is mapped to a vertex $b(e)$, as follows. Let $e = (r, v)$ be the edge outgoing the root $r$ to the rightmost vertex (to which there is a VP; we call this a *1-level* edge). This edge $e$ is mapped to the root $b(r)$ of $T$. Remove $e$ from $\Psi$. As a consequence, two VPLs remain: $\Psi_1$ with root $r$ and $\Psi_2$ with root $v$, when their roots are located at the leftmost vertices of both VPLs. Recursively, the left child of vertex $b(e)$ will be $b(\Psi_1)$ and its right child will be $b(\Psi_2)$. If any of the VPLs $\Psi$ is empty, so is its image $b(\Psi)$ (in other words, we can stop when a $\Psi$ that consists of a single edge is mapped to a binary tree that consists of a single vertex).

**Step 2**: Build a binary tree $T'$, which is a reflection of $T$ (that is, we exchange the left child and the right child of each vertex).

**Step 3**: We transform back the binary tree $T'$ into the (unique) VPL $\Psi'$ such that $b(\Psi') = T'$.

An example is depicted in Figure 3, where the layouts for $\mathcal{L} = 2, \mathcal{H} = 3$ and $\mathcal{L} = 3, \mathcal{H} = 2$ are shown, together with the corresponding trees $T(2, 3)$ and $T(3, 2)$, and the corresponding binary trees constructed as explained above. The edge $e$ in the layout $T(3, 2)$ is assigned the vertex $v(e)$ in this layout, and is assigned the vertex $b(e)$ in the corresponding tree $b(T(3, 2))$.

It is shown in [17] that for every $\mathcal{H}$ and $\mathcal{L}$, the trees $b(T(\mathcal{L}, \mathcal{H}))$ and $b(T(\mathcal{H}, \mathcal{L}))$ are reflections of each other. A closer look at these trees reveals more connections, that relate the multiset of the hop counts to the multiset of the loads (for more details see [17]).

This symmetry can be extended for constructing optimal layouts for a chain network, with a general stretch factor. For this, a new families of trees is needed, and the symmetry can be explored using ternary (rather than binary) trees.

Moreover, these correspondences immediately suggest optimal constructions for optimizing the average hop count or load. For more details on these correspondences and symmetries the reader is referred to [17].

## 4.2 Optimal Constructions Using Geometry

In [13] the all-to-all path layouts is studied using a geometric approach. Stated in graph-theoretic terms, these layouts are translated into embeddings (or linear arrangements) of the vertices of a graph with $N$ vertices onto the points $1, 2, \cdots, N$ of the $x$-axis. The authors look for a graph with minimum diameter $D_{\mathcal{L}}(N)$, for which such an embedding is possible, given a bound $\mathcal{L}$ on the cutwidth of the embedding. They develop a technique to embed the vertices of such graphs into the integral lattice points in the $\mathcal{L}$-dimensional $l_1$-sphere. Using this technique, they show that the minimum diameter $D_{\mathcal{L}}(N)$ satisfies

$$\mathcal{R}_{\mathcal{L}}(N) \leq D_{\mathcal{L}}(N) \leq 2\mathcal{R}_{\mathcal{L}}(N),$$

where $\mathcal{R}_{\mathcal{L}}(N)$ is the minimum radius of a $\mathcal{L}$-dimensional $l_1$-sphere that contains $N$ points. Extensions of the results to augmented paths and ring networks are also presented. For crossing-free augmented paths, they show a tight bound $D_{\mathcal{L}}(N) = 2\mathcal{R}_{\mathcal{L}}(N)$ or $2\mathcal{R}_{\mathcal{L}}(N) - 1$. Using geometric arguments, they derive analytical bounds for $\mathcal{R}_{\mathcal{L}}(N)$. It was shown that:

**Theorem 23.** *For every $\mathcal{L} \geq 1$,*

*1.* $1/2 \cdot (\mathcal{L}! \cdot N)^{\frac{1}{\mathcal{L}}} - \mathcal{L}/2 \leq \mathcal{R}_{\mathcal{L}}(N) \leq 1/2 \cdot (\mathcal{L}! \cdot N)^{\frac{1}{\mathcal{L}}} + 1.$
*2.* $1/2 \cdot N^{\frac{1}{\mathcal{L}}} - 1/2 \leq \mathcal{R}_{\mathcal{L}}(N) \leq \mathcal{L}/2 \cdot N^{\frac{1}{\mathcal{L}}} + 1.$
*3.* $\frac{\log N}{\log(2\mathcal{L}+1)} \leq \mathcal{R}_{\mathcal{L}}(N).$

Using this geometric approach, it was shown that:

**Theorem 24.** *For every $\mathcal{L} \geq 1$,*

$$\mathcal{R}_{\mathcal{L}}(N) \leq D_{\mathcal{L}}(N) \leq 2 \cdot \mathcal{R}_{\mathcal{L}}(N). \tag{3}$$

The results can be summarized as follows:

**Theorem 25.** *For an $N$-vertex chain and for every $\mathcal{L} \geq 1$,*

$$max\{\frac{1}{2}[(\mathcal{L}! \cdot N)^{1/\mathcal{L}} - \mathcal{L}], \frac{1}{2}[N^{1/\mathcal{L}} - 1], \frac{\log N}{\log(2\mathcal{L} + 1)}\} \leq D_{\mathcal{L}}(N) \leq$$

$$(\mathcal{L}! \cdot N)^{1/\mathcal{L}} + 2.$$

A detailed comparison with [32, 38, 2] follows:

– The analytical upper bound of [13] in Theorem 25 improves the one in [32] by a factor of $2e$, and generalizes the one in [38],

- The analytical lower bounds of [13] in Theorem 25 coincide with those of [2] and [32],
- The analytical upper bound of [13] in Theorem 25 for augmented rings improves the one in [2] by a factor of $e$, and
- The analytical lower bound of [13] in Theorem 25 for augmented rings significantly improves the one in [2] for large values of $\mathcal{L}$.

# References

1. S. Ahn, R.P. Tsang, S.R. Tong, and D.H.C. Du. Virtual path layout design on ATM networks. In *IEEE Infocom'94*, pages 192–200, 1994.
2. W. Aiello, S. Bhatt, F. Chung, A. Rosenberg, and R. Sitaraman. Augmented rings networks. *J. Math. Modelling and Scientific Computing*, 1997.
3. B. Awerbuch, A. Bar-Noy, N. Linial, and D. Peleg. Compact distributed data structures for adaptive routing. In *21st Symp. on Theory of Computing*, pages 479–489, 1989.
4. B. Awerbuch and D. Peleg. Routing with polynomial communication-space trade-off. *SIAM Journal on Discrete Math*, 5(2):151–162, May 1992.
5. L. Beccheti, C. Gaibisso, and G. Gambosi. Optimal layouts of virtual paths on a mesh, 1997. Private communication.
6. H.L. Bodlaender. A tourist guide through treewidth. Technical report RUU-CS-92-12, Utrecht university, Dept. of Computer Science, Netherlands, March 1992.
7. H.L. Bodlaender, G. Tel, and N. Santoro. Trade-offs in non-reversing diameter. *Nordic Journal of Computing*, 1:111–134, 1994.
8. J. Burgin and D. Dorman. Broadband ISDN resource management: The role of virtual paths. *IEEE Communicatons Magazine*, 29, 1991.
9. F. Chung, E. Coffman, M. Reiman, and B. Simon. The forwarding index of communication networks. *ieeeinfo*, 33:224–232, 1987.
10. I. Cidon, O. Gerstel, and S. Zaks. A scalable approach to routing in ATM networks. In G. Tel and P.M.B. Vitányi, editors, *The 8th International Workshop on Distributed Algorithms (LNCS 857)*, pages 209–222, Terschelling, The Netherlands, October 1994.
11. I. Cidon, O. Gerstel, and S. Zaks.S. A scalable approach to routing in atm networks. In *Proceedings 8th Intl. Workshop Distributed Algorithms (WDAG)*, pages 209–222, 1994.
12. R. Cohen and A. Segall. Connection management and rerouting in ATM networks. In *IEEE Infocom'94*, pages 184–191, 1994.
13. Y. Dinitz, M. Feighelstein, and S. Zaks. On optimal graphs embedded into paths and rings, with analysis using $l_1$-spheres. In *Proc. of the 23rd International Workshop on Graph-Theoretic Concepts in Computer Science (WG), Berlin, Germany*, jun 1997.
14. Ye. Dinitz. Subset connectivity and edge-disjoint trees, with an application to target broadcasting. Technical Report TR-822, Technion, Haifa, Israel, 1994.
15. Ye. Dinitz, T. Eilam, S. Moran, and S. Zaks. On the total$_k$-diameter of connection networks. In *Proceedings of the fifth Israel Symposium on Theory of Computing and Systems*, pages 96–106, 1997.
16. T. Eilam, M. Flammini, and S. Zaks. A complete characterization of the path layout construction problem for atm networks with given hop count and load. In *Proc. 24th International Colloq. on Automata, Languages, and Programming*, 1997.

17. M. Feighelstein and S. Zaks. Duality in chain atm virtual path layouts. In *sirocco97*, pages 24–26, 1997.
18. M. Flammini, E. Nardelli, and G. Proietti. Atm layouts with bounded hop count and congestion. In *wdag11*, pages 24–26, Saarbrüecken, Germany, September 1997.
19. G.N. Frederickson and R. Janardan. Separator-based strategies for efficient message routing. In *27th Symp. on Foundations of Computer Science*, pages 428–437, 1986.
20. G.N. Frederickson and R. Janardan. Designing networks with compact routing tables. *Algorithmica*, 3:171–190, 1988.
21. O. Gerstel. *Virtual Path Design in ATM Networks*. PhD thesis, Technion, Israel Inst. of Technology, December 1995.
22. O. Gerstel, I. Cidon, and S. Zaks. Efficient support for the client-server paradigm over ATM networks. In *IEEE Infocom'96*, pages 1294–1301, 1996. Submitted to IEEE/ACM Trans. on Networking.
23. O. Gerstel, I. Cidon, and S. Zaks. The layout of virtual paths in ATM networks. *ACM/IEEE Transactions on networking*, 4(6):873–884, December 1996. A preliminary version of this paper appears in [10].
24. O. Gerstel and A. Segall. Dynamic maintenance of the virtual path layout. In *IEEE INFOCOM'95*, April 1995.
25. O. Gerstel, A. Wool, and S. Zaks. Optimal layouts on a chain ATM network. *3rd Annual European Symposium on Algorithms (ESA), (LNCS 979)*, Corfu, Greece, September 1995, pages 508-522. To appear in *Discrete Applied Mathematics*.
26. O. Gerstel and S. Zaks. The virtual path layout problem in fast networks. In *The 13th ACM Symp. on Principles of Distributed Computing*, pages 235–243, Los Angeles, USA, August 1994.
27. R. Händler and M.N. Huber. *Integrated Broadband Networks: an introduction to ATM-based networks*. Addison-Wesley, 1991.
28. M-C. Heydemann, J-C. Meyer, and D. Sotteau. On forwarding indices of networks. *Discrete Applied Math*, 23:103–123, 1989.
29. ITU recommendation. I series (B-ISDN), Blue Book, November 1990.
30. L. Kleinrock and F. Kamoun. Hierarchical routing for large networks; performance evaluation and optimization. *Computer Networks*, 1:155–174, 1977.
31. L. Kleinrock and F. Kamoun. Optimal clustering structures for hierarchical topological design of large computer networks. *Networks*, 10:221–248, 1980.
32. E. Kranakis, D. Krizanc, and A. Pelc. Hop-congestion tradeoffs for high-speed networks. In *7th IEEE Symp. on Parallel and Distributed Processing*, pages 662–668, 1995.
33. F.Y.S. Lin and K.T. Cheng. Virtual path assignment and virtual circuit routing in ATM networks. In *IEEE Globecom'93*, pages 436–441, 1993.
34. C. Partridge. *Gigabit Networking*. Addison Wesley, 1994.
35. D. Peleg and E. Upfal. A trade-off between space and efficiency for routing tables. *Journal of the ACM*, 36:510–530, 1989.
36. K.I. Sato, S. Ohta, and I. Tokizawa. Broad-band ATM network architecture based on virtual paths. *IEEE Transactions on Communications*, 38(8):1212–1222, August 1990.
37. Y. Sato and K.I. Sato. Virtual path and link capacity design for ATM networks. *IEEE Journal of Selected Areas in Communications*, 9, 1991.
38. L. Stacho and I. Vrt'o. Virtual path layouts for some bounded degree networks. In *Proc. 3rd International Colloquium on Structural Information and Communication Complexity (SIROCCO), Siena, Italy, June, 1996.*

# The Mobile Agent Technology

S. Covaci and T. Magedanz

GMD FOKUS
Competence Center for Intelligent Mobile Agents (IMA)
Kaiserin-Augusta-Allee 31, D-10589 Berlin, Germany

Fon: + 4930 3463 7000, Fax: + 4930 3463 8000
Email: [covaci | magedanz]@fokus.gmd.de, WWW: http://www.fokus.gmd.de/ima

**Abstract.** Within the last five years the paradigm of *Intelligent* and *Mobile Agents* has gained momentum in the field of software engineering. Mobile Agents (MAs) are autonomous, asynchronous and intelligent software entities which, in order to fulfil their tasks, can migrate to and reside in a number of networked nodes. A variety of agent languages, architectures and platforms have been developed within the last three years in academia and industry, including AgentTCL, Telescript, Tabriz, Active X, Java and a variety of Java enhancements, such as Aglets, Java-To-Go, etc. However, all of these developments are incompatible with each other and designed for specific environments and application domains, due to the lack of agent standards.

The application of the agent technology, however, has already a history and spreads over many areas, including user interface/personal assistance, mobile computing, information retrieval & filtering, data mining, smart massaging, the electronic marketplace, and telecommunication services control and service / network management. The last two areas represent still new application fields for agent technology, now rapidly gaining importance in the problem area addressed by the emerging global *Information Infrastructure*.

The heterogeneity, distribution, scale and dynamic nature of the emerging Information Infrastructure, referred to as NII, EII, or GII, is calling for new paradigms for the *control* and *management* of the *open resources*. In this context, the *agent-based technology* seems to offer a promising solution to cope with the complexity of this environment. Within such an open environment, an agent-based solution can:

- reduce the requirement of traffic load and the availability on the underlying networks (via the autonomy and asynchronous operations of the agents);
- reduce the requirement on customer intelligence during the installation, operation and maintenance of the resources (via the intelligence and autonomy of the agents);
- enable "on demand" provision of customized services (via dymanic agent downloading from the provider system to the customer system and further on back to the provider system or directly to the resources);
- increase the flexibility, reusability and effectiveness of the software-based problem solutions;

– allows for a more decentralized realisation of service control and management software, by means of bringing the control or management agent as close as possible or even onto the resources.

Agent platforms could be considered as enhancements of todays distributed object-oriented platforms, e.g. DPEs. Hence the long term target should be to head for an appropriate enhancement of standard object-oriented service and management platforms, such as CORBA and the TINA DPE in order to allow for more flexible service realisations. However, in face of existing "legacy" environments, such as IN, TMN and Internet based platforms, the short to medium term integration of agent technology into these systems could enhance the flexibility of these systems to a large extend and thus offer new market opportunities.
Key to the success for the integration of agent technology in particular environments is the adoption of standardised solutions. The standardization of agent technologies has started in 1996 within three fora, namely the Object Management Group (OMG), the Foundation of Physical Intelligent Agents (FIPA), and the Agent Society. Fortunately all these activities are aligned due to mutual agreements.

This talk provides an overview of MA basics, technologies, standards and applications and thus is partitioned into four parts:

– *Part 1* looks at MA basics, covering mobility aspects, the remote programming paradigm, benefits, problems, platform requirements and the relationship with distributed object computing.

– *Part 2* provides an overview of existing MA technologies, such as General Magic's Telescript, Sun's Java, IBM's Aglets Workbench (AWB) and AgentTcl form Dartmouth University. In addition a brief introduction to OMG's Common Object Request Broker Architecture (CORBA) is given.

– *Part 3* reviews relevant Agent standardisation activities, namely OMG's Mobile Agent Facility (MAF) specification, the Foundation of Intelligent Physical Agents (FIPA) and the Agent Society.

– *Part 4* addresses the application of MA technology in the field of telecommunications with emphasis on service control (IN) and network management (TMN).

For more information contact: *http://www.fokus.gmd.de/ima*

# Theory and Practice in Interactionally Rich Distributed Systems

David A. Duce

Rutherford Appleton Laboratory, Chilton, DIDCOT, OX11 0QX, UK

**Abstract.** This paper explores the notion of richness in interactive systems, at the device level, the application level and the theoretical level. The paper attempts to show something of the breadth of recent research and practice, with a somewhat unintentional leaning towards the domain of scientific visualization in a broad sense. Theoretical work which may lay the foundations for a systematic understanding of interactionally rich systems is also discussed, drawing on the literature of psychology and system modelling. The paper concludes with some thoughts on future directions for both theory and practice.

## 1 Introduction

The phrase "Interactionally Rich Systems" is the title of a cooperative network funded by the Human Capital and Mobility Programme of the European Union, coordinated by the University of York, with 6 other partners including Rutherford Appleton Laboratory. One of the motivations behind this project is the wide range of devices for input and output that are becoming available, for example input is possible through devices that register gesture in three dimensions, eye movements, hand configurations and speech. Output may be presented as high quality three-dimensional images, possibly stereoscopic and with sound, animation and tactile force display. The class of interactionally rich distributed systems encompasses multimodal, multimedia, distributed VR and CSCW systems.

The concept of 'richness' in interactive systems is quite hard to characterize. Two pertinent meanings of the adjective 'rich' are "abounding in desirable qualities" and "of choice or superior quality". In the IRS network we tried to elicit a deep sense in which a system may be interactionally rich, the sense that leverage and flexibility in interactive, conjoint, behaviour between user and computer, may be achieved in some minimal way, e.g. with respect to the cognitive load imposed by the interaction per se. In mathematics one can think of group theory as a very rich structure, yet it is a structure generated from just four axioms; a rich structure with a minimal basis. Two examples were suggested from other domains. The hammer is a highly flexible tool in woodworking and metalworking. Skilled use of a simple tool by, say, a blacksmith, can generate a phenomenal range of both functional and decorative artefacts. Similarly the pencil is an enormously flexible device. A system consisting of two people, two pencils and a large sheet of paper can be considered interactionally very rich in this sense.

A hammer (instrument) may be used by a human (agent) to knock (action) a nail (object1) into a piece of wood (object2). A hammer *affords* the action on a pair of objects, but to use a hammer successfully also requires skill on the part of the user - for example where exactly to hit the nail, how hard to hit the nail, how exactly to hold the hammer etc. (*Affords* is a technical term due to J.J. Gibson. Loosely speaking it is the meanings an environment has for an observer, e.g. an apple affords eating.) A ball-point pen (Biro) may be used to write on a piece of paper, to dial a telephone fitted with an old-style rotary dialling device, or to access the concealed button on a Macintosh Powerbook which reboots the device. A single instrument may be capable of performing many different actions. The number of actions an instrument can perform could be considered as a measure of its 'richness'. The instrument is a tool, with no autonomous behaviour of its own. Some computer systems fall into this category; systems that perform a precise task under the total control of a single user.

Again taking an analogy from woodworking, a carpenter might say to his assistant, nail these two pieces of wood together, taking care not to mark the top surface. This is a form of task delegation and leaves the assistant with a number of decisions to make. This theme is taken up in intelligent agent systems, see, for example [72]. A system to which the user could delegate a task might be thought of as more interactionally rich than a system over which the user has to exercise complete control at every point.

This idea of negotiating the task to be performed is taken further by Genter and Nielson [31]. It also echoes the ideas for "man-computer symbiosis" of Licklider in the 1960s (quoted in [5], page 43).

> However, many problems that can be thought through in advance are very difficult to think through in advance. They would be easier to solve, and they would be solved faster through an intuitively guided trial-and-error procedure in which the computer cooperates, turning up flaws in the reasoning or revealing unexpected turns in the reasoning ... One of the major aims of man-computer symbiosis is to bring the computing machine effectively into the formulative parts of technical problems ... To think in interaction with a computer in the same way that you think with a colleague whose competence supplements your own will require much tighter coupling between man and machine ... than is possible today.

A recent issue of IEEE Computer Graphics and Applications includes a preliminary report of a workshop on Human-Centred Systems: Information, Interactivity and Intelligence, organised by the National Science Foundation in the USA [68]. The aim of the workshop was to formulate recommendations for one of the components of the next High Performance Computing Initiative. The focus for the workshop was how to achieve synergy between human and computer. Human-centred systems were seen as a key way forward and four main goals in developing human-centred systems were identified: (a) scale up current technology to reliably and cost-effectively support human-centred activities: (b) develop revolutionary technology that expands the space of current human activities; (c) expand the understanding of human behaviour and needs in view of changing environments; (d) deepen the understanding of how technology affects human life.

Six general characteristics of human-centred systems are identified:

– They take into account human perceptual and motor capabilities and limitations.
– They support actual practice (real behaviour in real tasks) effectively.

- They are flexible rather than rigid - that is, they can be used in many ways and do not unnecessarily constrain users.
- They are context-sensitive and adapt to users' changing needs.
- They are open and inspectable so that users can understand them.
- They are engaging and enjoyable.

From the discussion above we can discern a number of dimensions to the quality of richness.

- At the device level, the flexibility of the effectors and sensors through which the interaction is mediated. Flexibility may also include the sense of achieving the same result in several ways, e.g. by menu item selection and control keys.
- The richness of the range of tasks which can be performed effectively by the system.
- Taking a more holistic view of user(s) and system(s), the character of the synergy realised between users and system.

The next part of this paper will explore some examples of input and output device technology, followed by some examples of interactive systems that use some of these devices in rich ways. Virtual reality seems to provide much impetus for research, both experimental and theoretical, into systems and devices that fully engage and exploit human perceptual and motor capabilities [28]. The final part of the paper will explore attempts that are being made to model such systems theoretically. No attempt is made to give a complete coverage of the areas discussed in this paper; rather the paper aims to give a flavour of some research directions, in both practical systems and theoretical analysis.

## 2 Technology

### 2.1 Visual

In this section we describe some developments in visual display devices that address a number of issues: (a) devices that support the creation of a sense of immersion in the virtual environment in which user(s) and system(s) are cooperating; (b) devices that exploit the stereo viewing capabilities of the visual perceptual system, in non-intuitive ways; (c) devices that emulate or augment the normal working environment of users, for example, workbench, desk or cockpit.

This section just gives a flavour of pertinent developments; it is not exhaustive, for example technologies such as wide angle displays and holographic displays are not covered.

**Head-mounted Displays** have become a familiar sight in game arcades, exhibitions, advertising, scientific and engineering establishments, etc. Ellis [28] discusses the origins of some of the technology used in virtual environments. Philco and Argonne National Laboratory worked on teleoperation displays using head-mounted, closed circuit television systems in the early 1960s. Sutherland wrote in 1970 [67] "My idea was very simple: Mount miniature cathode ray tubes on the user's head, one tube in front of each eye, so that the computer can control exactly what he sees. Measure the position of the user's head and compute a perspective picture appropriate to that viewing position ...". Such a head-mounted display was developed at the University of Utah. Two CRTs were mounted in goggles. Tracking the position of the head was done by mechanical linkages. Since then the engineering aspects of head-mounted displays have improved tremendously, but the principle remains much the same.

Head-mounted displays are also commonplace in augmented reality applications, where the requirement is to overlay computer generated information on the real-world scene being viewed by the user. Aircraft cockpits, maintenance and repair environments and surgery are examples of application areas for this technology. Girolamo et al. [32] discuss display technology being investigated for 21st century warriors. Warriors are no strangers to head-mounted displays as an illustration of a 1916 patent for a head-mounted look-and-shoot system, reproduced in [30] shows.

The CCSV (Counterbalanced CRT-based stereoscopic Viewer) [58] was developed at NASA Ames Research Center in the late 1980s as a supplementary viewing device for a virtual environment project. This overcomes some of the limitations of head mounted displays; the device consists of a wide field-of-view stereoscopic viewer using 2 CRTs, mounted on a 6 degree of freedom counterbalanced arm.

**The Cambridge Autostereo 3D Display.** Two-view stereo photographs became popular in the late 19th century. Most approaches to 3D display systems are based on the two-view stereoscopic principle which usually require the viewer to wear glasses of some kind, for example passive glasses with red green or polarizing filters or active glasses with a liquid crystal shutter in front of each lens, synchronized with the display so that left and right eyes receive different views. Such displays are called stereoscopic displays, in contrast to autostereoscopic displays [27] which generate images that can be seen without optical aids.

The display under development at the University of Cambridge [70, 41] requires multiple distinct pictures of the object to be viewed, taken from different viewing positions. There is a set of shutters in front of the cathode ray tube. When a picture is displayed, one of the shutters is opened, making the picture visible to part of the area in front of the display. By placing the shutters between a pair of lens, it turns out that each image in the final focal plane can be seen from only one direction. The effects of stereo and movement parallax combine to give an illusion of real depth in the image. A 25 inch, 28 view, full colour display prototype now exists.

The significance of the Cambridge Autostereo display in the context of this paper is that it is non-intrusive : the viewer does not have to wear glasses or other special equipment. This approach also allows several people to view stereo at one time.

**The CAVE** (CAVE Automatic Virtual Environment) was developed at the Electronic Visualization Laboratory, University of Illinois at Chicago [19]. The CAVE is described as a theatre, typically a cube of dimension 3m, made up of three rear-projection screens for walls and a down-projection screen for the floor. Full colour screen images (1280 x 512 stereo) at 120 Hz are projected onto the screen. The CAVE is also equipped with a multispeaker audio capability. A user's head and hands are tracked with electromagnetic sensors. Users have to wear LCD stereo shutter glasses. Groups of people can enter a CAVE and experience the virtual environment, but there are some limitations caused by interference between users, leading to collapse of stereo vision.

**The Responsive Workbench** was developed at GMD, German National Research Centre for Information Technology[53]. A lower cost device, the Virtual Workbench, has been developed in Singapore [61]. The Responsive Workbench grew out of experiments a range of device hardware in applications including medical imaging, molecular design, fluid dynamics visualization, autonomous systems and architecture. The Responsive Workbench is centred on a real bench with a special glass plate as a bench top. Images are projected onto the glass plate from below. The responsive environment consists of powerful graphics workstations, tracking systems, cameras, projectors and microphones. Tools such as voice and gesture recognition systems, off-axis stereoscopic rendering and a stylus system to simulate surgical equipment are also included. In multiple user scenarios, the users have to move as a group around the workbench. When in use, virtual objects and control tools are located on the workbench; objects are displayed as stereoscopic images and appear to be above the surface of the worktop.

It was found that head movement needs the fastest possible visual feedback; incorrect perspective severely impacts the visual appearance of virtual objects. Fast directional sound feedback further enhances the realism of interaction. Tactile feedback via a glove was found to be very important. Where such a device was not available, collisions between the hand and an object were denoted by specific sounds and a change to the colour of the object touched. The main drawback was found to be object distortion resulting from the single user perspective rendering.

## 2.2 Auditory

In recent years low cost audio cards for PCs and workstations have become commonplace. High-end speech and non-speech sound synthesisers have also advanced significantly in capabilities and quality. The MIDI (Musical Instrument Digital Interface) standard, developed in Japan to provide a uniform method of interconnection between music hardware has made a significant impact on the availability and accessibility of audio hardware. For a review of this area, see [52].

## 2.3 Haptic

Webster's New International Dictionary defines haptics as "pertaining to sensation such as touch, temperature, pressure, etc mediated by skin, muscle, tendon or joint". Haptic displays permit a user to receive haptic feedback. Many of these devices are input/output devices, they display haptic feedback and position and also track position and measure contact forces. The Touch Lab at MIT [36] is a good source of information on haptic devices as is the haptic finger web site [45]. The paper by Mark et al. [56] addresses issues and solutions in adding force feedback to graphics systems.

Devices for position tracking, especially hand positions, are now well-known. Sturman and Zeltzer have written an excellent review of the area [66]. Early devices were based on developments of master-slave manipulator arms (for example for manipulating hazardous materials). Hand position is characterized by the location of the hand in space and the orientation of the palm. Three basic technologies are used for tracking hand position: optically based, magnetically based and acoustically based.

There are two common methods of optical tracking. The first uses small markers on the body detected by cameras surrounding the user. The second uses a single camera to capture a silhouette image of the subject, which is analysed to determine positions of parts of the body and user gestures. The non-intrusive nature of the computer vision approach makes it a very attractive approach, though computer vision techniques and processor performance do not yet fully combine to give a system capable of interpreting complex visual scenes (including dynamic and static features) in real-time. Blake and Isard [10] recently described some new algorithms that take an important step in this direction. These enable "the curved silhouettes of moving non-polyhedral objects" such as hands and lips to be tracked at full video frame rates without any special hardware beyond a desktop workstation, video-camera and framestore. Using these algorithms they have demonstrated the use of a hand as a 3D mouse, rigid motion being used to control 3D position and attitude and nonrigid motion to signal button pressing.

Magnetic tracking is based on a source element radiating a magnetic field and a sensor that reports its position and orientation with respect to the source. The Polhemus 3-space tracking sensor is a well-known example of this technology. Accuracies of better than 0.1 inches in portion and 0.1 degrees in rotation can be achieved.

Acoustic tracking uses a high-frequency sound emitter (for example mounted on the hand) and precisely placed microphones to detect the sound. Positional accuracies of a few millimetres can be achieved. Reflections from walls can have an adverse effect.

Glove-based devices are one approach to measuring the shape of the hand as the fingers and palm bend. Sturman and Zeltzer [66] describe the main developments in glove technology in research groups and in the market place. The main problem in glove technology is to device a reliable and robust technology for measuring the shape of hand and fingers. One early example was by DeFanti and Sandis at the University of Illinois at Chicago, which used flexible tubes mounted along each finger with a light source at one end and photocell at the other. When a tube bent, the amount of light passing between source and photocell decreased; the light received by the photocell could be correlated with the amount by which the finger was bent. This device was used to give multi-dimensional control, for example emulating a set of potentiometers.

The DataGlove, commercialized by UPL Research, passed into common usage in the 1980s. This device monitored 10 finger joints, and the hands' positions and orientation (using a 3-space magnetic tracker). The accuracy of the glove was sufficient for simple gestural input, but not for fine manipulations or complex gestural recognition.

The Dextrous Hand Master, marketed by EXOS, was originally developed for controlling a robot hand. It is an exoskeleton-like device worn on the hand and fingers which can accurately measure the bending of the three joints of each finger as well as more complex motions. The device is very accurate.

Similar devices have been extended to provide haptic display, usually to the fingers and arm [74]. EXOS Inc., for example, have commercialized a 5 degree-of-freedom force feedback device that tracks the motions of, and provides force feedback to, the shoulder, elbow and forearm.

The user interacts with the PHANToM (TM) Haptic Interface, developed initially at MIT by Massie and Salisbury, now marketed by SensAble Technologies [47] by placing a finger in a thimble. The key to the operation of the device is that by exerting a precisely controlled force on a fingertip sensations of touching and interacting with imaginary objects can be invoked. The device is a desktop device, the thimble is mounted on an arm that can be moved around in space. Researchers at MIT have explored the use of two such devices to perform two fingered manipulations of virtual objects.

Akamatsu and Sato [1] describe a mouse with tactile and force feedback. Tactile stimulation is generated by pulsing a small pin which projects slightly from the surface of a mouse button. Force feedback is generated by an electromagnet inside the mouse in conjunction with an iron mousepad.

Hirota and Hirose at the University of Tokyo have developed a device called the surface display (described in [74]). The prototype consists of a set of rods arranged in a square 4 x 4 lattice at intervals of 20 mm covered by a foam sheet. Each rod can be moved vertically through ±25mm by a servo motor. The rods take up an initial configuration corresponding to the shape of the surface of a virtual object. The device senses the user's touch and reacts to the force exerted moving the rods which deform the surface of the sheet, thus simulating the behaviour of the object.

Medical applications have led to the development of a wide range of special purpose input devices (for example devices used in training systems [23]), The University of Virginia, for example, have developed a two handed interface, the user holds a small doll's head in one hand and points to it with various tools (for example a plastic cross-sectioning plate) held in the other [33].

The Fraunhofer Institute for Computer Graphics in Darmstadt, Germany, have developed a system that includes thermal perception along with other display modalities. They generate thermal display using computer controlled fans and infrared lamps and small Peltier element devices, which can be placed at different locations on the body's surface to cool or heat the skin locally [23, 22].

### 2.4 Wearable Computers and Direct Interfacing

Wearable computers are seen by some as opening up new application areas, for example in maintenance applications. Early versions, in common with most new technologies, were cumbersome in the extreme (c.f. the suits worn in the early days of diving!), but these problems are now being overcome through technology advances [55] and attention is being given to the interface design challenges they present [7].

There are also research projects looking at direct neural and brain interfaces, e.g. [37, 43]. Clearly there are very important applications of this technology in prosthetic systems; whether the general user would want to be so equipped is an issue on which readers will have their own views.

## 3 Applications

### 3.1 Auditory Interfaces and Sonification

The use of sound in human computer interfaces poses some interesting problems. The physics of sound is relatively simple and well-understood, whereas the perception of sound is neither simple nor well-understood [4, 26]. We may distinguish between speech and non-speech sound. In this section we will only be concerned with the use of non-speech sound. Non-speech sound is used extensively in film and television; so-called sound effects, which are often exaggerations of real-world sounds to convey a particular meaning or impression.

One approach to the use of sound in interfaces is the idea of *auditory icons* initiated by Gaver. Auditory icons are based on real-world sounds and concentrate on the objects by which they are generated. Gaver used this approach in his *SonicFinder* for the Apple Macintosh.

An alternative approach, first proposed by Blattner in 1989, is called *earcons*, abstract structured musical tones that have no real-world associations. Information is encoded in the rhythm, pitch patterns and use of timbres; the user has to learn the structuring rules before information can be extracted, but having done so, their use can be very effective. See Brewster's web page for examples [42]. Earcons have been applied to the design of interfaces to support blind or partially sighted users. An algebra earcon, for example, has been developed at the University of York to give blind mathematics students the ability to glance at a mathematical expression [65]. Another use is in the enhancement of a graphical information display in order to overcome the problems caused by information overload. Brewster [14], for example, describes two experiments in which earcons were added to buttons and scrollbars. Results showed that sound improved usability by increasing performance and reducing the time to recover from errors. Subjective workload measures also showed a significant reduction.

The process of rendering information in an auditory perceivable form is called *sonification*. The Atlas computer installed at the author's institution in the 1960s had a loudspeaker attached to the CPU. Users learnt to recognise the behaviour of their programs from the sounds they made. It was not uncommon for operators of the machine to telephone users in the middle of the night, hold the telephone close to the speaker, and ask the user if their program was alright! Another early example is the computer generated sound track of the film *Hash Tables*, made by Bob Hopgood at this laboratory [35]. Sound is used to convey an impression of the work done in populating a hash table as the number of entries increases and conflicts arise.

Astheimer has written a very useful review of the sonification area [4]. Minghim and Forrest [59] have developed a system to support surface-based data presentation and analysis. Their work is based on properties of timbre perception and results in auditory scene analysis and sound grouping to control the mapping of surface properties to sound. Their sonifications can be constructed from up to 4 streams, each associated with a particular property of the surface, such as surface normal, gradient sign and gaussian curvature. Surface normals, for example, use frequency, stereo field and volume properties of sound. They have also developed an earcon representation to 'help catalogue and memorise surface information'. In the conclusions of their paper they write "The effectiveness of auditory display for conveying information and even complex structures has been confirmed by us... Also, with time and practice understanding is dramatically improved, so that often it is possible to understand a great deal of the data by sound alone. The expected improvements in memory and interpretation have been observed... Sound can be useful ... provided the sonic designs employed consider fundamentals of human hearing and correspondence with familiar analogies."

## 3.2 Applications of Haptic Interfaces in Visualization

In 1990 Brooks et al. at the University of North Carlonia atChapel Hill [15] published a paper describing progress on a project (called GROPE) which began in 1967 to develop a haptic display for 6D (3 forces, 3 torques) interacting molecular force fields, including molecular docking. The group set out to test, through controlled experiments, the hypothesis 'Do haptic displays demonstrably aid perception?'. They concluded that the hypothesis is true and demonstrate performance improvements of 2.2 times in a simple manipulation task and 1.7 times in a complex 6-D molecular docking task. They argue that haptic displays are most useful in applications where the force field is important to task performance, complex and adifficult to visualize optically.

Recently Hughes and Forrest [49] have explored the use of a tactile mouse in visualization applications. Their motivation stems in part from work done by Forrest in the early 1970s with a computer peripheral for machining models from rigid plastic foam. Examples were encountered of surfaces whose visual appearance was seemingly indistinguishable, yet local flattening of the surface could be discerned by running a finger over the model. They experimented with a range of tactile mice. Their first device consisted of a mouse with a stripped down audio speaker attached to a vibrotactile pad mounted on the side of the mouse. Their second device used two pads. They have also experimented with solenoids, dot-matrix printheads and with a transducer which stimulated transcutaneous nerves by administering a very mild electric shock. An application to Geographical Information Systems is described in their paper. Consider the problem of overlaying two or more sets of map data at the same time, e.g., 'one may wish to know simultaneously how many people live in an area and the radiation levels due to a power station near by'. The difficultly with conventional methods for displaying overlapping data is the avoidance of visual clutter. Instead they display one data set visually, and one using the vibrotactile mouse. The user can see one map and can explore the other using the mouse. They found that it was quite easy to, say, find an area in which there is both high population and high radiation. They note the difficulty of finding satisfactory touch generating devices. Their initial impressions are that touch has a place in the visualization of complex data spaces, though they note that there are open questions about how many information channels can be used simultaneously.

## 3.3 The Virtual Desk

A collaboration between the Rank Xerox Research Centre in Cambridge and the Rainbow Group [41] developed a system called the DigitalDesk (See Wellner's paper [71]) a real physical desk on which the user may stack papers in the usual way, enhanced to provide some of the characteristics of a workstation display space. A visual display is projected onto the surface of the desk, and video cameras pointing at the desk are linked to an image processing system that can sense what the user is doing. Wellner cites three important characteristics of the digital desk: (a) it projects electronic images down onto the desk and onto paper documents, (b) it responds to interaction with pens or bare fingers, and (c) it can read paper documents placed on the desk. The work is being progressed in other projects in the Rainbow group exploring the use of video in user interfaces.

### 3.4 Gesture Recognition

Many of the xamples of applications and pilot applications in the literature that use gesture for input, e.g. [8, 12, 66] are glove-based with a gesture recognition engine to extract gestures from the data from the glove's positional sensors. The main drawback to this approach, from the users' point of view, is the intrusive nature of the glove.

An alternative approach is to use a video camera and image recognition techniques. One such project at the Fraunhofer Institute for Computer Graphics in Darmstadt [46] is investigating approaches using standard PC hardware, frame grabber and video camera. Maggioni and Kämmerer [40] have described a video-based system called the Gesture Computer which is proving effective in a number of applications. The system can recognize static hand gestures. Such approaches are also being developed in the computer music community, and to support disabled users (for example, recognition of sign languages).

### 3.5 Applications of the CAVE and Responive Workbench

A range of applications of the Responsive Workbench are described in recent papers [62, 63]. These include Command and Control systems, production modelling and planning, visualization and a virtual wind tunnel. The latter, carried out at NASA Ames Research Center used a prototype duo display capability developed by Fakespace [44]. Earlier work with virtual wind tunnels at NASA Ames used a BOOM device [16]. The duo display allows two people to use the workbench, each with independent head-tracked points of view. The authors report that this has proven very useful in collaborative research, however no hint is given as to the extensibility of this technology beyond two users. Use of a CAVE in molecular modelling is reported in [2].

## 4 Cooperative Systems

The literature on cooperative systems is vast. McCarthy [57] gives a very readable review of the area, addressing social and organizational issues as well as technological issues. Most of the early applications of CSCW technology used audio, video, whiteboard and shared editing tools. This kind of technology is now *relatively* mature and desktop cooperative sessions are a practical possibility, subject to constraints imposed by the underlying network technology used [39].

There is growing interest in richer forms of CSCW, for example the distributed collaboratory projects [51], and the MANICORAL project [38]. British Telecom Laboratories have recently reported a large scale user trial of a shared space virtual environment in a recreational setting [3]. Two CAVEs have been linked together to support remote collaboration in the design of vehicles using a high bandwidth ATM network (the authors estimate approximately 1 Mbit/sec per site, low latency is required)[54]. Audio and video communications were included, video being displayed on the walls of the CAVE. The DIVE system developed at SICS in Sweden [48] is an Internet-based multi-user VR system which enables users to navigate in 3D space and observe, meet and interact with other users and applications. The system is a research prototype but it demonstrates what is possible with desktop workstations and the Internet. Xerox PARC also work extensively in this field. The advent and standardization of VRML (Virtual Reality Modeling Language) also opens the door to new multiuser applications.

## 5 Theory of Interactionally Rich Systems

Interactionally rich systems involve both human and computer agents, hence the body of theory that is relevant to such systems encompasses many different aspects: (a) theory of interactive systems which enables a system designer to model, analyse and reason about how such computer systems behave; (b) theory of human users which presents the designers of interactionally rich systems with an understanding of human beings, including co-operative behaviour between humans and cultural influences; (c) approaches that bring together system and user theory and enable a designer to reason about the conjoint behaviour of system and user(s). The following sections examine each of these areas in turn.

### 5.1 System Models

Modelling and specifying interactive systems is not a new activity. Dearden and Harrison [20] have written a recent review of the use of formal mathematical models in the design of interactive systems. The Esprit Basic Research Action project, Amodeus 2, enabled a lot of fundamental work and case studies to be done in this area [11, 9]. One of the approaches refined in that project was the *interactor* approach for representing the observations of a user interface or its components. An interactor consists of an internal state representing some facet of the application domain, and a presentation that describes the perceivable components of that state. The effect of actions on the state and presentation, and invariants between the observables, are described within mathematical structures in a formal language such as MAL or Z. We can reason about aspects of interactive systems expressed in this way, such as the relationship between state and presentation, that relate to user-system interaction and hence affect usability.

The author has recently come across work in the multi-agent and intelligent agent systems area which looks to have potential applications to interactive distributed systems with multiple human and computer agents [73, 17, 13, 21]. The papers cited all describe formal frameworks for addressing issues such as interactions between different goals and intentions, autonomous agents which cope with the inherent uncertainty in autonomous systems, fairness and selfishness, cooperative problem solving including notions such as team action. These traits of multi-agent systems also appear in interactive and CSCW systems and there is much interesting work to be done in drawing threads of research from these areas together.

### 5.2 Theory of Human Users

Designers of interactive systems look to the psychology and cognitive science literature for theory that will provide a packaged understanding of the human user in a way that is useful for understanding how the human behaves when engaged in interaction with a system. One is looking for theories that characterise the properties of the human perceptual and action systems in a way that enables designers to design systems that make effective use of the properties of these human systems and hence will be 'comfortable' and 'effective' for the human users. Extending this idea to systems which involve multiple human users (CSCW systems, for example), designers are also interested in theory that will enable analysis and synthesis of the multi-user aspects of such systems.

Theories of the human visual system are arguably the best well-known in computer science. Most modern text books on computer graphics and university courses on graphics include sections on the visual system; the main components of the eye, how we perceive colour, properties of colour perception, visual illusions, etc. Studying the way artists use colour and effects such as perspective can also give valuable insights to the system designer. Gordon's book [34] is a highly readable account of seven major theories of visual perception. He gives detailed introductions and critiques of each. The theories covered by Gordon are: Gestalt theory, probabilistic functionalism (Brunswik 1903-1955), neurophysiological approaches, empiricism, direct perception and ecological optics (J.J.Gibson), and Marr's computational approach.

It seems to this author, and indeed to other computer scientists, that theories of audio perception have not received as much attention as theories of visual perception and work in the audio area is not so well-known. One of the central problems in audition is dealing with mixtures of sounds. Bregman in a book entitled 'Auditory Scene Analysis' developed the notion that the human auditory system exploits the organisation inherent in complex signals. There are similarities between Marr's approach and Bregman's. Cooke [18] has developed a computational model for separating out multiple simultaneous sound sources, which is rooted in a model of the auditory system. The model is based on an initial processing step to decompose signals in a certain way, followed by a second stage which seeks coherent subsets of auditory objects in the decomposed representation, according to a set of auditory grouping principles.

In the field of haptics, there is work in places such as the MIT Touch Lab [36] to develop models of haptic phenomena, including the mechanisms of tactile sense [64].

There are many other areas of psychology, physiology, linguistics and cognitive science that provide important foundations for an understanding of human computer interaction, for example task analysis, cognitive complexity theory, programmable user models. Johnson's book [50] gives a good overview of these approaches, from the viewpoint that the user and computer work together to perform tasks and tasks are seen as the central focus of design. This is a vast area which is beyond the scope of this paper.

One approach to formulating theory is to develop models of restricted scope accounting for properties in particular task settings. Barnard has taken a different approach in Interacting Cognitive Subsystems (ICS) [6, 69].

ICS is a comprehensive model of human information processing that describes cognition by a collection of sub-systems that operate on specific mental representations or 'codes'. Although specialised to deal with specific codes, all subsystems have a common architecture, shown in Figure 1. Incoming data streams arrive at an input array, from which they are copied into an image record representing an unbounded episodic store of all data received by that subsystem. In parallel with the basic copy process, each subsystem also contains transformation processes that convert incoming data into certain other mental codes. This output is passed through a data network to other subsystems. If the incoming data stream is incomplete or unstable, a process can augment it by accessing or buffering the data stream via the image record. However, only one transformation in a given processing configuration can be buffered at any moment. Coherent data streams may be blended at the input array of a subsystem, with the result that a process can 'engage' and transform data streams derived from multiple input sources.

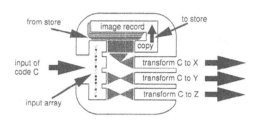

**Fig. 1.** Generic ICS Subsystem.

ICS assumes the existence of 9 distinct subsystems:

Sensory subsystems
**VIS** visual: hue, contour etc. from the eyes
**AC** acoustic: pitch, rhythm etc. from the ears
**BS** body-state: proprioceptive feedback

Structural subsystems
**OBJ** object: mental imagery, shapes, etc.
**MPL** morphonolexical: words, lexical forms

Meaning subsystems
**PROP** propositional: semantic relations
**IMPLIC** implicational: holistic meaning

Effector subsystems
**ART** articulatory: subvocal rehearsal, speech
**LIM** limb: motion of limbs, eyes, etc

The overall behaviour is constrained by the possible transformations and by several principles of processing, e.g. visual information cannot be translated directly into propositional code, but must be processed via the object system that addresses spatial structure. Although in principle all processes are continuously trying to generate code, only some will generate stable output that is relevant to a given task. Such a collection is called a *configuration*. The thick lines in Figure 2 shows the configuration of resources deployed while using a data glove to operate on some object within a visual scene.

The propositional subsystem (1) is buffering information about the required hand posture through its image record and using a transformation (written prop-obj) to convert propositional information into an object-level representation. This is passed over the data network (2), and used to control the positioning of the hand through obj-lim (3) and lim-hand (4) transformations. However, both obj and lim are also receiving information from other systems. The users' view of the rendered scene arriving at the visual system (5) is translated into object code that gives a structural description of the scene; if this is to be blended at obj with the users' propositional awareness of their hand position (from 2) the two descriptions must be coherent. A propositional representation of the scene is generated by obj-prop and passed to prop (6) where it can be used to make decisions about the actions that are appropriate in the current situation. In parallel with this 'primary' configuration, proprioceptive feedback from the hand is converted by the body-state system (7) into 'lim' code (8) in a secondary configuration.

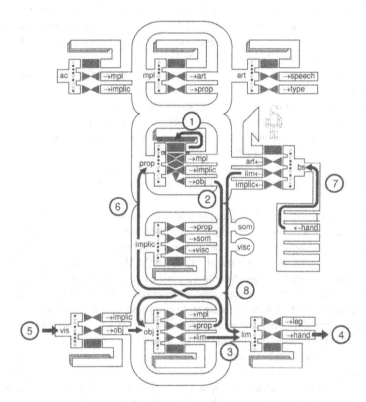

**Fig. 2.** ICS Resources.

Having built up an understanding of the resources needed to perform a particular task using a particular interface, it is then possible to reason about the suitability of that interface by, for example, looking at the conflicts that arise in the use of the identified cognitive resources. ICS provides a framework within which questions such as the one posed at the end of section 3.2 'how many information channels can be used simultaneously' can be addressed, using concepts such as stability of configurations, oscillation between configurations competing for resources, and the requirements for different data streams to blend. Such questions may require new experimental work with human subjects to be undertaken; ICS provides a framework within which such work can be understood and set in a wider context.

### 5.3    Combining User and System Modelling - Syndetic Modelling

User and system models are typically viewed as independent representations that provide insights into aspects of human-computer interaction. Within system development it is usual to see the two activities as separate, or at best loosely coupled, with either the design artefact or some third 'mediating' expression providing the context in which the results of modelling can be related. One endeavour in which the author is engaged, seeks to combine formal system models directly with a representation of human cognition to yield an integrated view of human-system interaction, called a *syndetic model* (from the Greek meaning 'bringing together').

The key observation underlying syndetic modelling is that the structures and principles embodied within ICS can be formulated as an axiomatic model in the same way that interactor models are formulated. This means that the cognitive resources of a user can be expressed in the same framework as the behaviour of a computer-based interface, allowing the models to be integrated directly. ICS and formal system models operate at a commensurate level of abstraction; both impose constraints on the processing of information within the overall system. There are models of cognition that are more operational than ICS, but these impose a level of detail that would make a syndetic model intractable.

Various formalisms have been, and continue to be, investigated for expressing ICS and the formal system model. Examples that have been published are expressed using Modal Action Logic ([24, 25]. The idea of the approach is to express both system model and user model as collections of axioms in MAL, using the notions of interactors to structure the descriptions. An interactor consists of an internal state representing some facet of the application domain, and a presentation that describes the perceivable components of the state. Changes to the state and presentation are effected by actions. The resulting formal model is an approximation of ICS as a cognitive theory; nevertheless it is an approximation that has helped to hone the presentation of some aspects of the theory *per se* and is rich enough to formalise some of the reasoning that takes place when applying ICS to interactive systems. Syndetic modelling has been applied to the access control subsystem of a media space system, an experimental multi-modal flight information system (where for example spoken information can be combined with gesture via a mouse to produce a single command for an application) and a system using gestural interaction to edit 3D scenes constructed from visual objects [24].

Faconti and Duke [29] have used ICS and syndetic modelling as a novel basis for comparing different input devices based on an analysis of the *cognitive* ergonomics of interaction, in terms of the mental resources needed to utilise a particular device for a specific task.

There are, of course, many other approaches drawn from experimental and theoretical psychology that enable aspects of interfaces and user performance to be evaluated, for example the GOMS model and the Keystroke-Level Model (see Newman and Lamming's book for a discussion of such models [60]. Obviously there is a place for such methods. What syndesis offers as a longer term prospect is the ability to bring the analytical power of mathematical logic to bear on problems whose complexity makes the use of less formal techniques problematic. Syndesis is also more abstract than many other techniques and broader, encompassing the whole gamut of human cognitive activity. These characteristics may enable the use of syndesis to filter interface designs before very specific design commitments are made, which might require the use of more specialized theories dealing with individual modalities for their detailed analysis.

## 6  Summary and Prospects

This paper has skimmed the surface of a wide range of topics: device technology, applications, formal system modelling, psychology and syndetic modelling. All (apart from the last) are long-established areas of study and research to which it is impossible to do justice in a paper of this kind. The author's aim in attempting to cover material of this breadth was to try to encourage a more holistic view of research and practice in interactive systems, emphasising the multi-disciplinary nature of the subject and the need to consider computer system and human users as 'equal partners' in some sense in determining the overall behaviour of an interactive system.

Referring back to the dimensions of the quality of richness listed at the end of section 1, at the device level we have tried to show the richness of the range of effectors and sensors now available and the ways in which these exercise and are exercised by the subsystems of the human user. We have given examples of intrusive and non-intrusive devices, for example stereographic displays that require the user to wear some kind of glasses versus autostereographic displays that free the user from such encumbrances, and glove-based gestural input devices versus video-based gesture recognition. Examples of more experimental haptic display devices have also been described.

In applications we have seen how auditory display is now being seriously investigated, both to complement visual display and as an alternative for visually impaired users. Developments such as the Virtual Desk and the Responsive Workbench aim to integrate interactive computer systems with a traditional physical workplace. The examples in section 3 have been chose to illustrate a rich range of tasks across diverse application areas.

The third dimension of richness, quality of synergy, is addressed in perhaps a rather indirect way, by the syndetic modelling work described in section 5.3.

As the work of Faconti and Duke on Device Models shows, work such as this makes it possible to characterise the resources (and hence, perhaps still in a crude sense, the degree of synergy) required to carry out a particular task, relating these directly to a system model.

The cooperative working dimension to interactive systems is a dimension that may be expected to grow in significance. Cooperative working may involve multiple users working in the same location, or users working individually at different locations, or working in small groups at different locations. Moving from one to many users introduces new challenges for the development of non-intrusive technology, for both effectors and sensors. Multiple users also highlight limitations of existing technology, for example, the Responsive Workbench displaying an object from a single viewpoint may be very limiting when used by groups of users, most of whom will receive incorrect stereo images for their locations.

On the theoretical front, work in formal systems modelling, intelligent agent systems and psychology offers increasingly firm foundations for reasoning about interactive systems and their design, but there is much more work needed to draw on the richness of the insights these areas offer and package them in such a way that designers can readily use them to analyse and refine designs at an early stage in the design process and also to have confidence that the insights offered are applicable within the boundaries defined by the intended use and users of the system. This is a vast undertaking, but it is through tentative steps such as syndetic modelling that progress is made. Packaging the results of such research is an equally important and equally difficult endeavour. As others have pointed out, it is often through the provision of well-engineered tools that progress is made in the transfer of theory to practice.

# References

1. M. Akamatsu and S. Sato. A multi-modal mouse with tactile and force feedback. *Int. J. Human-Computer Studies*, 40:443 – 453, 1994.
2. N. Akkiraju, H. Edelsbrunner, P. Fu, and J. Qian. Viewing Geometric Protein Structures from Inside a CAVE. *IEEE Computer Graphics and Applications*, 16(4):58–61, 1996.
3. B. Anderson and A. McGrath. Strategies for Mutability in Virtual Environments. In *From Desktop to Webtop: Virtual Environments on the Internet, WWW and Networks*. British Computer Society Computer Graphics and Displays Group, 1997.
4. P. Astheimer. Acoustic Simulation for Visualization and Virtual Reality. In *Eurographics '95, State of the Art Reports*. Eurographics Association, ISSN 1017-4656, 1995. For ordering information, see http://www.eg.org.
5. R.M. Baecker and W.A.S. Buxton, editors. *Readings in Human-Computer Interaction*. Morgan Kaufmann, 1987.
6. P.J. Barnard. Interacting Cognitive Subsystems: A psycholinguistic approach to short term memory. In A. Ellis, editor, *Progress in the Psychology of Language*, volume 2, chapter 6, pages 197 – 258. Lawrence Erlbaum Associates, 1985.
7. L. Bass, C. Kasabach, R. Martin, D. Siewiorek, A. Smailagic, and J. Stivoric. The Design of a Wearable Computer. In *CHI 97 Conference Proceedings*, pages 139 – 146. ACM Press, 1997.
8. T. Baudel and M. Beaudouin-Lafon. CHARADE: Remote Control of Objects using Free-hand Gestures. *Communications of the ACM*, 36(7):28–35, 1993.
9. V. Bellotti, A. Blandford, D. Duke, A. MacLean, J. May, and L. NIgay. Interpersonal Acess Control in Computer-Mediated Communications: A Systematic Analysis of the Design Space. *Human-Computer Interaction*, 11:357–432, 1996.
10. A. Blake and M. Isard. 3D position, attitude and shape input using video tracking of hands and lips. In *Computer Graphics Proceedings, Annual Conference Series, 1994*, pages 185 – 192. ACM Press, 1994.
11. A.E. Blandford and D.J. Duke. Integrating user and computer system concerns in the design of interactive systems. *Int. J. Human-Computer Studies*, 46:653 –679, 1997.
12. M. Bordegoni and M. Hemmje. A Dynamic Gesture Language and Graphical Feedback for Interaction in a 3D User Interface. *Computer Graphics Forum*, 12(3):C-1 – C-11, 1993.
13. S. Brainov. Altruistic Cooperation Between Self-Interested Agents. In W. Wahlster, editor, *ECAI 96, 12th European Conference on Artificial Intelligence*. John Wiley and Sons, 1996.

14. S.A. Brewster. Using Non-Speech Sound to Overcome Information Overload. See http: //www.dcs.gla.ac.uk/ stephen/, 1997.

15. F.P. Brooks, M. Ouh-Young, J.J. Batter, and P.J. Kilpatrick. Project GROPE - Haptic Displays for Scientific Visualization. *Computer Graphics*, 24(4):177–185, 1990.

16. S. Bryson and C. Levit. The Virtual Wind Tunnel. *IEEE Computer Graphics and Applications*, 12(4):25 – 34, 1992.

17. H.-D. Burkhard. Abstract Goals in Multi-Agent Systems. In W. Wahlster, editor, *ECAI 96, 12th European Conference on Artificial Intelligence*. John Wiley and Sons, 1996.

18. M. Cooke. *Modelling Auditory Processing and Oorganisation*. Distinguished Dissertations in Computer Science. Cambridge University Press, 1993.

19. C. Cruz-Neira, D.J. Sandin, and T.A. DeFanti. Surrond-Screen Projection-Based Virtual Reality: The Design and Implementation of the CAVE. In *Computer Graphics Proceedings, Annual Conference Series, 1993*, pages 135 – 142. ACM Press, 1993.

20. A.M. Dearden and M.D. Harrison. Abstract models for HCI. *Int. J. Human-Computer Studies*, 46:151 – 177, 1997.

21. M. d'Inverno and M. Luck. Understandning Autonomous Interaction. In W. Wahlster, editor, *ECAI 96, 12th European Conference on Artificial Intelligence*. John Wiley and Sons, 1996.

22. J. Dionisio. Virtual Hell: A Trip Through the Flames. *IEEE Computer Graphics and Applications*, 17(3):11–14, 1997.

23. Jose Dionisio, Udo Jakob, and Rolf Ziegler. New Dimensions of Human-Computer Interaction: Force, Temperature, and Motion Feedback in VR. *Computer Graphik Topics*, 9(1/97):31–33, 1997. Reports of the House of Computer Graphics, Wilhelminenstrasse 7, D-64283 Darmstadt, Germany.

24. D.J. Duke. Reasoning about gestural interaction. *Computer Graphics Forum*, 14(3):C–55 – C–66, 1995.

25. D.J. Duke, P.J. Barnard, D.A. Duce, and J. May. Systematic development of the human interface. In *Proceedings of the Second Asia-Pacific Software Engineering Conference, APSEC'95*, pages 313 – 321. IEEE Computer Society Press, 1995.

26. A.D.N. Edwards, P.C. Wright, and S.A. Brewster. Physical and perceptual properties of sounds and their use in human-computer interfaces. University of York, 1995.

27. J. Eichenlaub and A. Martens. 3D without glasses. *Information Display*, 8(3):9 – 12, 1992.

28. S.R. Ellis. What Are Virtual Environments? *IEEE Computer Graphics and Applications*, 14(1):17–22, 1994.

29. G.P. Faconti and D.J. Duke. Device Models. In F. Bodart and J. Vanderdonckt, editors, *Design, Specification and Verification of Interactive Systems '96*. Springer-Verlag/Wien, 1996.

30. R.E. Fischer. Optics for Head-Mounted Displays. *Information Display*, 10(7 & 8):12 – 16, 1994.

31. D. Gentner and J. Nielson. The Anti-Mac Interface. *Communications of the ACM*, 39(8):70 – 82, 1996.

32. H.J. Girolamo, C.E. Rash, and T.D. Gilroy. Advanced Information Displays for the 21st-Century Warrior. *Information Display*, 13(3):10 – 17, 1997.

33. J.C. Goble, K. Hinckley, R. Pausch, J.W. Snell, and N.F. Kassell. Two-Handed Spatial Interface Tools for Neurosurgical Planning. *IEEE Computer*, 28(7):20 – 26, 1995.

34. I.E. Gordon. *Theories of Visual Perception*. John Wiley and Sons, 1990.

35. F.R.A. Hopgood. Computer Animation as an Aid in Teaching Computer Science. In *IFIP 74 Proceedings*, 1974. Also appears in 'The Best Computer Papers of 1975', edited by Auerbach.

36. http: //touchlab.mit.edu.

37. http: //transducers.stanford.edu/stl/Projects/StanfordDVA.html.

38. http: //www-geomatics.tu graz.ac.at/manicoral/home.html.

39. http: //www mice.cs.ucl.ac.uk/merci.

40. http: //www.bath.ac.uk/Centres/MEDIA/SIEMENS/GestureComputerMainFrame.html.

41. http: //www.cl.cam.ac.uk/.

42. http: //www.dcs.gla.ac.uk/ stephen/.

43. http: //www.ee.ic.ac.uk/research/neural/bci/bci.html.

44. http: //www.fakespace.com.

45. http: //www.haptic.mech.nwu.edu/links/finger.html.

46. http: //www.igd.fhg.de/www/igd a7/Projects/Biosignals/gesture.html.

47. http: //www.sensable.com/phantom.htm.

48. http: //www.sics.se/dce/dive/.

49. R.G. Hughes and A.R. Forrest. Perceptualisation using a Tactile Mouse. In *Proceedings of IEEE Visualization '96*, pages 181 – 188, 1996.

50. P. Johnson. *Human-Computer Interaction, Psychology, Task Analysis and Software Engineering*. McGraw-Hill, 1992.

51. R.T. Kouzes, J.D. Myers, and W. Wulf. Collaboratories: Doing Science on the Internet. *Computer*, 29(8):40 – 46, 1996.

52. G. Kramer. *Auditory Display, Sonification, Audification and Auditory Interfaces*. Addison Wesley, 1994.

53. W. Krueger, C.-A. Bohn, B. Froehlich, H. Schueth, W. Strass, and G. Wesche. The Responsive Workbench: A Virtual Work Environment. *IEEE Computer*, 28(7):42 – 48, 1995.

54. V.D. Lehner and T.A. DeFanti. Distributed Virtual Reality: Supporting Remote Collaboration in Vehicle Design. *IEEE Computer Graphics and Applications*, 17(2), 1997.

55. S. Mann. Wearable Computing: A First Step Toward Personal Imaging. *IEEE Computer*, 30(2):25–32, 1997.

56. W.R. Mark, S.C. Randolph, M. Finch, J.M. Van Verth, and R.M. Taylor. Adding Force Feedback to Graphics Systems: Issues and Solutions. In *Computer Graphics Proceedings, Annual Conference Series, 1996*, pages 447 – 452. ACM Press, 1996.

57. J. McCarthy. The state-of-the-art of CSCW: CSCW systems, cooperative work and organization. *Journal of Information Technology*, 9:73 – 83, 1994.

58. I.E. McDowall, M. Bolas, S. Pieper, S.S. Fisher, and J. Humphries. Implementation and integration of a counterbalanced crt-based stereoscopic display for interactive viewpoint control in virtual environment applications. In *Proc. Stereoscopic Displays and Applications II*, volume 1256. SPIE, 1990.

59. R. Minghim and A.R. Forrest. An Illustrated Analysis of Sonification for Scientific Visualisation. In *Proceedings of IEEE Visualization '95*, pages 110 – 117, 1995.

60. W.M. Newman and M.G. Lamming. *Interactive System Design*. Addison-Wesley, 1995.

61. T. Poston and L. Serra. Dextrous Virtual Work. *Communications of the ACM*, 39(5):37 – 45, 1996.

62. L.J. Rosenblum. Applications of the Responsive Workbench. *IEEE Computer Graphics and Applications*, 17(4), 1997.

63. L.J. Rosenblum, R. Doyle, and J. Durbin. The Virtual Reality Responsive Workbench: Applications and Experiences. In *From Desktop to Webtop: Virtual Environments on the Internet, WWW and Networks*. British Computer Society Computer Graphics and Displays Group, 1997.

64. M.A. Srinivasan and K. Dandekar. An Investigation of the Mechanics of Tactile Sense Using Two-Dimensional Models of the Primate Fingertip. *Transactions of the ASME Journal of Biomechanical Engineering*, 118:48 – 55, 1996.

65. R.D. Stevens, S.A. Brewster, P.C. Wright, and A.D.N. Edwards. Design and evaluation of an auditory glance at algebra for blind readers. In G. Kramer, editor, *Auditory Display: The Proceedings of the Second International Conference on Auditory Display*, 1995.

66. D.J. Sturman and D. Zeltzer. A Survey of Glove-based Input. *IEEE Computer Graphics and Applications*, 14(1):30–39, 1994.
67. I.E. Sutherland. Computer Displays. *Scientific American*, 222(6):56 – 81, 1970.
68. N. Talbert. Toward Human-Centered systems. *IEEE Computer Graphics and Applications*, 17(4), 1997.
69. J.D. Teasdale and P.J. Barnard. *Affect, Cognition and Change: Re-modelling Dpressive Thought*. Lawrence Erlbaum Associates, 1993.
70. A.R.L. Travis, S.R. Lang, J.R. Moore, and N.A. Dodgson. Time-Manipulated 3-D Video. *Information Display*, 13(1):24 – 29, 1997.
71. P. Wellner. Interacting with Paper on the Digital Desk. *Communications of the ACM*, 36(7):87–96, 1993.
72. M.D. Wilson. Metaphor to Personality: the role of animation in intelligent interface agents. In *IJCAI '97 Workshop in Intelligent Animated Agents*, 1997. to appear.
73. M. Wooldridge and N. Jennings. The cooperative problem solving process: A formal model. Technical report, Department of Computing, Manchester Metropolitan University, Chester St., Manchester M1 5GD, UK, 1994.
74. Rolf Ziegler. Haptic Displays - How can we feel Virtual Environments? In *Eurographics '96, State of the Art Reports*. Eurographics Association, ISSN 1017-4656, 1996. For ordering information, see http://www.eg.org.

# Configuration-Based Programming Systems

Valérie Issarny

INRIA/IRISA, Campus de Beaulieu, 35042 Rennes Cédex, France

**Abstract.** This paper provides an overview of configuration-based programming, focusing primarily on associated formal methods aimed at easing the development of correct distributed applications. Briefly stated, configuration-based programming consists of describing an application using a configuration that defines interconnections between software components through connectors characterizing communication protocols. Existing work in the field of formal specification of configurations allows to verify that the interconnections of components are correct with respect to both the functional behaviors of components and the use of protocols within components. In addition, we discuss a way to specify non-functional properties provided by configuration elements, hence allowing to reason about the application behavior from the standpoint of resource management policies implemented by the underlying (distributed) runtime system.

## 1 Introduction

The ever increasing complexity of distributed applications calls for methods and tools easing their development. In that framework, industrial consortia have emerged so as to provide application developers with standard distributed software architectures. In general, such a standard specifies a base distributed system for communication management, a set of services for distribution management (e.g., naming service), and a set of tools for application development (e.g., Interface definition language). The definition of a standard architecture then serves as a guideline for the implementation of a programming system. A well known example of standard distributed architecture is the OMA (*Object Management Architecture*) [19] from the OMG (*Object Management Group*) that is aimed at the development of distributed applications relying on client-server type interactions. Another example of standard distributed architecture is TINA (*Telecommunication Intelligent Network Architecture*) [29] that is being specified by the TINA-C consortium and that is aimed at the development of telecommunication applications.

A more ambitious approach to the development of complex distributed applications is the one undertaken in the software architecture research field of software engineering. Instead of concentrating on the definition of a specific architecture, the ongoing research work aims at providing a sound basis for the specification of various styles of software architectures (e.g., see [1, 28]). An architectural style identifies the set of patterns that should be followed by the system organization, that is, the kinds of components to be used and the way they interact. Although the software architecture field is continuously evolving, it is now accepted that the description of an application architecture, qualified as *configuration-based software* in the following, decomposes into at least three abstractions (e.g., see [28]):

(*i*) *Components* that abstractly define computational units written in any programming language,

(*ii*) *Connectors* that abstractly define types of interactions (e.g., pipe, client-server) between components,

(*iii*) *Configuration* that defines an application structure (i.e., a software architecture or configuration-based software) in terms of the interconnection of components through connectors.

Programming systems that are based on the software architecture paradigm then integrate a configuration language that allows application specification in terms of the three above abstractions, together with runtime libraries that implement base system services (including primitive connectors). Such an application description fosters software reuse, evolution, analysis and management. Furthermore, it enables formal verification of the application correctness, including the correct use of connectors with respect to the protocols associated with them [2].

In the light of research results in the software architecture field, configuration-based description of distributed applications constitutes a promising approach for facilitating the development of correct, complex distributed applications. The next section gives an overview of configuration-based programming, focusing primarily on existing work on formal specification of configuration elements. Existing proposals roughly subdivide into two categories depending on whether they address formal specification of functional properties provided by components, or of communication protocols provided by connectors. Section 3 discusses a complementary work that we are carrying out in the framework of the Aster project[1], within the Solidor research group at IRISA/INRIA. Our proposal deals with formal specification of non-functional properties (e.g., security, availability, timeliness) provided by connectors. The resulting framework allows to reason about the behavior of a distributed application with respect to the resource management policies that are transparently implemented by the underlying distributed runtime system. There is still much research work to be undertaken in the field of configuration-based programming to cope with the overall requirements of complex distributed applications such as multimedia services; some open issues are addressed in Section 4 that offers conclusions.

## 2  Formal Specification of Configuration Elements

This section focuses primarily on research work in the area of formal specification of configuration elements, illustrating provided means for verifying the semantic

---

[1] Information about the Aster project may be found at the URL: http://www.irisa.fr/solidor/work/aster.

correctness of configuration-based applications. In a first step, we briefly introduce the main principles underlying the description of a configuration-based software.

## 2.1  Describing a configuration-based software

The description of a configuration-based software relies on the use of a *configuration language* that allows the description of an application in terms of the interconnection of software components. Existing configuration languages differ depending on whether they allow the description of connectors (i.e., communication protocols) or not (e.g., see [5]):

- The former type of language, qualified as an ADL (*Architecture Description Language*), supports the most general description of configurations since various types of communication protocols can be specified for component interactions.
- The latter type of language, qualified as a MIL (*Module Interconnection Language*), subsumes in general a specific communication protocol for component interactions due to the absence of connectors. However, as shown by the Polylith configuration-based programming system [24], this shortcoming can be alleviated through the introduction of a notion (called *software bus* in Polylith) similar to connector, at the level of the programming system rather than within the configuration language.

In the following, we concentrate on the definition of ADL languages, the one of MILs being straightforward from it.

An ADL introduces a set of notations for the description of components, connectors and configurations.

A component may be either primitive or complex: a *complex component* is equivalent to a configuration; and a *primitive component* corresponds to a program that may be written in any programming language. The description of a primitive component gives the component interface, that is to say, the list of operations (or services) provided for other components and the list of operations required from other components. The component description further states the corresponding implementation file.

In the same way, a connector may be either primitive or complex, a *complex connector* being described in terms of existing components and connectors. However, to our knowledge, the description of complex connectors has not yet been integrated within an existing ADL. A *primitive connector* corresponds to a communication protocol of the target execution environment. For instance, in a Unix framework, there will be a connector describing a pipe. The description of a primitive connector includes the connector interface and a reference to the implementation of the corresponding protocol.

Finally, the description of a *configuration* consists in interconnecting a set of components so as to bind the operations that are required by some configuration components to the corresponding operations that are provided by other components of the configuration. These interconnections are further realized through connectors, hence specifying the communication protocols that are used for the resulting interactions between components.

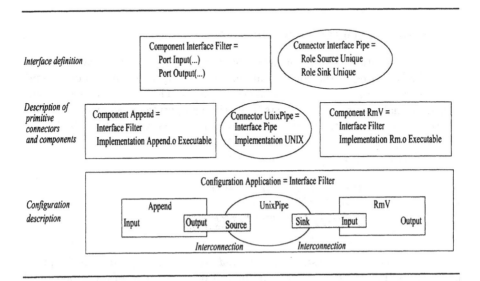

**Fig. 1.** Application description using an ADL

For illustration purpose, let us give the description of a simple configuration that is taken from [5]. The proposed *Application* configuration describes an application that takes a character string as input and that displays it on a screen after vowel removal (see figure 1). The configuration implements the interface named *Filter*, which declares the *Input* and *Output* operations using the **port** keyword. Furthermore, the configuration is composed of two primitive components that also implement the *Filter* interface: the *Append* component collects characters and the *RmV* component removes vowels. The configuration components interact through the interconnection of the *Output* operation of *Append* with the *Input* operation of *RmV*. This interaction is achieved using the *Unix-Pipe* primitive connector that implements the pipe-type communication protocol provided by the Unix system. The *UnixPipe* connector interface, named *Pipe*, declares two operations using the **role** keyword: *Source* and *Sink*.

The above configuration description is given in terms of the informal specification of its constituents. There is ongoing research work on formal specification of components and connectors, hence allowing some verifications about the semantic correctness of configuration-based applications. For instance, formal specification of components enables to check for the correct interconnection of components with respect to the behaviors specified for the required and provided operations. Specification of connectors further allows to verify the correct use of the corresponding communication protocols within components. The two following subsections discuss proposed solutions to the formal specification of components and connectors, respectively.

## 2.2 Formal specification of components

Formal specification of components amounts to the formal specification of the operations declared in the components interfaces, that is to say, the operations that are required and provided by the components. Thus, an approach to component formal specification consists in specifying operations behaviors in terms of pre- and post-conditions using the Hoare logic [8] as proposed in [23, 30].

For illustration purpose, let us consider the *create* operation of a file system. A specification for this operation may be:

> create (F: **FileName**) : (P: **FilePtr**)
> *pre:* **not** *exist(F)*
> *post: exist(F)* **and** *open(F)* **and** *valid(P)*

where the meanings of *exist, open* and *valid* are straightforward from the predicate names. Let us now consider an application constituted of a component $C$ such that $C$ accesses a file system and creates at least two copies of each accessed file for availability reason. From the specification of *create*, it can be inferred that the application configuration composed of the $C$ component and of a file system component implementing *create*, is not correct with respect to the application requirement except if the $C$ component assigns distinct names to each copy of a file.

In addition to the benefit of component specification for verifying the correctness of component interconnection, component specification favors software reuse and evolution: a component may be retrieved from a component database using the component specification; the correctness of component substitution within a configuration can be checked with respect to the specifications of involved components. The aforementioned verifications lie in the definition of relations over specifications. These relations set conditions upon which software interconnection, re-use, and substitution are correct [23, 17, 30]; and they are defined in terms of specification matching. Examples of such relations, which are taken from [30], are given hereafter.

**Specification matching.** A component specification is the set of specifications of the component's (required and provided) operations. Thus, specification matching over components is defined in terms of matching between operation specifications.

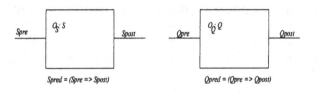

**Fig. 2.** Specifications of the $O_Q$ and $O_S$ operations

Let us first examine matching between the specifications $Q$ and $S$ of two operations $O_Q$ and $O_S$ (see figure 2). There is an *exact match* between two operation specifications if the two specifications are equivalent. We get:

$$exact\text{-}match(S, Q) = (Q_{pre} \Leftrightarrow S_{pre}) \wedge (Q_{post} \Leftrightarrow S_{post})$$

A weaker matching relation, called *plug-in match,* can be defined. It allows to verify that the operation $O_Q$ may be replaced by the operation $O_S$ whilst not entailing an error. We have:

$$plug\text{-}in\text{-}match(S, Q) = (Q_{pre} \Rightarrow S_{pre}) \wedge (S_{post} \Rightarrow Q_{post})$$

Let us now examine matching between the specifications $S_A$ and $S_B$ of components $A$ and $B$. There is an *exact match* between these specifications if for each operation of $B$: ($i$) there is a corresponding operation within $A$, and conversely, and ($ii$) there is an exact match between the specifications of $B$'s operation and of the corresponding operation of $A$. We get:

$$C\text{-}exact\text{-}match(A, B) = \exists U_{op} : S_B \rightarrow S_A \mid$$
$$U_{Op} \text{ is one-to-one and onto, and}$$
$$\forall Q \in S_B : exact\text{-}match(U_{op}(Q), Q)$$

where the $U_{op}$ function returns the specification of an operation of $A$ that matches the specification $Q$ of an operation of $B$, for each operation of $B$. There is a *plug-in match* between the specifications of $A$ and $B$ components if for each operation of $A$: ($i$) there exists an operation of $B$ that corresponds to it, and ($ii$) there is a plug-in match between the specifications of $A$'s operation and of the corresponding operation of $B$. We get:

$$C\text{-}plug\text{-}in\text{-}match(B, A) = \exists U_{Op} : S_A \rightarrow S_B \mid$$
$$U_{Op} \text{ is one-to-one and}$$
$$\forall Q \in S_A : plug\text{-}in\text{-}match(U_{op}(Q), Q)$$

**Configuration correctness from the standpoint of component interconnections.** From a practical point of view, matching between component specifications enables to verify the correctness of software evolution, and of component interconnections within configurations. It further provides a sound basis for software reuse. In addition, the above facilities may be integrated within a configuration-based programming system through their implementation by means of a theorem prover.

From the standpoint of configuration evolution, a component $A$ may be replaced by a component $B$ if either $C$-$exact$-$match(B, A)$ or $C$-$plug$-$in$-$match(B, A)$ holds. The former case corresponds to software evolution without modifying the application's functional properties; for instance, this occurs when the algorithm that is implemented by the component is modified so as to improve the application performance. The latter case takes place when the functional properties of the application are enriched.

From the perspective of software reuse, matching between component specifications constitutes the basis for the definition of a search function over a set of components. We have:

$$Search: \qquad S \times P \times B \to B$$
$$Search(s,\ C\text{-}match,\ b) = \{c \in b:\ C\text{-}match(c.spec, s)\}$$

where $S$, $P$, and $B$ respectively denote the sets of component specifications, of component specification matching relations, and of components. Furthermore, an element of the set $B$ of components is a pair of the form $(impl, spec)$ where $spec$ is the specification of the implementation $impl$.

Finally, the correctness of an interconnection between a component $c_1$ and a component $c_2$ can be defined using the following function:

$$Interconnection: \qquad B \times B \to Bool$$
$$Interconnection(c_1, c_2) = C\text{-}plug\text{-}in\text{-}match(c_2, c_1') \wedge protocol\text{-}match(c_{2_p}, c_{1_p})$$

where $c_1'$ is the set of $c_1$ operations that are required from $c_2$, that is to say, the operations required by $c_1$ that are bound to the operations provided by $c_2$ within the embedding configuration. Furthermore the *protocol-match* predicate refers to interconnection correctness with respect to the communication protocols that are used by the interacting components. The predicate definition is given in terms of the formal specification of interactions, which is addressed in the next paragraph.

### 2.3 Formal specification of interactions

Formal specification of component interactions has been examined in [2]. This proposal concentrates on the specification of connectors from the standpoint of provided communication protocols, using a process algebra that is a subset of CSP [9]. For illustration purpose, let us give the resulting formal specification of the components and connector that compose the *Application* configuration (see figure 1) [5]. We first introduce the notations used for defining processes.

**Notations.** A process is defined in terms of the following notations: **skip** denotes the process that does nothing; $e \rightarrow P$ denotes the engagement of a process in event $e$; $P \square Q$ denotes the deterministic choice between processes $P$ and $Q$ by the environment; $P \sqcap Q$ denotes the non-deterministic choice between processes $P$ and $Q$ by the embedding process; $P = $ *definition* denotes process naming; and **let** $Q = exp$ **in** $P$ defines a scoped process name.

**Specifying the interactions performed by a component.** Using the above notations, the specification of a component from the standpoint of performed interactions decomposes into the specification of a process for each component operation (or port), and the specification of the process defining coordination among ports. We get the following enhanced description of the *Filter* interface:

> **component interface** Filter =
>    **port** Input = read ? x → Input $\square$ readEnd → **skip**
>    **port** Output = send ! x → Output $\sqcap$ termination → **skip**
>    **coordination** =
>       **let** end = Input.readEnd → Output.termination → **skip in**
>       end $\square$ (Input.read ! x → (Output.send ? x $\sqcap$ **skip**) → **coordination**)

The specification of the interactions performed by a component is then the parallel composition of the processes defining ports and of the coordination process.

**Specifying the interactions performed by a connector.** We now introduce the specification of the *Pipe* connector from the standpoint of the provided communication protocol. This specification defines a process for each role of the connector, and for the coordination among these roles. We get:

> **connector interface** Pipe =
>    **role** Source **unique** = send ! x → Source $\sqcap$ termination → **skip**
>    **role** Sink **unique** =
>       **let** end = termination → **skip in**
>       **let** communication = (read ? x → Sink $\square$ *eof* → end) **in**
>       communication $\sqcap$ end
>    **coordination**
>       **let** readOnly =
>          Sink.read ! x → readOnly $\square$
>          Sink.*eof* → Sink.termination → **skip in**
>       **let** sendOnly =
>          Source.send ? x → sendOnly $\square$
>          Source.termination → **skip in**
>       Source.send ? x → **coordination** $\square$ Sink.read ! x → **coordination** $\square$
>       Source.termination → readOnly $\square$ Sink.termination → sendOnly

The specification of the interactions performed by a connector is then the parallel composition of the role and coordination processes.

**Configuration correctness from the standpoint of performed interactions.** Given the *Application* configuration depicted in figure 1, the actual behavior of the *UnixPipe* connector is given by the substitution of the role processes by the bound port processes. Process specification in terms of a subset of CSP then allows to verify that the behavior of a port is correct with respect to the expected behavior for the role to which it is bound [2].

To our knowledge, formal specification of connectors has only been addressed from the standpoint of provided communication protocols. Hence, existing approaches do not allow to take into account the resource management policies that are transparently implemented by the underlying distributed system and that influence the application behavior from the standpoint of criteria such as performance or dependability. A solution to this issue is discussed in the next section.

## 2.4  Discussion

This section has introduced existing solutions to the formal specification of configuration elements. There is much ongoing work in this area and there exist other complementary solutions that together contribute to the development of correct distributed applications. For instance, let us mention the work of [18] and [13]. However, to our knowledge, only results in the field of specification matching have been integrated within configuration-based programming systems.

Existing configuration-based programming systems can be classified into two categories depending on whether configurations are declared using a MIL or an ADL (see § 2.1); examples of systems belonging to the former category are Conic [16], Inscape [23], Durra [3], and Polylith [24] while examples of systems belonging to the latter category are Unicon [28], Darwin [15], Rapide [14] and Aesop [7].

## 3  Non-functional Properties of Configuration Elements

The previous section has shown that the configuration-based description of an application enables to verify the application correctness from the standpoint of component interconnections with respect to both the behavior of the components' provided and requested operations, and the behavior of communication protocols used for achieving the corresponding interactions. On the other hand, the formal specifications of configuration elements that have been introduced do not cope with non-functional properties provided by a connector, i.e., the resource management policies that are transparently implemented by the underlying distributed runtime system. This section discusses a solution that we have proposed in the framework of the Aster project so as to deal with the aforementioned issue. The following subsection details specification of non-functional properties, giving examples in the framework of the OMA architecture [19]. It is then shown how these specifications may be used to either select an existing connector or to implement a connector that fits the application requirements in terms of distribution management.

## 3.1 Specifying non-functional properties

A non-functional property characterizes a resource management policy that is transparently implemented by the underlying (possibly distributed) runtime system. For instance, we identify properties relating to the location of the system entities, to fault-tolerance and to the response time that is guaranteed for the system functions. In a way similar to the work on formal specification of configuration components (see § 2.2), non-functional properties are specified in terms of the first order logic. However, due to the non-functional nature of considered properties, specifications relate to operations that are not explicitly stated within the configuration description but that are instead transparently implemented by the distributed system (or connector) managing component interactions. This leads us to specify any non-functional property in terms of a unique predicate instead of pre- and post-conditions. Furthermore, predicates describing non-functional properties are defined in terms of *base predicates* that characterize the communication functions provided by the connector. These functions are named **send** and **receive** in the following; they respectively denote the emission and the reception of a message occurring during component interactions.

**Base predicates.** Let $C_1$ and $C_2$ be two components and $m$ be the message sent from $C_1$ to $C_2$ during an interaction between these two components. The **send** and **receive** functions provided by any connector are then defined using the following predicates:

- $send(C_1, C_2, m)$ corresponds to the emission of $m$ by $C_2$ to $C_1$.
- $receive(C_1, C_2, m)$ specifies that $C_2$ receives $m$, *once and only once*.
- $receive^+(C_1, C_2, m)$ specifies that $C_2$ receives $m$, *at least once*.
- $failure(C_1, C_2, m)$ denotes the occurrence of a failure between the emission of $m$ by $C_1$ and the end of the treatment of $m$ by $C_2$.

The above predicates hold subsequently to the execution of the **send** and **receive** functions. As discussed in [11], this can be specified using the Hoare logic [8].

**Non-functional properties.** There exists a wide diversity of non-functional properties. Up to now, we have examined the specification of properties provided by RPC systems, covering properties relating to failure semantics [11], interoperability [26], and server trading [12]. Examples of the corresponding specifications are given in the remainder. In addition, we are studying the specification of non-functional properties in a broader way by considering classes of non-functional properties such as properties relating to security [4] and fault-tolerance [27].

**Example: Non-functional properties provided by the OMA.** Let us illustrate specification of non-functional properties in the framework of the OMG's OMA architecture. Briefly stated, the OMA is an architecture that is based on the object paradigm. It defines three types of objects:

- *Objects services* are objects providing non-functional properties [20].
- *Common facilities* provide facilities that are commonly used within distributed applications (e.g., window manager).
- *Application objects* are objects, which are specific to user applications.

*The CORBA Architecture.* Interactions between the OMA objects rely on the CORBA (*Common Object Request Broker Architecture*) architecture [22] that defines the RPC-based communication system named ORB. More precisely, any ORB implements a set of interfaces for achieving interactions between objects:

- The *Static* and *Dynamic Invocation Interfaces* (SII and DII, respectively) are provided for issuing requests within client objects using respectively *Client stubs* generated by the OMG IDL compiler, and ORB functions.
- The ORB *standard interface* enables to access ORB services.
- The IDL *Skeleton* and *Dynamic Skeleton Interfaces* are provided for handling requests within server objects.
- The *Object Adapter* allows to access a given kind of objects (e.g., database objects); it provides functions such as the interpretation of objects references.

Prior to give some non-functional properties provided by the OMA, let us compare a programming system complying with the CORBA specification, with a configuration-based programming system that offers only a connector corresponding to the ORB. The main difference between these two types of systems then lies in the expression of component (or object) interconnection. Using CORBA, the application programmer has to code calls to server objects within the implementation of client objects while this may be achieved automatically from the configuration description in a configuration-based system (e.g., see [6]). The main advantage of the OMA comes from the fact that it is a standard for distributed computing. In particular, one may expect the implementation of various object services and common facilities, hence promoting software reuse. It follows that the design of a configuration-based system enabling to use the ORB as one connector allows to combine the advantages of both systems [10].

*Non-functional properties provided by the ORB.* With respect to the CORBA specification [22], any ORB implements various RPC semantics.

Semantics that are offered when using the SII are the following. By default, the ORB provides *synchronous* calls with the *at most once* failure semantics. Should an object service operation be qualified with the **oneway** attribute, the operation is called *asynchronously* with the *best effort* failure semantics. The use of the DII allows a larger choice of RPC semantics. In addition to the aforementioned semantics, the following types of calls may be requested:

asynchronous with synchronization on the receipt of result, and non-ordered multicast. Furthermore, these synchronization and topology semantics may be combined with either the *best effort* or *at most once* failure semantics, depending upon the presence of the **oneway** attribute.

Let us now give the specification of the non-functional properties characterizing the failure semantics provided by the ORB. We get:

$Best\text{-}Effort(C_1, C_2, req) \equiv$
  $send(C_1, C_2, req) \wedge (\neg failure(C_1, C_2, req) \Rightarrow receive^+(C_1, C_2, req))$
$At\text{-}Most\text{-}Once(C_1, C_2, req) \equiv$
  $send(C_1, C_2, req) \wedge (\neg failure(C_1, C_2, req) \Rightarrow receive(C_1, C_2, req) \wedge$
  $(failure(C_1, C_2, req) \Rightarrow (\neg receive^+(C_1, C_2, req) \vee receive(C_1, C_2, req)))$

Given the above properties, it is straightforward to specify a non-functional property that is either required or provided by a configuration element [11]. It amounts to set the properties parameters. Let $Fail(C_1, C_2, m)$ denote any property characterizing a failure semantics. Let us further assume that a component $C$ of a configuration requires the $Fail$ property for all the operations it invokes. We get:

$$\forall C' \in \mathcal{I}_C, \ \forall req \in R_{C,C'}, \ Fail(C, C', req),$$

where $\mathcal{I}_C$ denotes the set of components providing operations called by $C$, and $R_{C,C'}$ denotes the set of call messages that are exchanged between $C$ and $C'$.

*Non-functional properties provided by the OTS service.* Let us now consider non-functional properties provided by OMA object services. As an example, we focus on OTS (*Object Transaction Service*) [21] that implements failure atomicity for transactional systems based on the flat and/or the nested transaction models. The operations of the OTS service are called through the ORB and allow to create, commit and abort transactions. In addition to failure atomicity, a transactional system provides two other types of non-functional properties, which relate to concurrency control and persistence. In the OMA architecture, the provision of concurrency control and persistence rely respectively on the CCS (*Concurrency Control Service*) and POS (*Persistent Object Service*) services.

Prior to give the specification of the non-functional properties provided by OTS, let us consider a simple transaction model that provides failure atomicity at the level of a single operation. Committing the effects of the operation execution depends upon the execution of the **commit** operation. We thus introduce a base predicate, noted $validate(C_1, C_2, m)$ that holds if the effects of the call from $C_1$ to $C_2$ are committed using **commit**. We get the following property to characterize failure atomicity at the level of a single operation:

$Atomic(C_1, C_2, m) \equiv$
  $send(C_1, C_2, m) \wedge$
  $((\neg failure(C_1, C_2, m) \Rightarrow (receive(C_1, C_2, m) \wedge validate(C_1, C_2, m) \ ) \wedge$
  $(failure(C_1, C_2, m) \Rightarrow \neg receive^+(C_1, C_2, m)))$

The *Atomic* property does not specify the properties of the calls issued during the treatment of the $m$ request by the $C_2$ component. We introduce a stronger property, noted *Atomic-Rec*$(C_1, C_2, m)$, that specifies the properties of nested calls. This corresponds to the property defining failure atomicity under the flat transaction model. The *Atomic-Rec* property specifies that the effects of a transaction are committed only if all the operations that are called in the scope of the transaction are committed. On the other, the effects of all these operations are aborted if a failure occurs during the execution of one of them. We get:

$$
\begin{aligned}
\textit{Atomic-Rec}(C_1, C_2, m) &\equiv \textit{send}(C_1, C_2, m) \;\wedge \\
&\quad ((\textit{Receive-Rec}(C_1, C_2, m) \Rightarrow \textit{validate}(C_1, C_2, m)) \;\wedge \\
&\quad (\neg \; \textit{Receive-Rec}(C_1, C_2, m) \Rightarrow \textit{Undo}(C_1, C_2, m))) \\
\textit{Receive-Rec}(C_1, C_2, m) &\equiv \textit{receive}(C_1, C_2, m) \;\wedge\; \neg \; \textit{failure}(C_1, C_2, m) \\
&\quad \wedge_{\forall (C,n) \in \textit{called}_m} \; \textit{Receive-Rec}(C_2, C, n) \\
\textit{Undo}(C_1, C_2, m) &\equiv \neg \; \textit{receive}^+(C_1, C_2, m) \; \wedge_{\forall (C,n) \in \textit{called}_m} \; \textit{Undo}(C_2, C, n)
\end{aligned}
$$

where $\textit{called}_m$ gives the requests that are issued by $C_2$ during the treatment of $m$. Notice that due to the use of the *validate* predicate, *Atomic-Rec* specifies only the validation of effects for the top-level operation. However, specifying the validation of effects for all the nested calls is straightforward; it amounts to provide a recursive definition of *validate* in a way similar to the one of *Undo*. We may now give the specification of the failure atomicity property provided by the nested transaction model. The *Atomic-Rec* property must be enriched so as to cope with the specifics of the nested transaction model in the presence of failure: a nested transaction may either terminate successfully or be aborted. We get:

$$
\begin{aligned}
\textit{Nested-Rec}(C_1, C_2, m) &\equiv \textit{send}(C_1, C_2, m) \;\wedge \\
&\quad ((\textit{Nested-Rcv}(C_1, C_2, m) \Rightarrow \textit{validate}(C_1, C_2, m)) \;\wedge \\
&\quad (\neg \; \textit{Nested-Rcv}(C_1, C_2, m) \Rightarrow \textit{Undo}(C_1, C_2, m))) \\
\textit{Nested-Rcv}(C_1, C_2, m) &\equiv \textit{receive}(C_1, C_2, m) \;\wedge\; \neg \; \textit{failure}(C_1, C_2, m) \\
&\quad \wedge_{\forall (C,n) \in \textit{nested}_m} \; (\textit{Nested-Rcv}(C_2, C, n) \;\vee\; \textit{Undo}(C_2, C, m)) \\
&\quad \wedge_{\forall (C,n) \in \textit{called}_m - \textit{nested}_m} \; \textit{Nested-Rcv}(C_2, C, n)
\end{aligned}
$$

where $\textit{called}_m$ has the same meaning as before and the set $\textit{nested}_m$ gives the requests of $\textit{called}_m$ that are nested transactions.

## 3.2 Using non-functional properties to set up connectors

The previous subsection has discussed our approach to the specification of non-functional properties. A non-functional property is defined as a first order logic formula in terms of base predicates characterizing functions provided by the underlying runtime system (i.e., connector). In this subsection, we show how non-functional properties may be used to either select or build a connector that meets the application requirements from the standpoint of non-functional properties.

In Subsection 2.2, we addressed the use of specification matching to retrieve an application component from its specification. We undertake the same approach for the selection of connectors, and of components providing non-functional properties.

Let us recall that at least two matching relations may be defined over specifications: *exact-match* (noted $\lhd_{exact}$ in the following) requires specification equivalence for matching while *plug-in-match* (noted $\lhd_{plug-in}$ in the following) allows weaker specification matching. Let $\mathcal{P}_1$ and $\mathcal{P}_2$ be two sets of behaviors where a behavior is defined as a conjunction of non-functional properties. We have:

$$\mathcal{P}_1 \lhd_{exact} \mathcal{P}_2 \quad \equiv \forall P_2^i \in \mathcal{P}_2 : \exists P_1^j \in \mathcal{P}_1 \mid P_1^i = P_2^i$$
$$\mathcal{P}_1 \lhd_{plug-in} \mathcal{P}_2 \equiv \forall P_2^i \in \mathcal{P}_2 : \exists P_1^j \in \mathcal{P}_1 \mid P_1^j \Rightarrow P_2^i$$

**Selecting a connector.** Given the above specification matching relations, the selection of a connector meeting application non-functional requirements is direct. Let $\mathcal{R}_C$ be the set of behaviors required by the components of a configuration $C$ for their interactions, and let $\mathcal{P}_C$ be the set of behaviors provided by $C$'s components. A connector $S$ providing the set of behaviors $\mathcal{P}_S$ and requiring the set of behaviors $\mathcal{R}_S$ may be used for achieving interactions between $C$'s components if and only if:

$$(\mathcal{P}_S \cup \mathcal{P}_C) \lhd_{plug-in} (\mathcal{R}_S \cup \mathcal{R}_C).$$

**Building a connector.** Should the components of an application require some non-functional properties that are not provided by available connectors, a connector customized to the application needs must be built. As depicted in figure 3, such a connector can be built through the interconnection of a base connector (e.g., an ORB) with system-level components providing non-functional properties (e.g., OMA object services such as OTS). Specification matching then allows to retrieve the constituents of a customized connector that meets the application requirements.

Let $B$ be a base connector and $S_i$, $i = 1..n$, be a set of system-level components composing a customized connector. The behavior of the customized connector is:

$$\mathcal{P}_B \bowtie (\cup_{i=1..n} \mathcal{P}_{S_i}) = \cup_{\forall P \in \mathcal{P}_B} P \wedge_{i=1..n} \wedge_{j=1..m} P_{S_i}^j,$$

where $\mathcal{P}_B$ is the set of behaviors provided by $B$; $\mathcal{P}_{S_i}$ is the set of non-functional properties provided by the $S_i$ component and $P_{S_i}^j$ is one of these properties.

Let now $\mathcal{S}_E$ be the set of system-level components that are available within the development environment and $\mathcal{B}$, be the set of available base connectors. A connector customized to the requirements of a configuration $C$ can be built if and only if the following predicate holds:

$$
\begin{aligned}
&system(C) \equiv \\
&\quad \exists B \in \mathcal{B}, \exists \{S_1, ..., S_n\} \subset \mathcal{S}_E \mid \\
&\qquad ((\mathcal{P}_B \bowtie (\cup_{i=1..n} \mathcal{P}_{S_i})) \cup \mathcal{P}_C) \lhd_{plug-in} (\mathcal{R}_C \cup \mathcal{R}_B \cup_{i=1..n} \mathcal{R}_{S_i})
\end{aligned}
$$

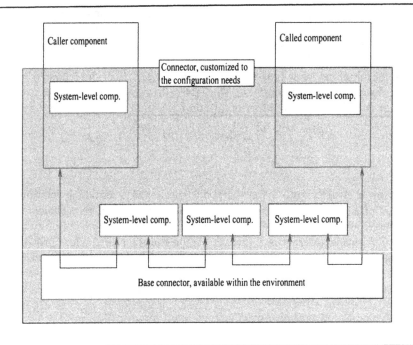

**Fig. 3.** Architecture of a customized connector

where $\mathcal{P}_B$ and $\mathcal{P}_C$ are the sets of behaviors provided by $B$ and $C$, respectively; $P_{S_i}$ is the set of properties provided by the $S_i$ component; and $\mathcal{R}_C$, $\mathcal{R}_B$ and $\mathcal{R}_{S_i}$ are the set of requirements specified for $C$, $B$ and $S_i$, respectively.

The retrieval of the connector and of system-level components that are needed for $system(C)$ to hold can be automated by means of a tool using a theorem prover. Examples of such tools can be found in [25, 17]. Furthermore, we have developed a tool that is dedicated to the implementation of our customization method.

### 3.3 Discussion

This section has sketched formal specification of non-functional properties provided by configuration elements. From the practical standpoint, this proposal has been integrated in the Aster configuration-based programming system [10]. The Aster system integrates a configuration language that allows to specify non-functional properties, specified properties being used to make up connectors matching application requirements. The Aster prototype currently supports RPC-based connectors, including the Orbix implementation of the ORB. We are now examining the enrichment of the Aster prototype so as to deal with various types of connectors as in ADL-based systems (see § 2.4).

## 4 Conclusion

This paper has given an overview of configuration-based programming, concentrating primarily on associated formal methods to ease the development of correct distributed applications. The proposed methods allow to reason about the correctness of distributed configurations from the standpoint of both functional and non-functional properties. Furthermore, these methods foster software reuse, evolution, analysis and management. Although existing proposals provide much help in the development of complex distributed software, there are still open issues to be addressed to cover the requirements of emerging distributed applications. In particular, dynamic evolution of configuration elements must be dealt with as this constitutes a key requirement of applications based on either multimedia data or mobile agents. In the case of multimedia services, dynamicity relates to the adaptability of the service with respect to resource availability so as to provide the largest number of users with the best quality of service. In the case of agent-based applications, dynamicity relates to the variety of execution environments within which a single agent may execute.

**Acknowledgment:** The author would like to acknowledge Christophe Bidan and Titos Saridakis who are contributors to the Aster research results addressed in this paper. She also would like to thank Jean-Pierre Banâtre for fruitful discussions about the topics discussed in this paper.

## References

1. G. Abowd, R. Allen, and D. Garlan. Using style to understand descriptions of software architecture. In *Proceedings of the ACM SIGSOFT'93 Symposium on Foundations of Software Engineering*, pages 9–20, 1993.
2. R. Allen and D. Garlan. Formalizing architectural connection. In *Proceedings of the Sixteenth International Conference on Software Engineering*, pages 71–80, 1994.
3. M. R. Barbacci, D. L. Doubleday, M. J. Gardner, R. W. Lichota, and C. B. Weinstock. DURRA: A Task-Level Description Language – Reference Manual (Version 3). Technical Report 18, Software Engineering Institute, 1991.
4. C. Bidan and V. Issarny. Security Benefits from Software Architecture. In *Proceedings of* COORDINATION '97: Coordination Languages and Models, 1997. LNCS.
5. J. Bishop and R. Faria. Connectors in configuration programming languages: Are they necessary. In *Proceedings of the Third International Conference on Configurable Distributed Systems*, pages 11–18, 1996.
6. J. R. Callahan and J. M. Purtilo. A packaging system for heterogeneous execution environments. *IEEE Transactions on Software Engineering*, 17(6):626–635, 1991.
7. D. Garlan, R. Allen, and J. Ockerbloom. Exploiting style in architectural design environments. In *Proceedings of the ACM SIGSOFT'94 Symposium on Foundations of Software Engineering*, pages 175–188, 1994.

8. C. A. R. Hoare. An axiomatic basis for computer programming. *Communications of the ACM*, 12:576–580, 1969.

9. C. A. R. Hoare. *Communicating Sequential Processes*. Prentice-Hall International, 1985.

10. V. Issarny and C. Bidan. Aster: A CORBA-based software interconnection system supporting distributed system customization. In *Proceedings of the Third International Conference on Configurable Distributed Systems*, pages 194–201, 1996.

11. V. Issarny and C. Bidan. Aster: A framework for sound customization of distributed runtime systems. In *Proceedings of the Sixteenth IEEE International Conference on Distributed Computing Systems*, pages 586–593, 1996.

12. V. Issarny, C. Bidan, and T. Saridakis. Designing an open-ended distributed file system in Aster. In *Proceedings of the Ninth International Conference on Parallel and Distributed Computing Systems*, pages 163–168, 1996.

13. D. Le Metayer. Software architecture styles as graph grammars. In *Proceedings of the ACM SIGSOFT'96 Symposium on Foundations of Software Engineering*, pages 15–23, 1996.

14. D. C. Luckham, J. J. Kenney, L. M. Augustin, J. Vera, D. Bryan, and W. Mann. Specification and analysis of system architecture using Rapide. *IEEE Transactions on Software Engineering*, 21(4):336–355, 1995.

15. J. Magee and J. Kramer. Dynamic structure in software architecture. In *Proceedings of the ACM SIGSOFT'96 Symposium on Foundations of Software Engineering*, pages 3–14, 1996.

16. J. Magee, J. Kramer, and M. Sloman. Constructing distributed systems in Conic. *IEEE Transactions on Software Engineering*, 15(6):663–675, 1989.

17. A. Mili, R. Mili, and R. Mittermeir. Storing and retrieving software components: A refinement based system. In *Proceedings of the Sixteenth International Conference on Software Engineering*, pages 91–100, 1994.

18. M. Moriconi, X. Qian, and R. A. Riemenschneider. Correct architecture refinement. *IEEE Transactions on Software Engineering*, 21(4):356–372, 1995.

19. OMG. Object Management Architecture Guide (OMA Guide). Technical Report 92.11.1, OMG, 1992. ftp: omg.org.

20. OMG. Common Object Services Specification, Volume 1 – Revision 1.0. Technical Report 94.1.1, OMG Document, 1994. ftp: omg.org.

21. OMG. Object Transaction Service. Technical Report 94.8.4, OMG Document, 1994. ftp: omg.org.

22. OMG. The Common Object Request Broker: Architecture and Specification – Revision 2.0. Technical report, OMG Document, 1995. ftp: omg.org.

23. D. E. Perry. The Inscape environment. In *Proceedings of the Eleventh International Conference on Software Engineering*, pages 2–12, 1989.

24. J. M. Purtilo. The Polylith software bus. *ACM Transactions on Programming Languages and Systems*, 16(1):151–174, 1994.

25. E. J. Rollins and J. M. Wing. Specifications as search keys for software libraries. In *Proceedings of the Eighth International Conference on Logic Programming*, pages 173–187, 1991.

26. T. Saridakis, C. Bidan, and V. Issarny. A programming system for the development of TINA services. In *Proceedings of the Joint IFIP International Conference on Open Distributed Processing and Distributed Platforms*, pages 3–14, 1997.

27. T. Saridakis and V. Issarny. Towards formal reasoning on failure behaviors. In *Second European Research Seminar on Advances In Distributed Systems (ERSADS)*, 1997.

28. M. Shaw, R. DeLine, D. Klein, T. Ross, D. Young, and G. Zelesnik. Abstractions for software architecture and tools to support them. *IEEE Transactions on Software Engineering*, 21(4):314–335, 1995.
29. TINA-C. Overall Concepts and Principle of TINA – Version 1.0. Technical Report TB_MDC.018_1.0_94, TINA-C Document, 1995. URL: http://www.tinac.com.
30. A. M. Zaremski and J. M. Wing. Specification matching of software components. In *Proceedings of the ACM SIGSOFT'95 Symposium on Foundations of Software Engineering*, pages 6–17, 1995.

# Automatic Generation of Parallelizing Compilers for Object-Oriented Programming Languages from Denotational Semantics Specifications

Prakash K. Muthukrishnan and Barrett R. Bryant

Department of Computer and Information Sciences
University of Alabama at Birmingham
Birmingham, Alabama 35294-1170, U. S. A.
{prakash, bryant}@cis.uab.edu

**Abstract.** The denotational semantics of object-oriented programming languages (OOPL's) are used to derive the control and data dependencies that exist within programs and this information is then used to produce parallel object code by a compiler. This approach is especially suited for OOPL's because of the concurrency inherent in their semantics. A denotational semantics of an OOPL called SmallC++ is developed with some specialized operations that facilitate the automatic generation of a parallelizing compiler for SmallC++. The result is a compiler which generates code for shared or distributed memory multi-processors, thereby achieving a language implementation which realizes fully transparent parallel object-oriented programming.

## 1 Introduction

The object-oriented (OO) programming paradigm provides linguistic support for objects, classes and inheritance [18]. It supports the software engineering principles of data abstraction, information hiding, modular design, and code reuse, and it allows the modeling of real world objects in a natural fashion. Since objects are independent entities that communicate through message passing, the object-oriented paradigm lends itself naturally to the development of distributed systems. This has been pursued primarily by developing object-oriented languages with explicit concurrency constructs called concurrent object-oriented programming languages. For surveys of general principles, the reader is referred to [17]. The various concurrent object-oriented programming systems provide some explicit concurrency constructs which the programmer has to use to obtain parallelism. The development of concurrent object-oriented programming systems have been complemented by efforts in defining the formal semantics of such systems [20, 16]. There has been only very little work done in automatic parallelization of OO languages [19, 4, 2].

We use formal semantics specification in deriving the parallelism that is available in object-oriented programs, which are inherently parallel, and propose a tool which a software developer can use to develop parallel applications by just writing sequential object-oriented code. Such technology has not been applied for OOPL's. We propose a methodology that extends existing techniques in automatic compiler generation [9] and parallelizing compilers [21]. One immediate benefit of such a tool is that it could be used to enable an existing object-oriented programming language, such as Smalltalk[5] or C++ [15], to be directly compiled into parallel code.

The paper is structured as follows. In Section 2 we present SmallC++ [11], a subset of C++ that we believe to be tractable for developing a prototype of a parallelizing compiler. In Section 3 we present the background information on Denotational Semantics and how it can be used in generating compilers. This is illustrated with an example using the semantics of SmallC++ in Section 4, where we present the semantics of SmallC++. The use of denotational semantics to produce a parallelizing compiler and the generation of the parallel target code is explained in Section 5. And finally in Section 6 we present the summary of our project and future work.

## 2   SmallC++

SmallC++ is a distilled version of the object-oriented programming language C++. The object-oriented features of SmallC++ include classes, objects, message-passing and inheritance with dynamic binding. In our subset, we can create classes and instances of a class (objects). A user programming with our subset can declare a member (data or function) to be *private, protected* or *public*. Since we do not support pointers, dynamic creation of objects is not supported in our language. Our subset allows only public derivation; public derivation in C++ models the *is-a* relationship and thus supports inheritance. Like Smalltalk, we do not support multiple inheritance. We have retained the keyword *virtual* in our subset and also the reference feature of C++. Combining these two features, polymorphism and late binding is achieved. Figure 1 is an example program written in SmallC++.

## 3   Formal Semantics of Programming Languages and Semantics-Directed Compilers

A programming language has two main characteristics, namely *syntax*, which describes the appearance and structure of its sentences, and *semantics*, which describes the assignment of meanings to the sentences. The syntactic specification of languages has been standardized using the Backus-Naur Form (BNF), or context-free grammar, and there are many automated tools which support the syntax analysis phase of a compiler (e.g. lex and Yacc [1]). In semantic specification, there is no such standard, although of the various methods for defining the formal semantics of programming languages, denotational semantics is the most commonly used because of its unique combination of mathematical properties and formalization of execution models [14].

A denotational semantics specification of a programming language typically consists of the following components:

```
class Point
{
    private:
      int x;
      int y;
    public:
      Point (int a, int b) { x = a; y = b; }
      int GetX () { return x; }
      int GetY () { return y; }
};

void main ()
{
  Point obj1 (3,4);
  Point obj2 (5,6);
  int d;
  // Note that all four method invocations can execute in parallel
  d = sqrt(square(obj1.GetX() - obj2.GetX()) +
           square(obj1.GetY() - obj2.GetY()));
  cout << d;
}
```

**Fig. 1.** SmallC++ example program

1. *Abstract syntax* gives the abstract syntactical details of the language, which can be derived from its *concrete syntax*, the syntax described by the BNF grammar. It is typically defined as a set of *syntax domains*, each of which describes an abstract syntax representation of the corresponding concrete syntax construction.

2. *Semantic algebra* describes the various *semantic domains* and the operations associated with the elements of those domains. Basic domains include numbers, Boolean values, lists, etc. More complex domains may be structured in three ways:

   (a) *Product domains* are domains of tuples whose components consist of other semantic domains. These are typically denoted by $D_1 \times D_2 \times ... \times D_n$, where $D_1$, $D_2$, ..., $D_n$ are the component domains. The main product domain operation is a *selector* to extract a component, for example, by its position in the tuple.

   (b) *Sum domains*, also called disjoint unions, are unions of sets of values, each value being tagged with the type of the domain it belongs to. These are typically denoted by $D_1 + D_2 + ... + D_n$, where $D_1$, $D_2$, ..., $D_n$ are the component domains. Sum domain operations include *injection* operations to insert a value into the sum with the appropriate type tag, and *projection* operations to extract a value which is tagged with a certain type.

(c) *Function domains* are mappings of one domain into another, expressed as $D_1 \rightarrow D_2 \rightarrow ... \rightarrow D_n$, which denotes a function over domain $D_1$ whose range is $D_2 \rightarrow ... \rightarrow D_n$. That is, the $\rightarrow$ operator associates to the right. Function domains are usually modelled using $\lambda$-calculus and may be used to represent mappings such as the mapping of an identifier into its denotation, or the mapping of a memory location to its value. Examples of $\lambda$-calculus expressions are *addone* $= \lambda y.y + 1$ and *apply* $= \lambda f.\lambda x.fx$, equivalent to the more standard function notations $addone(y) = y + 1$ and $apply(f, x) = f(x)$, respectively. A detailed discussion of $\lambda$-calculus is beyond the scope of this paper (see [6] for an introduction).

3. *Semantic functions* express the denotation of each abstract syntax phrase in terms of the denotations of its components. By applying the semantic functions, it is possible to elaborate the denotation of a complete program as the composed denotation of its various components. Through this process, all syntax domains are translated into semantic domains. At the highest level of this mapping, programs are translated into their denotations, in the form of a $\lambda$-calculus expression that maps input into output. It is this mapping which allows compilers to be constructed from denotational semantics specifications.

The process of compiling a program according to a denotational semantics specification is called *semantics-directed* compiling and the theory behind this process dates back over thirty years. The typical process is to use the semantic functions of the denotational semantics to translate a program $P$ into a functional form, e.g. in $\lambda$-calculus, which maps input $I$ to output $O$. That is, a denotational semantics may be regarded as a function over domain $P \rightarrow I \rightarrow O$. If we have $P$, then we can get $I \rightarrow O$. If we have both $P$ and $I$, we can get $O$. Since a denotational semantics defines both *static* and *dynamic* semantics, i.e. type checking as well as run-time behavior, it is usual for such a compiler to have at least two phases, one to evaluate the semantic functions involved with the static semantics at compile-time, and the other to perform the translation into the final functional form which represents only the functions which should be executed at run-time. We shall refer to the functional form produced by this process as "denotational code." This code may be executed by any number of different functional programming language interpreters, such as ML [10], or it may be compiled into actual machine code using functional programming language implementation techniques [13]. This process is illustrated in Figure 2.

Since this approach may be used to produce a compiler from a language specification, it is called *automatic compiler generation*. The most usual approach for producing target code in such systems is not to compile the denotational code as a functional program but rather to consider it as representing a fixed set of operations called *combinators*. A *combinator* is a $\lambda$-expression without free variables, essentially a function without global variables (*addone* and *apply*

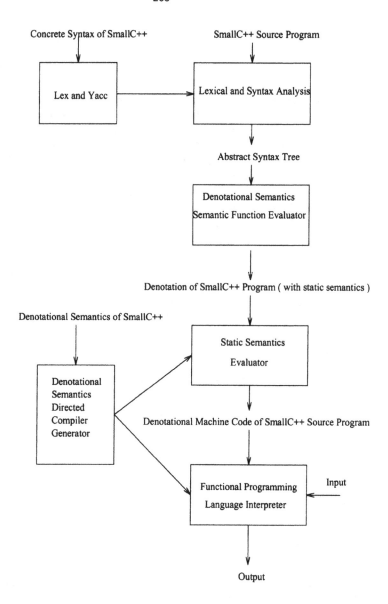

**Fig. 2.** Overview of semantics-directed compilation process

mentioned previously are examples of combinators). Combinators generated as denotational code are typically denotational expressions which have direct representations in conventional target machine languages (e.g. arithmetic and logical operations, conditional and iterative constructions, etc.). This greatly facilitates the automatic generation of target machine code.

# 4 Semantics of SmallC++

Denotational semantics functions typically have two styles, *direct* and *continuation*. Direct semantics models the semantic representation of a program as a standard composition of functions, e.g. $f_n(...(f_2(f_1))...)$, where $f_1$ is the first function denoted by the program, $f_2$ the second, etc. Continuation semantics, on the other hand, models this composition as a sequence, each function in the sequence taking the "rest of the sequence" as an argument, e.g. $f_1(f_2(...(f_n)...))$, where $f_2(...(f_n)...)$ is the *continuation* of $f_1$. Not only is this approach more like a sequence of machine instructions when each function is a combinator, but also it is useful in modelling more powerful control-flow constructions than direct semantics, such as function call and return, loop exits, etc. We present the denotational semantics of SmallC++ using the continuation style approach.

## 4.1 Abstract Syntax

We begin our denotational semantics specification with the abstract syntax for SmallC++ shown in Figure 3. Each of the indicated syntax domains may be thought of as a model of an abstract syntax tree, whose children are themselves abstract syntax trees defined by other domains. For example, an Assignment Statement, AS, is a syntax tree with operator =, and two children, defined by VAR and E, respectively, each of which denotes an additional abstract syntax tree defined by their abstract syntax rules. From our SmallC++ example, it can be seen that the single statement in main matches the AS syntax domain, with d being the VAR and the sqrt expression corresponding to E. The syntax domain rules for E break this down further. VAR syntax domian models identifiers that could be class-identifiers, object-identifiers, identifiers to denote ground types or function identifiers.

## 4.2 Semantic Algebras

In this section we present some of the semantic algebras used in modeling our language. The basic semantic domains are Boolean Truth Values, Identifiers, which are strings of characters, and Natural numbers. These basic domains may be composed into larger semantic domains to represent denotations of program entities. The Denotable-Values domain models the values that are denoted by identifiers in a program. SmallC++ identifiers may denote memory locations (Loc) if they are call-by-reference formal parameter variables, procedures (Proc) if they are

```
Prog := P
P    := D_1, D_2, D_3, ..., D_n
D    := SD | CD | FD
SD   := int I_1 | I_c I_1 | int I_1 [n_1]...[n_n] | I_c I_1 [n_1]..[n_n]
CD   := class I_c ML | class I_c : public I_c ML
FD   := TS I_f AL BD
TS   := int | void | I_c | int I_c :: | void I_c :: | I_c I_c ::
ML   := AS ML_1, ML_2,..., ML_n | SD | FD | virtual FD = 0
AS   := private | public | protected
AL   := AL_1, AL_2 | int I_1 | I_c I_1 | int & I_1 | I_c & I_1
BD   := SD CSL | SD CSL


CSL  := C_1;C_2
C    := AS | IF | FOR | OUT | IN | RET
AS   := VAR = E
IF   := if E CSL else CSL
FOR  := for AS_1 E AS_2 CSL
OUT  := cout ≪ E
IN   := cin ≫ E
RET  := return E


EL   := E_1, E_2
E    := I | N | I_f (EL) | UOP VAR | UOP N | UOP I_f(EL) | E_1 BOP E_2
        | E_1.E_2 | OID.I_f (EL) | this → I_f (EL)
VAR  := I | VAR.I | this → I | VAR [EL]
```

**Fig. 3.** Abstract syntax of SmallC++

functions, as well as classes, arrays, system variables (like **cin** and **cout**), objects, and error values (e.g. if the identifier doesn't have a declaration). Since these are all different types, a sum domain is used to model the union of the various types. The Storable-Values domain is the set of values that can be stored in locations in the store, either integers or files, in the case of **cin** and **cout**, and Expressible-Values is the domain that is used to model the values produced by expression evaluations.

**Denotable-Values** Domain $d \in$ **Dv** = Loc + Proc + Class + Array + System-Var + Object + ErrValue

**Storable-Values** Domain $v \in$ **Sv** = Nat + File

**Expressible-Values** Domain $e \in$ **Ev** = Nat + Tr + Dv

The symbol table which models bindings of identifiers is defined using the Environment domain with basic symbol table operations. The environment is regarded as a mapping of identifiers to denotations, so is modelled as a function. Some of the operations defined on this domain are *initialenv* to set up the initial environment for a local scope, *accessenv* which accesses an identifier in the scope, *updateenv* to update the symbol table with a new definition, and a composition operator to combine symbol tables. Environment is a static semantics domain which would not be present in the denotational machine code.

**Environment** Domain $r \in$ **Env** $=$ Id $\rightarrow$ Dv

The principal dynamic semantics domain of any object-oriented language is the memory store, which is modelled after the memory store of a conventional stored-program computer, organized as a stack of activation records, where every address is computed from a relative location within a stack frame. Like Environment, Store consists of a mapping function which associates locations with the Storable-Values that are stored in the locations. A Location is simply an address, so may be modelled by natural numbers. The store is denoted by a triple where the first component represents the stack frame, the second component the top of the stack and the third component the map of Location to Storable Values.

**Store** Domain $s \in$ **Store** $=$ Loc $\times$ Loc $\times$ (Loc $\rightarrow$ Sv)
Operations:
*access* : Store $\rightarrow$ Loc $\rightarrow$ [Sv + {Errvalue}]
access $= \lambda$ (sp,top,map) . $\lambda$ l . l $<$ top $\rightarrow$ map(l) ⫿ ErrValue
*update* : Store $\rightarrow$ Loc $\rightarrow$ Sv $\rightarrow$ [Store + { errorStore }]
update $= \lambda$ (sp,top,map) . $\lambda$ l . $\lambda$ v . l $<$ top $\rightarrow$ ([l$\mapsto$v](map,top)) ⫿ errorStore
*mark-locn* : Store $\rightarrow$ ( Loc $\times$ Loc $\times$ Store )
mark-locn $= \lambda$ (sp,top,map) . ( sp, top, (sp+top,0,map) )
*alloc-locn* : Store $\rightarrow$ Nat $\rightarrow$ (location $\times$ Store)
alloc-locn $= \lambda$ (sp,top,map) . $\lambda$ n . (sp+top, (sp,locn + 1,map))
*dealloc-locn* :Store $\rightarrow$ Loc $\rightarrow$ Loc $\rightarrow$ Store
dealloc-locn $= \lambda$ (sp,top,map). $\lambda$ $l_1$. $\lambda$ $l_2$. ($l_1$,$l_2$,map)
*initial-store* : File $\rightarrow$ Store
initial-store $= \lambda$ f. let $s' = empty\text{-}store$ in ([cin $\mapsto$ f]$s'$)
*empty-store* : Store
empty-store $= (0,0,\lambda$ l.inErrValue())

A continuation is formally a function that maps intermediate results ( that are expected by the "rest of the program" ) to their final answers. We use three standard continuations in our formal semantics: Command continuations, Expression continuations and Declaration continuations. A command modifies a store and passes this modified store to the rest of the program following it. So we define the command continuations as follows:

**Command Continuations** Domain $c \in$ **Cc** $=$ Store $\rightarrow$ Ans

A Declaration modifies the Environment and possibly the store. Since Declarations pass the modified environment and the store to the rest of the program following them we define Declaration Continuations as:

**Declaration Continuations** Domain $u \in$ **Dc** $=$ Env $\rightarrow$ Store $\rightarrow$ Ans

Since Expressions pass their values and the modified store to the rest of the program following them, we define Expression Continuations as following:

**Expression Continuations** Domain $k \in$ **Ec** $=$ Ev $\rightarrow$ Store $\rightarrow$ Ans

## 4.3 Semantic Functions

We illustrate the semantic functions for SmallC++ using the example program presented earlier. The top-level function is **PROG** which maps a program, Prog, into a functional mapping an input file, File, into an Answer, Ans. The domain Ans is the final answer which will be a finite string of values (of domain Nat) ending either with a **stop** (in case of a successful completion) or **error** (in case of a failed execution).

**PROG** : Prog $\rightarrow$ File $\rightarrow$ Ans
**PROG** $[\![$ Prog $]\!]$ = $\lambda$ f. **P** $[\![$ Prog $]\!]$ *initialenv initial-store* f

**P** : ListOfDefinitions $\rightarrow$ Env $\rightarrow$ Cc
**P** $[\![ D_1, D_2, \ldots D_n ]\!]$ r :=
$\qquad$ **D** $[\![ D_1 ]\!]$ r ( $\lambda r_1$ . $\lambda s_1$ . **D** $[\![ D_2 ]\!]$ r$[r_1]$ ( $\ldots$
$\qquad$ ( $\lambda r_{n-1}$ . $\lambda s_{n-1}$ **D** $[\![ D_n ]\!]$ $r_n$ ( $\lambda r_n$ . $\lambda s_n$ .
$\qquad$ **R** $[\![$ *main* $]\!]$ ( $r_f$ ( $\lambda e_{final}$. $\lambda s_{stop}$. result $s_{stop}$ "stop" ) $s_n$ ) )
$\qquad$ $s_{n-1}$ ) $\ldots$ ) $s_1$ )
$\qquad$ where $r_n$ = r$[r_1[\ldots [r_{n-1}]\ldots]]$

A program is a list of declarations and the denotation of a program is a function to construct an environment from the list of declarations and then invoke the function **main** to start the execution. The static semantics of the declarations will be simplified during static semantics evaluation and we will be left with a denotation as a pure dynamic semantic function. For our example program, there is only a declaration of the **Point** class and the **main** function.

Of the various static semantic functions, we would like to concentrate our discussion on the semantics of a SmallC++ function definition, since it must also define the dynamic semantics of a function. A function definition binds the denotation of a function to the function identifier $I_f$ in the current environment r. The denotation of the function is a component of the Procedure domain, a type for the function and its denotation, and is injected into that domain using in-Proc. The denotation of the function is expressed as a $\lambda$-expression which takes a Command Continuation c, memory store s, and set of Expressible Values (the actual parameter values), and returns the result of evaluating the body of the function, BD, with the formal parameter list, AL, bound to the values of the actual parameters. This denotation also contains the environment r, since the body may contain declarations that statements that require resolving the semantics of identifiers from the environment (in the semantics of BD, these declarations are elaborated by the semantic function **SD** which we have not shown here). This environment would also be simplified in static semantics processing, leaving the purely dynamic semantics part of the denotation. Only at run-time will this denotation actually be executed, once the function is actually called and the actual parameter values are known. Here we merely associate this denotation with the function identifier in the environment. It is the capability of $\lambda$-calculus to denote such functions that makes it extremely suitable for our semantic representation.

**FD** : FunctionDefinition $\rightarrow$ Env $\rightarrow$ Dc $\rightarrow$ Cc
**FD** $[\![$ TS $I_f$ AL BD $]\!]$ r u :=
$\qquad$ let $t_1$ = TS $[\![$ TS $]\!]$ in
$\qquad$ let $r_1$ = $updateenv(I_f$, inProc ( $t_1$, p ), r ) in u $r_1$
$\qquad$ where p = $\lambda$ c. $\lambda$ s. $\lambda$ $e_1$ ... $e_n$.
$\qquad$ $BD$ $[\![$ BD $]\!]$ (**AL** $[\![$ AL $]\!]$ ( r s ($e_1$ ... $e_n$ ) ) c )

**BD** : Body $\rightarrow$ Env $\rightarrow$ Store $\rightarrow$ Cc $\rightarrow$ Ans
**BD** $[\![$ SD CSL $]\!]$ r c := **SD** $[\![$ SD $]\!]$ r ($\lambda$ $r_1$. $\lambda$ $s_1$. **CSL** $[\![$ CSL $]\!]$ $r_1$ ($\lambda$ $s_2$. c $s_2$) $s_1$)

After processing the declarations for our example program, we will have an environment consisting of the **Point** class and the **main** function. The declarations inside each will have been processed in a similar manner. Let us turn our attention to the body of **main**. The semantics of a CommandsList is defined as taking an environment and a command continuation and returning a command continuation.

**CSL** : CommandsList$\rightarrow$Env$\rightarrow$Cc$\rightarrow$Cc
**CSL** $[\![$ $C_1;C_2$ $]\!]$ r c := **CSL** $[\![$ $C_1$ $]\!]$ r ( $\lambda$ $s'$. **CSL** $[\![$ $C_2$ $]\!]$ r c $s'$ )

**CSL** $[\![$ C $]\!]$ := **C** $[\![$C $]\!]$

The first command in the CommandsList, $C_1$, is evaluated with the current environment r and the continuation is the evaluation of the second command, $C_2$, which may actually be another CommandsList. Individual statements are evaluated using the **C** function.

**C** : Command$\rightarrow$Env$\rightarrow$Cc$\rightarrow$Cc
**C** $[\![$ I = E $]\!]$ r c := **L** $[\![$ I $]\!]$ r ($\lambda$ l. $\lambda$ $s_1$. isLoc l $\rightarrow$
$\qquad$ **R** $[\![$ E $]\!]$ r ($\lambda$ e. $\lambda$ $s_2$. c ( $update$ $s_2$ l e ) ) $s_1$ $[\![$ error )

**C** $[\![$ cout $\ll$ E $]\!]$ r c := **R** $[\![$ E $]\!]$ r ( $\lambda$ e. $\lambda$ $s_1$. c ( $updatecout$ $s_1$ e ) )

Here we show the semantics of the assignment and output statements which are used in the example program. An assignment denotation requires the location of the identifier, I, on the left side of the assignment, which is achieved by evaluating the **L** function which looks this identifier up in the environment and returns its denotation. If this denotation is not a location, then we have an error denotation. Note that the way this works is to pass an expression continuation along with the environment to **L**. This expression continuation contains a call to the expression evaluation function **R** which takes en expression E, environment and expression continuation for updating the location of I with the value of E. The output function similarly evaluates its expression using **R**, passing an expression continuation to update the output stream component of the store. The **L** and **R** rules are further evaluated below.

**R** : E $\rightarrow$ Env $\rightarrow$ Ec $\rightarrow$ Cc
**R** $[\![$ I $]\!]$ r k := let d = $accessenv($ I , r ) in isLocn ( d ) $\rightarrow$ deref k d $[\![$ error

$\mathbf{L} : E \rightarrow \text{Env} \rightarrow Ec \rightarrow Cc$

$\mathbf{L} [\![ \; I \; ]\!] \; \mathbf{r} \; \mathbf{k} := \text{let } d = accessenv \; (\; I \;, \mathbf{r} \;) \text{ in } \mathbf{k} \; d$

The expression obj1.GetX() - obj2.GetX() in our program is evaluated using the following semantic function where $E_1$ represents the expression obj1.GetX() and $E_2$ represents the expression obj2.GetX(). Here first we evaluate Expression $E_1$ to get the value $e_1$ and the modified store $s_1$. We then evaluate $E_2$ using the new store and get the value $e_2$ and the modified store $s_2$. Values $e_1$ and $e_2$ are passed to the BOP function which applies the binary operation on the values and returns a new values which is passed along with the store $s_2$ to the continuation k.

$\mathbf{R} [\![ \; E_1 \; \text{BOP} \; E_2 \; ]\!] \; \mathbf{r} \; \mathbf{k} :=$

$\qquad \mathbf{R} [\![ \; E_1 \; ]\!] \; \mathbf{r} \; (\; \lambda \; e_1 \; . \; \lambda \; s_1 \; . \; \text{isNat}(e_1) \rightarrow$

$\qquad \mathbf{R} [\![ \; E_2 \; ]\!] \; \mathbf{r} \; (\; \lambda \; e_2 \; . \; \lambda \; s_2 \; . \; \text{isNat}(e_2) \rightarrow$

$\qquad \mathbf{k} \; \mathbf{BOP} \; [\![ \; \text{BOP} \; ]\!] \; (\; e_1, \; e_2 \;) \; s_2 \; [\!] \; \text{error} \;) \; s_1 \; [\!] \; \text{error} \;)$

The semantics of function calls are defined by the following equation where we first obtain the denotation of the function (main in our example program), create a new stack frame, then evaluate the arguments, and execute the body of the function in this new store. At the end of execution of the body the return value is stored on top of the stack and is returned to the caller of the function.

$\mathbf{R} [\![ \; I_f(E_1, E_2, \ldots E_n) \; ]\!] \; \mathbf{r} \; \mathbf{k} :=$

$\qquad \text{let } d = accessenv(I_f, \mathbf{r}) \text{ in let } l_s \; l_t \; s_{old} = mark\text{-}locn \; s \text{ in}$

$\qquad \mathbf{R} [\![ \; E_1 \; ]\!] \; \mathbf{r} \; (\; \lambda \; e_1. \; \lambda \; s_1. \; \mathbf{R} [\![ \; E_2 \; ]\!] \; \mathbf{r} \; (\; \ldots$

$\qquad \lambda \; e_{n-1}. \; \lambda \; s_{n-1}. \; \mathbf{R} [\![ \; E_n \; ]\!] \; \mathbf{r} \; (\; \lambda \; e_f. \; \lambda \; s_f.$

$\qquad d{\downarrow}2 \; ((\lambda \; s_{ret}. \; \text{let } s_{new} = dealloc\text{-}locn \; l_s \; l_t \; s_{ret} \text{ in}$

$\qquad \mathbf{k} \; (access \; s_{new} \; (s_{new}{\downarrow}1 + s_{new}{\downarrow}2) \; s_{new} \;),$

$\qquad s_f, \; (e_1 \ldots e_n) \;) \;) \;) \; s_{n-1} \;) \; s_1 \;)$

The result of elaborating these denotations would be a $\lambda$-expression which maps any input file (there being no input in the example program) to the distance between the two points, expressed as an integer.

# 5 Semantics Based Parallelizing Compilation

In this section we outline the process of generating distributed code for SmallC++ using semantics based compilation techniques. We define a two step process to generate the parallel code. First SmallC++ is translated into a sequential combinator language which provides a sequential interpratation. Next we perform semantics preserving transformations on the combinator representation that replace all potential parallelism by parallel combinators and communication primitives. We use the Tuple Space model of concurrency [3] as the underlying model for object interaction. The result of the transformations is an executable parallel program with well understood semantics.

## 5.1 Sequential Semantics

A set of sequential combinators is developed using the semantics of SmallC++ defined in Section 4. We have defined a combinator for every major semantic action such as looping, conditionals, arithmetic and boolean operations, memory access and type checking. Figure 4 summarizes the set of sequential combinators we designed to implement SmallC++. Of these combinators *message-send* and *create-object* are directly related to the semantics of objects. An object is encoded as a store-like function that returns the denotations of instance variables and methods for that object. Figure 5 is the denotation of the SmallC++ program introduced in Section 2 after compilation. For simplicity sake we do not change the identifier names into their actual denotations.

| | |
|---|---|
| `access-array, update-array` | access and update array elements |
| `access-field, update-field` | access and update components of structures |
| `access-member, update-member` | access and update instance variables of objects |
| `create-object` | create an object |
| `copy-block` | copy structures and arrays |
| `assign` | define an assignment operation |
| `compose` | sequential composition |
| `if` | conditional (must have "then" and "else" parts) |
| `for` | for-loop (must have an index, initial value, termination condition and loop body) |
| `call` | procedure call |
| `message-send` | send a message to an object (similar to "call") |
| `return` | return a value |
| `fix` | the fix-point combinator, represents recursively defined functions |
| `ref` | call-by-reference parameter |
| `deref` | access a call-by-value parameter |
| `plus, minus, times, slash, equal, not-equal, less, less-equal, greater, greater-equal, and, or, not` | standard arithmetic, relational, and logical operators |

**Fig. 4.** Sequential Combinators

## 5.2 Parallel Semantics

The second step in defining the semantics is applying a set of semantics preserving parallelizing transformations that encode SmallC++ by a parallel combinator language augmented with communication primitives. This encoding is executable on a distributed platform. The transformation is performed using flow

```
(compose (create-object Point obj1 3 4)
         (create-object Point obj2 5 6) (create-object int d)
         (assign d (call sqrt (plus
           (call square (minus (message-send obj1 GetX)
                               (message-send obj2 GetX)))
           (call square (minus (message-send obj1 GetY)
                               (message-send obj2 GetY))))))
     (write d))
```

**Fig. 5.** Denotation of the example program in sequential combinators

analysis, dependence analysis, and parallel code generation. We describe the parallel combinators, the Tuple Space model of communication, and the proposed transformations below.

### 5.3 Parallel Combinators

Parallelism is encoded using parallel combinators. Parallel combinators were introduced by [7] to define the granularity of parallelism, and [8] to generate parallel functional code. Our approach is most like [8] in that the combinators themselves represent the basic instruction set of the parallel machine model being used, in our case the Tuple Space model. Some examples of parallel combinators are given in Figure 6. The set of parallel combinators is still under development. So, only a high-level set is illustrated here.

| | |
|---|---|
| parallel | argument expressions can be evaluated in parallel |
| sequence | argument expressions must be evaluated sequentially |
| distribute | argument objects can be distributed |
| cluster | argument objects should not be distributed |
| for-all | parallel for-loop (must have an index, initial value, termination condition and loop body) |

**Fig. 6.** Parallel Combinators

### 5.4 Tuple Space

We use the *generative communication model of distributed computing* or Tuple Space (TS) for the implementation of interprocess communication. TS represents communication among distributed processes by tuples. Formally, a tuple is an $n + 1$-tuple in which the first element of the tuple is a name, and the remaining $n$ elements are parameters, either formal or actual. The actual parameters represent data to be sent and the formal parameters represent place holders. The model is dynamic, with the tuple space of a program changing as the program executes. For example, if A and B are objects, A sending a message to B will generate a new tuple in the space. When B is ready to receive this message it removes the tuple and executes. These actions are encoded by the functions **out**, which puts a new tuple in the space; **in**, which removes a tuple from the space; and **read**, a non-destructive **in**.

TS provides a good interprocess communication model for an OO environment for several reasons. First, tuples, like objects, can have both data members and functions. Thus they provide a natural representation of objects. Second, a tuple can consist of a name and a function, facilitating the creation of a process which can execute in parallel. We call such tuples *message tuples* and they can be encoded using the **eval** operator that was introduced in [12]. Message tuples can return a value in the TS, that can be used by other processes, at the end of its execution. Thus, a method and data member can be treated equally in our implementation. Finally, using **eval**, a process may request the execution of a program (a method, for example) on a specific node or allow the system to select the node within the TS network which has the least number of active processes thus providing an automatic load balancing mechanism.

## 5.5 Generating the Parallel Semantics and Code

The basic structure of a parallelizing compiler for an object-oriented programming language is similar to that for more conventional languages [21]. However, there are some major differences in the nature of the components, caused by object-orientation. Apart from the regular loop-parallelism that is available in Fortran-like languages, there are at least three other forms of concurrency possible within an object-oriented system. Inter-object concurrency refers to different objects carrying out different activities at the same time. Intra-object concurrency refers to a single object executing several methods simultaneously. Thirdly each of these methods could themselves be carrying out several operations in parallel. Thus, an object may have several threads of control, each corresponding to different types of concurrent execution.

The primary components of the parallelization of OO programs are flow analysis, dependence analysis, and parallel code generation. A general overview is presented in Figure 7.

Flow analysis creates a graphical representation of the flow of data among components of a program. Unlike the flow analysis performed on non-OO programs where nodes represent simple statements, the flow graph we create has three types of nodes: (1) individual methods, (2) a collection of methods within a class or object, and (3) all the methods of an OO program. The primary complexity introduced by the OO paradigm is determining the precise instance

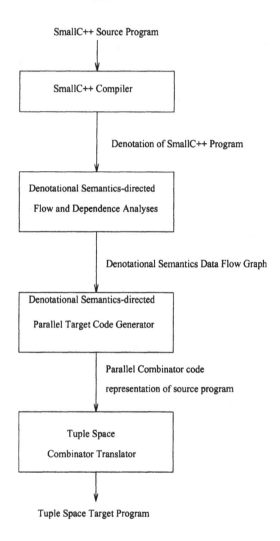

**Fig. 7.** Overview of Denotational Semantics-Directed Parallelization

variables and methods being referenced in the presence of polymorphism and inheritance.

Dependence analysis uses the flow graph to determine the data dependencies and the interconnectedness among the objects in the program. Because we are working with an OO programming paradigm, we must consider both the dependencies within methods and the dependencies that result from message passing. The first type is handled by constructing a standard dependence graph for the code within each method. We determine the interconnections and flow of data among objects using a second type of dependence graph we call a *message flow graph*.

Parallel code generation uses the flow and dependence information to partition the objects into processes in a manner that reduces communication costs. This includes determining whether an object process should be executed synchronously or asynchronously. Furthermore, we may generate code that automatically controls communication between processes by increasing the amount of data transferred in a single message so as to reduce the number of messages sent. This includes, for example, the restructuring of for-loops containing a message updating a database to a single message containing all of the updates. The result of the parallel code generation is a parallel combinator program which may be executed using Tuple Space.

The sequential combinator code produced for the example program in Section 5.1 is now transformed to the program shown in Figure 8 after performing flow and data dependence analysis. This program contains the parallel combinators identifying parts of the code that can be executed in parallel. The creation of the two point objects (and the distance variable) are done in parallel. Various portions of the calculation of the distance between the two points are also done in parallel.

The tuple space code corresponding to the parallel combinator code is shown in Figure 9. The Tuple Space representation **outs** two tuples, one for each **Point** object. The names **"Pointobj1"** and **"Pointobj2"** are keys that are used to identify the tuples, created by concatenating the Class Name and the Object Name (we would wish to use a more unique key creation algorithm in practice). The first **eval** forks four processes each executing one method and returning a value in the temporary variables. The second **eval** forks two more processes, each using the previously computed temporary values. The final **eval** forks a process which computes the distance and returns it in **d**. In essence, the computation in the above program is done in just three steps - one step for each **eval**, ignoring the **outs** in the first two statements.

# 6  Summary

We have implemented the sequential semantics generating sequential combinator code from SmallC++ programs. We are currently implementing the program dependence graphs from the sequential combinator code. We intend to investigate the graphs to discover more parallel combinators that could be used in generat-

```
(compose
 (parallel (create-object Point obj1 3 4)
           (create-object Point obj2 5 6) (create-object int d))
 (assign d
  (call sqrt
   (plus
     (parallel
      (call square (minus
                     (parallel
                       (message-send obj1 GetX)
                       (message-send obj2 GetX))))
      (call square (minus
                     (parallel
                       (message-send obj1 GetY)
                       (message-send obj2 GetY))))))))
 (write d))
```

**Fig. 8.** Denotation of the example program in parallel combinators

```
out ("Pointobj1", 3, 4); out ("Pointobj2", 5, 6);
eval ("dtmp1",tmp1 = GetX (Pointobj1), tmp2 = GetX (Pointobj2),
              tmp3 = GetY (Pointobj1), tmp4 = GetY (Pointobj2));
eval("dtmp2",tmp5 = square(tmp1 - tmp2), tmp6 = square(tmp3 - tmp4));
eval("d",d = sqrt (tmp5 + tmp6));
```

**Fig. 9.** Tuple Space code for example program

ing the parallel combinator code. This would be followed by an implentation of the mapping from the parallel combinator code to the tuple space code which would be run on a network of sun workstations. Our future research goal is to extend SmallC++ and provide a complete semantics specification for a larger subset of C++. Also once the tool is developed, we are planning to apply this technique to SmallTalk.

# References

1. Aho, A.V., Sethi, R., Ullman, J.D., *Compilers Principles, Techniques, and Tools*, Addison-Wesley, Reading, MA, 1986.
2. Bik, A. J. C., Gannon, D. B., "Automatically Exploiting Implicit Parallelism in Java," in *Concurrency: Practice and Experience*, vol. 9, no. 6, 1997, pp. 579-619.
3. Gelernter, D., "Generative Communication in Linda," in *ACM Transactions On Programming Languages and Systems*, vol. 7, no. 1, Jan. 1985, pp. 81-112.

4. Genjiang, Z., Li, X., Zhongxiu, S., " A Path-Based Method of Parallelizing C++ Programs," in *SIGPLAN Notices*, vol. 29. no. 2, Feb 1994, pp. 19-24.

5. Goldberg, A.J., Robson, A.D., *Smalltalk-80: The Language and its Implementation*, Addison-Wesley, Reading, MA, 1983.

6. Hindley, J., R., Seldind, J., P., *Introduction to Combinators and λ-calculus*, Cambridge University Press, 1986.

7. Hudak, P., Goldberg, B., "Distributed Execution of Functional Programs using Serial Combinators," in *IEEE Transactions on Computers*, vol. C-34, no. 10, Oct. 1985, pp. 881-891.

8. Knox, D.L., Wright, C.T., "Combinators as Control Mechanisms in Multiprocessing Systems," in *Proceedings of the International Conference on Parallel Processing*, 1987, pp. 158-161.

9. Lee. P., *Realistic Compiler Generation*, MIT Press, Cambridge, MA, 1989.

10. Milner, R., Tofte, M., and Harper, R., *The Definition of Standard ML*, MIT Press, Cambridge, MA, 1990.

11. Muthukrishnan, P.K, Bryant, B.R, "The Syntax and Semantics of SmallC++," *Technical Report, Department of Computer and Information Sciences, University of Alabama at Birmingham*, 1995.

12. Patterson, L., *Fault Tolerant Tuple Space*, Ph.D. thesis, Department of Computer and Information Sciences, University of Alabama at Birmingham, Birmingham, AL, 1992.

13. Peyton Jones, S., *The Implementation of FunctionalProgramming Languages*, Prentice Hall, Englewood Cliffs, NJ, 1987.

14. Schmidt, D.A., *Denotational Semantics A Methodology for Language Development*, Allyn and Bacon, Inc., Boston, MA, 1986.

15. Stroustrup, B., *The C++ Programming Language*, Addison-Wesley, Reading, MA, 1986.

16. Tokoro, M., Nierstrasz, O., Wegner, P., eds., *Object-Based Concurrent Computing, Proceedings of ECOOP '91 Workshop*, Springer- Verlag, 1991.

17. Tomlinson, C., Scheeval, M., "Concurrent Object-Oriented Programming Languages," in *Object-Oriented Concepts, Databases, and Applications*, ACM Press/Addison-Wesley, Reading, MA, 1989, pp. 79-124.

18. Wegner, P., "The Object-Oriented Classification," in *Research Directions in Object-Oriented Programming*, MIT Press, Cambridge, MA, 1987, pp. 479-560.

19. Yin, M., Bic, L., Ungerer, T., " Parallel C++ Programming on the Intel iPSC/2 Hypercube," in *Proceedings of the 4th Annual Parallel Processing Symposium*, 1990, pp. 380-394.

20. Yonezawa, A., Tokoro, M., eds., *Object-Oriented Concurrent Programming*, MIT Press, Cambridge, MA, 1987.

21. Zima, H., Chapman, B., *Supercompilers for Parallel and Vector Computers*, ACM Press/Addison-Wesley, Reading, MA, 1990.

# A Formal Software Engineering Paradigm: From Domains via Requirements to Software — Formal Specification & Design Calculi —

Dines Bjørner

Software Systems Group
Dept. of Information Technology, Technical University of Denmark
Bldg.345/167–169, DK–2800 Lyngby, Denmark
E–Mail: db@it.dtu.dk — Fax: +45-45.88.45.30

**Abstract.** We postulate that a development process is possible in which three major stages:

- Domain Engineering

- Requirements Engineering

- Software Design

"smoothly" connect to one another: That from Domain Models we can develop Requirements Models and relate them, and that from Requirements Models we can develop Software Architecture Models, and subsequently Program Organisation and Refinement Models and relate them.

In this paper we will show, by a simple example, what we mean by stage and stepwise refinement (or development). And we will therefore be able to make plausible our claim that the three stages itemised above can indeed be formally expressed and related.

That is: If we could not show a relatively "smooth" progression from usually non-computable abstractions via computable ones to executable specifications (designs), then what we are claiming becomes a bit academic! Now we can indeed show this transition, transformation as it were, from abstractions to concretisations. And therefore we can also — however superficially — illuminate what we mean by abstractions and concretisations. We can perhaps convince the reader not only that small abstractions can be beautiful, but that the ability to first capture Domain abstractions, and then, in two more stages and, within these, in three–five steps or more, show real, efficiently executable concretisations related strongly to the abstractions, can be even more beautiful!

This paper is is an extract of a larger, more consistent and comprehensive 'opus'. We presently show the use of the **RAISE** Method [2] and the RAISE Specification Language **RSL** [1].

# 1 Definitions

1. Method:

   By a Method we understand a set of *principles* of *analysis,* and for *selecting* and *applying techniques* and *tools* in order *efficiently* to *construct efficient artifacts* — here Software.

2. Methodology:

   By Methodology we understand the study and knowledge about Methods. Since we can assume that no one Software Development Method will suffice for any entire construction process we need be concerned with Methodology.

3. Software Development:

   To us Software Development consists of three major components: Domain Engineering, Requirements Engineering and Software Design. Together they form Software Engineering.

4. Domain Concepts:

   Two approaches seem current in today's 'domain engineering': one which takes its departure point in model–oriented, Mathematical Semantics specification work (and which again basically represents the 'Algorithmic' school), and one which takes its departure point in knowledge engineering — an outgrowth from AI and Expert Systems. The latter speaks of Ontologies. For now we focus on the former approach.

   (a) Domain = System + Environment + Stakeholders:

   By Domain we roughly understand an area of human or other activity. We "divide" the Domain into System, Environment and Stakeholder. All are part of a perceived world.

   (b) System:

   By System we understand a part of the Domain. The System is typically an enterprise. Once the Machine has been installed in the System then it becomes a part of a new Domain wrt. future Software Development.

   (c) Environment:

   By Environment we understand that part of the perceived world which interacts with the System. Thus the System complement wrt. "the perceived world", i.e. the Environment, together with the System and Stakeholder makes up the Domain of interest.

(d) **Stakeholder = Clients + Staff:**

By Stakeholder we mean any of the many kinds of people that have some form of "interest" in the (delivered) Machine: enterprise owners, managers, operators and customers of the enterprise: within the System or in the Environment.

(e) **Client:**

By Client we understand the legal entity which procures the Machine to be developed. The Client is one of the Stakeholders, and must be considered a main representative of the System.

(f) **Staff:**

By Staff we understand people who are employed in, or by, the System: who works for it, manages, operates and services the System. Staff are a major category of Stakeholders.

(g) **Customer:**

By Customer we understand the legal entities (people, companies), within the System, who enter into economic contracts with the the Client: buys products and/or services from the Client, etc. Customers form another main category of Stakeholders: outside the System, but within the Domain.

(h) **Domain Engineering = Capture $\mapsto$ Model $\gg$ Analysis $\mapsto$ Theory:**

Domain Engineering, through the processes of Domain Acquisition and Domain Modelling, establishes models of the Domain. A Domain Model is — in principle — void of any reference to the Machine, and strives to describe (i.e. explain) the Domain <u>as</u> <u>it</u> <u>is</u>. Domain Analysis investigates the Domain Model with a view towards establishing a Domain Theory. The aim of a Domain Theory is to express laws of the Domain.

(i) **Domain Capture = Acquisition $\mapsto$ Modelling:**

*Discussion: We make a distinction between the "soft" processes of Domain Acquisition: linguistic and other interaction with Stakeholders, and Domain Modelling: the "hard" processes of writing down, in both informal and formal notations, the Domain Model.*

*The Domain Capture process, when actually carried out, often becomes confused with the subsequent Requirements Capture process. It is often difficult for some Stakeholders and for some Developers, to make the distinction. It is an aim of this report to advocate that there is a crucial distinction and that much can be gained from keeping the two activities separate. They need not be kept apart in time. They may indeed be pursued concurrently, but their concerns, techniques and documentation need be kept strictly separate.*

(j) Ontology:

What we call Domain Models some researchers call Ontology — almost!

In the 'Enterprise Integration and in the 'Information Systems communities ontology means: "formal description of entities and their properties". Ontological analysis is applied to modelling the Domain of (manufacturing) enterprises and such systems (typically management systems) whose implementation is typically database oriented.

5. Requirements Concepts:

(a) Requirements = System + Interface + Machine:

Requirements issues are either such which concern (i) Machine support of the System, (ii) human (and other) interfaces between the System and the Machine, or (iii) the Machine itself.

Requirements describes the System as the Stakeholders **would like to see it**.

(b) Requirements Engineering = Capture ↦ Model ≫ Analysis ↦ Theory:

Requirements Engineering, through the process of Requirements Capture, establishes Models of the Requirements. The "conversion" from Requirements information obtained through Requirements Elicitation, via Requirements Modelling to Requirements Models is called Requirements Capture. Requirements Models are formally derived from and extends Domain Models. Requirements Engineering also analyses Requirements Models, in order to derive further properties of the Requirements.

(c) Requirements Capture = Elicitation $\mapsto$ Modelling:

Remarks similar to those under Domain Capture — item 4i (page 3 of this paper)) apply.

6. Software Concepts:

(a) Software Design = Software Architecture Specification
$\rightarrow$ Program Organisation Specification
$\rightarrow$ Refinements
$\rightarrow$ Coding

Software Design, through the process of Design Ingenuity, proceeds from establishing a Software Architecture, to deriving a Program Organisation, and from that, in further steps of Design Reification, also called Design Refinement, constructing the "executable code". .

(b) Software Architecture:

A Software Architecture Description specifies the concepts and facilities offered the user of the software — i.e. the external interfaces.

(c) Program Organisation:

A Program Organisation Description specifies internal interfaces between program modules (processes, platform components, etc.).

(d) Refinement:

Design Refinement covers the derivation from the Requirements Model of the Software Architecture, of the Program Organisation from the Software Architecture, and of further steps of concretisations into Program Code.

7. Systematic, Rigorous and Formal Development:

The Software Development may be characterised as proceeding in either a systematic, a rigorous or even, in parts, a formal manner — all depending on the extent to which the underlying formal notation is exploited in reasoning about properties of the evolving descriptions.

(a) *Formal Notation:*

By a Formal Notation we understand a language with a precise syntax, a precise semantics (meaning), and a proof system. By "a precise ..." we usually mean "a mathematical ...".

(b) *Systematic Use of Formal Notation:*

By a Systematic Use of Formal Notation we understand a use of the notation in which we follow the precise syntax and the precise semantics.

(c) *Rigorous Use of Formal Notation:*

By a Rigorous Use of Formal Notation we understand a systematic use in which we additionally exploit some of the 'formality' by expressing theorems of properties of what has been written down in the notation.

(d) *Formal Use of Formal Notation:*

By a Formal Use of Formal Notation we understand a rigorous use in which we fully exploit the 'formality' by actually proving properties.

(e) *Formal Method ≈ Formal Specification ⊗ Calculation:*

We refer to item 1 (page 2 of this paper) for a definition of 'method'.

The methods claimed today to be formal methods may be formal, but are not methods in the sense we define that term! Since we do not believe that a method for developing Software: from Domains via Requirements, can be formal, but only that use of the notations deployed may be, we (now) prefer the terms: Formal Specification and Calculation.

(f) *Design Calculi — or Formal Systems:*

By a Design Calculus we understand a Formal System consisting of a Formal Notation and a set of precise Rules for converting expressions of the Formal Notation into other such, semantically 'equivalent' expressions.

8. Satisfaction = Validation ⊕ Verification:

The Domain Acquisition and Requirements Elicitation processes alternate with Domain Modelling and Requirements Modelling, respectively, and these again with securing Satisfaction.

(a) Validation:

In this report we are not interested in the crucial process of interactions between software developers (i.e. Software Engineers, which we see as Domain Engineers, Requirements Engineers and

Software Designers) and the Stakeholders. Validation is thus the act of securing, through discussion, etc., with the Stakeholders that the Domain Model correctly reflects their understanding of the Domain.

(b) Verification:

Let $\mathcal{D}$, $\mathcal{R}$ and $\mathcal{S}$ stand for the theories of the Domain, Requirements and Software. Then Verification:

$$\mathcal{D}, \mathcal{S} \models \mathcal{R}$$

shall mean that we can verify that the designed Software Satisfies the Requirements in the presence of knowledge (i.e. a theory) about the Domain. This proof obligation is well known [].

9. Software Engineering:

Software Engineering is the combination of Domain Engineering, Requirements Engineering and Software Design, and is seen as the process of going between science and technology. That is, of developing Descriptions on the basis of scientific results using mathematics — as in other engineering branches — and of understanding (the constructed Domain of) existing (software) technologies by subjecting them to rigorous Domain Analysis.

## 2 Domain Model

Items 4h (page 3 of this paper) and 4i (page 3 of this paper) briefly characterise the notions of *Domain Engineering* = *Capture* ↦ *Model* ≫ *Analysis* ↦ *Theory*, and *Domain Capture* = *Acquisition* ↦ *Modelling*. Modelling leads to Model. Analysis leads to Theory.

The domain concept to be illustrated is that of an abstract airline time-table (TT). Basic, further un-explained terms are: flight number (Fn), airport (A), arrival and departure times (T).

**object**
   ttt: **class type** A, Fn, T, TT **end**

One can view (view) an airline time-table, and for any recorded flight from and to which airports it flies, and when — that is: its journey (jour). First let us introduce the concept of gate-time: the time interval that a flight is at an airport gate:

**object** gt:
  **class**
    **type**
      GT = {| (at,dt) | at,dt:ttt.T • at < dt |}
    **value**
      bef: GT × GT → **Bool**
    **axiom**
      $\forall$ at,dt,at',dt':ttt.T •
        bef((at,dt),(at',dt')) ≡ (dt < at')
  **end**

Every journey involves at least two airports. The airports of a journey can be totally ordered such that we can speak of an airport of origin and one of final destination. Possibly other airports of a journey are called intermediate (stop) airports. Arrival and departure times (ground times) designate the time interval during which the aircraft is at the gate (ie. gate time). Thus arrival time for origin and departure time for final destination airports are sensible quantities.

**object** jrn:
  **class**
    **type**
      Journey
      Jrn' = ttt.A $\overrightarrow{m}$ gt.GT
      Jrn = {| jrn | jrn:Jrn' • iswfJrn(jrn) |}
    **value**
      obs_Jrn: Journey → Jrn

      iswfJrn: Jrn → **Bool**
      iswfJrn(j) ≡
        $\forall$ a,a' • a $\neq$ a' ∧ {a,a'} ⊆ **dom** j ⇒
          gt.bef(j(a),j(a')) ∨ gt.bef(j(a'),j(a) )
  **end**

The time–table (TT) is likewise an abstraction but shall "contain" many facets — i.e. enable the many uses people may have of time-tables.

**object** tt:
  **class**
    **type**
      time_table = ttt.Fn $\overrightarrow{m}$ jrn.Jrn

**value**
    obs_time_table: TT → time_table
**end**

Users sees time-tables as:

**scheme** TTU =
  **class**
    **value**
      view: ttt.TT → ttt.TT
      view(tt) ≡ tt.obs_time_table(tt)

      jour: ttt.Fn × ttt.TT $\xrightarrow{\sim}$ jrn.Jrn
      jour(fn,tt) ≡ (tt.obs_time_table(tt))(fn)
  **end**

**object** du:TTU

In addition there is the time-table as seen by those (airline staff, S) who construct time-tables. They start (init) with an empty time-table (isempty), and they add (add), change (chg) and delete (del) flight numbered journies.

**scheme** TTS =
  **class**
    **value**
      init: TT
      initi() **as** tt **post**: tt.obs_time_table(tt) = [ ]

      add: ttt.Fn × jrn.Journey × ttt.TT $\xrightarrow{\sim}$ ttt.TT
      add(fn,j,tt) **as** tt′
        **pre**:  fn ∉ **dom** tt.obs_time_table(tt)
        **post**: tt.obs_time_table(tt′) ≡
             tt.obs_time_table(tt) ∪ [ fn ↦ j ]

      chg: ttt.Fn × jrn.Journey × ttt.TT $\xrightarrow{\sim}$ ttt.TT
      chg(fn,j,tt) **as** tt′
        **pre**: fn ∈ **dom** tt.obs_time_table(tt)
        **post**: tt.obs_time_table(tt′) ≡
             tt.obs_time_table(tt) † [ fn ↦ j ]

del: ttt.Fn × ttt.TT $\xrightarrow{\sim}$ ttt.TT
del(fn,tt) **as** tt′
    **pre**: fn ∈ tt.**dom** obs_time_table(tt)
    **post**: tt.obs_time_table(tt′) = tt.obs_time_table(tt)\\{fn}
**end**

We need refer to only one staff 'entity':

**object** ds:TTS

Many different users and one staff can use the time-table system.

**object** idx: **class type** Index **end**

The total system consists of the ds and the

**object** dusers[ i:idx.Index ]:TTU

That's all! You may think of the above time-table as manifested in the form of a paper document. You, as a passenger (user), potential or actual, may browse through the document (view), and airline staff may initialise, add, change or delete (paper document) entries. The 'isempty' and 'domain' functions are technicalities — needed to properly define other functions. (Only the use of 'domain' is shown.)

Let us, just for the sake of argument and illustration, assume that our time-table is indeed in a very old-fashioned paper document form.

By not modelling the time-table, i.e. by only naming its sort, we are free — later on — to introduce other observation functions. Examples could be: inquire about distance between airports: in aeronautical miles, in flying time, etc., the price of a flight (economy, business, first class), etc.

## 3  Requirements Model

Items 5a (page 4 of this paper), 5b and 5c briefly characterise the notions of *Requirements = System + Interface + Machine*, *Requirements Engineering = Capture + Model + Analysis + Theory* and *Requirements Capture = Elicitation + Modelling*.

The domain model for time-tables was simple. It reflects **what there is**.

Now a set of requirements may be:

1. Projection of Domain onto Machine:

   "Transfer" all aspects of the previous paper document form to the computer and support all the human functions mentioned in the domain description by a suitable interface between passengers and staff — on one side — and the Machine on the other side.

   This means that the requirements specification extends DU and DS.

2. Additional Functionality:

   Provide additional support for inquiring about connections (cnns): given two airports find all one, and two, and (at most) three flight connections between those two airports — if any!

   This means that we extend DU with definitions concerning connections.

   **scheme**
      RU′ = **extend** TTU **with**
       **class**
        **type**
         Conns = ttt.Fn-**set**
              × (ttt.Fn×ttt.A×ttt.Fn)-**set**
              × (ttt.Fn×ttt.A×ttt.Fn×ttt.A×ttt.Fn)-**set**
        **value**
         cnns: (ttt.A×ttt.A) × ttt.TT $\overset{\sim}{\to}$ Conns
         cnns((fa,ta),tt) **as** rs
         **pre** {fa,ta} ⊆ **dom** jrn.obs_Jrn(tt)
         **post** ...
       **end**

3. User & Staff Computer Human Interfaces:

   (a) *Command Keyword Buttons:*

   The interface between users and the Machine shall be in the form of a menu-driven dialogue: A menu shall show the various query possibilities in the form of "command" menu buttons (B):

   – For users:

       i. View time-table                     (view_B)

       ii. Given flight number display journey     (jour_B)

iii. Find appropriate connections         (cnns_B)

– For staff:

   i. Initialise time-table         (init_B)

   ii. Add a journey         (add_B)

   iii. Change a journey         (chg_B)

   iv. Delete a journey         (del_B)

The user buttons:

**object**
  rbu:
    **class**
      **type**
        B_U == nil | view_B() | jour_B() | cnns_B()
    **end**

The staff buttons:

**object**
  rbs:
    **class**
      **type**
        B_S == nil | init_B() | add_B() | chg_B() | del_B()
    **end**

Button "clicks" can at most engage one button — the most recent "clicked"! Buttons are thus either 'on' or 'off', and at most one can be 'on'.

(b) *Command Prompt Menus:*

"Clicking" on one of these buttons shall then result — for all but the first "command" — in an input request menu being displayed, one which prompts the user for further information (to be read, M...r):

– For users:

   i. view_Mr:         nothing!

   ii. jour_Mr:         flight number

    iii. cnns_Mr:                   two airport names

&ndash; For staff:

    i. init_Mr:                                      nothing!

    ii. add_Mr:          flight number and journey

    iii. chg_Mr:         flight number and journey

    iv. del_Mr:                  flight number

The user prompts:

**object**
  rmur:
    **class**
      **type**
        M_Ur == nil
                | view_Mr()
                | jour_Mr(fn:ttt.Fn)
                | cnns_Mr(fa:ttt.A,ta:ttt.A)
    **end**

The staff prompts:

**object**
  rmsr:
    **class**
      **type**
        M_Sr == nil
                | init_Mr()
                | add_Mr(fn:ttt.Fn,j:jrn.Journey)
                | chg_Mr(fn:ttt.Fn,j:jrn.Journey)
                | del_Mr(fn:ttt.Fn)
    **end**

(c) *Execution Results:*

Filling these menus in — as appropriate — or just submitting the 'view' or 'init' commands, shall then result in an appropriate response:

&ndash; For users:

    i. view_R:                        time-table

      ii. jour_R:                                    the journey

      iii. cnns_R:                            three scrollable lists

– For staff:

      i. init_R:                      the update time-table

      ii. add_R:                   the update time-table

      iii. chg_R:                  the update time-table

      iv. del_R:                  the update time-table

User execution results:

**scheme**
  RRU =
    **class**
      **type**
        R_U == nil
                | view_R(t_t:tt.time_table)
                | jour_R(j:jrn.Journey)
                | cnns_R(cnns:TTSRCs.Conns)
    **end**

**object** rru:RRU

Staff execution results:

**scheme**
  RRS =
    **class**
      **type**
        R_S == nil
                | init_R(tt:ttt.TT)
                | add_R(tt:ttt.TT)
                | chg_R(tt:ttt.TT)
                | del_R(tt:ttt.TT)
    **end**

**object** rrs:RRS

4. The user requirements:

  First the 'generic' user:

**scheme**

  RU(du:DU,ru:RU'(du)) =

    **class**

      **value**

        user: (rbu.B_U × rmur.M_Ur × rru.R_U) × ttt.TT

                 $\overset{\sim}{\to}$ rbu.B_U × rmur.M_Ur × rru.R_U

        user((b,m,),tt) ≡

          (rbu.nil,rmur.nil,

            **cases** b:

              rbu.nil

                → rru.nil,

              rbu.view_B()

                → rru.view_R(du.view(tt)),

              rbu.jour_B()

                → rru.jour_R(du.jour(rmur.fn(m),tt)),

              rbu.cnns_B()

                → rru.cnns_R(ru.cnns(rmur.fa(m),rmur.ta(m),tt))

           **end**)

         **pre** ...

    **end**

Then the "set" of all users.

  **object** rus[ i:idx.Index ]:RU

The 'rus' object array form one part of the requirements definition.
The single staff requirements is now concluded:

5. The staff requirements:

**scheme**

  RS =

    **class**

      **value**

        staff: (rbs.B_S × rmsr.M_Sr × rrs.R_S) × ttt.TT

               $\overset{\sim}{\to}$ (rbs.B_S × rmsr.M_Sr × rrs.R_S) × ttt.TT

        staff((b,m,),tt) ≡

          **cases** b:

            rms.nil

              → ((rbs.nil,rmsr.nil,rrs.nil),tt),

rms.init_M()
 → **let** tt′ = ds.init() **in**
  ((rbs.nil,rmsr.nil,rrs.init_R(tt′)),tt′) **end**,
rms.add_M()
 → **let** tt′ = ds.add(rms.fn(m),rms.j(m),tt) **in**
  ((rbs.nil,rmsr.nil,rrs.add_R(tt′)),tt′) **end**,
rms.chg_M()
 → **let** tt′ = ds.chg(rms.fn(m),rms.j(m),tt) **in**
  ((rbs.nil,rmsr.nil,rrs.chg_R(tt′)),tt′) **end**,
rms.del_M()
 → **let** tt′ = ds.del(rms.fn(m),tt) **in**
  ((rbs.nil,rmsr.nil,rrs.del_R(tt′)),tt′) **end**

  **end**
  **pre** ...
 **end**

**object** rs:RS

6. The overall system requirements:

The requirements are now made up from the 'rus' object array and the 'rs' object, and, through them all the objects and schemes their definitions depend on.

## 4 Software Design

In the Requirements there was an object 'rs' related to staff and an object array 'rus' related to a set of users. With the staff object we defined a "next" state transition function 'staff', and likewise for each of the user objects, ie. 'user'.

The intention that the time-table is to be seen as a global entity, as was rather implicitly modelled in the Domain as an object 'ttt'. But that object is not an updateable quantity. The Domain Model, as well as the Requirements Model, skirted the notion of a 'state'. They only expressed that if you come with a time-table then they define how user and staff functions behave. The former by using the time-table, the latter by (using and by) "offering" to update it.

So what to do?

Let us, for the sake of argument, and so that we can illustrate issues of Software Architecture and Program Organisation, assume that resolution of these issues were left to the Software Designers.

## 4.1 Software Architecture Model

We refer to our definition of Software Architecture item 6b (page 5 of this paper).

**The Idea:** The basic idea is that the one staff and the many user functions "alternate" in some fashion, for example:

**scheme**
  SASys' =
    **class**
      **type**
        $\Sigma$ == b:rbu.B_U m:rmu.M_Ur r:rru.R_U
        $\Psi$ == b:rbs.B_S m:rms.M_Sr r:rrs.R_S
        $\Omega$ = idx.Index $\overrightarrow{m}$ $\Sigma$
      next: $\Psi \times \Omega \times$ ttt.TT $\overset{\sim}{\to} \Psi \times \Omega \times$ ttt.TT
      next($\psi,\omega$,tt) $\equiv$
        **let** ($\psi'$,tt') = rs.staff($\psi$,tt) **in**
        **let** $\omega'$ = [ i $\mapsto$ du.user($\omega$(i),tt')|i:idx.Index•i $\in$ **dom** $\omega$ ] **in**
        ($\psi'$,$\omega'$,tt')
    **end end end**

**object** sys:SASys'

The idea is that $\Sigma$ represents a Machine state reserved for users, whereas $\Psi$ represents a Machine state reserved for properly authorised staff. $\Omega$ represents the totality of all user states.

    The above definition has stated that there is some regulating force which "cycles" between the staff's manipulation of the time-table, if any, and each of all the users use of the time-table. The latter occurs in some non-deterministic fashion: the order in which each user is "selected" for possible access to the global time-table is not specified.

    The above is, however, only a sketch. We shall not pursue it further. Instead we shall next look at a process architecture for the time-table system.

*An Aside — on Human Computer Interface Development:* We could have chosen to detail — either already in the Requirements or in the Software Architecture — with the "look" of the user and staff menus (thus dealing also with user-friendliness issues of GUIs (graphic user interfaces)). But we have refrained. That is not to say that we still could'nt deal with it as part of Requirements or as part of Software Architecture. It is just a matter of refining the RBU, RMU, RRU, RBS, RMS and RRS schemes. Such can be done independently of the present development. Here we focus on the difficult interconnections between user and staff "processes" — and in the next section they will indeed become processes.

**The Process Architecture:** The "idea" model above said nothing about the processes that carry out the various functions and operations. Now we can meaningfully do so.

The Software Architecture designer will decide on such processes which seem relevant to the user, including their communication.

- We associate one process with the time-table, call it tt$\pi$.

- We associate one main process associated with each user, call it user$\pi$.

- And we associate one main process with the staff, call it staff$\pi$.

- Each main user process (user$\pi$) may be implemented in terms of core_$U\pi$, button_$U\pi$, menu_$U\pi$ and result_$U\pi$ processes, one per user process (user$\pi$).

- Similarly for the one staff process (staff$\pi$).

- We may finally distribute the time-table process (tt$\pi$), [3], ie. as a set of processes each taking care of a part of the time-table. This will not be shown explicitly in the specifications below.

The idea is that the total system of processes is as informally shown in figure 1 (page 19 of this paper).

In figure 2 (page 20 of this paper) we likewise informally detail the synchronisation and communication between processes:

1. **The System Process:**

    **scheme**
      SASys =
        **class**
          **value**

**Fig. 1.** A Program Organisation

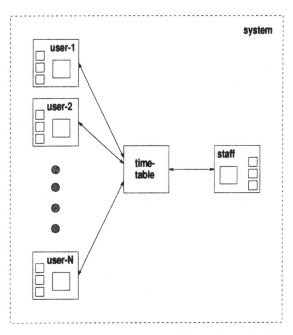

**System of N Users, one Timetable and one Staff**

system: **Unit** → **Unit**
system() ≡
    ttprocess.ttπ()
    || staffprocess.staffπ()
    || ||{ usr[ i ].userπ() | i:idx.Index }
**end**

**object sys:SASys**

2. Channels between time-table and a(ny) user process:

**object**
   ttuch[ i:idx.Index ]:
     **class**
       **channel**

**Fig. 2.** Program Flow

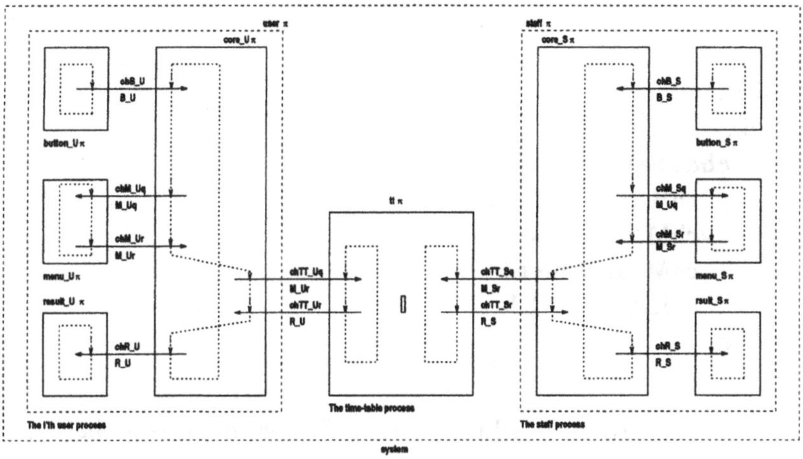

A Simple System of Processes: one User [i], Timetable and Staff

$$\text{chTT\_Uq : rmur.M\_Ur,}$$
$$\text{chTT\_Ur : rru.R\_U}$$
**end**

There is thus an array of user $\leftrightarrow$ time–table channels.

3. Channels between the time-table process and the staff process:

**object**
   ttsch:
      **class**
         **channel**
            chTT_Sq : rmsr.M_Sr, chTT_Sr : rrs.R_S
      **end**

4. Type of query and read messages over "internal" user channels:

**object**
   muq:
      **class**
         **type**
            M_Uq == nil | view_Mq() | jour_Mq() | cnns_Mq()
      **end**

5. Channels "internal" to user processes:

**object**
  uch[ i:idx.Index ]:
    **class**
      **channel**
        chB_U : rbu.B_U,
        chM_Uq : muq.M_Uq,
        chM_Ur : rmur.M_Ur,
        chR_U : rru.R_U
    **end**

We have chosen to make all the "internal" user channels global.

6. Type of query messages over "internal" staff channels:

**object**
  msq:
    **class**
      **type**
        M_Sq == nil | init_Mq() | add_Mq() | chg_Mq() | del_Mq()
    **end**

7. Channels "internal" to staff process:

**object**
  schs:
    **class**
      **channel**
        chB_S : rbs.B_S,
        chM_Sq : msq.M_Sq,
        chM_Sr : rmsr.M_Sr,
        chR_S : rrs.R_S
    **end**

The channels are likewise chosen to be global.

8. The time-table process:

```
object
  ttprocess:
    class
      value
        ttπ: TT → in { ttuch[i].chTT_Uq | i:idx.Index },
                      ttschs.chTT_Sq
                   out { ttuch[i].chTT_Ur | i:idx.Index },
                      ttschs.chTT_Sr Unit
        ttπ(tt) ≡
          let (c,i) = []{let c = ttuch[i].chTT_Uq? in
                          (c,i) end | i:idx.Index } in
          cases c:
            rmur.view_Mr()
               → ttuch[i].chTT_Ur!du.view(tt),
            rmur.jour_Mr(fn)
               → ttuch[i].chTT_Ur!du.jour(fn,tt),
            rmur.cnns_Mr(fa,ta)
               → ttuch[i].chTT_Ur!ru.cnns(fa,ta,tt)
          end end;
          ttπ(tt)
          []
          let cmd = ttschs.chTT_Sq? in
          let tt' =
          cases cmd:
            rmsr.init_Mr()    → ds.init(),
            rmsr.add_Mr(fn,j) → ds.add(fn,j,tt),
            rmsr.chg_Mr(fn,j) → ds.chg(fn,j,tt),
            rmsr.del_Mr(fn,j) → ds.del(fn,j,tt)
          end in
          ttschs.chTT_Sr!tt';
          ttπ(tt') end
    end end
```

9. The staff process and its "internal" processes:

```
object
  staffprocess:
    class
      value
        staffπ: Unit → in chTT_Sr out chTT_Sq Unit
```

staff$\pi$() $\equiv$
  core_S$\pi$() || button_S$\pi$() || menu_S$\pi$() || result_S$\pi$()

core_S$\pi$:
  Unit $\rightarrow$
    in  schs.chB_S,
        schs.chM_Sr,
        ttsch.chTT_Sr
    out schs.chM_Sq,
        ttsch.chTT_Sq,
        schs.chR_S
    Unit
core_S$\pi$() $\equiv$
  let p() $\equiv$
    (let cmd = schs.chM_Sr? in
     schs.chTT_Sq!cmd;
     schs.chR_S!chTT_Sr? end) in
    cases chB_S?
      rbs.init_B()
        $\rightarrow$ schs.chM_Sq!msq.init_Mq();p(),
      rbs.add_B()
        $\rightarrow$ schs.chM_Sq!msq.add_Mq();p(),
      rbs.chg_B()
        $\rightarrow$ schs.chM_Sq!msq.chg_Mq();p(),
      rbs.del_B()
        $\rightarrow$ schs.chM_Sq!msq.del_Mq();p(),
      _ $\rightarrow$ skip
    end end;
    core_S$\pi$()

button_S$\pi$: Unit $\rightarrow$ out schs.chB_S Unit
button_S$\pi$() $\equiv$
    (schs.chB_S!
        (rbs.nil$\lceil$rbs.init_B()$\lceil$rbs.add_B()$\lceil$rbs.chg_B()$\lceil$rbs.del_B()));
  button_S$\pi$())

menu_S$\pi$:
  Unit $\rightarrow$ in schs.chM_Sq out schs.chM_Sr Unit
menu_S$\pi$() $\equiv$
  let mq = schs.chM_Sq? in

```
            cases mq:
              msq.init_Mq() →
                  ...; schs.chM_Sr!rmsr.init_Mr(),
              msq.add_Mq() →
                schs.chM_Sr!(let fn:Fn,j:Journey ... in
                                  ...; rmsr.add_Mr(fn,j) end),
              msq.chg_Mq() →
                schs.chM_Sr!(let fn:Fn,j:Journey ... in
                                  ...; rmsr.chg_Mr(fn,j) end),
              msq.del_Mq() →
                schs.chM_Sr!(let fn:Fn ... in
                                  ...; rmsr.del_Mr(fn) end)
            end end;
            menu_Sπ()


        result_Sπ: Unit → in schs.chR_Sr Unit
        result_Sπ() ≡
            let res = schs.chR_Sr? in res end;
            result_Sπ()
      end
```

10. The user processes:

```
  object
    usr[i:idx.Index]:
      class
        value
          userπ:
              Unit → in ttuch[i].chTT_Ur out ttuch[i].chTT_Uq Unit
          userπ() ≡
                core_Uπ() || button_Uπ() || menu_Uπ() || result_Uπ()


          core_Uπ:
            Unit →
              in  uch[i].chB_U,
                  uch[i].chM_Ur
                  ttuch[i].chTT_Ur
              out uch[i].chM_Uq,
                  uch[i].chR_U Unit,
                  ttuch[i].chTT_Uq
```

```
core_Uπ() ≡
  let p() = (let cmd = uch[i].chM_Ur? in
              ttuch[i].chTT_Uq!cmd end;
              let res = ttuch[i].chTT_Ur? in
              uch[i].chR_U!res end) in
  cases uch[i].chB_U ?
    rbu.view_B()
      → uch[i].chM_Uq!muq.view_Mq(); p(),
    rbu.jour_B()
      → uch[i].chM_Uq!muq.jour_Mq(); p(),
    rbu.cnns_B()
      → uch[i].chM_Uq!muq.cnns_Mq(); p(),
    _ → skip
  end;
  core_Uπ()


button_Uπ: Unit → out chB_Ur Unit
button_Uπ() ≡
  (uch[i].chB_Ur!
    (dbu.nil⌐|dbu.view_B()⌐|dbu.jour_B()⌐|dbu.cnns_B());
  button_Uπ())


menu_Uπ:
  Unit → in  uch[i].chM_Uq out uch[i].chM_Ur Unit
menu_Uπ() ≡
  let mq = uch[i].chM_Uq? in
  cases mq:
    muq.view_Mq()
      → uch[i].chM_Ur!view_Mq()
    muq.jour_Mq()
      → uch[i].chM_Ur!
            (let fn:Fn ... in jour_Mq(fn) end)
    muq.cnns_Mq()
      → uch[i].chM_Ur!
            (let fa,ta:A ... in cnns_Mq(fa,ta) end)
  end end; menu_Uπ()


result_Uπ:
  Unit → in uch[i].chR_Ur Unit
result_Uπ() ≡
```

```
        let res = uch[i].chR_Ur? in res end;
        result_Uπ()
end end
```

The system flow and synchronisation, and also the non-deterministic external choice operator [], can be pictured as a Petri Net. See figure 3 (page 26 of this paper).

**Fig. 3.** A "Program" Petri Net

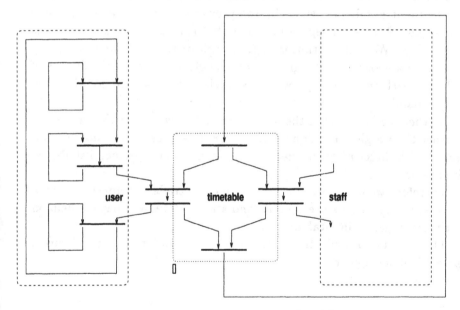

**Petri Net: User / Timetable / Staff Flow**

In this subsection we have only sketched a software architecture possibility. And we have only given the signature of possible processes. The process states, i.e. data structure signatures are then to be derived from the types given under Domain, Requirements and Software Architecture Models.

## 4.2 Program Organisation Model

We refer to our definition of Program Organisation item 6c (page 5 of this paper).

Finally we need illustrate the concept of a Program Organisation as being distinct from its related Software Architecture.

So far users and staff have had the impression that the time-table was a single, global entity. It is the rôle of a Software Architecture to give them that impression, and it is also the rôle of a Program Organisation to let them still believe so — even though that Program Organisation now decides to distribute the time-table.

What do we mean by 'distribution'? Well, abstractly we decide to have $n$ time-tables, each taking care of non-overlapping sets of flight numbers and such that the totality of such individual time-tables together cover all the flight numbers covered by the previous seemingly unique, global time-table. With each separate, distributed time-table we associate one staff. Why? Well, the intuition is that each distributed time-table may be that of a geographical region, or of a single airline or of an alliance of airlines. With these changes we can still claim that we are not changing the Domain!

Neither the $m$ users nor the $n$ staff should be aware that the time-table is other than a global, unique entity. Therefore they should not be concerned with directing their queries or updates to any specific, distributed time-table.

Therefore we need introduce a notion of a "trader": something (say a process) which accepts user queries and staff updates and passes them on to the right (sub) time-table.

The program organisation structure is as shown informally in figure 4 (page 28 of this paper).

We leave the formalisation of the program organisation to another time!

## 5 Discussion

What can we now conclude from this "brief" of a realistic example development?

- The Domain Model focuses only on:

  • **what there really is.**

**Fig. 4.** A Program Organisation Diagram

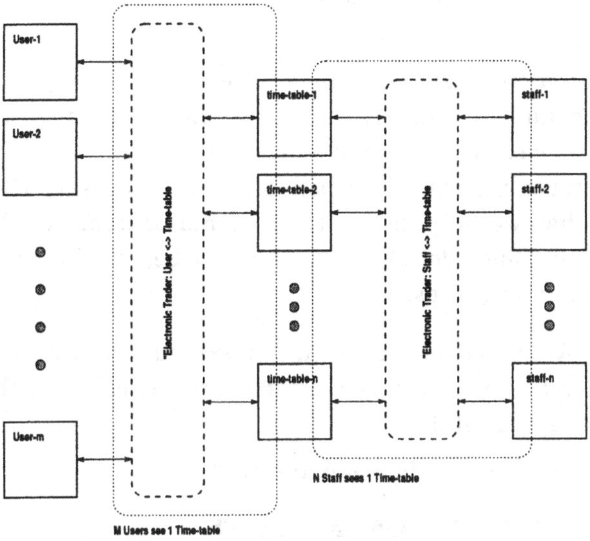

**Time-table System Program Organisation:**
**Distributed Time-table and Shared (Distributed) "Electronic Trader" Access**

It is descriptive.

We have kept apart the user and the staff models till the very end of this Domain Engineering stage where we "combine" them. But we do not really exploit this combination in the following!

Maybe browsing through a time-table, in the domain, allow users to "construct" connections, maybe not. We have assumed, but only for the sake of argument, that they cannot. This allows part of the next step:

- The Requirements Model, in this case extends the Domain Model.

The resulting model expresses:

- **what there ought to be.**

It is prescriptive.

It only specifies a next state transition, not the continuous evolution of user, time-table and staff states.

We have "extended" the Domain Model with the connections "feature". and we have kept separate the user and the staff models (until, as in the Domain Engineering stage, "the very end" of this Requirements Engineering stage.

We have left "vague" the treatment of the visual display units and their keyboard and icon input, as well as the formatting of menus and result displays. We could have detailed it in the Requirements Engineering stage, but have left it till later. We could say, with some justification, that we only deal with some major issues of Requirements, and that "finer details" are, or could be, "unfolded" in the Software Design stage steps.

– The Software Architecture Model further extends the Requirements Model by concretising the external interface to the users: In this case passengers and airline staff.

The Software Architecture Model expresses:

- **how the external interfaces behaves**.

This Software Design step, in consequence of detailing the architectural issues of external interfaces also takes up "unfinished" Requirements "business" by also modelling more detailed Requirements issues.

We have included among the external interfaces those "within" the user, respectively the staff processes. But we could hide them.

– The Program Organisation Model now emphasises the internal process realisation and the distribution of a data structure.

The Program Organisation Model expresses:

- **how the internal interfaces behave**.

– The overall proof of corretness responsibility remains:

- $D, SA \models R$

- $D, R, PO \models SA$

– The boundaries between Domain (D), Requirements (R), Software Architecture (SA) and Program Organisation (PO) are "smooth". We do not wish to dogmatically impose too strict adherence to the definitions of what a Domain Model, a Requirements Model, a Software Architecture Model, and a Program Organisation Model is. That would constrain the Domain Engineering, the Requirements Engineering and the Software Design "processes" too much. As long as we afterwards, once the development is over, can precisely point to which parts of which documents reflect the Domain, which the Requirements, which the Software Architecture, etc., then we are satisfied. To be satisfied also means that the developer clearly understands the differences between these stages and steps and their models.

# 6 Acknowledgement

Thanks to Dr. Bo Stig Hansen for valuable comments.

# References

1. The RAISE Language Group. *The RAISE Specification Language.* The BCS Practitioner Series. Prentice-Hall, Hemel Hampstead, England, 1995.
2. The RAISE Method Group. *The RAISE Method.* The BCS Practitioner Series. Prentice-Hall, Hemel Hampstead, England, 1992.
3. Chris George. *A Theory of Distributing Train Rescheduling.* Research Report 51, UNU/IIST, P.O.Box 3058, Macau, December 1995.

# The Whole Picture to Software Process Improvement

Annie Kuntzmann-Combelles, OBJECTIF TECHNOLOGIE, France
28, Villa Baudran, F - 94742 Arcueil Cedex
Tel. +33 1 49 08 58 00
Fax +33 1 49 08 95 88
e-mail akc@objectif.fr

*Abstract*: The paper focuses on a systematic method to ensure the success for process improvement as part of the strategic plan of an organisation to be and stay ahead of competitors. The entire program is driven by the business goals: CMMTM is the road map and ami® provides guidelines to select actions to address weaknesses and difficulties of the software processes and progressively match primary goals.

## 1 Introduction

Software process assessment and improvement are based on some simple concepts:

- evolution is possible and takes time. There is a systematic approach to improve the way software is developed and maintained. This approach - a process view - is in contrast with the "silver bullet" one,

- there are stages of process maturity. The levels of CMM or the capability dimension of SPICE[1] are indicators of process maturity and, as a consequence, of decrease of risks and increase of performance,

- evolution implies that there is a recommended sequence to get control of the process depending on external constraints,

- maturity will erode unless sustained. Maturity regression may be observed in some cases.

Our experience in the process control business started as a conscious process in 1990 - and probably as an unconscious one before -. When we started, the CMM model from the Software Engineering Institute was available to the public and we used it as the basis for developing a methodology for software measurement [5]. Late 1993, we noticed some interest among our customers in process improvement and we caught the opportunity to promote CMM model and concepts.

---

[1] *Software Process Improvement and Capabiliy dEtermination, an ISO standard (15504) going-on for assessing and improving the software process*

In 1994, the company started to dedicate some efforts to the international ISO WG10 team - SPICE - and conducted trial experiments in 1995 according to the SPICE model. The ami® [2] method developed previously has inspired part of the informative guidelines contained the SPICE series, especially in the Process Improvement Guide.

After some experiences of leading assessments and monitoring process improvement for large and medium size companies, I want to discuss what I think are prerequisites to succeed, the main steps of the process and what type of difficulties might be on the road. Analysing the origin of these difficulties can greatly help not to repeat mistakes and will certainly result in more enthusiasm and benefit among teams.

## 2   The business performance perspective

"What results can my organisation expect as it matures?" This frequently asked question is probably one of the most difficult to answer simply.

Most of the Companies I started with were at the initial level (level 1 of the maturity scale); they are usually populated with software heroes: engineers who are doing their best to come out in time with pieces of code which of course will be error free because they, all of them, are very good programmers spending evenings and week-ends trying to match milestones fixed by managers without proper estimates.
I have no intention to blame any of these practices but obviously continuous stress, change of priorities, overtime are not adequate conditions to perform software business and will obviously not result into better competitivity. Generally, when an organisation decides to start process improvement, the decision has been made by management and a clear commitment exists. However, I have also had bad surprises in assuming such a commitment, and have discovered that external constraints - i.e. pressure from customers or prerequisites to bid or simply fad - were commonly used drivers as well. When business performance is the main target - the normal situation - all the difficulty is to capture which are the main indicators management will use to analyse software process improvement impact on the overall business. At the same time it is crucial to understand how the software business is perceived regarding cost analysis and risk management: is software component on the critical path to deliver the final product in time? does the cost of the software development prevail compared to other development costs? will software cost reduction impact the sale price of the final product? These are examples of questions which should be debated.

Senior management is asked about qualitative and quantitative data regarding the current situation and evolution of:

- who are the customers and how does the division interact with them,
- what added value is brought to the customer through the software components of the product,

---

[2]   *The ami method (Application of Metrics in Industry) is a four phases loop based on Assessment, Goal/Question/Metrics Analysis, Metrication and Improvement paradigm*

- responsiveness to the market needs and the time required to respond,
- the size and complexity of the software components of the product,
- the cost of development and the cost of maintenance,
- areas of emphasis for the final product and the specific ones for the software,
- some strategic aspects such as architecture, product decomposition, sub-contracting.

The information collected and the understanding of the business context will then be used all along the next steps: process assessment, project risks analysis, recommendations, actions planning and follow up. Any data from the projects considered or practice should be analysed against adequacy and efficiency criteria to meet the business goals.

The types of "business drivers" considered during the process are illustrated by the following examples drawn from real goals:

- Time to Market: the time between Request for Tender and delivery of the final product must be less than 6 months,
- Cost: Reduce cost to develop a feature to 1/2 of the level of 1/1/97,
- Quality: Reduce number of defects discovered by the customer during Acceptance Testing to 1/10 of the level of Release N.

Having highlighted the importance of business orientation in the improvement initiative is probably not enough to convince organisations that SPI cannot be started as "yet another quality attempt to get a better final product". Software process is definitively part of the organisation asset and contributes to the overall risk maturity level and entrepreneurship.

Finally, most of the organisations who really succeeded and demonstrated business impact of the initiative have not only considered software process improvement but included other processes in their analysis such as product management/marketing, purchasing and sales or exploitation. It is namely difficult to get benefit of a software improvement initiative if the other organisational processes do not reach similar maturity level.

## 3 General concepts

Process improvement is not specific to software activities and the cyclic approach followed is common to any discipline. This cycle has been extensively developed by Deming - Plan, Do, Check, Act - or in the ami® framework - Assess, Analyse, Metricate, Improve -. In other words, improvement is based on the results of an appraisal, a snap-shot of the real practices of development teams, at a given time, in a given context. Obviously, the success of the actions derived from the assessment is tightly linked with the quality of the observations. Continuous improvement happens by repeating the same steps of Initiating (agreeing on the motive and strategy for undertaking change), Analysing what to change and prioritising, Implementing the changes, Monitoring the progress and Leveraging (capturing and capitalising lessons learned). Capability models such as the CMM, play an important role in the Analysis step: it is the reference against which the development process strengths and weaknesses are diagnosed.

## 3.1 Overview of the CMM

The CMM consists of the five levels of process maturity where each level has an associated set of Key Process Areas (KPAs). At the Initial level of maturity (level 1), software projects rely on the skills and theoric efforts of individual engineers. there are no KPAs associated with level 1. Fire fighting is prevalent and projects tend to leap from one emergency to the next.

The Repeatable (level 2) maturity level has six KPAs associated with it. These KPAs relate to requirements management, project planning, tracking and oversight, subcontracts management, software quality assurance and configuration management. Projects under a level 2 organisation are repeatable and under basic management control.

At the Defined (level 3) maturity level, the software development organisation now defines common processes, develops training programs, focuses on intergroup co-ordination and performs peer reviews. The result is the development of tailorable software processes and other organisational assets so that there is a certain level of consistency across projects.

At the Managed (level 4) maturity level, the software development organisation implements a quality and metric management program and monitors both project and organisational performance.

At the Optimising (level 5) maturity level, quantitative data are used for process improvement and defect prevention. in addition, technology changes are introduced and evaluated in an organised, systematic process.

## 3.2 Initiating the assessment

Apart of the effort dedicated to understand the business objective, preparation of the assessment covers three activities: training at various levels, selecting projects and planning. Based on my experience of running assessments in different parts of the world, I have learned that cultural differences have to be considered when organising the collection of data. There are usually three types of data collection: interviews based on prepared questions, open discussions and documents analysis - i.e. project documents -. In some cases, open discussions are difficult to manage because the principle does not exist in the country's culture or company's culture. As a consequence, preparing the assessment includes this aspect; discussions are necessary with the sponsor and the assessment team to decide which procedure will be the most efficient and not strange to people and managers.

### 3.3 The on-site period

The on-site period is a very exciting period not only for the assessment team but also for the teams being assessed. A lot of expectations are raised and one danger is the dissatisfaction which may happen if actions are not decided later on. The team building step, at the very early beginning of an assessment, is generally a strong point. The lead assessor has to identify quickly the possible difficulties that may endanger the consensus; talkative people or non talkative people, aggressiveness, or any other psychological behaviour which may arise are risks against the success of the initiative. In some cases, for example, participants have not well captured the difference between this type of appraisal and an audit; as a consequence, the information provided is not what one is looking for or remains simply hidden.

Interviews are conducted with individuals with decision-making authority. This could include Project Managers and Product Managers, but also Chief Architects, and Functional Managers for services such as Integration, Validation, and Product Support. The interviews are led from a script of questions adapted to the role of the individual. Although the subjects are based on the CMM, the questions are posed in the frame of reference of the interviewee, e.g. "Would you please describe how you estimate the work to be performed." Of course, follow-up questions will be posed to look for documented project or organisational procedures when not mentioned.

The group discussions assemble multiple practitioners who perform similar work on different projects. There is no script for the discussions, but the participants are asked to comment on their day to day activities. Consolidation of information is performed by the appraisal team at the end of each day of interviews. The objectives are twofold:

- to continuously compare the observations with business drivers as expressed by the senior management, in order to identify the suitability of the practices to the business needs,
- to continuously compare the observations with KPA goals, in order to evaluate the appropriateness of practices against established standards. As the questions are posed in the frame of reference of the projects, a cross reference table has been created which maps the questions to KPA goals.

At each consolidation point, the team must decide if the information received is sufficient for achieving these objectives. If not, the appraisal team has to take corrective actions such as adding a set of additional questions to some of the remaining interviews or open discussions, or performing follow-up interviews.

One danger is that assessors judge conformance rather than adequacy and effectiveness. Omitting adequacy and effectiveness means that business goals identified at the very early beginning are no longer considered. This will directly impact on the nature of the improvement actions derived from the assessment results: if actions are defined outside the company's strategy, keeping levels as only targets, success may not be guaranteed.

### 3.4 Results of an assessment

An assessment brings conformance results which are more or less detailed. Final results include:

- a profile of each KPA goal with a general rating (Not Satisfied/Partially Satisfied/Fully Satisfied),
- an analysis of the coherence between the business drivers and the existing practices,
- a detailed section describing observations made, consequences assessed and recommendations.

The practices profile will allow teams to perform comparisons with the previous assessment results in a continuous improvement framework.
For each of the critical areas pointed out in the assessment, recommendations will be derived from these results in terms of specific practices needing to be introduced. In the context of an on-going improvement program, this will be a combination of new actions and adaptation, redesign or re-engineering of current actions.

There is an important aspect of assessment which I have not yet mentioned: assessment procedure helps to capture and reuse lessons learned. In a way, a company initiating an improvement and organising appraisals of various organisational units becomes experienced in internal benchmarking. The best practices in place somewhere can be identified first - through the assessment - and then disseminated widely within the company. There is a huge advantage in tracking internal best practices: the cost/benefit analysis is easier within internal teams, the training to new practices can be organised based on real examples which are more familiar to practitioners than any others, and finally you create a challenging climate that is productive and controllable. But again this "best practices" research exercise cannot be entirely successful and benefiting if you only consider the conformance dimension.

## 4  From assessment to improvement

Deriving an action plan from an assessment profile is one of the most difficult exercise in the whole approach. I feel that there are two reasons for that: a social one and a technical one. The social difficulty comes from the fact that the on-site period is a very strong stage in the entire process, raising a lot of expectations; many people participate, senior managers show commitment and promise changes and finally, because of the appraisal procedure itself, there is a sort of dramatic atmosphere. During the few weeks following an assessment, there might be a danger of loosing all the positive aspects raised before, mainly due to the technical difficulty of defining improvement actions.
Three main points  have to be addressed: the role of business goals, deciding priorities and finally establishing plans based on the very early recommendations provided by the assessment team. It  is important to highlight here that capability models do not provide any specific help to address these points; only some guidelines are provided in additional technical reports. Who should achieve this work? How

should it be managed? What form should the documents take? All these questions and others have been largely debated by experts but most of the time, you have to adapt to what you have read according to the company's/country's culture, practitioners willingness to participate or not etc..But don't underestimate the importance of organisational aspects in the success and efficiency of the whole initiative and in the motivation of teams.

## 4.1 The importance of business goals

Again, business goals are extremely important to consider at that stage. Improvement actions should help to achieve business goals with a minimum risk. Too often I have been faced with cases where the profile result show weaknesses in different areas, among which configuration management; the immediate action taken was to investigate configuration management tools which obviously would solve all the problems!! In nine cases out of ten, other weak processes were far more urgent to work on to minimise project risks.

So you should dedicate some time to decompose your business goals and analyse how the various processes identified during assessment would contribute to each individual sub-goal. Doing so, you would probably discover that there are logical sequences to address weaknesses and that there is no way to attack all fires at the same time. This brings us to the second point: putting priorities on actions.

## 4.2 How to prioritise actions

Priorities are necessary, because a)the improvement budget is not infinite and b)the resources are generally limited. There is another important reason to consider hierarchy within the list of actions which I call "the time dimension". As soon as you enter the continuous improvement loop, you need to show the impact of the decisions taken on the projects and processes. The demonstration of real changes is absolutely necessary to sustain the initiative, at senior management level as well as at the practitioners' level. Therefore, actions which will bring visible benefit quickly have to be organised first. A good example relates to "Peer reviews"; this tactic to detect early defects, usually helpful and extremely efficient is performed with intelligence. However the Peer Reviews process is part of Level 3 of the CMM model and many groups do not decide to install peer reviews due to their low maturity.

One difficulty remains: to select these priorities. There is no absolute rule nor algorithm to establish them. Expertise, business goals and also effectiveness measures, risks associated with some improvement scenarios, those are the main factors to reason upon.

## 4.3 Define action plan

The action plan is the workproduct of this stage; it can be considered as the development plan of the improvement initiative. Therefore, it should at least include a clear definition of actions, an action owner - a group of people or an individual -, measurable targets of each actions, and a set of metrics to monitor their achievement, effort estimates and planning.

My experience is that measurable targets and follow up metrics are mandatory for many reasons: to communicate about progress, to show results of the improvement, to be sure that the actions are adequate to remove weaknesses and risks identified during the assessment and cost/benefit analysis. The ami® approach is a technique to quantify targets and define metrics.

From my experience, 90% of organisations who defined actions without writing a complete plan and without installing measurements have failed within the 18 months period following the assessment. There were no clear signs of improvement, no regular follow up, and demotivation and frustration were quite systematically observed.

## 5 Cultural aspects

In the previous sections, I have indicated several times that cultural or social factors were important. Let's take cultural aspects first. Some studies have been done regarding management styles which are to some extent applicable to organisation styles. In Europe, from north to south and from east to west there are differences, and obviously some other cultural factors have to be considered if you visit an organisation in India or in Asia. These factors have an extremely sensitive impact on the way the initiative will be motivated, and later on sustained and on the way work will be carried out during assessment and improvement.

When I talk about Senior management support, it is not just assigning dedicated resources to work on improvement or just attending the kick off meeting and final findings presentation. Middle management plays a main role too in the whole process; in some organisations and countries, their power is huge and they will greatly resist to change instead of freeing up individuals to work on and implement the improvements. In some cases, middle managers have very precise business targets to achieve and will systematically put priorities on operations - resources will be kept on on-going projects - instead of on improvement plans. Therefore, senior management should clearly communicate the importance of improvement and express their understanding of the impact on projects or products.

The results of a study conducted by Cindi Wise [3] on US organisations are interesting; she reports that Senior management should be taking action in the following areas to ensure success:

- articulate the role of software process improvement in the organisation's strategy and business objectives,
- provide visible demonstration of their commitment,
- provide and sustain resources for improvement,
- provide and sustain funding,
- create and communicate rewards that encourage teams,
- provide continuous monitoring of improvement activities,
- ensure that organisation processes supporting improvement are established.

All these areas might be covered differently depending on culture and history.

The second aspect I want to emphasise is the one related to social characteristics. The SPICE series highlights topics such as values, attitudes and behaviour as much as education and training. I have observed in some cases that the improvement programme initiated goals and messages varying deeply within the usual organisation's context. This results in discrepancies between the image given to external customers and the internal status; individuals will very shortly feel uncomfortable. One extreme example I faced is the following: an improvement programme had been started by a company 18 months ago when suddenly the senior management decided to decrease the number of employees by 10% without sound explanations. In this case, improvement actions were strongly altered by the decision not because resources failed but because of the internal climate resulting from senior management attitude.

In order to conclude this chapter, improvement loop is an approach that often results into deep changes: technical or technological changes on one hand, behavioural and managerial changes on the other hand. Visibility, communication, performance management, knowledge of weaknesses are the key factors of software process improvement: the same values will be needed in the rest of the company's processes otherwise improvement will not survive and investments will be quickly wasted.

## 6 Report on real experiences

The existing literature on process improvement results [7] shows that CMM based process improvement is making a difference in the organisations that are committed to improving. Raytheon yielded a twofold increase in its productivity and a ratio of 7.7 to 1 return on its improvement expenditures for a savings of $4,48 million during 1990 for a $0,58 million investment. Over a four and one half year period, from mid 1988 to the end of 1992, the company eliminated $15,8 million in rework costs. Hughes Aircraft has computed a ratio of 5 to 1 return for its process improvement initiatives, based on changes in its cost performance index. The company has experienced an annual savings of approximately $2 million over its process improvement expenditures.

As discussed by the SEI in "a Systematic Survey of CMM experience and Results", results from industry show that there are many positive results from CMM based improvements. Examples are:

Ability to meet Schedule evolving from 40% (level 1) to 58% (level 2),

Product Quality evolving from 77% (level 1) to 90% (level 2),

or Productivity increasing from 53% (level 1) to 68% (level 2).

Motorola GED reports [8] that each level of SEI CMM maturity reduces defect density by a factor of 2 and emphasises the importance of return on investment analysis during process improvement. "It would be easy to set up a high SEI maturity organisation that would suffer in cycle time and productivity if process were followed for process's sake".

The following example is interesting because it demonstrates how to use the ami® concepts to build an improvement action plan after a CMM based appraisal and how to monitor it for success. One strength of this experiment is the continuously documented link between SPI and company business goals.

The company is a defence market actor in Europe named C (for confidentiality reasons, I am not able to give more details and even allowed to, it will not add any value to the information reported). The software development staff is about 70 people, less than 10% of the total number of employees including commercial people, system engineers and hardware design people as well. Software is one of the growing components of the systems developed.

We met many organisations sharing the same characteristics within the two last years. The software part increase is about 20% each year and the software business starts to be considered as one of the key factors for success in the next decade. But software development still provides surprises and is sometimes considered as a group of artists who may either make significant profits or loose a lot of money without clear explanation why. Therefore, I really believe that mastering the software process is one of the major challenges of the near future. Software process improvement might be one of the key success factors of the competition if organisations ensure that the impact of decided actions is able to provide benefits quicker than the external environment evolves. In other words, any SPI initiative is able to win if and only if improvement actions are addressing in a timely manner the strategic marketing objectives of the organisation.

The C company performed a professional CMM appraisal in 1994 and defined an actions plan by beginning of 1995 using well recognised and efficient methods to do it. Discussions to convince the improvement team that the ami® approach was able to contribute positively to the actions already in place lasted about 6 months ; confidence was obtained after I had the opportunity to meet top managers (executive VP, Product strategy manager and QA manager) and discuss strategic goals for SPI. These people were enthusiastic about the idea to document the link between their vision of the business and the practices used to develop software. They had obviously no deep knowledge of software development issues and difficulties but they accepted to explain what were the key drivers for the next 2 to 3 years in their specific business. It was our job to ask the right questions in order to be able to derive improvement goals and make the transformation very clear between business strategy and software process strategy. These questions were prepared based on the G/Q/M paradigm on one hand and on the assessment findings on the other hand.

Examples of questions used are the following:

| # | Question |
|---|----------|
| 1 | Would you please explain which products should be strategic in the next 2 to 3 years? Why? |
| 2 | Considering this market strategy, would you say that software business is a key business (software component a key component)? Why? Which is the added value provided by software? |
| 3 | What major changes do you expect in the near future? Would they impact the software characteristics and how? technology, competitors, new customers, new types of users, standards, internationalisation etc. |
| 4 | Would you please explain the cost structure for the development of a new product (or new version of an existing product)? maintenance cost? |
| 5 | Would you please explain what type of goals you defined regarding costs and if software is part of them? |
| 6 | Would you please tell if you would need a different level of visibility in the costs and why? |
| 7 | Would you please explain how you define and measure the quality of the product? |
| 8 | What is the cost of non quality? How is it calculated? |
| 9 | What are the most critical factors for your customers? |
| 10 | How will you characterise a successful business now? in 2 years? |
| 11 | How will you manage trade off between cost and quality? |

For example, discussions with top managers made very clear that maintenance cost - evolutive and corrective - of the applications was very high and prevented teams to do all the work they were supposed to do in order to satisfy the customers with new features. At least top managers had this feeling but at the same time were unable to produce any quantitative data to support their judgement. Half of the project leaders shared this view and half did not ; they had other good explanations of the overload. Therefore, it was decided to start with an improvement goal of decreasing the maintenance cost.

Going further with G/Q/M paradigm, I discovered that assessment findings had been related to actions of the plan but that nobody had checked their impact on the business performance. At that stage, I decided to restart the complete process of defining actions based on the assessment findings using ami® concepts. The very first task was to train all the people supposed to be involved - from middle managers, project leaders to development teams -. The 4 phases of the ami® loop, the importance of measurements to be pro-active and the G/Q/M paradigm were extensively explained. This one day course has later on been extremely useful, helping all the participants to share the same language and understanding.

Furthermore, a certain number of brainstorming meetings were organised. I choose to play the role of a catalyst for the organisation. I thought that it was a more efficient way of transferring the technology and to help the organisation to acquire the capacity to reproduce this process later on. Meetings were scheduled over a 3 months period.

The first meeting was dedicated to analyse business goals as expressed by senior managers, to put priorities among them and to translate them in terms of project goals or software related goals. Brainstorming and complementary interviews with customer related teams were used to do this. The result was a first level of decomposition of primary goals into more technical goals. Each time figures were easily available they were used to consolidate the thinking process. The results was the goals decomposition (1 step). Improvement team had to validate this decomposition with all the participants and involved people before the second meeting.

**Reduce cost**

**1. Manage the stability of requirements**

**1.1 Reduce the number of changes requests**

**2. Acquire the knowledge of cost decomposition**

**2.1 Analyse the reliability of initial estimates**
2.1.1 Identify the deviations and origin

**2.2 Identify how the effort is split along the life cycle**
2.2.1 Define effort/task
2.2.2 Identify rework effort
2.2.3 Identify the availability of resources

During the second meeting an impact analysis was performed based on the validated sub-goals. Each assessment finding and set of recommendations was carefully discussed to identify which type of recommendations could drastically improve the situation. This type of analysis and judgement is greatly based on assessment team wide experience. It is not always easy because there are very few 1 to 1 relations. In other words, several weaknesses, if corrected, might have a positive impact on the same goal ; as a result, it is extremely difficult to draw the map and this map is certainly not applicable in various contexts. This linkage between weaknesses and business goals is not repeatable from project to project and even from organisation to organisation. Environmental constraints, cultural aspects are as many explanatory variables as you can find in the process. Again, here, G/Q/M is extremely helpful in structuring the brainstorming for more efficiency and better quality.

The following table gives an overview of one of the interim result obtained during this second meeting. Weaknesses observed during the assessment were formalised based on the top level goals and sub-goals. Only the weaknesses which when corrected would make improvement were further considered.

| CMM KPA | Weaknesses |
|---|---|
| Requirements Management | • change requests insufficiently formalised<br>• impact analysis not systematically achieved and poorly documented<br>• decision to implement changes not necessarily taken at the right level<br>• absence of process measurement: e.g. # changes addressed/quarter, average # days necessary to make the change, lead time between request and installation of the new release |
| Project Management | • lack of precision of the effort tracking system<br>• no historical data available (project leaders have their own data)<br>• sw development plan exists but is not updated regularly<br>• risk evaluation and management informally performed<br>• corrective actions taken late; most of the time result into an increase of resources hired from other projects or software houses<br>• overall planning (synchronisation between projects) poorly achieved |

After the second meeting, I left the improvement team validate the results obtained and list the top few actions which should bring the highest benefit in order to improve the business performance quickly. I reviewed their conclusions based on my experience and knowledge of the context.

The third meeting was dedicated to identify actions (the "how") to implement the recommendations (the "what") which could really help to match targets. It is clear that the impact analysis and the goals decomposition maintain continuously the traceability with top level objectives. At the same time actions were allocated to specific roles in the organisation and I verified that it was meaningful to assume that these key roles were involved in the achievement of the business goals.

The fourth meeting concentrated on the definition of metrics to track actions defined. In between meetings 3 and 4, a huge effort was allocated to the precise definition of each action, budget estimation, definition of mechanisms to pilot it and finally to measure efficiency and adequacy. This has been mainly performed by the organisation with some inspections from me in order to verify overall consistency and time frame for results. These progress metrics were aligned with business targets stated by top managers. Wherever they had been able to quantify, quantitative metrics were systematically defined; when not, either quantitative or qualitative indicators were suggested depending on the experience of Sw process group.
Some examples of these indicators are given below.

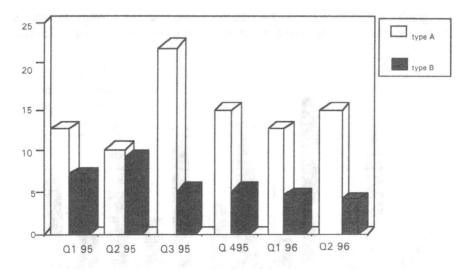

**Fig. 1. Deviation rate for an homogeneous family of products**
**Y axis is % of deviation against estimates**

Type A projects appeared quite unstable; Q3 and Q4 95 recorded heavy deviations for this population mainly due to a major change in technology. This family - decision making systems - moved to OO design at that time; deviations origin were two folds: a bad estimate when the projects started because estimations procedures had not been reviewed according to new technology and some technical difficulties faced at because of the lack of experience of the teams.

Type B, on the contrary, embedded systems in a rather stable environment, showed a decrease of the deviation rate.

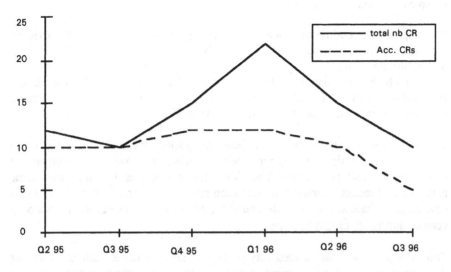

**Fig. 2. Ratio of change requests accepted to be included into the final product**
**compared to the total number of requests.**

The graph shows a decrease of this ratio over time based on a policy to address change requests at a unique point: the change control committee, meeting each two weeks.

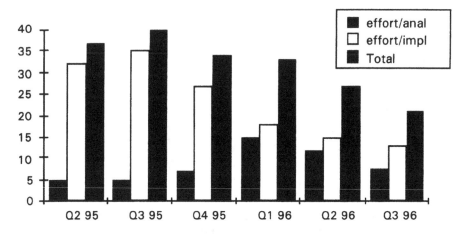

**Fig. 3. Effort spent in man days for performing**
**Impact analysis of the CRs before making the decision**
**Implementation of the CRs accepted for the final product**

The diagram shows an important decrease of effort along time. A change review board was institutionalised during Q1 96 and its impact is clearly visible on the maintenance effort. The results of the boards were regularly hand over to customers and the reasons for postponing or ignoring some of his requests explained to him based on real and technical data.

It is important to outline that along with meetings 3 and 4 (end of 1995), existing data started to be recorded and exploited by the organisation; therefore, better understanding of the weaknesses and action impact measurement were achieved concurrently. No new data to collect were defined but the ones which were already collected for other organisational reasons were used. It might be a very common situation and it has to be considered as not risky at all.

The improvement/measurement plan being completed, the metricate stage could start. The plan was first deployed on 3 pilot projects to validate the mechanisms for data collection and exploitation. The observation period was 3 months with some data collected regularly, some collected when reviews were organised at the end of a development phase and some by the end of the project (this last category has not been verified during the prototype phase).

This pilot phase was also judged very positive as far as it raised motivation among the participants who then communicated their enthusiasm to the rest of the software community. Some minor adaptations were necessary in the measurement plan in order to get a completely adapted collection mechanism and to avoid duplication with the existing global system for cost management.

# 7  Conclusions

This experiment let me observe what worked well and what could be improved. Models to be reproduced are the one keeping a continuously trace of the business goals and top level managers expectations or the introduction of simple metrics. What could be improved was some CMM coverage i.e. following more closely the underlined road map. Because there was a potential risk that the company C tried to implement each activity of the CMM and forget business goals, I took exactly the opposite approach; there could be opportunities for a more efficient one.

When the actions have been decided and allocated to groups of people, I observed that (1) the motivation was high - people wanted to reach their goals - (2) the justification of why to do it and how to do it were better understood by a majority of people because the links with the organisation strategy was documented. Because the understanding was better, responsible people for actions implementation were more strong in promoting them among the development teams.

I reproduced the approach in other more mature organisations with similar benefits. But because the initial starting point was a higher level of maturity, I decided to involve product line managers in the assessment and in the improvement planning in order to better consider the gap between market strategy and software process improvement. The result was a stronger emphasis on customer relationships weaknesses - inappropriate, not at the right time etc.... - as well, which is not well covered in CMM V1.0.

Over the last 3 years, I visited more than 20 sites where software process improvement was the key investment. The most successful ones, whatever the approach followed to support the continuous improvement programme, had applied the same following key rules:

• business goals were considered as the key driver of the programme,

• senior and middle management were committed and supportive,

• effectiveness measurement was installed.

Without satisfying these prerequisites, no software process improvement will succeed.

## Bibliography

[1]  Key Practices of the Capability Maturity Model, Version 1.1, SEI-93-TR-025
[2]  The level 4 Software process from the Assessor's Viewpoint, Ken Dymond, ISCN Conference 96

[3] C. Wise, Senior Management Actions Critical for Successful Software process Improvement, the 7th annual Software Technology Conference, April 1995

[4] K. Dymond, A Guide to the CMM, Process Inc US, ISBN 0-9646008-0-3

[5] K. Pulford, A. Kuntzmann-Combelles, S. Shirlaw, A quantitative approach to Software Management, the ami® Handbook, Addison-Wesley 1995

[6] Software process Assessment Part 7: guide for use in process improvement Working Draft V1.00

[7] Judith G. Brodman and Donna Johnson, Return on Investment from Software Process Improvement as Measured by US Industry

[8] Michael Diaz and Joe Sligo, Cost/Benfit Analysis of Software Process Improvement and the SEI CMM, Motorola GED Technical Report

# Object-Oriented Design Patterns

Wolfgang Pree

Applied Computer Science
University of Constance, D-78457 Constance, Germany
Voice: +49-7531-88-44-33; Fax: +49-7531-88-35-77
E-mail: pree@acm.org

**Abstract.** There is an undeniable demand to capture already proven and matured object-oriented design so that building reusable object-oriented software does not always have to start from scratch. The term *design pattern* emerged as buzzword that is associated as a means to meet that goal. This paper starts with an overview of relevant design pattern approaches. It goes on to discuss the few essential design patterns of flexible object-oriented architectures, so-called frameworks. The paper sketches the relationship between these essential design patterns and the design pattern catalog by Erich Gamma et al. [8]. The implications for finding domain-specific design patterns are outlined.

**Keywords.** Design patterns, object-oriented design, object-oriented software development, frameworks, reusability

## 1    Introduction

Over the past couple of years (design) patterns have become a hot topic in the software engineering community. In general, patterns help to reduce complexity in many real-life situations. For example, in many situations the sequence of actions is crucial in order to accomplish a certain task. Instead of having to choose from an almost infinite number of possible combinations of actions, patterns allow the solution of problems in a certain context by providing time-tested combinations that work. What does this mean in the realm of software construction?

Programmers tend to create parts of a program by imitating, though not directly copying, parts of programs written by other, more advanced programmers. This imitation involves noticing the pattern of some other code and adapting it to the program at hand. Such imitation is as old as programming.

The design pattern concept can be viewed as an abstraction of this imitation activity. In other words, design patterns constitute a set of rules describing how to accomplish certain tasks in the realm of software development. As a consequence, books on algorithms also fall into the category of general design patterns. For example, sorting algorithms describe how to sort elements in an efficient way depending on various contexts. If the idea sketched above is applied to object-oriented software systems, we speak of design patterns for object-oriented software development.

Currently, patterns comprise a wide variety of activities in the software development process, ranging from high-level organizational issues such as project management and team organization to low-level implementation issues such as tips and tricks regarding the use of a particular programming language. The pattern conference proceedings reflect this wide spectrum, for example, [6]. Due to the origin of the pattern movement in the framework community, numerous patterns still focus on how to construct frameworks. Section 3 discusses these approaches in detail.

## 2    History and Overview of Pattern Approaches

The roots of object-oriented design patterns go back to the late 1970s and early 1980s. The first available frameworks such as Smalltalk's Model-View-Controller (MVC) framework [13] and MacApp [18, 21] revealed that a framework's complexity is a burden for its (re)user. A framework user must become familiar with its design, that is, the design of the individual classes and the interaction between these classes, and maybe with basic object-oriented programming concepts and a specific programming language as well. This is why (framework) cookbooks have come to light:

**Framework Cookbooks.** Cookbooks contain numerous *recipes*. They describe in an informal way how to use a framework in order to solve specific problems. The term framework usage expresses that a programmer uses a framework as a basis for application development. A particular framework is adapted to specific needs. Recipes usually do not explain the internal design and implementation details of a framework. Cookbook recipes with their inherent references to other recipes lend themselves to presentation as hypertext.

Cookbooks exist for various frameworks. For example, Krasner and Pope present a cookbook for using the MVC framework [13]. The MacApp cookbook [5] describes how to adapt the GUI application framework MacApp in order to build applications for the Macintosh. Ralph Johnson wrote a cookbook [12] for the HotDraw framework, a system developed by Kent Beck and Ward Cunningham for implementing various kinds of graphic editors. ParcPlace-Digitalk Smalltalk provides an extensive cookbook for adapting the corresponding framework library.

**Coding Styles & Idioms.** These patterns form a quite different pattern category. C++ is a representative example of an object-oriented programming language whose complexity requires coding patterns to tame its montrosity. The principal goals of coding patterns are

- to demonstrate useful ways of combining basic language concepts

- to form the basis for standardizing source-code structure and names

- to avoid pitfalls and to weed out deficiencies of object-oriented programming languages, which is especially relevant in the realm of C++.

Coplien's C++ Styles & Idioms [7] and Taligent's guidelines for using C++ [19] fall into this category.

**Formal Contracts.** In the early 1990s more advanced design pattern approaches gained momentum that focus on framework development. Available object-oriented analysis and design methods appeared to be insufficient to construct reusable software architectures. In order to better understand object-oriented architectures, they should be described on an abstraction level that is higher than their implementation language. For example, Richard Helm *et al.* described interactions between objects of different classes in a framework in a formal way [11].

**Gamma's Description of ET++.** Pioneering work was accomplished by Erich Gamma in his doctoral thesis [9] which presents patterns incorporated in the GUI application framework ET++ [20, 1]. Gamma was inspired by Helm's formal contracts. Instead of using a formal notation he decribed ET++ by means of informal text combined with class and interaction diagrams. The design pattern catalog [8], also known as Gang-of-Four (GoF) book, resulted from Gamma's PhD thesis.

**Influence of Building Architecture.** At the 1991 Object-Oriented Programming Systems, Languages and Applications (OOPSLA) Conference—a major forum for researchers and practitioners in the field of object-oriented technology—Bruce Anderson headed the workshop "Towards an Architecture Handbook". Participants were encouraged to describe design patterns in a manner similar to the descriptions of architecture patterns presented in Christopher Alexander's books *A Pattern Language* [4] and *The Timeless Way of Building* [3]. These books show non-architects what good designs of homes and communities look like. One pattern, for example, recommends placing windows on two sides of a room instead of having windows only on one side. Alexander's patterns cover different levels of detail, from the arrangement of roads and various buildings to the details of how to design rooms. In general, this architecture handbook workshop inspired the pattern community, in particular the writing of the GoF book.

Workshops on object-oriented patterns were organized at subsequent OOPSLA conferences. A separate conference on design patterns (PLoP; Pattern Languages of Program Design) started in the U.S. in August 1994, the European pendant, called EuroPLoP, was first held in 1995.

# 3 Essential Framework Design Patterns

Frameworks are well suited for domains where numerous similar applications are built from scratch again and again. A framework defines a *high-level language* with which applications within a domain are created through *specialization* (= adaptation). Specialization takes place at points of predefined refinement that we call *hot spots*. We consider a framework to have the quality attribute *well designed* if it provides adequate hot spots for adaptations. For example, Lewis et al. present various high-quality frameworks [14].

## 3.1 Flexibility Through Hooks

Methods in a class can be categorized into socalled hook and template methods: Hook methods can be viewed as place holders or flexible hot spots that are invoked by more complex methods. These complex methods are usually termed template methods [8, 15]. Note that template methods must not be confused with the C++ template construct, which has a completely different meaning. Template methods define abstract behavior or generic flow of control or the interaction between objects. The basic idea of hook methods is that overriding hooks through inheritance allows changes of an object's behavior without having to touch the source code of the corresponding class. Figure 1 exemplifies this concept which is tightly coupled to constructs in common object-oriented languages. Method t() of class A is the template method which invokes a hook method h(), as shown in Figure 1(a). The hook method is an abstract one and provides an empty default implementation. In Figure 1(b) the hook method is overridden in a subclass A1.

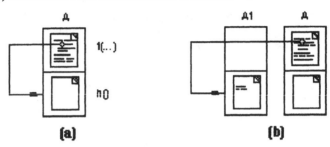

**Fig. 1.** (a) Template and hook methods and (b) hook overriding.

Let us define the class that contains the hook method under consideration as *hook class* H and the class that contains the template method as *template class* T. A hook class quasi parameterizes the template class. Note that this is a context-dependent distinction regardless of the complexity of these two kinds of classes. As a consequence, the essential set of flexibility construction principles can be derived from considering all possible combinations between these two kinds of classes. As template and hook classes can have any complexity, the construction principles discussed below scale up. So the domain-specific semantics of template and hook classes fade out to show the clear picture of how to achieve flexibility in frameworks.

### 3.2 Unification versus separation patterns

In case the template and hook classes are unified in one class, called TH in Figure 2(a), adaptations can only be done by inheritance. Thus adaptations require an application restart.

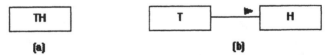

**Fig. 2.** (a) Unification and (b) separation of template and hook classes.

Separating template and hook classes is equal to (abstractly) coupling objects of these classes so that the behavior of a T object can be modified by composition, that is, by plugging in specific H objects.

The directed association between T and H expresses that a T object refers to an H object. Such an association becomes necessary as a T object has to send messages to the associated H object(s) in order to invoke the hook methods. Usually an instance variable in T maintains such a relation. Other possibilities are global variables or temporary relations by passing object references via method parameters. As the actual coupling between T and H objects is an irrelevant implementation detail, this issue is not discussed in further detail. The same is true for the semantics expressed by an association. For example, whether the object association indicates a *uses* or *is part of* relation depends on the specific context and need not be distinguished in the realm of these core construction principles.

A separation of template and hook classes also forms the precondition of *run-time adaptations*, that is, subclasses of H are defined, instantiated and plugged into T objects while an application is running. Gamma et al. [8] and Pree [16] discuss some useful examples.

### 3.3 Recursive combination patterns

The template class can also be a descendant of the hook class (see Figure 3(a)). In the degenerated version, template and hook classes are unified (see Figure 3(b)). The recursive compositions have in common that they allow building up directed graphs of interconnected objects. Furthermore, a certain structure of the template methods, which is typical for these compositions, guarantees the forwarding of messages in the object graphs.

The difference between the simple separation of template and hook classes and the more sophisticated recursive separation is that the playground of adaptations through composition is enlarged. Instead of simply plugging two objects together in a straightforward manner, whole directed graphs of objects can be composed. The implications are discussed in detail in [15, 16, 17].

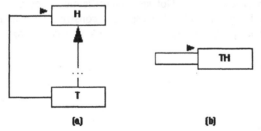

**(a)**                    **(b)**

**Fig. 3.** Recursive combinations of template and hook classes.

### 3.4 Hooks as name designators of GoF pattern catalog entries

Below we assume that the reader is familiar with the patterns in the pioneering Gang-of-Four catalog [8]. Numerous entries in the GoF catalog represent small frameworks, that is, frameworks consisting of a few classes, that apply the essential construction patterns in various more or less domain-independent situations. So these catalog entries are helpful when designing frameworks, as they illustrate typical hook semantics. In general, the names of the catalog entries are closely related to the semantic aspects that are kept flexible by hooks.

**Patterns based on template-hook separation.** Many of the framework-centered catalog entries rely on a separation of template and hook classes (see Figure 1(b)). The catalog pattern Bridge describes this construction principle. The following catalog patterns are based on abstract coupling: Abstract Factory, Builder, Command, Interpreter, Observer, Prototype, State and Strategy. Note that the names of these catalog patterns correspond to the semantic aspect which is kept flexible in a particular pattern. This semantic aspect again is reflected in the name of the particular hook method or class. For example, in the Command pattern "when and how a request is fulfilled" [8] represents the hot spot semantics. The names of the hook method (Execute()) and hook class (Command) reflect this and determine the name of the overall pattern catalog entry.

**Patterns based on recursive compositions.** The catalog entries Composite (see Figure 3(a) with a 1: many relationship between T and H), Decorator (see Figure 3(a) with a 1:1 relationship between T and H) and Chain-of-Responsibility (see Figure 3(b)) correspond to the recursive template-hook combinations.

# 4    How to find domain-specific patterns

Hot spot identification in the early phases (eg, in the realm of requirements analysis) should become an explicit activity in the development process. There are two reasons for this: Design patterns, presented in a catalog-like form, mix construction principles and domain specific semantics as sketched above. Of course, it does not help much, to just split the semantics out of the design patterns and leave framework designers alone with bare-bone construction principles. Instead, these construction principles have to be combined with the semantics of the domain for which a framework has to be developed. Hot spot identification provides this information. Figure 4 outlines the synergy effect of essential construction principles paired with domain-specific hot spots. The result is design patterns tailored to the particular domain.

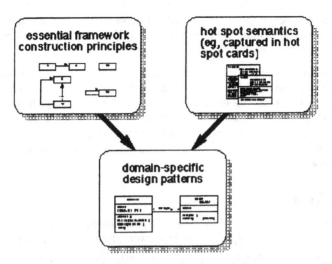

**Fig. 4**. Essential construction principles + hot spots = domain-specific design patterns

Hot spot identification can be supported by hot spot cards, a communication vehicle between domain experts and software developers. Pree [16, 17] presents the concept of hot spot cards and detailed case studies where they are applied.

A further reason why explicit hot spot identification helps, can be derived from the following observations of influencing factors in real-world framework development: One seldom has two or more similar systems at hand that can be studied regarding their commonalities. Typically, one too specific system forms the basis of framework development. Furthermore, commonalities should by far outweigh the flexible aspects of a framework. If there are not significantly more standardized (= frozen) spots than hot spots in a framework, the core benefit of framework technology, that is, having a widely standardized architecture, diminishes. As a consequence, focusing on hot spots is likely to be more successful than trying to find commonalities.

## 5    Outlook

Are patterns just a hype? Lewis *et al.* [14] view the pattern movement from the perspective of frameworks as part of the evolution of this technology: "Patterns ... is one of the most recent fads to hit the framework camp. ... Expect more buzzwords to appear on the horizon." Because patterns have become the vogue in the software engineering community, the term is used now wherever possible, adorning even project management or organizational work. So the genericity of the term pattern might be the reason that patterns are found everywhere, a fact which is regarded as a clear indication of a hype.

Nevertheless, we view pattern catalogs, in particular the GoF-catalog, as important first step towards a more systematic construction of flexible object-oriented architectures. There is no doubt that organizational measures are at least equally important to be successful as framework development requires a radical departure from today's project culture. Goldberg and Rubin [10] discuss this in detail, without using the term pattern for these management issues.

Probably, we should read the recent books published by the building architect Christopher Alexander [2] in order to predict what will happen to software patterns. He states that the cataloging of architectural styles did not really help architects to come up with buildings that have what he calls a quality without a name. A reduction to very few principles, all related to the concept of "center", allow the generation of this quality without a name. Let us wait and see what this means in the realm of software.

## 6 References

1. Ackermann P. (1996). *Developing Object-Oriented Multimedia Software—Based on the MET++ Application Framework.* Heidelberg: dpunkt.Verlag

2. Alexander C. (1997). *The Nature of Order.* New York: Oxford University Press

3. Alexander C. (1979). *The Timeless Way of Building.* New York: Oxford University Press

4. Alexander C., Ishikawa S., Silverstein M., Jacobson M., Fiksdahl-King I. and Angel S. (1977). *A Pattern Language.* New York: Oxford University Press

5. Apple Computer (1989). *MacApp II Programmer's Guide*; Cupertino, CA: Apple Computer, Inc.

6. Coplien J. and Schmidt D. (eds.) (1995). *Pattern Languages of Program Design.* Conference Proceedings. Reading, Massachusetts: Addison-Wesley

7. Coplien J.O. (1992). *Advanced C++ Programming Styles and Idioms.* Reading, Massachusetts: Addison-Wesley

8. Gamma E., Helm R., Johnson R., Vlissides J. (1995) *Design Patterns—Elements of Reusable Object-Oriented Software*; Reading, MA: Addison-Wesley.

9. Gamma E. (1991). *Objektorientierte Software-Entwicklung am Beispiel von ET++: Klassenbibliothek, Werkzeuge, Design*; doctoral thesis, University of Zürich, 1991; published by Springer Verlag, 1992.

10. Goldberg A., Rubin K. (1995). *Succeeding with Objects—Decision Frameworks for Project Management.* Reading, Massachusetts: Addison-Wesley

11. Helm R., Holland I.M. and Gangopadhyay D. (1990). Contracts: specifying behavioral compositions in object-oriented systems. In *Proceedings of OOPSLA '90*, Ottawa, Canada

12. Johnson R.E. (1992). Documenting frameworks using patterns. In *Proceedings of OOPSLA '92*, Vancouver, Canada

13. Krasner G.E. and Pope S.T. (1988). A cookbook for using the Model-View-Controller user interface paradigm in Smalltalk-80. *Journal of Object-Oriented Programming*, **1**(3)

14. Lewis T., Rosenstein L., Pree W., Weinand A., Gamma E., Calder P., Andert G., Vlissides J., Schmucker K. (1996) *Object-Oriented Application Frameworks*. Manning Publications/Prentice Hall

15. Pree W. (1995). *Design Patterns for Object-Oriented Software Development*. Reading, MA: Addison-Wesley/ACM Press

16. Pree W. (1996). *Framework Patterns*. New York City: SIGS Books

17. Pree W. (1997). *Komponentenbasierte Softwareentwicklung mit Frameworks*. Heidelberg: dpunkt.Verlag

18. Schmucker K. (1986). *Object-Oriented Programming for the Macintosh*. Hasbrouck Heights, NJ: Hayden

19. Taligent (1994). *Taligent's Guide to Designing Programs*. Reading, Massachusetts: Addison-Wesley

20. Weinand A., Gamma E., Marty R. (1988). *ET++ - An Object-Oriented Application Framework in C++*; OOPSLA'88, Special Issue of SIGPLAN Notices, Vol. 23, No. 11.

21. Wilson D.A., Rosenstein L.S. and Shafer D. (1990). *Programming with MacApp*. Reading, Massachusetts: Addison-Wesley

# Object-Oriented DBMS and Beyond

Klaus R. Dittrich and Andreas Geppert

Department of Computer Science, University of Zurich
{dittrich,geppert}ifi.unizh.ch

**Abstract.** Over the past 10+ years, object-oriented database systems have gone a long way from research prototypes to commercial products to real-life mission-critical applications. Currently, we also witness the extension of relational systems with salient object features, resulting in so-called object-relational DBMS.

In this paper, we introduce and review the salient features of both approaches, discuss their merits and shortcomings, and for which kinds of applications they are best suited. We also elaborate on further necessary improvements of the current state of the art. Furthermore, we will speculate about several upcoming areas of database research in a broad sense (like global information systems, workflow management, component technology) where object-orientation and object-relational and object-oriented database systems in particular might (and should!) play a leading role.

## 1 Introduction

Database management systems have a long tradition of research and successful deployment in computing practice. The "classical" systems (hierarchical, network, and relational) are established and viable platforms for data management in a large variety of application areas. With the advent of object-orientation and new requirements imposed by advanced application domains, research has proposed object-oriented database systems beginning in the mid-80s. Since the beginning of the 90s, such systems have been commercially available. After a period of discussion (some may even say "religious warfare") about whether object-oriented DBMSs are actually needed or whether relational systems can meet these new requirements just as well, relational vendors have recently reacted by incorporating object-oriented features into their systems. These extensions are also considered in the forthcoming SQL3-standard.

In this paper, we review object-oriented concepts in database systems and hypothesize about future trends in database systems with respect to object-orientation. Particularly, we argue that database technology will play an increasingly important role in new and emerging areas, such as workflow management and information integration. These areas definitely need database services, and object-oriented database technology is well-suited to meet their requirements such as modeling power and seamless integration. Nevertheless, database functionality as required by these areas can no longer be sufficiently provided by single, monolithic systems; we therefore hypothesize that, in order to be successful, database systems will have to adapt a component-oriented view that allows to bundle services into systems dedicated to the intended application domain.

The remainder of this paper is structured as follows. First, we review and survey object-oriented concepts in database systems. We speculate about future trends in section 3 and the impact of object-orientation and related paradigms (e.g., component technology) on database applications and information systems in section 4. Section 5 identifies challenges and open issues, and section 6 concludes the paper.

# 2    Object-Orientation and Database Systems

Object-oriented database systems try to blend the object-oriented paradigm with database technology. We therefore shortly introduce both areas and then present object-oriented database systems.

## 2.1    Object-Orientation

The general characteristics of the object-oriented paradigm [24] are objects and object identity, classes, and inheritance/specialization. They should apply to all models, systems, methods, etc. that claim the label of being object-oriented.

An *object* is a concept to represent a real or virtual entity; each object has a state and exhibits a certain kind of behavior. In addition, each object has an identity that distinguishes it from all other objects. Complex object structures can be built by using references among objects.

*Classes* define the structure and behavior of objects. A class therefore prescribes an intensional set of objects that are similarly structured and exhibit the same behavior. Individual objects can be constructed by instantiating a class. In this way, a class acts as an object factory, and the set of instances of a class existing at a certain point of time is called the *class extension*.

Similarities between (objects of multiple) classes can be expressed by *specialization*. The semantics of this association is *inclusion polymorphism* [5]: an object can be an instance of multiple classes, i.e., each instance of a more specific class (the subclass) is also an instance of the general class (the superclass). Furthermore, an instance of a (direct or indirect) subclass can be used wherever an instance of a general class is required (*substitutability*). Specialization is related to *inheritance* in that all the properties (structure and behavior) defined for the superclass also apply to the subclass—the subclass inherits structure and behavior from the superclass. Subclasses can extend the inherited class definition or can override parts of it.

Nowadays object-orientation is a widely used paradigm, e.g., for user interfaces, system modeling and design (e.g., UML [28]), and object-oriented programming languages are in widespread use (e.g., C++ or Java).

## 2.2 Database Systems

*Database management systems* [12] (DBMSs) are software systems for the long-term, reliable, and persistent storage of and concurrent access by multiple users to large sets of data items. *Databases* are collections of data items whose structure is defined by a (database) *schema*. A *data model* specifies data definition constructs (i.e., restricts the set of definable schemas) as well as operators that can be applied to data items constructed under the schema. Ultimately, a *database system* (DBS) is a DBMS together with a concrete database (including its schema).

Popular data models include the hierarchical, network, and relational model. In the relational data model [8], all data items are represented as relations (tables) whose elements are tuples, which in turn are sets of attributes. The popularity of the relational model is mainly due to its query language SQL, which allows to retrieve sets of tuples in a declarative manner. In terms of number of installations, relational DBMSs (RDBMS) are currently by far the most successful DBMSs in the marketplace.

## 2.3 Object-Oriented Database Systems

During the last decade, a major part of database research has focused on *object-oriented database management systems* (OODBMSs) to combine database technology with object-orientation. Motivation behind this trend are mainly the impedance mismatch between object-oriented languages and RDBMSs and requirements of then new applications such as CAD/CAM, software engineering environments, etc. In the meanwhile, several products are commercially available and occupy a still small but growing market share.

An OODBMS is *a DBMS with an object-oriented data model* [1, 10]. The first requirement refers to the usual database functionality, i.e., persistent data storage, management of large sets of data, secondary storage management, transaction management, and a query language are indispensable. This means that an object-oriented programming language with (only) support for persistent objects would not classify as an OODBMS.

The requirement to support an object-oriented data model [1] refer to:

- objects with identity,
- complex objects,
- classes,
- encapsulation,
- inheritance and specialization,
- computational completeness (the data manipulation language should allow to express any computable algorithm),
- overriding, overloading, and late binding,
- extensibility (classes should be user-definable).

OODBMS-products are available commercially since the beginning of the 90s. These systems have in the meantime overcome several of their initial weaknesses (e.g., additional functionality such as schema modification, better performance, etc.). Also, a standard has been defined [6] so that customers are now less dependent on single products.

## 2.4 Object-Relational Database Systems

*Object-relational DBMSs* (ORDBMSs) are DBMSs whose data model extends the relational model [8] with object-oriented features. The object-relational extensions [22, 23] are part of the SQL3-standard which is currently under specification by ANSI and ISO. Note that apart from introducing object-oriented concepts, SQL3 defines several other extensions; however, we focus on the object-relational ones.

The object-relational model still considers relations/tables as the fundamental sort of data modeling construct, i.e., every real-world entity to be stored in a database must be represented by a tuple of the appropriate relation. The most important object-oriented extensions refer to support for tuple types, tuple identifiers, abstract data types (ADTs), and inheritance

An ADT is characterized by an (optional) list of supertypes, an internal representation, comparison operators, a constructor, and further type-specific operators. Tuple types (*row types*) can be defined separately from relations. A row type defines a set of attributes, where references to other tuples (of any type), collection-valued types, and atomic types are permitted as attribute types. Tuple types can inherit from other tuple types. Tuples types can be assigned to relations upon relation creation. For each relation, an attribute can be designated as tuple identifier in that the values of this attribute are specified to be system-generated. Relations can be defined as specializations (i.e., subsets) of other relations, with the semantics that each tuple of the specialized relation is also an element of the more general relation.

Although the SQL3-standard is still under definition, vendors have already started to include object-relational features into their products. Illustra (now available as Informix Universal Server) [35] supports user-definable types and inheritance. Additionally, nonstandard functionality can be packaged into so-called *DataBlade modules* that can be linked to a DBMS. Each such module can be regarded as a library of types. Along with each type, functions, operators, and access methods of the type are given. Examples for DataBlade modules include geometric data types, images, and time series. Extensions similar to Data Blades are supported by Oracle8 (cartridges) and DB2 (extenders [20]).

Most relational and object-relational DBMSs also support constraints and/or triggers. *Triggers* or *event-condition-action rules* (ECA-rules) [11] allow to specify situations in which the DBMS has to react automatically by executing a pre-defined action. The semantics of such a rule is that whenever the event occurs and the condition holds, then the action is executed. Possible event types are updates of a tuple or object, conditions are typically specified as queries (with the semantics that the condition holds if the query returns a non-empty result), and actions are specified in the data manipulation language of the DBMS. ECA-rules are useful constructs to implement a variety of tasks, including consistency constraints and materialized view maintenance.

## 2.5 Strengths and Weaknesses

When identifying the strengths and shortcomings of relational, object-oriented, and object-relational DBMSs for specific types of applications, meaningful criteria include

1. whether real-world structures can be modeled (more or less) naturally,
2. whether there is a design method how to use the respective type of DBMS,
3. whether the DBMS is efficient for typical applications.

For a classification of strengths of the various DBMS-types, the DBMS-matrix [35] has been proposed using complexity of data structures and the need for complex queries as the two dimensions. Following this matrix,

- RDBMSs are well-suited for simple data structures and queries,
- OODBMSs can handle complex data structures, but do not support queries that well,
- ORDBMSs are well-suited for applications characterized by complex data structures and complex queries.

From our point of view, this classification is too schematic. For instance, the ODMG-standard also specifies a query language (OQL) which is for quite some time already implemented in some OODBMS-products (e.g., $O_2$ [2]). Thus, given that the data models of ORDBMSs and OODBMSs are quite similar, in principle OODBMSs might handle queries equally well as ORDBMSs. Moreover, the classification seems to assume that ORDBMSs can increase the modeling power of RDBMSs (which is then comparable to that of OODBMSs) and at the same time preserve the performance of query processing as present in RDBMSs.

The strength of OODBMSs is that they are well-suited for complex object structures and that they integrate very smoothly with object-oriented programming languages. Complex structures can be more or less directly modeled in the database, and there is no need to convert data structures in application programs to/from simple-structured tuples. In other words, the so-called *impedance mismatch* vanishes. This is particularly important for applications written in object-oriented languages; they can map their objects directly onto database objects. To some extent, the impedance mismatch will continue to exist in OR-DBMSs, as queries will return sets of tuples which then have to be converted into programming language objects. Due to complex ADTs and type-specific behavior definable for them, the impedance mismatch will however become smaller in ORDBMSs than in RDBMSs.

*Versions* and *configurations* [21] represent causal and/or historical developments of objects which for instance are needed in design environments (e.g., CASE). A versioning concept can be nicely integrated into an object-oriented model, so OODBMSs are a natural starting point for adding version support. Many OODBMS-products already support some sort of version or configuration functionality, along with design transactions.

A limitation of current OODBMS-products—but not of OODBMS-technology in general—seems to be that they are less adequate for the management of huge amounts of data (in the range of many giga or even terabytes). Such amounts of data are typically encountered in data warehousing [7] or data mining applications. The critical factor here is efficiency of query processing. RDBMSs are seen at an advantage, because they can rely on a much simpler model as well as on much more experience and research in query processing and optimization than OODBMSs. Moreover, SQL can be easily parallelized [9], which is still an open issue for query languages of OODBMSs.

Another area in which RDBMSs rely on sound methods and a bulk of experiences (again, as a consequence of the data model simplicity) is *database design*. Both, the notation and the process are well understood, while this is not really the case for OODBSs (see [34] for a survey). Ideally, a design method and notation supports the conceptual database design and allows to transform the conceptual schema into the logical schema of the used database system. In other words, an equivalent to the entity-relationship-model for the relational case is needed for OODBS-design. Such a design method should be an extension or at least be compatible with existing design methods, such as UML [28].

Another weakness of OODBMSs is that they provide only very limited support for specifying and enforcing consistency constraints (OODBMSs following the principle of persistence by reachability ensure referential integrity, and the ODMG-standard allows to define user-defined keys). Similar to the relational case, consistency constraints can be implemented as ECA-rules [16]; however, since OODBMSs do not (yet) support ECA-rules, explicit constraints mostly have to be enforced by method implementations and application programs.

Summarizing, OODBMSs are a natural choice for applications that need to manage complex object structures and for object-oriented applications that need to avoid the impedance mismatch. It is still to be seen whether they are feasible platforms for emerging application classes such as data warehousing and data mining. Concerning ORDBMSs, we feel that they are still too young in order to assess whether they can live up to their claims. In particular, it remains to be seen whether ORDBMSs can extend their modeling power while preserving the query processing performance of RDBMSs. In fact, we expect that—as data structures become as complex as those managed by OODBMSs—complexity of query processing will be the same in OODBMSs and ORDBMSs.

## 2.6 And the Winner is...

object-orientation. The advent of ORDBMSs shows that relational vendors accept the the need for object-oriented concepts in databases. Thus, the DBMS of the future will incorporate object-oriented concepts, regardless whether its origin is relational or "pure" object-orientation.

It remains to be seen to which extent and how smoothly ORDBMSs can incorporate object-oriented extensions. Extending an RDBMS to an ORDBMS might imply to re-design major parts of the DBMS, and it is likely that ORDBMS-builders will encounter the same problems and difficulties as OODBMS-designers did. Particularly, efficient object storage and management and efficient query processing and optimization will turn out as major hurdles. Users will also expect performance to remain the same for "traditional" relational data and queries, and it will be interesting to see whether object-relational extensions will hamper the performance of conventional queries.

In other areas, both OODBMSs and ORDBMSs will share their destiny, for instance in database design. Apparently, once abstract data types, inheritance, triggers, user defined functions and so forth are supported, the traditional database design methods must be extended and adapted to the new functionality (which we know from OODBMSs to be a severe problem).

Unfortunately, the academically clean solution will *not* be among the winners: a clear-cut, comprehensive, object-oriented data model and system that would include the "relational world" as a special case. Whereas OODBMS-vendors are too small to engage in and promote such an approach, RDBMS-vendors do not have any interest in it since this would incur huge investments.

## 3   Beyond OODBMS: OBMS

Given the diversity of developments and different types of DBMSs, we now turn toward some of the major directions in which database technology will evolve. Particularly, we speculate about the future prospects of object database systems.

We believe that the future will show two trends:

1. database technology will be even more ubiquitous in computing,
2. the successful DBMSs will no longer be monolithic systems, but will instead allow to "bundle" user-defined or external components.

In order to accomplish these trends, object-orientation in databases will be a prerequisite. In this respect, "object-orientation" can mean object-oriented or object-relational DBMSs; we thus henceforth subsume both types of systems under the term *objectbase systems* (OBSs and OBMSs).

Typical users work with data that is stored in a variety of data stores, whereby only a fraction of the data is actually managed by database systems. Integrating all these diverse data stores will provide users a uniform view of all their relevant data, will avoid redundancies and inconsistencies, and as a consequence ease the daily tasks of many users significantly. Moreover, although the data stored and managed by tools cannot be migrated to a database, users would benefit from database services, such as query facilities or transactions in their tools [36]. Thus, in the not too distant future we will see database systems expand and provide their services, e.g., to desktop applications, in a componentized form [4, 37].

Data stores managed by these tools and other data repositories will have to be integrated into large-scale information systems, and database services must be available for them. These information systems will be operated over the Internet or Intranets, and users will query them, without willing to bother where the desired information is actually stored (here, "where" means both, the type of data store containing the required information as well as the network node where the data store is located). In other words, the established query languages will be leveraged to much more powerful *universal query services* that are able to retrieve and browse the contents of potentially any kind of data stores anywhere in a network.

In addition to queries, databases will be accessed and manipulated in complex, long-lasting processes. The information system of the future can no longer be idealized as a database and an attached collection of application programs. Instead, in the future we will see databases as one part of large-scale cooperative environments. The behavior of these environments will be defined by process models, and databases, users, and all sorts of software systems will interact to execute processes (or synonymously, *workflows*).

DBS-services will thus be a crucial component of quite varied scenarios, and it is to be expected that no single DBMS will meet all requirements equally well and provide for all its tasks in an efficient manner. Successful solutions thus will be characterized by a component-oriented architecture which allows to add new components needed, to exchange existing components by others better suited for certain tasks, and to remove components whose functionality is not needed for specific applications [37]. In other words, we suggest that the real next great wave will be *bundable DBMSs* that allow to configure application-specific DBMSs out of pre-fabricated components.

Bundling can certainly be accomplished by traditional (purely relational) database systems and procedural programming techniques. Nevertheless, a really powerful component-oriented approach to database management will require object-orientation on all levels:

- only object-oriented modeling facilities provide the expressive power that is needed, e.g., for integrating all sorts of data,
- only object-oriented (software) design principles will allow for the required extensibility and customizability of DBMSs.

Below we discuss several areas which will be influenced by object database technology in more detail.

## 4 Four Cases for Objectbase Technology

In this section, we elaborate on several areas of objectbase technology which might benefit from object-orientation. "Benefit" can hereby bear different (not mutually exclusive) meanings:

- benefit in terms of object-oriented modeling concepts,
- benefit of using object-oriented architectures and implementations,
- benefit of using database services for the construction of (information) systems.

### 4.1 Component Technology

*Component technology* is an approach to build software systems out of pre-existing software pieces [25]. The attractive aspect of component technology is that it allows to build new systems and extend existing ones at rather little cost, provided that reusable adequate components are available (see Fig. 1). Moreover, existing components often exhibit a higher degree of maturity and robustness than newly developed software modules. However, in order to put component technology to work, several issues must have been settled:

- a software architecture model [32] is needed that allows to specify componentized systems and to add components in a controlled way,
- systems must be dynamically reconfigurable so that new components can be added without redesigning or recompiling the whole system,
- components must be designed and implemented in such a way that they are effectively reusable (e.g., do not make assumptions about the context in which they are going to be used).

A component certainly is more abstract and of larger granularity than an object (it is rather comparable to a module in software engineering). However, object-orientation lends itself to component technology in a very natural way:

- a prerequisite for component technology (as for any kind of software reuse) is information hiding, i.e., encapsulation of object state and behavior,
- specialization, inheritance, and polymorphism provide the appropriate means to specialize and customize services/functionality in subclasses.

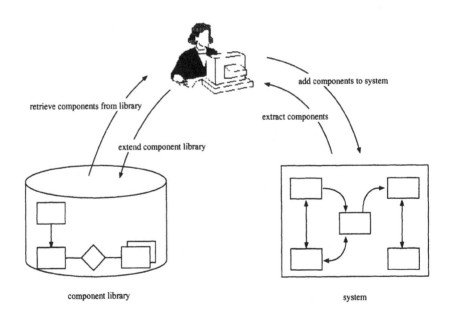

**Fig. 1.** System Construction based on Component Libraries

Component technology will influence databases in two ways. First, it will allow to build application systems out of components (and DBMSs will be one prominent type of component). Second, given that DBMSs are conceived as componentized software systems, component technology will allow to build DBMSs themselves out of components.

Component technology will be of paramount importance for applications development, especially for heterogeneous and distributed systems [27]. The problem here is to integrate a collection of heterogeneous systems distributed over a network into a coherent and homogeneous application system, using the various systems as components and middleware services [27] as the glue that abstracts from heterogeneity and distribution details.

The prominent representative of component technology in this sense is OMG's Common Object Request Broker Architecture (CORBA) [26]. CORBA defines an object model and an interface definition language (IDL) that is used to specify service signatures. Any functionality offered by some server and/or needed by some client must be specified in IDL. In addition, CORBA defines so-called *Common Object Services* (COS) most of which directly represent database services (e.g., persistence, transaction, query, and relationship services). CORBA has been assessed as a feasible platform for applications that are more oriented towards operation shipping, while it seems to be less beneficial for efficient data-intensive applications [31].

Another representative is Microsoft's *Object Linking and Embedding* (OLE) [27]. In OLE, related functions are packaged into interfaces. Interfaces can be defined using the IDL of DCOM (*Distributed Component Object Model*). An OLE-component is defined by a class that implements one or more interfaces and a class factory (which knows how to create instances of the class).

One step further, component technology is not only used to build application objects, but also to build DBMSs themselves [15]. A DBMS is *unbundled* into a collection of services and potential components implementing them. The notions of "component" and "services" are defined by an underlying (software) architecture model; i.e., this model defines how services and components can be defined and interrelated. Consider concurrency control, object storage, query optimization or versioning as sample (database) services (note that persistence, concurrency control and queries are also elements of OMG's COS). Then, a strict two-phase lock manager would be one possible component implementing the concurrency control service. Given unbundled ODBMS-services and components, a DBMS can be extended or customized by adding new components when new services are needed, or exchanging components when service requirements change.

The appealing aspect of database component technology is that DBMSs can be tailored to the needs of their applications: every DBMS provides just the functionality that is effectively needed, and applications do not have to pay—in terms of performance—for functionality that they do not use. A first step towards component technology are the DataBlades (or extenders, cartridges) in ORDBMSs, and the user-definable classes in OODBMSs. However, these mechanisms solely extend the set of data types. We envision the need to push component technology in DBMSs one step further, namely in that the DBMS itself can be extended or customized. For instance, it does not need a rocket scientist to define an ADT for multimedia objects whereas the critical task is to enable efficient management and retrieval of instances of such types. Thus, it might be necessary to add advanced access paths and methods, and it might even be necessary to extend the query processor and optimizer. Otherwise, instances of ADTs can be modeled and stored, but cannot be retrieved and queried with acceptable performance.

Summarizing, component technology will become an even more fundamental prerequisite of database technology for the construction of both, database applications and DBMSs. Object-orientation is the underlying paradigm that makes component technology practical in database technology.

### 4.2 Workflow Management

Workflow management [14] is concerned with the modeling and execution of automated production/design/administration processes. *Workflow management systems* (WfMSs) are software systems that provide for the specification and execution of workflows (see Fig. 2). A workflow type specifies the structures of workflows (i.e., subworkflows and/or atomic activities), processing entities (software systems or humans) responsible for carrying out these activities, and execution constraints (e.g., ordering constraints, deadlines, data flow). Workflow types can be instantiated; the workflow instances can be executed under control of the WfMS.

*Control data* comprises all the data elements that are manipulated or accessed by processing entities and routed between the various processing stations within a workflow. As an example, consider an insurance claim workflow. In this case, relevant data includes all the information about the insurance policy, the insurance case, the claimed amount, etc. Some of this data (or handles to it) will flow through a workflow, while other data will be external to the WfMS and managed by some sort of data stores (e.g., a DBMS). During workflow specification, control data must be specified, and upon workflow execution, it must be routed through the various processing stations.

Workflow management can benefit from object-oriented data modeling facilities for two kinds of data:
- control data,
- internal data structures.

Object-oriented modeling of control data yields several advantages: first, the modeling power of object-orientation can be exploited. For instance, an insurance claim is a quite complex structure, and therefore could be modeled naturally as a complex object. Second, the WfMS execution engine and the processing entities (which in turn can be software systems) need to agree on common data formats, so that using a standard object model would make sense (e.g., ODMG's [6] or OMG's [26] object models would be reasonable choices).

In addition to control data, a WfMS needs to store information about workflow types and workflow executions. The WfMS needs to retrieve this information

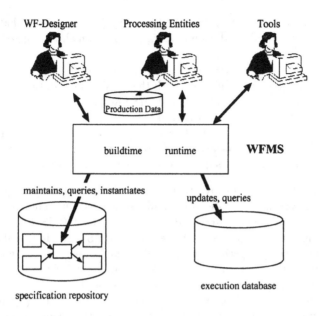

WF-Designer     Processing Entities     Tools

Production Data

buildtime    runtime     **WFMS**

maintains, queries, instantiates

updates, queries

execution database

specification repository

**Fig. 2.** Workflow Management Systems

in order to properly enact the workflow. For instance, parts of workflow specifications must be retrieved in order to determine whether and which further activities are ready to be executed and which processing entities are eligible for executing the activity. Again, all these various types of information can be modeled much more naturally in an object-oriented way. Currently, most WfMSs still use RDBMSs as platforms (e.g., InConcert [30]), and thus have to suffer the impedance mismatch in case their modeling language is object-oriented.

WfMSs must be extensible, need to be integratable with other components, and should be customizable for all sorts of platforms and environments. For instance, a WfMS should be able to operate in a CORBA-environment, which would open up the possibility to use the common object services and to integrate application objects into a coherent workflow. This integration is eased if the WfMS relies on component technology and object-orientation (e.g., as in ObjectFlow [19]).

These requirements are best met if the WfMS is built out of components. In this way, the WfMS is also flexible to use different DBMSs as platforms, or to implement parts of its functionality using database services. If available, a WfMS can greatly benefit from these services, at all the various stages of workflow specification and execution:

- data modeling and storage service: specifications of workflows information need to be modeled and stored persistently for further inspection and reuse,
- query service: workflow specifications and workflow execution states might have to be inspected, which is best supported by a query service,

- transaction service: during execution, the WfMS must ensure that its operations are recoverable and atomic (e.g., information about current execution states must not get corrupted),
- concurrency control service: typically many clients (e.g., processing entities, workflow designers) will interact with the WfMS and will thereby update its internal data structures. Thus, the WfMS must perform concurrency control in order to ensure that concurrent clients are isolated form each other,
- access control service: the WfMS has to prevent the access to and manipulation of workflow specification and execution information by unauthorized users, which can be best supported by a flexible access control service.

Hence, database services are crucial for a WfMS to manage required information about workflow specifications and executions. One step further, more advanced database services are used to leverage the DBMS to a full-fledged WfMS. In other words, given a componentized DBMS and additional services, a WfMS can be plugged together out of database and other services. For instance, ECA-rules (see above) are used for the implementation of the workflow logic. Every occurrence in the workflow system is expressed as an event, and ECA-rules define what should happen in reaction to this event [17] (for instance, a typical rule might define to execute a successor activity as soon as two preceding activities have terminated with a certain result). This style of workflow execution makes use of further database services, namely ECA-rule definition, event detection, and rule execution. These services must in turn be flexible and customizable in order to tailor their semantics to the need of WfMSs (e.g., specific event types are needed, and rule execution semantics might be different from those found in active DBMSs).

Summarizing, workflow management will become prevalent in application systems and will allow to define and enact processes executing in cooperative and distributed environments. Object-oriented database technology enables WfMSs to be designed and implemented in an object-oriented way on all levels. Advanced OBMS-functionality such as ECA-rules allows to implement certain aspects of workflow execution within the OBMSs.

### 4.3 Information Integration and Global Information Systems

*Information integration* refers to the provision of a unified and uniform view of information stemming from varied data sources. A typical example of information integration is data warehousing [7]. Aside from data warehousing, information integration can be seen in a much broader context, e.g., in terms of the data sources that are considered. Particularly, information from all sorts of sources might be desired to be integrated, such as voice mail, email, video clips, X-ray images, and so forth. Furthermore, data can be structured (as in relational databases), semi-structured (e.g., email messages), or unstructured (e.g., texts).

The real value of information integration and global information systems lies in the ability to *browse* through and *query* the integrated data. Hereby, "query" does not necessarily mean pure, traditional database queries, i.e., exact queries that return only those data elements that fully match the query. In addition, probabilistic queries as typical for information retrieval should be specifiable which return "imprecise" results that are somehow related to the query. This type of query is of particular importance for semi- or unstructured data sources (e.g., texts, images, Web-pages). Furthermore, it should be possible to specify queries on the global level, without having to worry about which data sources contains the data and how the information is formatted there. These considerations are undertaken by the global information system that decomposes the query, forwards the subqueries to the data sources, receives answers, and combines them according to the user's initial query.

The critical issues in information integration are a uniform and powerful model to define the (integrated) data from the various sources and query facilities to retrieve and combine the data from the sources.

Data warehouses usually use the relational model, because their data sources are relational and so the mapping from the data source to the warehouse is simple. This is no longer the case when general data sources are considered; then an object-oriented model will be much more feasible. Generic behavior (e.g., of "information objects") can be described in abstract classes, and specialized semantics can be captured in subclasses using inheritance. Classes and class-specific operations will be especially important when multimedia data needs to be integrated. Information objects can be complexly structured, so the ability to model complex objects and aggregations are needed. Moreover, especially when objects from different sources are associated to more complex structures, then object identity is a prerequisite [18]. Thus, for the general case of information integration, an object-oriented global model will be useful (e.g., the Garlic system [29] uses and extends the ODMG-model).

Another difference between data warehouses and federated DBMSs [33] and the kind of information integration considered here is that the former maintain copies of data (in the warehouse) and that updates and transaction management are critical issues in federated DBMSs. In contrast, here we focus on the integration of information and on queries against the integrated data sources. All the benefits of an object-oriented model apply to such a query service as well (e.g., representation of complex results at the end-user interface using object references via object identifiers, or presentation of results according to the class-specific behavior). In addition, given that the global model is object-oriented, then an object-oriented query language is a natural choice for formulating queries (i.e., OQL [6] or SQL3).

A universal query service (Fig. 3) requires an architecture model that allows to incorporate any kind of data source. This architecture model in turn will be object-oriented and based on component technology. *Wrappers* are one possibility [29], where a wrapper encapsulates a specific data source and defines a predefined behavior, namely to "understand" queries and translate them into the query language of its data source. Another possibility is OLE/COM, which is used in OLE DB [4] to define and implement components. Typical classes of components in OLE DB are "service providers" and "data providers". A data

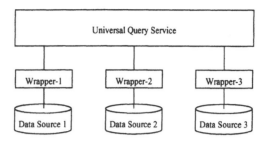

**Fig. 3.** Universal Query Service

provider is a component that can export so-called rowsets (i.e., sets of rows, which in turn are sets of attributes), and a service provider can take rowsets as input, transform them, and produce new rowsets as output. Thus, provided that information is given in the form of rowsets, any data source can be queried, and different queries can be combined to more complex ones, again producing rowsets as results. Note that the restriction of OLE DB to row-set-based structures might be too strong for some current types of data.

Summarizing, information integration and global information systems will gain more and more importance in Intranets and the Internet. Object database technology can beneficially support these systems and tasks with respect to powerful modeling and querying facilities.

### 4.4 Repositories

Future information systems will integrate much more diverse information than they do today, they will provide access to these information sources using universal query services, and complex, distributed workflows will execute within them. In order to achieve this vision, database services must be available in a very flexible form. The price for this much higher degree of flexibility is complexity. How can we manage all these componentized database services and configurations of them into complex information systems?

To that end, organizations need "meta databases" that store all the relevant information about their information systems. In order to configure and operate the kind of system we outlined above, knowledge about database services, their usage, integrated data sources, involved applications and so forth must be available. In other words, we need database technology in order to construct the powerful information systems of the future: the construction of these systems must be assisted and guided by repository systems [3].

A *repository manager* supports the management of information about design artifacts (in our case, "artifacts" would represent any kind of information about database services, components, etc.). Users interact with the repository manager through *repository tools*, i.e., editors, browsers, and so forth that allow to store new artifacts in the repository and to extract old ones from it. Aside from the storage of artifacts, a repository manager implements further tasks, such as version and configuration management to represent design histories and to model compositions of artifacts, or workflow control to model and enact design processes.

Repositories are related to database technology in two ways. First, repository support is crucial to methodize DBMS-development and usage. Second, a repository system can in turn be built using objectbase technology.

As for the first relation, OBMSs and their application systems will be manageable only if *information about them* is available in an appropriate form. For instance, assume an ORDBMS with a large number of pluggable ADTs (or Data-Blades, extenders). Information about these artifacts is crucial for managing and maintaining the ORDBS. Lack of this information will lead to redundancies and inconsistencies, and chaos will be the consequence. Similarly, imagine a global information system that integrates data from a large number of heterogeneous sources. This information system will be manageable only if meta data about each of the sources is available (which data they contain, how they can be accessed, how regularly they are updated and how frequently updates are propagated, etc). Moreover, wrappers may have to be built for each of these sources, and both, the design and the implementation of the wrappers should be reusable during the integration of further sources. Or, assume a workflow system with a large number of (potentially versioned) workflow specifications, processing entities, etc. Again, all this information must be kept accessible to manage and maintain the workflow system and to reuse artifacts for future specifications.

Hence, while the increased flexibility of component technology and extensibility of DBMSs are certainly to be appreciated, users and administrators need support to master the equally increasing complexity of DBMSs. Otherwise, if sufficient support to systematically develop and maintain these systems is not provided, users will probably refrain from using all these advanced features.

Fortunately, OBMS-technology itself can help overcome this problem (in some sense, OBMSs can bootstrap themselves). This is the second relationship between repositories and object database technology: OBMSs can be leveraged to repository managers when adequate OBMS-services are available. Version management facilities are required to define and manage the complex artifacts of a repository. OQL and SQL3 are feasible languages to query repositories and to retrieve artifacts. An ECA-rule service can be used to specify and enforce the complex constraints typical for repositories and to implement notification mechanisms. Workflow management and design transaction support can be provided as discussed in section 4.2 above, and by customizing a concurrency control and transaction service. The latter case is also a nice example for the customization of internal DBMS-components, i.e., adding a new ADT is not sufficient here.

Summarizing, repository technology is mission-critical for users to master the complexity of OBMS-technology and its applications. On the other hand, repositories are also typical applications that can be built using OBMS-functionality and component technology.

# 5    Challenges to Future Work

In this section we present a set of challenges to research and practice that have to be met in order to turn OBMSs into an even more useful technology and to master the complexity of advanced OBMS-applications.

As for the *functionality of OBMSs*, additional concepts need to be integrated into existing systems and products. For instance, support for ECA-rules is required as a basic mechanism to implement a variety of tasks such as consistency constraint enforcement. ORDBMSs already support at least simple forms of ECA-rules (they consider, e.g., only a restricted set of event types), while for OODBMSs only research prototypes are available (e.g., [13]). In order to convince vendors that ECA-rules are a viable technology, users should urge them to give support for consistency constraints, notification, etc., and researchers should come up with models that are powerful yet usable and that can be implemented in a robust and efficient manner. Another area which requires more work are query languages and query processing techniques for global information systems. In this respect, research is necessary to develop systematic information integration techniques (e.g., easy-to-build and reusable wrappers). The existing and well-understood query languages and query processing techniques should be extended to cover also distributed and heterogeneous data sources, including semi- and unstructured data, and exact as well as probabilistic queries.

Concerning the usage of OBMSs, a better understanding of and more experience in how to develop OBMS-applications is desperately needed. Design methods and corresponding tools should be developed. These methods and tools should built upon the existing methods and tools established in software engineering, since those are already in use and designers will not be willing to use multiple, completely different methods/tools in parallel. While database design at first glance is an item on the research agenda, it also needs input from practitioners, because only they have real experience in building real-life, mission-critical large systems.

In a broader sense, appropriate design and development support is the decisive factor for OBMS-usage. As we have outlined above, OBMSs lend themselves in a natural way to play important roles in a variety of scenarios, ranging from information integration and global information systems to workflow management. Component technology and OBMSs are viable base technologies for these scenarios, but users will rely on the appropriate support of how to develop these applications in a systematic way.

Therefore, as a first challenge in this context, component object models must be developed that are powerful enough to describe componentized systems and OBMSs. Existing and new models (e.g., CORBA, COM, wrappers) need to be evaluated with respect to which scenario they support well, and experiences in using them are needed. Second, a method how to bundle components together into powerful OBMS-applications is required. This method should be systematic (i.e., rely on established techniques from software engineering) and should enable reusability of designs and openness of constructed systems.

# 6 Conclusion

In this paper, we have reviewed and surveyed object-oriented concepts in database systems and have hypothesized about (some) future trends and challenges in this area. We foresee that the trend towards large information systems that are built out of smaller units and that are required to (inter)operate in distributed environments will continue. DBMSs will have to exhibit a much higher degree of openness and flexibility in order to smoothly integrate into such large-scale systems and contribute their services in an optimal way.

From an abstract point of view, OBMSs (but not necessarily products as they are available today) are powerful enough to meet these requirements. However, even for them much more work remains to be done, e.g., with respect to methods for how to systematically unbundle and construct DBMSs. Approaches and experiences might—and should—be adapted from other research areas, such as component technology and software architecture. If the "componentization" efforts succeed, then OBMS-services will turn into commodities met in all sorts of future information systems and tools.

## Acknowledgements

We thank Dimitrios Tombros, Markus Kradolfer, and Ruxandra Domenig for their helpful comments on an earlier version of this paper.

## References

1. M. Atkinson, F. Bancilhon, D.J. DeWitt, K.R. Dittrich, D. Maier, and S.B. Zdonik. The Object-Oriented Database System Manifesto (a Political Pamphlet). In *Proc. 1$^{st}$ Intl. Conf. on Deductive and Object-Oriented Databases*, Kyoto, Japan, December 1989.
2. F. Bancilhon, C. Delobel, and P. Kanellakis, editors. *Building an Object-Oriented Database System*. Morgan Kaufmann Publishers, 1992.
3. P.A. Bernstein and U. Dayal. An Overview of Repository Technology. In *Proc. 20$^{th}$ Intl. Conf. on Very Large Data Bases*, Santiago, Chile, September 1994.
4. J.A. Blakeley. Data Access for the Masses through OLE DB. In *Proc. ACM-SIGMOD Intl. Conf. on Management of Data*, Montreal, Canada, June 1996.
5. L. Cardelli and P. Wegner. On Understanding Types, Data Abstraction, and Polymorphism. *ACM Computing Surveys*, 17(4), 1985.
6. R.G.G. Cattell and D. Barry, editors. *The Object Database Standard: ODMG 2.0*. Morgan Kaufmann Publishers, San Francisco, 1997.
7. S. Chaudhuri and U. Dayal. An Overview of Data Warehousing and OLAP Technology. *ACM SIGMOD Record*, 26(1), March 1997.
8. E. Codd. A Relational Model for Large Shared Data Banks. *Communications of the ACM*, 13(6), 1970.
9. D.J. DeWitt and J. Gray. Parallel Database Systems: The Future of High Performance Database Systems. *Communications of the ACM*, 35(6), June 1992.

10. K.R. Dittrich. Object-Oriented Data Model Concepts. In A. Dogac, T.M. Özsu, A. Biliris, and T. Sellis, editors, *Advances in Object-Oriented Database Systems*, volume 130 of *Computer and System Sciences*. Springer, 1994.

11. K.R. Dittrich, S. Gatziu, and A. Geppert. The Active Database Management System Manifesto: A Rulebase of ADBMS Features. In *Proc. $2^{nd}$ Intl. Workshop on Rules in Database Systems*, Athens, Greece, September 1995. Springer.

12. R. Elmasri and S.B. Navathe. *Fundamentals of Database Systems*. Benjamin/-Cummings, $2^{nd}$ edition, 1994.

13. S. Gatziu, A. Geppert, and K.R. Dittrich. Integrating Active Mechanisms into an Object-Oriented Database System. In *Proc. $3^{rd}$ Intl. Workshop on Database Programming Languages (DBPL)*, Nafplion, Greece, August 1991.

14. D. Georgakopoulos, M. Hornick, and A. Sheth. An Overview of Workflow Management: From Process Modeling to Workflow Automation Infrastructure. *Distributed and Parallel Databases*, 3(2), April 1995.

15. A. Geppert and K.R. Dittrich. Constructing the Next 100 Database Management Systems: Like the Handyman or Like the Engineer? *ACM SIGMOD Record*, 23(1), March 1994.

16. A. Geppert and K.R. Dittrich. Specification and Implementation of Consistency Constraints in Object-Oriented Database Systems: Applying Programming-by-Contract. In *Proc. Datenbanken in Büro, Technik und Wissenschaft (BTW)*, Dresden, Germany, March 1995. Springer.

17. A. Geppert, M. Kradolfer, and D. Tombros. Realization of Cooperative Agents Using an Active Object-Oriented Database Management System. In *Proc. $2^{nd}$ Intl. Workshop on Rules in Database Systems*, Athens, Greece, September 1995. Springer.

18. M. Härtig and K.R. Dittrich. An Object-Oriented Integration Framework for Building Heterogeneous Database Systems. In *Proc. IFIP DS-5 Conf. on Semantics of Interoperable Database Systems*, Lorne, Australia, November 1992.

19. M. Hsu and C. Kleissner. ObjectFlow: Towards a Process Management Infrastructure. *Distributed and Parallel Databases*, 4(2), 1996.

20. DB2 Relational Extenders. White Paper (http://www.software.ibm.com/pubs/-papers/), IBM Corp., May 1995.

21. R. Katz. Toward a Unified Framework for Version Modeling in Engineering Databases. *ACM Computing Surveys*, 22(4), 1990.

22. K. Kulkarni, M. Carey, L. DeMichiel, N. Mattos, W. Hong, M. Ubell, A. Nori, V. Krishnamurthy, and D. Beech. *Introducing Reference Types and Cleaning Up SQL3's Object Model*. International Organization for Standardization, August 1995.

23. J. Melton. A Shift in the Landscape. Assessing SQL3's New Object Direction. *Database Programming & Design*, 9(8), August 1996.

24. O. Nierstrasz. A Survey of Object-Oriented Concepts. In W. Kim and F.H. Lochovsky, editors, *Object-Oriented Concepts, Databases, and Applications*. ACM Press, New York, 1989.

25. O. Nierstrasz and L. Dami. Component-Oriented Software Technology. In O. Nierstrasz and D. Tsichritzis, editors, *Object-Oriented Software Composition*. Prentice Hall, London, 1995.

26. The Object Management Group. *The Common Object Request Broker: Architecture and Specification. Revision 2.0*, July 1995.

27. R. Orfali, D. Harkey, and J. Edwards. *The Essential Client/Server Survival Guide*. John Wiley & Sons, $2^{nd}$ edition, 1996.

28. Rational Software Corp., Santa Clara, CA. *Unified Modeling Language: Notation Guide*, 1997.

29. M.T. Roth and P. Schwarz. Don't Scrap it, Wrap it! A Wrapper Architecture for Legacy Data Sources. In *Proc. $23^{rd}$ Intl. Conf. on Very Large Data Bases*, Athens, Greece, August 1997.

30. S.K. Sarin. Object-Oriented Workflow Technology in InConcert. In *Proc. of the IEEE COMPCON Spring*, Santa Clara, February 1996.

31. J. Sellentin and B. Mitschang. Möglichkeiten und Grenzen des Einsatzes von CORBA in DB-basierten Client/Server-Anwendungssystemen. In *Proc. GI-Fachtagung Datenbanksysteme in Büro, Technik und Wissenschaft*, Ulm, Germany, March 1997. Springer.

32. M. Shaw and D. Garlan. *Software Architecture: Perspectives on an Emerging Discipline*. Prentice Hall, 1996.

33. A.P. Sheth and J.A. Larson. Federated Database Systems for Managing Distributed, Heterogeneous and Autonomous Databases. *ACM Computing Surveys*, 22(3), September 1990.

34. I.-Y. Song and E.K. Park. Object-Oriented Database Design Methodologies: A Survey. In T.W. Finin, C.K. Nicholas, and Y. Yesha, editors, *Selected Papers from 1st Intl. Conf. on Information and Knowledge Management*, volume 752 of *Lecture Notes in Computer Science*. Springer, 1992.

35. M. Stonebraker and D. Moore. *Object-Relational DBMSs*. Morgan Kaufmann Publishers, 1996.

36. D. Vaskevitch. Database in Crisis and Transition: A Technical Agenda for the Year 2001. In *Proc. ACM-SIGMOD Intl. Conf. on Management of Data*, Minneapolis, May 1994.

37. D. Vaskevitch. Very Large Databases. How Large? How Different? In *Proc. $21^{st}$ Intl. Conf. on Very Large Data Bases*, Zurich, Switzerland, September 1995.

# On Integration of Relational and Object-Oriented Database Systems*

Maria E Orlowska[1] and Hui Li[1] and Chengfei Liu[2]

[1] School of Information Technology, The University of Queensland,Brisbane, Qld 4072, Australia
[2] CRC for Distributed Systems Technology, Brisbane, Qld 4072, Australia

**Abstract.** The converging trend of relational database technology and object-oriented database technology results in object-relational database systems which extend the relational systems with add-on object features. Efforts from both academic research and industry have been directed into extended data type processing by fully functioned database systems. In this paper, we put forward an approach for realizing object-relational systems by deploying current distributed database technology. The paper first discusses the object-relational data model and its query language. Then a hetrogeneous database architecture is presented for realizing object-relational technology by integrating relational and object-oriented database systems. Two main functions of the architecture: schema transformation and query translation are further discussed and correspondent algorithms are proposed.

## 1 Introduction

Traditional relational database systems (RDBMSs) have demonstrated a strong capability for most database applications, but suffer from almost no support for complex data. All data are stored in tables, and every attribute in the table is defined on one of the few atomic basic types (e.g. integer, float, boolean, character, string). It is currently being accepted that RDBMSs should allow type extensibility, i.e., new data types can be added to the system without significant changes to any part of the existing code. Unfortunately, in traditional RDBMSs, complex data can only be stored as uninterpreted BLOBs (Binary Large Objects), and the interpretion of this data relies solely on the application. However, many specialised databases, such as engineering DBs, spatial DBs, multimedia DBs, scientific and statistical DBs, require more complex structures for data, and nonstandard application-specific operations. It is desirable to extend the relational model to accommodate these features.

About a decade ago, researchers began to investigate general methods to introduce objects into database systems. A number of different ways were explored:

---

* The work reported in this paper has been funded in part by the Cooperative Research Centres Program through the Department of the Prime Minister and Cabinet of the Commonwealth Government of Australia.

extended relational database systems, object-oriented (OO) databases [7, 3], toolkits for constructing special-purpose database systems, and persistent programming language. As claimed by Carey [2], the extended relational database systems, as they are called object-relational (OR) database systems now, appear likely to emerge as the ultimate winner in terms of providing objects for mainstream enterprise database applications.

OR systems start with the relational model and its query language, SQL, and are built from there. As such, it has a strong base: conceptual simplicity, open standard query language, powerful transaction management and recovery facilities, efficiency, availability, scalability, rich set of applications. All of these are what an OO database system lacks. Besides, OR systems also accommodate many object features. Among them are abstract data types(ADTs), row types, references, multivalued attributes and inheritance. *ADTs* are user-defined base types. Their role is to enable the set of built-in data types of the DBMS to be extended with new data types such as text, image, polygon, etc. *Row types* are a direct and natural extension of the type system for tuples. In addition to base type attributes, *row objects* are permitted to contain *reference-valued attributes* and *multivalued attributes*. The introduction of reference-valued attributes [8] will not only provide shareability of type instances, but also will enable us to support OO capabilities for rows in existing tables. A multivalued attribute value can be a set, a bag, a list or an array of base type elements. Finally, *inheritance* is also supported to enable natural variations among row types or ADTs to be captured in the schema.

Currently, different approaches are being taken by vendors to provide OR database products: native implementation such as Informix's Illustra – commercialized version of Postgres [12], Fujitsu's ODB II – commercialized version of Jasmine [5], Omniscience and UniSQL; incremental evolution taken by CA-Ingres, DB2/6000 C/S, Oracle 8, etc.; wrapper approach taken by HP's Odapter. Building an ORDBMS is a complex and time-consuming task, requiring hundreds of man years of effort. It is always psychologically difficult for people to discard their investment in an old system.

For the purpose of providing new technology without giving up old systems, we put forward a new approach [10], utilising a heterogeneous database architecture as a vehicle for OR database system implementation. A relational DBMS and an OODBMS are integrated together to provide an ORDB environment. This approach preserves an enterprise's current investment on relational database systems and applications, while still offering the benefits of new add-on OO features. Compared with the wrapper and gateway approaches discussed by Stonebraker [13], our approach is more biased towards using existing resources.

In our proposed heterogeneous database architecture, all the data are actually stored as underlying relational tables and objects. The OR table on the top is virtual. When an OR schema of a database application is defined, it is necessary to translate the schema into local relational and OO database schemas. When an query is issued against global OR schema, it must be translated to a set of local queries against corresponding local relational schema and OO schema for execution. These local queries are either pure relational queries or pure OO queries. The local partial results are then sent back to the top level. The RDB engine at the global level is responsible for merging the partial results into the final result. In the rest of the paper, the object-relational data model and its query language is discussed in section 2. In section 3, a heterogeneous database architecture is presented for implementation of OR systems. Two kernel functions of the architecture, the schema transformation and query partition, are further studied in section 4 and section 5, respectively. Section 6 concludes the paper.

## 2 Object-Relational Databases

In this section, an object-relational data model is introduced, with emphasis on type extensibility, and corresponding extended query features.

### 2.1 Object-relational data model

There is still no agreement on how the relational model should be extended to have the modelling power of object-oriented systems while keeping the simplicity of relational systems. Current SQL3 draft [14] only supports unnamed row types, ADTs and collection types. In the separate "SQL/Object" part of SQL3 [8], *named row types(NRTs)* are introduced with *polymorphism, identity, no inheritance*, and *no encapsulation*. Reference types are also introduced but only references to row types are allowed. In contrast, an ADT supports *polymorphism,inheritance,encapsulation*, but *no identity*. Beech [1] suggests that a possible future simplification of SQL3 is a combination of ADTs and NRTs. However, this may compromise the simplicity of relational systems. In this paper, we prefer to keep the relational flavour in OR systems, i.e., to keep a distinction between ADTs and NRTs.

The basic components of our object-relational model are *types*. An OR database schema consists of a set of row types, and each attribute in a row type is defined on a certain type, which can be a built-in type, an abstract data type(ADT), a collection type, a reference type or another row type. Therefore, the types can be defined recursively as follows:

**Base types** are the system built-in types including integer, float, date, string, boolean and day-time, which are supported by SQL92.

**Abstract Data Types (ADTs)** are user defined and implemented types. The implementation of objects, and their attributes and behaviours, is invisible to the query system. All accesses to the instances of an ADT are through the interface defined for the type. Therefore, for the purpose of our discussion here, we can describe an ADT by a collection of named functions, $T(f_1 : T_{f_1}, \cdots, f_n : T_{f_n})$, where $T$ is the name of the ADT, $f_i (1 \leq i \leq n)$ represents a method of the ADT, $T_{f_i}$ is the function type of $f_i$. A function type has the form $Fun(T_0, T_1, \cdots, T_n)$ where $T_1, \cdots, T_n$ is a list of input types, and $T_0$ is the output type of the function.

**Row types** have the form $T(A_1 : T_1, \cdots, A_n : T_n) : (T_r, \cdots, T_m)$, where $T$ is the name of the row type, $A_i (1 \leq i \leq n)$ is the name of an attribute, $T_i$ is the data type of $A_i$, and $T_j (r \leq j \leq m)$ is a supertype of $T$. A row type has a name and a set of attributes. An instance of a row type $T$ is an element of the Cartesian product of the value sets of the types that define $T$. A row type may be defined with or without a name. The former is called a named row type(NRT), and the latter is called unnamed row type.

**Reference types** have the form $ref(T)$ where $T$ is a row type. In current OR model, tables are the only top-level named entities that can be stored persistently. In other words, only tuples in a table are treated as independent objects with identity and thus can be referenced.

**Collection types** have the form $C(T)$ where $T$ can be a base type, ADT, row type, reference type or another collection type. $C$ represents one of the built-in collection type constructs, including set, bag, list, tree, etc.. They represent different ways to group up $T$'s instances. For the briefness of discussion, we only discuss set type in the paper.

## 2.2 Query language extension

Due to type extensibility, SQL92 is no longer suffcient for querying OR databases. In supporting type extensibility, we extend SQL92 with the following query features which are supported in our SQL3-like language.

**method invocations** – Since ADT have methods defined, method invocations are allowed to appear in the select-clause and where-clauses. A method invocation may take zero or more values as input and one value as output. The data type of every value is either a base type or an ADT.

**path expressions** – Since NRTs and reference types are introduced, path expressions are used to navigate the complex structures of objects and their relationships to other objects via references. A $deref()$ function is used to dereference an object identity to get its object. In the query, we adopt the dot notation to represent a path expression. For example, $V.A_1.A_2.\cdots A_k$ is a path expression in the query where $V$ is a tuple variable, $A_i (1 \leq i \leq k - 1)$ is either an attribute or a dereferenced attribute in the OR table of $V$, and $A_k$ can be an attribute or a dereferenced attribute of $type(A_{k-1})$, or a method invocation of $A_{k-1}$.

**set operations** – Since collection types are introduced, set operations are also allowed in the where-clause. The predicates such as membership $(IN)$ and inclusion $(ISSUB)$ are allowed.

The followings are examples of OR schema definition and queries.

*Example 1.* In the following we define an OR database schema for a company. It consists definitions of one ADT *point*, two NRTs *emp_t* and *dept_t*, and two tables *emp* and *dept*.

```
create ADT point (
      x_coordinate float,
      y_coordinate float;
      distance(point) float);
```

```
const CENTRAL_POINT = new point(0, 0);
create NRT emp_t (              create NRT dept_t (
      name varchar(30),               dname varchar(30),
      salary decimal(9,2),            budget float,
      interest set(varchar(40)),      location point,
      location point,                 manager ref(emp_t));
      dept ref(dept_t),
      friend SET(REF(emp_t)));
create table emp of emp_t       create table dept of dept_t
      scope for dept is dept,         scope for manager is emp;
      scope for friend is emp;
```

*Example 2.* Find the names and research interests of all employees who have interest in ORDB and work for the department which is located in central area (within 2 kilometers from the central point) and has budget more than 1 million dollars.

```
select e.name,e.interest
from emp e
where e.deref(dept).budget > 1,000,000 and
      "ORDB" in e.interest and
      d.location.distance(CENTRAL_POINT) < 2;
```

*Example 3.* Find the names of managers who have friends woking in the same department and their salaries are more than $100,000.

```
select d.deref(manager).name
from dept d, emp e
where d = e.deref(dept) and e in d.deref(manager).friend
      and e.salary >= 100,000;
```

Notice in the above example queries, we have used method invocations, set operations and path expressions to represent the queries. All these query features are used to support the type extensibility, therefore, they can not be found in SQL92.

## 3    HDB Architecture for ORDBMS

In this section, we address how a heterogeneous database (HDB) architecture can be used as a vehicle for realizing OR technology. Given a SQL3-like language, we need an ORDB engine to implement it. The HDB engine shown in Figure 1 can be used for this purpose. In fact, the HDB engine is a virtual ORDB engine, it is built based on local RDB engine and OODB engine. We provide so-called *functional transparency*. In the architecture, we are only interested in how a global ORDB request is functionally transformed into local RDB and OODB subrequests. By functional transparency we mean users need not know such functional transformation. Given a SQL3-like request, the HDB engine interprets the request in terms of supporting RDB and OODB engines. In the following, we describe each component in the architecture.

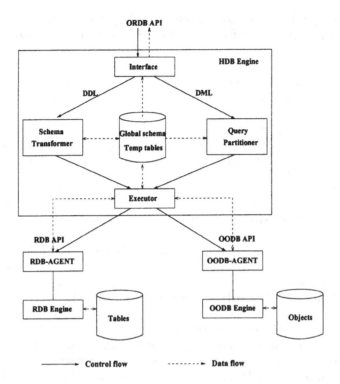

**Fig. 1.** HDB Architecture for Mixing Objects with Tables

**Interface** – It receives users' SQL3-like requests, does syntactical check and hands them to corresponding components for processing. It is also responsible to return the results of the requests back to users.

**Schema Transformer** – It handles SQL3-like requests for schema definition. The main function of this component is to keep the information of OR schema into the global directory and to transform the schema into schemas of both local RDB and OODB engines. A transformation plan is generated by this component which is delivered to the executor for execution. The schema mapping information is also kept in the global directory. The detail of this component is further discussed in section 4.

**Query Partitioner** – This component is responsible for translating SQL3-like queries which are issued against global OR schema into local queries on both local RDB and OODB engines. The schema mapping information is used for such transformation. All local queries form a query plan which is handed over to the executor. This component will be discussed in detail in section 5.

**Executor** – The executor is responsible for coordinating the distributed execution of the execution plan (both transformation and query plan). There is a RDB engine employed at the global level to hold temporary results which may pass between local relational system and OO system, to merge the results. Since users are on the top of an ORDB interface, they may require the result in a form which is more than a flat table. Therefore, some functions are added for this purpose, such as, *nest* for reconstructing set-valued column values, *deref* for obtaining the objects, etc.

**RDB-Agent** – The RDB-agent is responsible for monitoring the execution of local queries against transformed local relational schema and returning the results.

**OODB-Agent** – The functions of this agent include: (a) translating SQL3-like queries to ODMG-93 OQL [3] queries which are supposed to be supported by the OODB engine; (b) monitoring the execution of local queries against transformed local OO schema; (c) returning results to the executor.

## 4 Schema Transformation

C. Yu et al. [15, 11] have studied translations between relational systems and object-oriented systems. In providing relational (OO) frontend to an OO (relational) system, two tasks with opposite directions are necessary: one is the transformation from OO (relational) schema to its relational (OO) equivalent schema, the other is translation of relational (OO) queries to OO (relational) equivalent. Our work is different, we are supporting top-down OR database design, both schema transformation and query translation tasks are one-directional, i.e., from OR to both relational and OO. Besides, in [11], only structural part of OO schema is translated to relational. In this approach we consider both structural and behavioral aspects of an OR system.

In this section, we show how a global OR schema can be transformed into corresponding local relational and OO schemas. Our criteria is to maximise the transformation of OR schema to relational schema as a trivial transformation always exists from OR schema completely to OO local schema.

In the following, we first informally describe the steps of an algorithm for mapping OR schema to its corresponding relational schema and OO schema. Then, we summarize with a correspondence between an OR table and its transformed relational tables and OO classes.

### 4.1 Transformation Algorithm

**Input** An OR schema.
**Output** A relational schema and an associated OO schema.
We use two lists to keep useful intermediate information. $\mathcal{NRT}$ for definitions of NRTs, $\mathcal{TRAN}$ for table structures to be transformed. Initially, they are empty.
**Step 1**: For every NRT defined in the OR schema, create a structure in $\mathcal{NRT}$ to record the NRT name and its field definitions.
**Step 2**: For every ADT, output a definition of a class called *deputy class* which is used to implement the ADT. The definition of the deputy class is formed by including definitions of attributes and methods of the ADT definition and the definition of an extra object identity attribute *oid*.

**Step 3**: For table named $ORT$ defined in the OR schema, create a (frame) structure in $\mathcal{TRAN}$ to record the table name, copy the field definitions of the NRT in $\mathcal{NRT}$ on which it is defined as the field definitions of the table, and if scopes of referenced fields are defined with the table, keep the names of scope tables within the field definition. A table called *frame table* with name $ORT\_R$ will be created later in the relational schema based on this frame structure.

**Step 4**: Loop for each table structure in $\mathcal{TRAN}$ until $\mathcal{TRAN}$ becomes empty.

**Step 5**: For each table structure in $\mathcal{TRAN}$, let its table name $TBL$, create a relational table definition with table name $TBL\_R$. For each field definition $F : T$ in $TBL$, we classify 8 cases for different processing. Remove the structure of $TBL$ after all its field definitions have been processed.

**Case 1**: If $T$ is a built-in type, output a column definition $F : T$ for $TBL\_R$.

**Case 2**: If $T$ is an ADT, output a column definition $F : long$ for $TBL\_R$, we use type *long* to represent object identity.

**Case 3**: If $T$ is a NRT, replace the field definition with all field definitions of the NRT $T$ for processing. The newly added field definitions can be found in $\mathcal{NRT}$, their field names need to be renamed with the prefix $F\_$.

**Case 4**: If $T$ is a reference type $ref(RNRT)$, there should be a scope clause declaring the table, say $RTBL$, it really references to. Let $RPK : T_{RPK}$ its primary key definition, then output a column definition $F : T_{RPK}$ for $TBL\_R$.

**Case 5**: If $T$ is a set type $set(ET)$ and $ET$ is a built-in type, output a definition for a table called *auxiliary table* $TBL\_F\_R(PK : T_{PK}, TBL\_F : ET)$, where $PK : T_{PK}$ is the primary key definition for $TBL\_R$.

**Case 6**: If $T$ is a set type $set(ET)$ and $ET$ is an ADT, output a definition of a class called *auxiliary class* $TBL\_F\_C(oid : long, TBL_F : set(ET))$.

**Case 7**: If $T$ is a set type $set(ET)$ and $ET$ is a NRT, create an auxiliary structure in $\mathcal{TRAN}$ with $TBL\_F\_R$ as table name and $(PK : T_{PK}, EF_1 : ET_1, \cdots, EF_m : ET_m)$ as its field definition, where $PK : T_{PK}$ is the primary key definition for $TBL\_R$ and $(EF_1 : ET_1, \cdots, EF_m : ET_m)$ is the definition for $ET$ which can be retrieved from $\mathcal{NRT}$.

**Case 8**: If $T$ is a set type $set(ET)$, $ET$ is a reference type $ref(RNRT)$, there should be a scope clause declaring the table, say $RTBL$, it really references to. Let $PK : T_{PK}$ and $RPK : T_{RPK}$ be primary key definitions of $TBL$ and $RTBL$, respectively, then output a definition of a table called *auxiliary table* $TBL\_F\_R(PK : T_{PK}, F\_RPK : T_{RPK})$.

The above algorithm transforms an OO schema into a relational schema and an OO schema, if any. For every table in the OR schema, a frame table definition is always generated in the relational schema, together with definitions of auxiliary tables and classes, if any. A deputy class definition is always generated in the OO schema which implements an ADT in the OR schema. The definitions of NRTs in the OR schema are embeded in the transformed relational and OO schemas. In section 4.2 we will show how the transformed local relational and OO schemas correspond to the global OR schema. An example of schema transformation is given below.

*Example 4.* The transformed relational schema and OO schema of the OR schema of Example 1 are given below.

```
create class point (
      oid: long,
      x_coordinate: float,
      y_coordinate: float;
      distance(point): float);
create table emp_R (              create table dept_R (
      name varchar(30),                dname varchar(30),
      salary decimal(9,2),             budget float,
      location long,                   location long,
      dept varchar(30),                manager varchar(30),
      primary key is name);            primary key is dname);
create table emp_interest_R (     create table emp_friend_R (
      name varchar(30),                name varchar(30),
      emp_interest varchar(40),        friend_name varchar(30),
      primary key is name);            primary key is name);
```

## 4.2  Global and Local Schema Correspondence

As shown in above section, an OR table $TBL$ can be transformed into a frame relational table $TBL\_R$, together with a set of auxiliary relational tables and a set of auxiliary classes. An ADT can be transformed into a deputy class. Suppose $TBL$ is defined on the NRT $NRT_0(F_0 : BT_0, F_1 : BT_1, F_2 : ADT_2, F_3 : NRT_3, F_4 : ref(NRT_4), F_5 : set(BT_5), F_6 : set(ADT_6), F_7 : set(NRT_7), F_8 : set(ref(NRT_8)))$ where definitions of $F_i$ $(1 \leq i \leq 8)$ are in correspondence with 8 cases discussed above. Let $DC_2, DC_6$ are deputy classes for $ADT_2, ADT_6$; $AC_6$ is auxiliary class for $set(ADT_6)$; $TBL_5$, $TBL_7$, $TBL_8$ are auxiliary tables for $set(BT_5)$, $set(NRT_7)$, $set(ref(NRT_8))$; $RTBL_4$ and $RTBL_8$ are scope tables for $F_4$ and $F_8$. Suppose NRTs, existing tables and transformed tables have the following definitions: $NRT_3(F_0^3, \cdots, F_r^3), NRT_4(F_0^4, \cdots), NRT_7(RF_0^7, \cdots),$
$NRT_8(F_0^8, \cdots); RTBL_4(F_0^4, \cdots), RTBL_8(F_0^8, \cdots); TBL_5(F_0, BF_5),$
$TBL_7(F_0, RF_0^7, \cdots), TBL_8(F_0, F_0^8); AC_6(oid : long, AF_6 : set(DC_6)).$

The global table $TBL$ can be represented as
select $TBL\_R.F_1, TBL\_R.deref(F_2), TBL\_R.F_0^3, \cdots, TBL\_R.F_r^3, TBL\_R.F_4,$
$nest(TBL_5.BF_5), TBL\_R.deref(F_6), nest(TBL_7.RF_7), nest(TBL_8.F_0^8))$
from $TBL\_R, TBL_5, TBL_7, TBL_8, DC_2, AC_6, RTBL_4, RTBL_8$
where $TBL\_R.F_2 = DC_2.oid$ and $TBL\_R.F_4 = RTBL_4.F_0^4$ and $TBL\_R.F_0 = TBL_5.F_0$
and $TBL\_R.F_6 = AC_6.oid$ and $TBL\_R.F_0 = TBL_7.F_0$ and $TBL\_R.F_0 = TBL_8.F_0$ and
$TBL_8.F_0^8 = RTBL_8.F_0^8$

The predicates used in bridging relational side to OO side are called *bridge constraints*, e.g., $TBL\_R.F_2 = DC_2.oid$ and $TBL\_R.F_6 = AC_6.oid$ in the above *where* clauses are bridge constraints.

*Example 5.* The correspondence between OR table *emp* in Example 1 and its transformed representations in Example 4 is given below.

select $emp\_R.name, emp\_R.salary, nest(emp\_interest\_R.interest),$
$emp\_R.deref(point.oid), emp\_R.dept, nest(emp\_friend\_R.emp\_friend)$
from $emp\_R, emp\_interest\_R, emp\_friend\_R, point, dept\_R$

where $emp\_R.name = emp\_interest\_R.name$ and $emp\_R.name = emp\_friend\_R.name$
and $emp\_R.location = point.oid$ and $emp\_R.dept = dept\_R.dname$

# 5 Query Partition

Given a query against global OR schema, it must be translated to local queries against its transformed local relational and OO schemas for execution. We design a query partition algorithm which has three major steps: *substitution, decomposition* and final result processing. The result consists of a group of relational queries in SQL92 [6, 4] format and a group of OO queries in OQL [3] format.

To simplify the discussion, we assume that all the constraints in the where-clause are connected by only "AND" operators. This is reasonable because for any where-clause $C$ which contains an "OR", it can always be transformed into the disjunctive form, "$C_1$ OR $C_2$". The original query $Q$ can then be translated into "$Q_1$ UNION $Q_2$", where $Q_1$ and $Q_2$ take $C_1$ and $C_2$ respectively in the where-clause.

## 5.1 The substitution process

To process the OR query, the first task is to transform the OR style representations into suitable local forms. The OR table and attribute names should be substituted by local table and attribute names. As the set-valued attributes are flattened, some related predicates, including membership and inclusion, should be rewritten. More importantly, as the navigational access is not supported by relational systems, many path expressions $V.A_1.\cdots.A_n$ need to be translated into $V.A$ form and a set of join predicates to record the traversal information. After the substitution, all the data elements from the local relational tables have the strict $V.A$ form, and those from OO classes may have arbitrarily long paths starting from an OO variable.

An important concept used in the following algorithm is the *cluster* of path expressions. A cluster of path expressions in a query is a set of path expressions which start with the same tuple variable and have same first attribute. For example, $V.A_1$, $V.A_1.A_2$ and $V.A_1.A'_2$ are in the same cluster, while $V.A_1$ and $V.A'_1$ belong to two other clusters.

**Input** An OR query in the $select \cdots from \cdots where \cdots$ form, and there is no "OR" operation in where-clause.

**Output** An intermediate form. It is an integrated query against local schemes.

**Step 1:** (Initialisation) Define $k$ as the variable counter, and initialise with value 1. $k$ will be used to name the variables in the query. Each time a variable being renamed or a new variable being created, $k$ is increased by 1.

**Step 2:** (Translation of the from-clause and variable names) For each tuple variable $V$ of OR table $TBL$ in the from-clause of an OR query, replace the table name with the corresponding local frame table name $TBL\_R$ so that $V$ becomes a relational tuple variable on $TBL\_R$. Rename $V$ and its all occurrence in the query with "VAR_$k$" so that all the variable names have a standard format. In the rest of this algorithm we will still use $V$ to represent the variables wherever the name format is not concerned.

**Step 3:** (Translation of the path expressions) In select-clause and where-clause, for each cluster of path expressions in the form $V.A_1. \cdots$ or $V.deref(A_1). \cdots$, do the following process based on the type of $A_1$, say $type(A_1)$. Repeat this step until no more change can be applied.

**Case 1:** If $type(A_1)$ is a base type, $A_1$ should be the end of the path expressions. Nothing need to be done. Or if $V$ is defined on an OO class, the path expressions need not to be changed either.

**Case 2:** If $type(A_1)$ is an ADT, add a new variable definition in the from-clause, "$ADTC$ VAR_$k$", where $ADTC$ is the OO class of the ADT. In the where-clause, add a new predicate, "AND $(V.A_1 = $ VAR_$k$.OID)". Change the original path expressions into "VAR_$k. \cdots$".

**Case 3:** If $type(A_1)$ is a NRT, for each path expression in the cluster, there are following subcases.

**3.1** If there is a node, say $A_2$ after it, then based on the previous schema transformation algorithm, $A_2$ should now be an attribute of $TBL$_R and be renamed as $A_1$_$A_2$. Here $TBL$_R is the relation on which $V$ is defined. Therefore, we can translate the original path expressions to "$V.A_1$_$A_2. \cdots$".

**3.2** If $A_1$ is the last attribute in the path expression, and the expression appears in the select-clause, then change the path expression with a group of expressions, "$V.A_1$_$A^1, \cdots, V.A_1$_$A^n$", where $A^i, (1 \leq i \leq n)$ are the all attributes of $type(A_1)$.

**3.3** If $A_1$ is the last attribute in the path expression, and the expression appears in the where-clause, then change the path expression with "$V.A_1$_$PK$", where $PK$ is the primary key of $type(A_1)$.

**Case 4:** If $type(A_1)$ is a reference type $ref(RNRT)$, where $RNRT$ is a NRT, there should be a table $RTBL$_R to hold the referenced tuple. Add a new variable definition in the from-clause, "$RTBL$_R VAR_$k$". In the where-clause, add a new predicate, "AND $(V.A_1 = $ VAR_$k$.$RPK$)", where $RPK$ is the primary key of $RNRT$. For each path expression in the cluster, there are following subcases regarding the modification of the expression itself.

**4.1** If the form is like $V.deref(A_1).A_2 \cdots$ (there is another node after $A_1$), change the original path expressions into "VAR_$k$.$A_2 \cdots$".

**4.2** If the form is like $V.deref(A_1)$ ($A_1$ is the last node), and appears in the select-clause, then change the path expression with a group of expressions, "VAR_$k$.$A^1, \cdots$, VAR_$k$.$A^n$", where $A^i, (1 \leq i \leq n)$ are the all attributes of $RNRT$.

**4.3** If the form is like $V.deref(A_1)$, and appears in the where-clause, rewrite the path expression into $V.A_1$.

**4.4** If the form is like $V.A_1$, it should only appear the where-clause. No change need to be done.

**Case 5:** If $type(A_1)$ is a set type $set(ET)$ and $ET$ is a built-in type, there should be no more node after it in the path expressions. Add a new variable definition in the from-clause, "$TBL$_$A_1$_R VAR_$k$", where $TBL$_$A_1$_R is the relation that hold the set values. In the where-clause, add a new predicate, "AND $(V.PK = \text{VAR\_}k.PK)$", where $PK$ is the primary key for $TBL\_R$, on which $V$ is defined. Change the original path expressions into "VAR\_$k.TBL\_A_1$". If the path expression appears in the select-clause, add a new part in select-clause, "VAR\_$k.PK$".

**Case 6:** If $type(A_1)$ is a set type $set(ET)$ and $ET$ is an ADT, add a new variable definition in the from-clause, "$TBL\_A_1\_C$ VAR\_$k$", where $TBL\_A_1\_C$ is the OO class of the ADT set. In the where-clause, add a new predicate, "AND $(V.A_1 = \text{VAR\_}k.OID)$". Change the original path expressions into "VAR\_$k.TBL\_A_1.\cdots$".

**Case 7:** If $type(A_1)$ is a set type $set(ET)$ and $ET$ is a NRT, add a new variable definition in the from-clause, "$TBL\_A_1\_R$ VAR\_$k$", where $TBL\_A_1\_R$ is the table of the NRT set. In the where-clause, add a new predicate, "AND $(V.PK = \text{VAR\_}k.PK)$", where $PK$ is the primary key of $TBL\_R$, on which $V$ is defined. For each path expression in the cluster, there are following subcases regarding the modification the expression itself.

**7.1** If the form is like $V.A_1.A_2\cdots$ (there is another node after $A_1$), change the path expression into "VAR\_$k.A_2\cdots$". If the path expression appears in the select-clause, add a new part in select-clause, "VAR\_$k.PK$".

**7.2** If the form is like $V.A_1$ ($A_1$ is the last node) and appears in the select-clause, then change the path expression with a group of expressions, "VAR\_$k.A_1\_A^1, \cdots, \text{VAR\_}k.A_1\_A^n$", where $A^i, (1 \leq i \leq n)$ are the all attributes of $type(A_1)$.

**7.3** If the form is like $V.A_1$, and appears the where-clause, change the original path expression into "VAR\_$k.A_1\_PK$", where $A_1\_PK$ is the primary key of $type(A_1)$.

**Case 8:** If $type(A_1)$ is a set type $set(ET)$ and $ET$ is a reference type $ref(RNRT)$, there should be a table $RTBL\_R$ to hold the referenced tuples and another table $TBL\_A_1\_R$ to hold the set information. Add a new variable definition "$TBL\_A_1\_R$ VAR\_$k$" into from-clause. In the where-clause, add a new predicate, "AND $(V.PK = \text{VAR\_}k.PK)$", where $PK$ is the primary key of $TBL\_R$, on which $V$ is defined. For each path expression in the cluster, there are following subcases regarding the modification the expression itself.

**8.1** If the form is like $V.deref(A_1).A_2\cdots$ (there is another node after $A_1$), add a new variable definition "$RNRT\_R$ VAR\_$k'$" ($k' = k + 1$) into from-clause. In the where-clause, add a new predicate, "AND $(\text{VAR\_}k.RPK = \text{VAR\_}k'.RPK)$", where $RPK$ is the primary key of $RTBL\_R$. Change the original path expression into "VAR\_$k'.A_2.\cdots$".

**8.2** If the form is like $V.deref(A_1)$ ($A_1$ is the last node), and appears in the select-clause, add a new variable definition "$RNRT\_R$ VAR\_$k'$" ($k' = k + 1$) into from-clause. In the where-clause, add a new predicate, "AND $(\text{VAR\_}k.RPK = \text{VAR\_}k'.RPK)$", where $RPK$ is the primary key of $RTBL\_R$. Change the original path expression into a group of expressions, "VAR\_$k'.A^1, \cdots, \text{VAR\_}k'.A^n$", where $A^i, (1 \leq i \leq n)$ are the all attributes of $RNRT$.

**8.3** If the form is like $V.deref(A_1)$, and appears the where-clause, add a new variable definition "$RTBL\_R$ VAR\_$k'$" ($k' = k + 1$) into the

where-clause, add a new predicate, "AND $(VAR\_k.RPK = VAR\_k'.RPK)$",
where $RPK$ is the primary key of $RTBL\_R$. Change the original path expression into "$VAR\_k'.RPK$".

**8.4** If the form is like $V.A_1$, it should only appear in the where-clause. Change the original path expression into "$VAR\_k.A_1\_RPK$", where $RPK$ is the primary key of $RTBL\_R$.

**Step 4:** (Translation of where-clause) After Step 3, all the data elements in the query are either constants, in the $V.A$ form if $V$ is a relational tuple variable, or in $V.A \cdots$ form if $V$ is OO variable. This step will process the predicates in the where-clause, specificaly those have collection types involved. The following are the processes on the membership and inclusion predicates.

**Case 1:** (Membership of ADT set) The predicate should appear like $P_1$ IN $P_2$, where $P_1$ and $P_2$ are path expressions. The type of $P_1$ is an ADT and that of $P_2$ is the set of that ADT. In this case, nothing need be done.

**Case 2:** (Inclusion predicate between ADT sets) The predicate should appear like $P_1$ ISSUB $P_2$, where $P_1$ and $P_2$ are two path expressions. Both of their types are same, an ADT set. Change the predicate into "for all x in $P_1$ : x in $P_2$".

**Case 3:** (Other membership predicate) The predicate should appear like $E_1$ IN $E_2$. $E_1$ can be either (1). a constant; (2). path expression $V.A$, where $V$ is relational tuple variable and $A$ is a built-in typed attribute; or (3). $V.A_1. \cdots .A_n$ where $V$ is an OO variable and $A_n$ is a built-in typed attribute or an ADT method invocation returning a built-in typed value. $E_2$ is in the form $V'.A$ where $A$ is an attribute in the relational table, say $TBL\_R$, to hold the flattened information of the set. If $E_2$ has no other occurrence in the membership predicates of the query, simply change the operator "IN" into "=". Otherwise, create a new variable in from-clause, "$TBL\_R\ VAR\_k$". Change the predicate into "$E_1 = VAR\_k.A$". Add a new predicate "AND $(VAR\_k.PK = V'.PK)$" in the where-clause. Here $PK$ is the primary key of $TBL\_R$.

**Case 4:** (Other inclusion predicate between base type sets) The predicate should appear like $V_1.A_1$ ISSUB $V_2.A_2$, where $A_1$ and $A_2$ are attributes respectively in the relational tables, say $TBL_1\_R$ and $TBL_2\_R$ (not necessarily distinct), that hold the flattened information of the sets. Change the original predicate into following script.

```
not exists
select *
from TBL1_R VAR_k
where VAR_k.PK1 = V1.PK1 and not exists
  select *
  from TBL2_R
  where PK2 = V2.PK2 AND VAR_k.A1 = A2
```

*Example 6.* After substitution, the query in Example 2 is transformed into the following format.

```
select VAR_1.name, VAR_3.name, VAR_3.interest
from emp_R VAR_1, dept_R VAR_2, emp_interest_R VAR_3 VAR_5, point VAR_4
```

```
where VAR_2.budget > 1,000,000 and
      VAR_1.dept = VAR_2.dname and
      VAR_3.name = VAR_5.name and
      VAR_5.emp_interest = "ORDB" and
      VAR_1.name = VAR_3.name and
      VAR_4.distance(CENTRAL_POINT) < 2 and
      VAR_1.location = VAR_4.oid
```

## 5.2 Variable graph for decomposition

After the substitution process, the query is formulated on the local schemas. However, we need to decompose the integrated form into several parts so that each part can be executed by the relative local db engine. To decompose the query into local queries, we need to identify the boundary between the relational system and the OO system. A *variable graph* is used to assist the process. The major job is to find the *bridge constraints*. A bridge constraint is either a predicate (called *bridge predicate*) in the where-clause that involves both OO variables and relational tuple variables, or a method invocation (called *bridge invocation*) that takes relational elements as input parameters. The where-clause is split based on the definition of the variables. In general, all the predicates that have only relational variables involved are transfered into one local relational query, all the predicates that have only OO variables are translated into a set of local OO queries, all the predicates that belong to bridge constraints are translated into one top level query. The select-clause and from-clause are also split accordingly. The decomposition result includes a local relational query, a top level relational query, and a set of local OO queries.

The variable graph is an extension of the relational predicate graph in [11]. Not only predicates but also method invocations are taken into consideration. Because the methods may appear in select-clause, we draw the graph from the whole query instead of the where-clause.

**Definition 1.** *For a given query* $Q$ *, we define its variable graph:* $VG(Q)$ *as an annotated undirected graph:* $VG(Q) = (V, E)$. *Each vertex* $v$ *in* $V$ *represents a (relational or OO) variable used in* $Q$, *and each edge* $e$ *between vertices* $V_1$ *and* $V_2$ *in* $E$ *represents either a predicate in* $Q$ *that involve* $v_1$ *and* $v_2$, *or there is a method invocation* $v_i. \cdots .method()$ *that takes* $v_j. \cdots$ *as an input parameter* $(i, j \in \{1, 2\} \wedge i \neq j)$. *Each edge is annotated with the predicate or the method invocation.*

Compared with predicate graph, variable graph emphasises on the relationship among variables, and does not contain all the constraints in Q. Rearrange the vertices so that two disjoint circles can be drawn to enclose all the relational tuple variables and OO variables respectively. All the edges that go cross the circles' borders are called bridge edges, which correspond to the bridge constraints. An example is shown in Figure 2, which is the variable graph of Example 6. Three nodes are in the relational side, VAR_1, VAR_2 and VAR_3. One node is in the

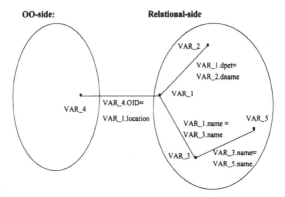

**Fig. 2.** The variable graph of example 6

OO side, VAR_4. The only bridge constraint is the predicate, VAR_1.location = VAR_4.OID.

By removing all the bridge edges, we get a disconnected graph $VG'$, called *local partition graph*. All the vertices of relational variables are always in one connected component of $VG'$. On the other hand, those OO variables may distribute in several connected components, called *connected OO components(COC)*. For each connected component, there should be a corresponding local query. In the example there is only one COC that contains one node, VAR_4.

### 5.3 The decomposition and final result processing

OR queries may have set valued attributes in the select-clause, which are flattened in relational side. Unfortunately, the top relational query engine is incapable of restoring the flat data back into the nested format. Therefore, an external procedure is needed to do the final job. We slightly extend the syntax of the top level relational query to include a special function call *nest()*.

**Input** An integrated query against local schemas

**Output** A set of relational queries in SQL92 form and a set of OO queries in OQL form.

**Step 1:** (Query decomposition) Draw the variable graph $VG(Q)$ of the substituted query Q. Identify the bridge constraints. Identify the connected OO components. If there is no bridge constraints, no decomposition need to be done and jump to the last step.

**Step 2:** (Create local relational query) Create the local relational query $LRQ$ in the "create table TEMP_R as select $\cdots$ from $\cdots$ where $\cdots$." form, where "TEMP_R" is the table to hold the partial result. In the from-clause, copy all the relational tuple variable defintions from from-clause of Q. In the select-clause, copy all the relational attributes from select-clause of Q. For each relational attribute mentioned in bridge constraints, add it into select-clause unless it is already there. In the where-clause, copy all the predicates that have no OO variable involved.

**Step 3:** (Create local OO queries) Initialize the local OO query counter $i$ and new attribute counter $j$ to 1. These counters work just like the variable counter $k$. For each COC(if any), create the local OO query $LOQ_c$ in the form "define TEMP_O_i as select distinct struct($\cdots$) from $\cdots$ where $\cdots$". In the from-clause, copy the definitions of all OO variables in the COC. In the where-clause, copy all the predicates mentioning only the variables in the COC. In the select-clause, for each data element "$V\cdots$" in the select-clause of Q or in the bridge predicates, where $V$ is in COC, add a new element "$name : V\cdots$". The naming rule is, if the data element is in "$V.A$" form, then use $A$ as the name, otherwise create a new name A_j. If the element is in "$V.oid$" form, change it into "$\&V$" which represents the identity of the object referenced by V. In the select-clause, if there is any method invocations that take an attribute from relational side as input parameter, then add "VR in TEMP_R" in from-clause and replace each occurrence "$V.A$" with "$VR.A$" for each relational variable $V$.

**Step 4:** (Create top level query) Create the top level relational query $TRQ$. The from-clause has the format "FROM TEMP_R VR, TEMP_O_1 VO_1, $\cdots$, TEMP_O_n VO_n", supposing there are $n$ local OO queries. In the select-clause, include all the contents in the select-clause of Q. In the where-clause, copy all the predicates acting as bridge constraints. Change all the data elements in select-clause and where-clause into right form. For each variable $V$, find out the connected component it belongs to in $VG'(Q)$ and then the corresponding local query. Suppose the query results in a temporary table TEMP_T, replace all occurrence of $V$ with $VT$. Any method invocation long path expression should be assigned to a new attribute in a certain local OO query, and therefore substitute the invocation or long path with that attribute name.

**Step 5:** (Final result processing) To restore the flattened set values, the standard SQL syntax need to be extended a little. The nest operation is expressed as "nest $A_1, \cdots, A_n$ on PK", where PK is the key regarding to the operation. The result is a set, $\{(A_1, \cdots, A_n)\}$. This expression can be nested. Therefore, we can apply this operation in the select-clause of TRQ when the set valued attributes need be restored.

*Example 7.* To continue Example 6, we have the following queries as final result.

```
/* Relational local query:
create table temp_R as
   select VAR_1.name as name, VAR_1.location as location,
          VAR_3.name as name_2, VAR_3.emp_interest
   from emp_R VAR_1, dept_R VAR_2, emp_interest_R VAR_3, VAR_5
   where VAR_1.dept = VAR_2.dname and VAR_2.budget > 1,000,000 and
         VAR_1.name = VAR_3.name and VAR_3.name = VAR_4.name and
         VAR_5.emp_interest = "ORDB";
/* OO local query:
create table temp_O as
   select distinct struct(oid: &VAR_4)
   from VAR_4 in point
   where VAR_4.distance(CENTRAL_POINT) < 2;
```

```
/* Relational top query:
select VR.name, nest VR.emp_interest on VR.name_2
from temp_R VR, temp_O VO_1
where VR.location = VO_1.oid;
```

Generally speaking, an OR query can be partitioned into three parts, a relational local query, a set of OO local queries, and a relational top query. Temporary tables are created to store the results of the local queries. The top query is applied on the temporary tables to merge the partial results. The following example shows a special situation. All the information used by the query is stored in the relational side. Therefore, the OO local queries are not created, and neither is the the the top query. The original query is translated into one relational query applied on the local relational database.

*Example 8.* The queries in Example 3 can be transformed into the following query (only local relational one):

```
select VAR_3.name
from dept_R VAR_1, emp_R VAR_2 VAR_3
where VAR_1.dname = VAR_2.dept and VAR_2.salary >= 100,000 and
      VAR_3.name = VAR_1.manager
```

## 6   Conclusion

The Object-relational data model opens up type system of traditional relational model, allowing more complex data structures. This requires new facilities to manage the data and handle the queries. Instead of building an object-relational DBMS from scratch, we proposed an approach to build the OR system by integrating existing relational and OO database systems. The heterogeneous database architecture which is used for this purpose has been presented in this paper. In particular, we focused on two kernel components of the architecture, schema transformer and query partitioner. Algorithms for implementing these two components have been proposed. Up to now, a prototype has been built at DSTC/UQ (CRC for Distributed Systems Technology/University of Queensland). Currently, we are improving the prototype with other functions such as query optimization, integrity control [9].

## References

1. D. Beech. Can SQL3 be simplified? *Database Programming and Design*, pages 46–50, January 1997.
2. Michael J. Carey and David J. DeWitt. Of objects and databases: A decade of turmoil. In *Proceedings of the 22nd International conference on VLDB*, pages 3–14, Mumbai (Bombay), India, September 1996. VLDB, Morgan Kaufmann.
3. R. G. G. Cattell, editor. *The Object Database Standard: ODMG-93*. Morgan Kaufmann Publishers, 1994.

4. C. J. Date and H. Darwen. *A Guide to The SQL Standard, 3rd ed.* Addison-Wesley, Reading, MA, 1993.

5. H. Ishikawa, Y. Yamane, Y. Izumida, and N. Kawato. An object-oriented database system Jasmine: Implementation, application, and extension. *IEEE Transactions on Knowledge and Data Engineering*, 8(2):285–304, 1996.

6. ISO/IEC. *ISO/IEC 9075:1992, Database Language SQL- July 30, 1992*, july 1992.

7. A. Kemper and G. Moerkotte. *Object-Oriented Database Management: Application in Engineering and Computer Science*. Pretice-Hall, 1994.

8. ISO DBL LHR-077 and ANSI X3H2-95-456 R2. *Introducing Reference Types and Cleaning up SQL3's Object Model*, November 1995. by K. Kulkarni, M. Carey, L. DeMichiel, N. Mattos, W. Hong, M. Ubell, A. Nori, V. Krishnamurthy and D. Beech.

9. C. Liu, H. Li, and M. E. Orlowska. Supporting update propagation in object-oriented databases. *Data & Knowledge Engineering*, 1997. Accepted.

10. C. Liu, M. E. Orlowska, and H. Li. Realizing object-relational databases by mixing tables with objects. In *Proceedings of 4th International Conference on Object-Oriented Information Systems*. Springer-Verlag, Nov. 1997.

11. W. Meng, C. Yu, W. Kim, G. Wang, T. Pham, and S. Dao. Construction of a relational front-end for object-oriented database systems. In *Proceeding of 9th International Conference on Data Engineering*, pages 476–483, 1993.

12. M. Stonebraker, L. A. Rowe, and M. Hirohama. The implementation of POSTGRES. *IEEE Transactions on Knowledge and Data Engineering*, 2(1):125–142, February 1990.

13. Michael Stonebraker. *Object-Relational DBMSs: The Next Great Wave*. Morgan Kaufmann, 1996.

14. ISO DBL YOW-004 and ANSI X3H2-95-084. *(ISO/ANSI Working Draft) Database Language SQL3*, March 1995. Jim Melton (ed).

15. C. Yu, Y. Zhang, W. Meng, W. Kim, G. Wang, T. Pham, and S. Dao. Translation of object-oriented queries to relational queries. In *Proceeding of 11th International Conference on Data Engineering*, pages 90–97, 1995.

# From OO Through Deduction to Active Databases –
# ROCK, ROLL & RAP

M.H.Williams* and N.W.Paton+
* Dept. of Computing and Elec. Eng., Heriot-Watt University, Edinburgh
+ Dept. of Computer Science, Manchester University, Manchester

**Abstract:** One important thread within advanced database systems research is the notion of rule-based database systems. The power of definite rules coupled with relational technology has led to the emergence of deductive databases. However, while this type of database system provides more advanced functionality, it suffers from other limitations of relational database systems such as the lack of data structures. The realisation that the object-oriented approach is complementary to the deductive one and that the two can be combined to produce deductive object-oriented databases with all the benefits of both represents an important breakthrough for rule-based database systems.
An alternative to the deductive rule approach is the active rule approach. Active rules are more powerful than deductive rules but lack the advantages of the sound theoretical foundation of the latter. The two ideas can be combined to produce an active DOOD provided that the integration is treated with care.

## 1 Introduction

The past two decades has seen a growth in activity in the area of advanced database systems. The limitations of relational systems were clearly recognised and alternative database paradigms have been investigated in order to widen the range of applications which can be handled with databases, or provide improved solutions to existing applications.

As a result of this activity, object-oriented database systems are now well established as one such alternative. The other main alternative consists of rule-based database systems which are the focus of this paper. These include both deductive databases and active databases, as well as the more recent deductive object-oriented database systems.

The following section provides some background to deductive and deductive object-oriented databases. Section 3 gives an overview of one particular deductive object-oriented database system, ROCK & ROLL. Section 4 introduces some aspects of active databases and as an example describes RAP, an active extension of ROCK & ROLL .

# 2 Deductive and DOO

## 2.1 Deductive Databases

The emergence of the deductive database model (or deductive relational database - DRDB) was a natural development from the relational model, providing a logical extension of the capabilities of relational databases and founded on a firm theoretical foundation. However, the provision of an upwardly compatible extension to the basic relational model is not sufficient in itself to account for subsequent interest in the deductive database model.

During the 1980s considerable interest and attention was focused on knowledge based systems, fuelled by the Japanese Fifth Generation programme and matching initiatives in Europe and the USA. In terms of database systems, this interest extended to combining the storage and manipulation of tuple data with the storage and manipulation of rules. This led to ideas such as *knowledge stores* [1] in which different rule bases might be stored in a common repository for access by different expert systems, and *expert database systems* [2] which combine the ideas of expert systems with those of database systems to provide some degree of "intelligence" in the database system. This latter extended to the application of rules to improving the user interface, e.g. [3], as well as to their use in improving the internal operation of the database, e.g. in controlling optimisation or identifying opportunities for restructuring [4].

In this context the deductive database system provides a natural tool for realising such applications. Initial research focused on simple implementations based on the coupling of an existing logic programming system (generally Prolog) to an existing relational database system. A typical example of this approach is EDUCE [5]. However, the performance limitations of such systems rapidly became apparent and the only practical solution was perceived to lie in purpose-built systems such as [ 6].

## 2.2 OO, OR and DOO

In parallel with the development of deductive database systems, the object-oriented database system began to emerge, offering a wider range of data types and control over behaviour. Just as with the deductive database, interest in this approach was enhanced by other factors. On the one hand, some applications requiring database capabilities were clearly not well suited to map onto the relational model, e.g. GIS, CAD, etc. On the other hand, interest in the object-oriented methodology for analysis and design of software and systems was growing.

However, this approach was not without its problems. The absence of an agreed object-oriented model meant that object-oriented systems lacked compatibility with each other. The substantive body of theory and techniques applicable to relational database systems could no longer be applied. Performance became an issue.

At this point it became clear that, while these different paradigms offered different functionality, they are not mutually exclusive. It was realised that considerable advantage might lie in combining different approaches to obtain the best of each. Object relational databases were an obvious start, combining the concepts of object-oriented databases with those of relational databases. These have now reached the stage of incorporation into standard products offered by companies such as Informix, IBM, Oracle, etc. Deductive object-oriented databases (DOODs) were another obvious development.

As with deductive databases, different approaches have been adopted to realising DOODs. In particular, three general classes of approach can be identified:

(1) Extensions of deductive languages, including
    (a) OO extensions of Datalog. These are further extensions of Datalog, incorporating notions of identity, inheritance, etc. They include proposals such as COL [7] (an extension of Datalog with complex typed terms) and IQL+ [8] (a further extension with object identity), as well as implementations such as ConceptBase [9] and ROL [10].
    (b) OO extensions of Prolog. Much work has been done on extending Prolog with object-oriented concepts, although generally this has been from a programming language point of view rather than a database one. An example of a database approach is ADAM [11].
    (c) OO and deductive extensions of SQL. An extension of SQL2, called ESQL2, which supports both object-oriented and deductive capabilities was developed as part of the EDS project [12].

(2) Changes to the underlying logic. Instead of building on first order predicate logic, this approach attempts to develop an object logic which incorporates essential characteristics of the object-oriented paradigm, and then to build a DOOD based on this logic. The idea behind this is that simple language extensions are inadequate for providing a clean and efficient combination of deductive and object-oriented capabilities, and that a more fundamental development at the foundation level is required. However, there are problems with the implementation of such systems. A classic example of such a logic is the O-logic [13], C-logic [14], F-logic [15] family. FLORID [16] is a partial implementation of ideas of F-logic.

(3) Integration of a declarative (rule-based) language with an imperative (procedural) language based on a common object-oriented data model. The two different sub-languages offer complementary facilities, the declarative language providing simple but powerful querying capabilities while the imperative language provides support for updates and procedural programs. ROCK & ROLL [17, 18] and Chimera [19] are examples of DOODs supporting not only imperative and deductive sub-languages but also active sub-languages. Coral++ [20] provides an integration of the Coral language with C++, a marriage which suffers from problems of impedance mismatch. Validity [21] provides declarative and imperative constructs in the language DEL in the only commercially available DOOD.

While the development of deductive databases (DRDB) was founded on a common starting point (the underlying model) and hence the resulting systems displayed similar behaviour, DOODs show much greater variety. This leads to the question of what constitutes a DOOD. Fernandes [22] presented a framework for characterising and comparing DOODs based on the following desired properties:

(1) Formal basis - the availability of a mathematical characterisation.
(2) User empowering - the ability to create, manipulate and query data without incurring excessive penalties through, e.g. optimisation.
(3) Implicit information - the ability to express more complex kinds of implicit information (e.g. derived data)
(4) Integrity constraints - the ability to express general constraints on the set of possible states of the database
(5) Recursion - the ability to use recursion to model structural and behavioural knowledge.
(6) Identity - the ability to distinguish the existence of an object from its current state at any point in time.
(7) Complex structures - the ability to model properties which have nonatomic values.
(8) Semantic orientation - the ability to model relationships between types.
(9) Behaviour orientation - the ability to assign specific behaviour to particular types, and to model dynamic aspects of interaction using notions of encapsulation, overloading, overriding and late binding.

He provides a detailed discussion of a number of different systems and shows how they relate to this framework.

DOODs may also be classified according to the strategy used in their design. One way of viewing this is as follows [23]:

(1) Language extension - in which an existing deductive language (Datalog, Prolog or SQL) is extended with OO features (such as identity, inheritance, etc.).
(2) Language integration - in which a deductive language and an imperative language are integrated (either unilaterally or bilaterally).
(3) New object logic language - in which a new logic language is created that incorporates essential features of the oo paradigm.

Another comparison paper [24] provides a similar comparison of different DOOD systems and extends this to additional deductive and object-oriented features.

# 3 ROCK & ROLL

The ROCK & ROLL database system is a specific example of a DOOD. It has two important features, i.e.:

(1) The system is based on the development and integration of a pair of languages:
- an imperative programming language ROCK which provides facilities to create and manipulate database objects in a conventional algorithmic style associated with object-oriented programming, and
- a declarative language ROLL which is used for expressing queries and methods, and provides the deductive capability as it allows derived properties of an object to be inferred from data stored explicitly through the use of ROLL methods.

The two languages are based on a common object-oriented data model (OM) which enables them to be integrated in a seamless fashion without any impedance mismatch.

(2) The object-oriented data model OM and the declarative language ROLL are formally specified in first order logic in such a way that the essential object-oriented features are captured in a simple logical model.

This section provides a brief introduction to ROCK & ROLL and concludes with a summary of the advantages of this approach.

## 3.1 The Object Model OM

The data model on which ROCK & ROLL is based is a simple object-oriented model, and a brief informal description is sufficient here. Atomic values are referred to as *primary objects*, all other abstractions are called *secondary objects*. Each object is assigned an *object type* and must conform to the structure associated with that type. Primary objects are instances of primitive types, secondary objects are instances of secondary types.

Each secondary object has a unique object identifier. Furthermore, each secondary type can define an object in terms of one or both of the following components:
(1) A set of type names used to model the properties of the type - the attributes of the type.
(2) Either

> (a) A structure which models its construction from other objects as a set (association), list (sequentiation) or tuple (aggregation). This is used to distinguish between the fundamental structural characteristics of a type and its other stored properties.

or

> (b) A reference to another object type with which the current object type is associated by inheritance. Specialisation denotes inheritance from previously defined supertypes, whereas generalisation does so from previously defined subtypes.

For example, the following type definitions indicate that a road has two attributes, roadName and town (both of type string) and consists of a list of roadSegments whereas a privateRoad is a specialisation of road.

```
type road
        properties:
                roadName: string;
                town: string;
                [roadSegment];
                interface:
                    ROCK:length(): real;
            ...
    end-type

type privateRoad:
        specialises: road;
            ...
    end-type
```

The definition of a secondary type may also include a behavioural interface. This consists of the definition of a *signature* for each operation associated with the type. For example, the definition of *road* includes the specification of the signature of the ROCK method *length* which returns a real result.

Associated with every type is a single class which has the same name as the type. Whereas the type definition contains sufficient information for a user to use the type, a class contains the actual code associated with a method together with other implementation details. For example

```
class road
        length(): real
        begin
                var l: real
                foreach s in self do
                        l := l + length@s;
            l
        end
            ...
    end-class
```

specifies that the method length applied to an object of type road adds together the lengths of each of the component roadSegments making up the road and returns this sum.

This separation of the *operation signature* from the *operation implementation* allows *overloading* of operation names, and *overriding* is supported through *late binding* which is used to choose the most specialised definition applicable to the message recipient.

## 3.2 The imperative language ROCK

The database programming language ROCK is a strongly typed imperative language based on the OM data model. It combines a data definition language for schema declarations with a data manipulation language which provides the ability to access and manipulate objects stored in the database.

The type definitions supported by the data definition language are those of the data model described in the previous section. The data manipulation language provides the means to create and manipulate objects. The facilities for doing this are similar to those provided by other object-oriented database languages.

A method consists of a sequence of statements optionally followed by an expression. A statement may be either a simple statement or a control statement. Simple statements include assignment, I/O (read, write,...), method invocation and object creation (new). Control statements include conditional statements (if then else), loops (while, foreach) and blocks (begin end). Expressions include simple expressions (which return an object) and control expressions (similar to control statements except that they return an object or set of objects).

Objects may be created by invoking the operation new for a given class name. Variables are set up through declarations such as

> var r : road

which indicates that the variable with object-name r may reference objects of class road, or

> var r := new road()

which creates a new road object which is referenced by r.

Operations on objects follow the messaging style where the symbol "@" is the message sending operator and message sends can be nested. The message recipient is an object expression (an expression which evaluates to an object). Generally this is simply an object although inside a method it can also be self or super. For example,

> n := get_roadName @ r

will assign to the variable n the result of sending the message get_roadName to the road r.

A number of *system generated* (or *built-in*) operations are provided. For each property of an object type or field in a tuple, the system generates a pair of methods whose names are the names of the corresponding property/field prefixed by get- and put-. These are used to retrieve or update the particular property/field. In addition a number of operators are available for manipulating constructions (e.g. add-member, remove-member, etc.).

In addition to these, the user may define his/her own operations in the form of methods. These may be defined as private (i.e. not accessible from outside the definition of a class) or public. The signatures of public methods are included in the interface part of the type definition.

For example,

```
class roadSegment
        length(): real
        begin
                var st := get_startJunction@self;
                var en := get_endJunction@self;
                var x1 := get_xCoord@st;
                var y1 := get_yCoord@st;
                var x2 := get_xCoord@en;
                var y2 := get_yCoord@en;
                sqrt((x1-x2)**2 + (y1-y2)**2);
        end
end-class
```

### 3.3 The declarative language ROLL

ROLL is a deductive query language which is based on the same data model OM. It is a function-free Horn clause language, the syntax of which is as follows.

The ROLL *alphabet* consists of variables, constants and predicate symbols. A constant symbol is simply the value of a primitive type (e.g. "Main Street", 127, true). A predicate symbol is an operation name declared in an operation interface. A variable is a name or a name preceded by "!" (if it is a ROCK variable).

A ROLL *term* is either a variable or a constant. If b is a predicate symbol, a is a ROLL term denoting an object and $t_1,...,t_n$ are ROLL terms, a ROLL *atom* can be defined as either

$$b(t_1,...,t_n)@ a$$

which is interpreted as "send the message b with arguments $t_1,...,t_n$ to the object a", or

$$b(t_1,...,t_{n-1})@ a == t_n$$

which is interpreted as "send the message b with arguments $t_1,...,t_{n-1}$ to the object a, expecting as result $t_n$". The first form is used when one does not need to distinguish between input and output parameters, the second when one does. If b requires no arguments then '$(t_1,...,t_n)$' can be omitted.

A ROLL atom is *ground* iff a and b are ground and $t_1,...,t_n$ are either ground or omitted. A ROLL *literal* is an atom (positive literal) or a negated atom (negative literal). A ROLL *clause* is a disjunction of literals with at most one positive literal. A ROLL clause with a single positive literal is a ROLL *fact*. A clause with one positive and at least one negative literal is a ROLL *rule*. A clause containing only negative literals is a ROLL *query*. The usual convention is followed of rewriting a clause as a reverse implication (i.e. head :- body) and replacing conjunctions with commas.

There are a number of operations in ROLL which are written as infix operators. An example is "<>" which denotes inequality.

To illustrate this, consider an example of a set of rules defining an operation *connected* associated with the class roadSegment which checks for direct or indirect connectivity between road segments. This may be written as:

```
class roadSegment
    public:
        touching(roadSegment)
        begin
          ...
        end
    public:
        connected(roadSegment)
        begin
        connected(RoadSegA)@RoadSegB :-
                touching(RoadSegA)@RoadSegB.
        connected(RoadSegA)@RoadSegB :-
                touching(RoadSegC)@RoadSegB,
                connected(RoadSegA)@RoadSegC.
        end
```

The first set of clauses (omitted for sake of brevity) define the operation *touching* which checks whether two road segments have a common junction. The remaining two clauses define the transitive closure of *touching*. An example of a ROLL query might be:

```
get_roadName@R == "Main Street", connected(R2)@R,
get_roadName@R2==R3
```

which will find the names of all roads which are connected directly or indirectly to Main Street.

ROLL is a strongly typed language which is integrated with ROCK through the OM. Thus there is no notion of a ROLL program as such but ROLL clauses are used to define methods in the context of classes in OM.

Within a ROLL clause one may invoke any ROCK method provided it is side-effect free (i.e. it does not update, directly or indirectly, non-local data). A ROCK variable may be represented by prefixing the object name by "!".

Within ROCK, ROLL may be used to query data and to define methods. The strong compile-time typing supported by ROCK extends to ROLL in this context and a type inferencing mechanism is supported which derives the type of an expression in ROLL. An example of such a construct is

var v := [ALL X | get_roadName@Y : road == "Main Street",
                    connected@Y == X]

The query is written between square brackets on the right hand side of the assignment to the ROCK variable v. It retrieves all instances of class *road* which are connected to the road with roadName "Main Street". The query body on the right hand side of "|" specifies that X is a road object which is connected to the road object Y whose roadName property has the value "Main Street". The projection on the left hand side of "|" specifies that all bindings for X should be collected and returned as the result.

### 3.4 Advantages of ROCK & ROLL Approach

The two major features of this approach are that it is based on a simple logic model and that it integrates two different languages in a clean way without mismatch. These two features and the advantages that spring from them are discussed below.

### (1) The Logic Model

A representation of the object model OM in clausal form logic is easy to derive with clauses representing the schema of the database, e.g.

    primitive_type(roadName, string) <-
    secondary_type(road) <-
    described_by(road, roadName) <-

as well as the data in the database, e.g.

    instance_of(!5, road) <-
    property( !5, roadName, "Main Street") <-

where !n denotes the unique object identifier of the object in class road whose current *roadName* is "Main Street". Similarly, ROLL clauses may be easily mapped into clausal form logic. For example

    connected(RoadSegA, RoadSegB) <- touching(RoadSegA, RoadSegB).
    connected(RoadSegA, RoadSegB) <- touching(RoadSegC, RoadSegB),
                    connected(RoadSegA, RoadSegC).

The properties of the system itself such as inheritance (including overloading and overriding) are described by a separate set of axioms. These describe attribution, specialisation, association, sequentiation and aggregation, including, for example, the transitivity of the is-a relations, and the propagation of classification through is-a links. They go on to impose a discipline on how structure and behaviour are inherited under conditions of overloading and overriding. The complete set of clauses is referred to as an object theory. This set of clauses is supplemented by a rewriting approach which maps ROLL into a language in which any ambiguity has been resolved deterministically. This is used to model both late binding and encapsulation.

Object theories for ROCK & ROLL constructed in this way capture all the relevant mandatory features of the OODB Manifesto [25] whereas most other proposals for DOODs capture fewer structural aspects and almost no behavioural ones. Consequently the semantics can be reduced to those of definite clause logic programs. Hence it can be shown that the semantics of such object theories can be defined by a mapping onto Datalog. From this point of view this approach is characterised as reductionist.

Not only does this simplify the proof- and model-theoretic characterisation of object theories, but it also ensures that most of the theory of deductive databases can be applied directly to such DOODs. In the case of ROLL this means that the techniques developed for optimisation in deductive databases can be used in implementing a query optimiser for ROLL. It also means that a precise characterisation of query evaluation is possible which has desirable termination properties.

### (2) Mismatch-free Integration

The second major feature of the ROCK & ROLL approach is the integration of a persistent imperative object-oriented programming language (ROCK) and a declarative deductive object-oriented query language (ROLL) in a way which involves no mismatch between the two.

From the current state of database systems it is clear that, depending on the circumstances, explicit control may or may not be desirable. For many applications it is unnecessary to specify control, and by keeping the query language declarative one can take advantage of various opportunities for optimisation. For some it is desirable to rely on explicit control in order to produce the most efficient implementations of particular queries (e.g. those involving geometric manipulations). One solution lies in the integration of a declarative query language with an imperative programming language.

However, the main problems with integrating two different languages lie in the impedance mismatch between them. In particular, the two most important areas of mismatch are in the type system and in the evaluation strategy. By defining a common model which characterises completely the structural and behavioural aspects, this fixes the type system and the evaluation strategy on which the two languages are based, and removes any impedance mismatch between them.

## 4 Active Databases

An active database system provides a different and more powerful form of rule-based capability from that of deductive databases. Whereas deductive rules are purely declarative, active rules are concerned with performing actions in response to events. As a result deductive rules lend themselves to a range of optimisation techniques, and their evaluation is well understood with clear termination properties; active rules on the other hand have more complex semantics, are more difficult to optimise and their termination is more difficult to predict.

Active rules are generally expressed in the form of event-condition-action (ECA) rules and may be incorporated into either relational or object-oriented databases. There are a number of different facets to active behaviour and a means of classifying active rule systems is given in [26].

In recent years there has been interest in combining deductive and active rule behaviour. One approach [27] argues that the two types of rule form the extremes of a spectrum, and leads to the conjecture that active rule systems could be constrained to provide deductive rule capability. A second viewpoint [28] suggests that a fixpoint interpretation extending the semantics of deductive rules can provide a common view of deductive rules and active rules. However, neither of these successfully accounts for a proper merger of these two types of rules.

A third view [29] is that the deductive aspects of rules should be kept separate from other parts of active rule behaviour. Thus instead of trying to integrate the two types of rule in such a way that one subsumes the other, it suggests an approach in which the two can be cleanly integrated while respecting the differences between them. It presents a logic-based formalisation of three languages - for events, conditions and actions - and shows how an active rule language can be assembled from them with an operational semantics that does not interfere with that of the deductive rules.

This approach has been used to design an active rule language which integrates with ROCK & ROLL. This is described in the next subsection.

## 4.1 The Active Rule Language RAP
The declarative language ROLL and the imperative language ROCK together provide an excellent starting point for the development of a separate active rule component for the DOOD. The active rule language RAP (ROCK & ROLL Active Programming language) provides support for ECA rules. In defining RAP [30] one needs to define three separate component languages - the event language, the condition language and the action language. However, ROLL is a natural candidate for the condition language and ROCK for the action language. This leaves only the event language.

The simplest events are primitive events. These include any new and delete operations, get_ and put_ operations, accesses to components of a constructed type or method invocations (to built-in or user defined methods). All of these operations are written as message events using the same syntax as in ROCK & ROLL. Thus the event

get_roadName()@R == "Main Street"

will match every retrieval from the attribute roadName which returned the result "Main Street". If the keyword SENT is appended to the event, the event will be triggered before the operation itself is carried out. Thus

get_roadName()@R == "Main Street" SENT

would trigger the event before the attribute roadName is fetched and checked. Other primitive events include transaction start and end, commit and abort.

Primitive events can be combined to produce complex events using an event algebra with the following operators:

[ ALL | EARLIEST | LATEST] event AND event [ WITHIN t]

[ ALL | EARLIEST | LATEST] event THEN event [ WITHIN t]

event OR event

event REPEATED n TIMES [WITHIN t]

event WITHOUT event [WITHIN t]

( event )

The keywords ALL, EARLIEST and LATEST are used to specify a consumption mode. For example, consider the composite event A AND B, and suppose that two occurrences of A are followed by an occurrence of B. If the consumption mode specified is EARLIEST, the first occurrence of A will be combined with the occurrence of B to trigger the event. If the consumption mode is LATEST, the second occurrence of A will be used. If the mode is ALL, both occurrences will be used and the event will be triggered twice.

The operator THEN specifies an ordered conjunction. The operator REPEATED requires an event to be repeated a specified number of times before it triggers the rule. The operator WITHOUT provides for specifying the non-occurrence of an event. The keyword WITHIN is used to restrict triggering to cases where the composite event is satisfied within a given number of seconds.

In addition to the event specification one also needs to specify both the frequency with which events are triggered and how event arguments are to be bound for use during condition or action evaluation. This is handled by a projection preceding the event specification. For example,

WHEN [ EACH | get_roadName()@R == "Main Street"]

will cause the event to be triggered each time the particular access is performed, with no parameters to be passed to the rest of the rule. It is also possible to use the keywords ALL or ANY if execution is deferred.

If parameters are to be passed from the event to the condition or action, these must be bound to an event variable. For example

WHEN r<==[ EACH | get_roadName()@R == "Main Street"]

will not only trigger the event whenever the particular access occurs, but will also bind the road object concerned to the variable r to make it accessible to the condition and action parts of the rule. Once again it is possible to use keywords ALL or ANY.

The condition part of a rule consists of a ROLL query or the keyword TRUE. Within the query an event variable can be accessed by prefixing the event variable name by "!". As with events, a projection may be used to pass values to the action component of the rule using the keywords ALL or ANY. For example,

WHEN r <== [ ALL R | get_roadName()@R == "Main Street"]

IF rt <== [ ALL RT | RT is_in !r, get_town()@RT == "Edinburgh"]

The action part of a rule may specify an operation to be performed in addition to any operations which trigger the rule or, if SENT is specified in the event part, an alternative operation may be specified using DO INSTEAD. The ROCK code which forms the action part may access the bindings produced by the event or condition by mentioning the relevant variable which gives access to the current state of these objects.

A complete ECA rule is encapsulated by RULE <rulename> END-RULE and followed by a specification of the coupling modes and priority associated with it. For example, suppose that Main Street in Edinburgh has been renamed High Street, and that any query which accesses this object via its original name must be replaced by an equivalent access to the object under its new name. In this case one might have

RULE newstreetname

WHEN r <== [ EACH R | get_roadName()@R == "Main Street" SENT]

IF get_town()@!r == "Edinburgh"

DO INSTEAD get_roadName()@R == "High Street"

END_RULE

COUPLING newstreetname CONDITION IMMEDIATE

ACTION IMMEDIATE END_COUPLING

PRIORITY newstreetname 4.0 END_PRIORITY

The execution model for RAP can be summarised as follows. The *Condition* and *Action modes* may each be either *Immediate* or *Deferred* (*Detached* is not possible because of the limited transaction facilities). If ALL is specified in the event specification, the *Transition Granularity* corresponds to *Set*, if EACH or ANY are used, it corresponds to *Tuple*. The *Cycle Policy* is {*Iterative, Recursive*} since rules with immediate conditions and actions are executed using a recursive depth-first strategy, while deferred rules are executed following an iterative strategy. In scheduling rules a numerical priority scheme is used. The scheduling policy is *All Sequential*. Errors are either ignored or may cause the system to *Abort*.

## 4.2 Advantages of RAP

The approach adopted in the design of RAP has three major advantages, i.e.

### (1) Clean Integration

RAP offers a clean integration of deductive and active rule behaviour which minimises the amount of new expertise required of the user. Provided the user is familiar with ROCK & ROLL, the new knowledge required consists mainly of the event language and coupling modes.

**(2) Opportunities for Optimisation**
The declarative features of the underlying ROCK & ROLL system are exploited in both the event and condition languages. The event language in combination with the consumption modes provides a powerful yet concise means of specifying a wide range of composite events. The condition language provides the full power of ROLL. Since the two are based on simple logic models, they have enabled the development of comprehensive rule analysis and optimisation features for RAP.

**(3) Reduction in implementation effort**
The decision to use ROLL for the condition language and ROCK for the action language clearly has benefits for the implementor in that it makes maximum use of existing software thereby reducing the amount of effort required to implement it.

# 5 Conclusion

This paper presents a brief outline of work done on combining deductive and object-oriented paradigms to produce a deductive object-oriented database and on adding an active component which integrates with the other two.

The work on ROCK & ROLL has shown how a declarative query language can be integrated with an imperative programming language in the context of a deductive object-oriented database system. The approach is strengthened by the fact that it is based on a logic model equivalent to a deductive database model and can therefore take advantage of deductive database techniques in its implementation. Although the two languages are distinct, they are so tightly integrated that users are not disadvantaged by this.

The work on RAP follows a similar approach. Here two different forms of rules are to be integrated and once again instead of trying to unify the two, the approach concentrates on producing an integrated whole. This is achieved by capitalising on ROCK & ROLL and by building a logical model to describe the concept set underlying active databases. Consequent advantages in terms of optimisation and analysis accrue from this.

## Acknowledgements
Some of the work reported here has been funded by the UK Engineering and Physical Sciences Research Council and their support is duly acknowledged. The authors would also like to thank Alvaro Fernandes, Andrew Dinn, Maria Luisa Barja and Alia Abdelmoty for their contributions in the development of the system. Thanks are also due to Dr. Michael Quinn of ICL, Neil Smith of Ordnance Survey and Prof. Keith Jeffery of Rutherford Appleton Laboratories for their assistance and support.

# References

[1] S.Salvini and M.H.Williams, Central Knowledge Management for Expert Systems, *Math. Comput. Modelling*, 16, pp. 137-144, 1992.

[2] F.Manola and M.L.Brodie, On knowledge based systems architectures, in *On Knowledge Base Management Systems*, Ed. M.L.Brodie and J.Mylopoulos, Addison-Wesley, pp. 87-92, 1986.

[3] J.C.F.M.Neves and M.H.Williams, Towards a co-operative data base management system, *Proc Logic Programming Workshop '83*, Ed. L.M.Pereira, Universidade Nova de Lisboa, Lisbon, pp. 341-370, 1983.

[4] M.H.Williams, I.M.Pattison and J.C.F.M.Neves, Reorganisation in a simple database system, *Software Practice and Experience*, 16, pp. 719-729, 1986.

[5] J.Bocca, EDUCE a marriage of convenience: Prolog and a relational DBMS, *Proc 3rd Symp Logic Programming*, Salt Lake City, Utah, USA, 1983.

[6] M.H.Williams, G.Chen, D.Ferbrache, P.Massey, S.Salvini, H.Taylor and K.F.Wong, Prolog and deductive databases, *Knowledge Based Systems*, 1, pp. 188-192, 1988.

[7] S.Abiteboul and S.Grumbach, COL: A Language for Complex Objects Based on Recursive Rules, *Proc Workshop on Database Programming Languages*, pp. 253-276, 1987.

[8] S.Abiteboul, Towards a Deductive Object-Oriented Database Language, *Data & Knowledge Engineering*, 5, pp.263-287, 1990.

[9] M.Jarke, R.Gallersdorfer, M.Jeusfeld and M.Staudt, ConceptBase - a deductive object base for meta data management, *Journal of Intelligent Information Systems*, 3, pp. 167-192, 1994.

[10] M.Liu, Rol: A deductive object base language, *Information Systems*, 21, pp. 431-457, 1996.

[11] P.M.D.Gray, K.G.Kulkarni and N.W.Paton, *Object-Oriented Databases: A Semantic Data Model Approach*, Prentice-Hall, 1992.

[12] G.Gardarin and P.Valduriez, Esql2 - extending sql2 to support object-oriented and deductive databases, EDS Project Technical Report, INRIA, 1992.

[13] D.Maier, A Logic for Objects, Technical Report CS/E-86-012, Oregon Graduate Center, Beaverton, OR, 1986.

[14] W.Chen and D.S.Warren, C-logic of Complex Objects, *Proc 8th ACM SIGACT-SIGMOD-SIGART Symposium on Principles of Database Systems*, ACM Press, pp. 369-378, 1989,

[15] M.Kifer and G.Lausen, F-logic: A Higher-Order Language for Reasoning about Objects, Inheritance and Schema, in *Proc ACM SIGMOD Conf*, Eds. J.Clifford, B.Lindsay and D.Maier, pp. 134-146, 1989.

[16] J.Frohn, R.Himmeroder, P.Kandzia and C.Schlepphorst, How to Write F-Logic Programs in FLORID: a Tutorial for the Database Language F-Logic, Institut fur Informatik, Freiburg University, Germany, 1996.

[17] M.L.Barja, N.W.Paton, A.A.A.Fernandes, M.H.Williams and A.Dinn, An Effective Deductive Object-Oriented Database Through Language Integration, in *Proc 20th VLDB Conf*, Eds. J.Bocca, M.Jarke and C.Zaniolo, Morgan-Kaufmann, pp. 463-474, 1994.

[18] M.L.Barja, A.A.A.Fernandes, N.W.Paton, M.H.Williams, A.Dinn and A.I.Abdelmoty, Design and Implementation of ROCK & ROLL: A Deductive Object-Oriented Database System, *Information Systems*, 20, pp. 185-211, 1995.

[19] S. Ceri and R. Manthey, Chimera: a model and language for active dood systems, *Proc of the East/West Database Workshop*, pp. 3-16, 1994.

[20] D.Srivastava, R.Ramakrishnan, P.Seshadri and S.Sudarshan, Coral++: Adding object-orientation to a logic database language, *Proc 19th VLDB Conf*, 1993.

[21] O.Friesen, A.Lefebvre and L.Vieille, VALIDITY: Applications of a DOOD System, *Proc EDBT*, Springer-Verlag, pp. 131-134, 1996.

[22] A.A.A.Fernandes, An Axiomatic Approach to Deductive Object-Oriented Databases, PhD thesis, Department of Computing and Electrical Engineering, Heriot-Watt University, Edinburgh, Scotland, 1995.

[23] A.A.A.Fernandes, M.H.Williams, N.W.Paton and A.Bowles, Approaches to Deductive Object-Oriented Databases, *Information and Software Technology*, 34, pp. 787-803, 1992.

[24] P.R.F.Sampaio and N.W.Paton, Deductive Object-Oriented Database Systems: A Survey, *Proc Third Int Workshop on Rules in Database Systems*, 1997.

[25] M. Atkinson, F. Bancilhon, D. DeWitt, K. Dittrich, D. Maier and S.B. Zdonik, The Object-Oriented Database System Manifesto, *Proc First Int Conf DOOD*, Elsevier Science Press, pp.223-240, 1990.

[26] N.W.Paton, O.Diaz, M.H.Williams, J.Campin, A.Dinn and A.Jaime, Dimensions of Active Behaviour, *Proc First Int Workshop on Rules in Database Systems*, Eds. N.W.Paton and M.H.Williams, Springer-Verlag, pp. 40-57, 1994.

[27] J.Widom, Deductive and Active Databases: Two Paradigms or Ends of a Spectrum?, Proc First Int Workshop on Rules in Database Systems, Eds. N.W.Paton and M.H.Williams, Springer-Verlag, pp. 306-315, 1994.

[28] C.Zaniolo, A Unified Semantics for Active and Deductive Databases, *Proc First Int Workshop on Rules in Database Systems*, Eds. N.W.Paton and M.H.Williams, Springer-Verlag, pp.271-287, 1994.

[29] A.A.A.Fernandes, M.H.Williams and N.W.Paton, A Logic-Based Integration of Active and Deductive Databases, *New Generation Computing*, 15, pp. 205-244, 1997.

[30] A.Dinn, N.W.Paton, M.H.Williams and A.A.A.Fernandes, An Active Rule Language for ROCK & ROLL, *Proc 14th BNCOD*, Eds. R.Morrison and J.Kennedy, Springer-Verlag, LNCS 1094, pp. 36-55, 1996.

# An Introduction to
# Virtual Reality Modeling Language

Jiří Žára

Czech Technical University, Faculty of Electrical Engineering
Department of Computer Science & Engineering
Karlovo nám. 13, 121 35 Praha 2, Czech Republic
e-mail: *zara@cs.felk.cvut.cz*

**Abstract.** An overview of a widely used description format for Virtual Reality on the Web (VRML) is presented. Basic principles as well as useful tricks for creating dynamic virtual worlds are illustrated by a number of examples. Other competitive languages for VR (QuickTimeVR, ActiveVRML) are compared with VRML. Possible future extensions and expectations are discussed.

## 1 Introduction

One of the main tasks in Computer Graphics is to create images from their symbolic description. When the description deals with spatial objects, the process of image creation is called *rendering*. Due to the limitation of computing power, two types of rendering are currently distinguished:

*photorealistic rendering*
> where the aim is to achieve the best quality image. The rendering speed need not play a primary role. The results of such a process remind photo snapshots of a real scene (world) and involve effects such as shadows and mirrored objects on complex surfaces. Two most popular methods from this area are *ray tracing* and *radiosity* algorithms.

*real time rendering*
> allows the user to quickly generate and present a high number of images giving feeling that the user (the viewer) is walking through the scene and/or objects move inside the scene. This feature represents the principal condition for the *virtual reality* (VR). Rendered images need not be highly realistic. The main effort is oriented around rendering speed, which should achieve about 30 images (frames) per second. Another important feature of virtual reality systems is the utilization of interactions, as well as the non-trivial dynamic behaviour of objects involved in a scene.

In this paper, we concentrate on the *Virtual Reality Modeling Language* (VRML) which represents an ASCII file format for description of 3D worlds which are to be rendered in real time. The following chapter gives an overview of the history of VRML, chapter 3 describes principles used in building 3D worlds and chapter 4 shows, how to optimize them for VR purposes. Chapter 5 describes methods for interaction and dynamic behaviour in VRML. Examples of alternative approaches to VR are mentioned in chapter 6.

## 2   Evolution of Virtual Reality Modeling Language

The history of VRML is coupled with the WWW and the Internet. Although a lot of VR systems existed before the WWW has become widely accessible and popular, VRML is the only VR language and file format designed especially for the Internet[1]. VRML allows the user to share and interchange interactive 3D worlds. It can be used in a variety of application areas such as scientific visualization, multimedia presentation, education, and entertainment.

The specification of VRML was designed and improved rapidly when compared with other successful standards. The design process was public and the authors extensively used Internet techniques. Public electronic conference (forum) was established and the development steps were discussed with a number of experts. For these reasons, the resulting version of VRML is now widely accepted by the Internet community. Notice, that this version is also one of a few ISO standards publicly available on the Internet.

The following list introduces a short history of VRML[2]:

**VRML 1.0** (November 95)
First descriptive language for static virtual worlds on the WWW. Based on Silicon Graphics *Open Inventor* ASCII file format [2, 3],
**VRML 2.0** (April 1996)
Common proposal from SGI & Sony called *Moving Worlds* accepted as the base for the new VRML version,
**VRML97** (April 1997)
Final version of VRML 2.0 that has been improved utilizing comments from the e-forum [7] organized by an independent consortium *VAG* (VRML Architecture Group) [8]. Accepted as the international standard ISO/IEC 14772.

VRML was designed to fulfill the basic criteria of VR systems. It offers a description of 3D worlds that is clear and simple, but allows to add new user-defined objects. It is possible to make file format translators from commonly used 3D modeling systems and formats. The authoring systems and viewers (browsers) can be implemented on a variety of computing platforms and executed regardless of a computer performance.

## 3   Building Static Scenes

We will use the abbreviation VRML instead of VRML97 for the remainder of this text. Building elements in VRML are called *nodes*. There are almost fifty

---

[1] MIME type for VRML files is `model/vrml` (or older `x-world/x-vrml`), file extension is `.wrl`

[2] All specifications are available on Internet [9].

nodes of various types and functions. They are combined into a tree or a multitree structure which is traversed top-down during rendering time. Nodes can be divided into several groups. Three common groups allow the user to define a static 3D object using its *geometry, property, and appearance*:

*Geometry nodes*

can be either simple solids (`Box`, `Cone`, `Cylinder`, `Sphere`) or more complex shapes consisting of polygonal facets (`IndexedFaceSet`). A terrain is defined by an `ElevationGrid` node, and sweeped shapes are described by an `Extrusion` node.

The geometry group is completed by nodes defining a set of spatial lines and points (`IndexedLineSet`, `PointSet`), and a `Text` node representing a banner in 3D space.

*Property nodes*

support and enhance certain geometry nodes, typically `IndexedFaceSet` and `ElevationGrid`. A `Coordinate` node defines a list of vertices in 3D space. For the purposes of the Gouraud shading algorithm, color and normal vectors are assigned to vertices using the `Color` and `Normal` nodes. Finally, the geometry of texture is described by a `TextureCoordinate` node.

*Appearance nodes*

add a natural look to the geometry of objects. A `Material` node contains color coefficients for the evaluation of the lighting (Phong) equation. They are needed in case of additional light sources in the scene.

Other nodes in this group deal with textures, which can create really realistic effects on the surfaces of objects. `ImageTexture` maps a 2D image (JPEG, PNG, or GIF) on the surface, `PixelTexture` defines a simple color pattern and the most powerful `MovieTexture` node allows the user to project a movie (MPEG) on a given surface.

Texture nodes represent the first example of WWW utilization – all textures can be referenced as links to the Internet (URL).

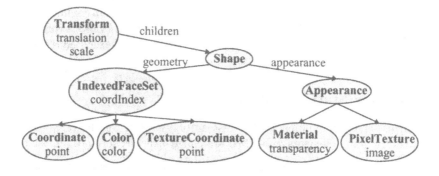

**Fig. 1.** VRML nodes arranged in a tree structure

**Example 1.** Simple VRML file with one object

A typical spatial object representation begins with a `Transform` node specifying position and scale of the object. The following example describes a solid triangle filled with interpolated colors. Its vertices are colored by red, green, and blue color respectively. In addition, the triangle is covered by a black and white checkerboard texture with a 50% transparency. The corresponding symbolic graph is shown in Fig. 1.

```
#VRML V2.0 utf8
Transform  {
  translation  0  2.0  0
  scale        1  2  0
  children  Shape  {
    geometry  IndexedFaceSet  {
      coord  Coordinate { point [ 0 0 0, 1 0 0, 1 0.5 0 ] }
      color  Color     { color [ 1 0 0, 0 1 0, 0 0 1 ] }
      texCoord TextureCoordinate { point [ 0 0, 3 0, 3 3, 0 3 ]}
      coordIndex [ 0, 1, 2, −1  ]
    }
    appearance  Appearance  {
      material Material { transparency  0.5  }
      texture  PixelTexture { image 2 2 1 0xFF 0x80 0x80 0xFF }
    }
  }
}  #   end of Example 1
```

### 3.1   Formal Node Specification

From the programmer's point of view, each VRML node can be seen as an object (or more accurately as a class) with its own data and methods. Private data are called *fields* and their values are set at the time of object definition (instantiation). An example of such data is the `radius` field in the `Sphere` node (see Example 2).

Public data are called *exposed fields*. They are accessible from other nodes using a special language construction called `ROUTE`. The ROUTE construction ensures sending and receiving events. A new value of an exposed field is set by an event, also any change of exposed field generates a new event. An example of an exposed field is the `translation` field in the `Transform` node (Example 1). Most fields in VRML nodes are exposed fields.

Special *eventOut* fields are designed for sending events, while *eventIn* fields only accept incoming events. These fields are suitable for interaction and dynamics. They will be discussed in detail in chapter 5.

VRML fields are of the data types listed in Table 1.

Instead of the commonly used term *array*, VRML uses a term *multiple field* for a sequence of values with the same data type.

| | |
|---|---|
| Bool, Float, Int32, String | – basic types |
| Time | – used for event passing |
| Node | – arbitrary VRML node |
| Color, Image | – structured types |
| Vec2f, Vec3f, Rotation | – geometrical data |

**Table 1.** Data types in VRML

## 3.2   Scene Graph Nodes

Single virtual objects are combined together and positioned by a `Transform` node. The same transformation is applied to all of its children. A `Transform` node is the basic element for creating a tree structure of the scene. A `Group` node has the same effect, but without performing any transformation. Child nodes can be dynamically removed or added through rendering. The only exception is a `Switch` node, which delivers just one of its children to the rendering process.

When a complex scene is built, many objects with the same shape are repeatedly defined and positioned. VRML allows the user to reuse previously defined objects, make copies and modify properties such as size or texture of these copies. Each VRML node can be named by a language construction `DEF`. The assigned user name is later used in the construction `USE`, which creates a new instance of the original node (see Example 2).

It should be stressed that the naming by `DEF` statement is also important for accessing node fields (chapter 5).

**Example 2.** Naming and instancing

An original object (a sphere) is named as `SOMETHING` and its second instance is firstly scaled and then moved. This order of transformation steps within one `Transform` node is obligatory. To change the order, a hierarchy of several `Transform` nodes must be defined.

```
#VRML  V2.0  utf8
Transform  {
   children   [
         DEF  SOMETHING  Shape { geometry Sphere { radius 0.5 } }
         Transform  {
                     translation  3  0  0
                     scale        1  2  1
                     children     USE  SOMETHING
         }
   ]
}  # end of Example 2
```

## 3.3 Behavioral Nodes

This group of nodes does not add "solid" parts to the virtual scene, but aids in increasing the user's feeling of "reality".

Light sources are of three types – DirectionalLight which shoots parallel rays from a large area from an infinite distance (e.g. a sky), PointLight which shoots rays with a limited radius and attenuated light energy, and the most complex SpotLight which has the characteristics of a reflector. Light sources have no geometrical representation. They influence only nodes on the same or lower level of the scene tree. This feature can be used for a sophisticated and efficient manipulation of light.

Next two nodes influence a scene globally. A Fog node defines a color blended with colors of spatial objects, a Background node allows the user to fill the background of a viewing window with either a set of interpolated colors or six images creating three-dimensional panoramic view.

The last node in this group is Sound, which is not only stereophonic but also spatial. Sound files downloaded from Internet can be in one of the following formats: WAV, MIDI or selected from a movie clip (MPEG).

## 3.4 User-defined Nodes

New nodes are defined in VRML as easily as a programmer defines a new procedure or a class in a programming language. A declaration of a new node is introduced by a PROTO statement followed by a name, a list of fields (parameters) with default values, and a body part – new VRML graph. Formal parameters are assigned to other fields by an IS construction. The first node in a body part specifies the type of the entire new node (shape, material, color, etc.)

```
PROTO  Cross  [  field          SFFloat  radius   1
                 field          SFFloat  radius2  0.2
                 exposed field  SFFloat  length   3   ]
{  Transform  {
        children  [
            Shape {geometry Sphere {radius IS radius}}
            DEF ARM Shape {geometry
                           Cylinder {  radius IS radius2
                                       height IS length
                            }
            }
            Transform {rotation 0 0 1 1.57
                       children USE ARM
            }
        ]
    }
} # end of Example 3
```

**Example 3.** New shape definition

Definition of a geometrical element `Cross` consists of three primitives - two crossing perpendicular cylinders and a small sphere in the middle. Obvious parameters (radii, length of arms) can be set when a `Cross` node will be instanced, moreover the later parameter can be changed dynamically.

A declaration does not create any instance. This is suitable for making libraries of prototypes. Instantiation then depends on the place from which the user-defined node is called. When this is from the same file, the original name is used, e.g.:

```
Cross { radius 0.5}
```

A call from an external file or library (here from library `myobjects.wrl`) is a little more complicated. A prototype can also be renamed:

```
EXTERNPROTO MyCross [ ] "http://myobjects.wrl#Cross"
MyCross { length 4 }
```

# 4 Special Features for VR and WWW

There are three groups of VRML nodes designed for optimization of real time rendering, for effective navigation through the virtual space, and for utilization of the WWW:

*VR optimization nodes*
> ensure, that VRML files can be sufficiently rendered even on slow machines such as a standard PC. This is achieved by a few "tricky" nodes.
>
> The first one is called `LOD` (Level Of Detail). It allows various representations for one object depending on distance from the viewer or on the rendering speed of the browsing machine. When near to viewer, a detailed object representation is rendered, when farther, simplified shape such as a small set of polygons is presented. Faraway and thus very small objects specified by `LOD` construction are not rendered at all. This feature represents the base condition for a constant frame rate needed in VR systems.
>
> The second tricky node is a `Billboard` node. All of its child nodes are automatically faced against a viewer (using rotation around a specified axis). This helps in the substitution of highly complex objects, such as a tree, by a texture placed on the billboard. Instead of modeling many small 3D parts, one image of the tree is presented face to face with a viewer independently on his or her position.

*Browser oriented nodes*
> facilitate navigation within a virtual world. The navigation inside virtual worlds is not simple, especially for beginners. VRML standardizes several features helping a user to go through and search virtual worlds. It is always good to define several `Viewpoint` nodes specifying a named viewer position and some camera characteristics. A browser then offers all predefined viewpoints to a user and performs automatic animated movement from current viewing position to selected new one.

A `NavigationInfo` node specifies the body proportions of a virtual visitor (a viewer), that one commonly called *avatar* in the field of VR. Fat avatar cannot go through a narrow door, tall avatar is not able to look under a table. Another important field specifies the method, how the information from an input device (typically a mouse) will be transformed (by a browser) to the movement within the scene. A user can `WALK` or `FLY` through the scene or `EXAMINE` virtual objects when looking around them.

Notice that avatar's movement in VRML is limited by the fact, that all virtual objects are "solid", i.e. the avatar cannot go through them. A `Collision` node has been designed to control collision detection of specified nodes. For instance, it is possible to go through a wall if its geometry is a child of an appropriately set `Collision` node.

*Web oriented nodes*

allow the user to use hyperlinks to other media such as sounds, images, and movies. As distinct from HTML, hyperlinks can be multiple, thus the chance of a successful download is increased. When the first hyperlink in the list fails, the next is tried. This is shown on Example 4, where the parent `Anchor` node specifies hyperlinks activated when a user picks a child node.

Other utilization of the WWW is ensured by an `Inline` node. It downloads virtual objects and includes them into a current scene. This is the way in which large scenes with elements taken from all over the world are built.

**Example 4.** Hyperlink to other VRML world

Activation of the first hyperlink causes the downloading of a VRML file from the Web and its presentation from the viewpoint `EntryView`. If the downloading fails, a current virtual world remains active and the viewpoint `AnotherView` is used.

```
Anchor {
  url ["http://www.ctu.cz/ctu.wrl#EntryView",
      "#AnotherView"
      ]
  description "U must visit this!"
  children    Shape { geometry Box { } }
} # end of Example 4
```

# 5   Dynamics in VRML

VRML nodes described in previous chapters are suitable not only for the modeling of static virtual worlds but also for dynamic ones. Common static worlds can be easily extended by a few additional specialized nodes so that the scenes become interactive and dynamic.

Symbolic scheme for dynamic actions in VRML scenes is shown in Fig. 2. Each individual action has to be initiated, controlled and targeted to a certain

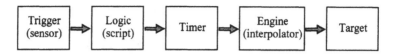

**Fig. 2.** Common event passing scheme for dynamic actions

node or to a group of nodes. Actions are driven by events represented as arrows in the figure. When the user activates a sensor, an event is sent to a logic part, where the conditions for the continuation of the action are checked. An action is then performed during a specified time interval (or a set of intervals) controlled by a timer. An animation is supported by interpolators connected with target objects.

The common symbolic scheme above is usually simplified. Many real dynamic actions in VRML consist of two or three nodes driven by one or two events only. The most simple dynamic action requires only one trigger and a target. Sophisticated interactive actions have to be constructed using all five node types from Fig. 2. The order of node types is not obligatory – nodes can be arranged into an even longer chain or into a graph with several logic controllers.

An event passing scheme is implemented by a `ROUTE` statement in VRML. The syntax is as follows:

```
ROUTE SourceNode.sourcefield TO TargetNode.targetfield
```

`SourceNode` and `TargetNode` are user names that were assigned earlier to VRML nodes by a `DEF` statement. An identifier `sourcefield` represents the name of either an *exposed* or *eventOut* field in a `SourceNode`. Similarly, a `targetfield` is the name of an *exposed* or *eventIn* field in a target node. Both a `sourcefield` and a `targetfield` must be of the same data type as specified in Table 1. An event is thus internally implemented as a couple of a time stamp and a typed data value. A time value is used by a browser to synchronize events and actions.

### 5.1  Manipulators

The most simple VRML nodes utilized in user interactions with the virtual world are *manipulators*, sometimes also called *drag sensors*. They allow the user to transform other nodes in a specific way. The movement of a pointing device is mapped to a position on certain spatial surface such as a planar, cylindrical or spherical surface. Using manipulators, virtual objects can be moved or searched from all directions.

Manipulators are of three types – **PlaneSensor** which allows movement on a plane, **CylinderSensor** which maps input 2D coordinates onto invisible cylinder aligned with the Y-axis ("up"-axis), and **SphereSensor** which rotates an object about the origin of the sensor local coordinate system.

Manipulators combine features of two actions – a trigger and an engine – together.

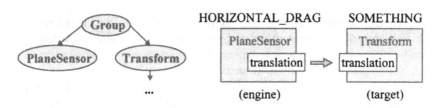

**Fig. 3.** VRML scene with **PlaneSensor** manipulator. Symbolic tree (left) shows static structure, event passing scheme (right) adds dynamics.

**Example 5.** Dragging object in one direction

**PlaneSensor** allows the user to specify the limits of possible manipulation in two dimensions. In the following example, the specified range is limited in one dimension, thus allowing movement only in the horizontal direction. A symbolic representation of this example is shown in Fig. 3.

```
#VRML   V2.0   utf8
Group  {
    children  [
                DEF  HORIZONTAL_DRAG  PlaneSensor  {
                        minPosition    0  0
                        maxPosition    5  0
                }
                DEF  SOMETHING  Transform  {
                        children  Inline  { url "object.wrl" }
                }
    ]
}
ROUTE  HORIZONTAL_DRAG.translation  TO  SOMETHING.translation
#  end of Example 5
```

## 5.2   Sensors

Manipulators work when the user directly activates the object (clicking) by an input device. In addition to them, VRML contains several sensors sensitive to various kinds of the user activity.

A simple **TouchSensor** node detects activity of pointing device. It sends events when a pointer is over a specified rendered object. It also checks the status of an input device button (see Example 6) and returns properties of the point on the touched object (coordinates, normal vector, texture coordinates).

Next two nodes are sensitive to avatar. A **Collision** node detects if avatar (not a pointing device!) is touching a specific object. A **ProximitySensor** node detects the changes of avatar's position and orientation within a specified spatial region. Careful design of virtual worlds with the use of proximity sensors efficiently uses a time when downloading from the WWW. One virtual room can be loaded quickly, while the other parts are loaded later when avatar leaves the room by either activating a door handle or jumping through the window.

A `VisibilitySensor` node is sensitive to avatar in other way. It outputs events if a specified object or a spatial region is visible from avatar's position, i.e if a specified object is at least partially rendered in a browser window. It is a useful tool for starting animation just when a user can see it.

The last sensor in this group is a `TimeSensor` node. It does not depend on the user's activity, but only on time. It generates events as time passes, possibly in a loop. `TimeSensor` is a basic building element for animations and other dynamic actions (see Example 7).

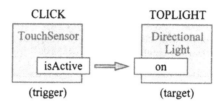

**Fig. 4.** Event passing scheme from Example 6 – Touch sensor

**Example 6.** Touch sensor switches a directional light

The scene is constructed from a group of three nodes. An `Inline` node downloads the shape of a virtual object, `TouchSensor` waits for the activation of that object, and the light source is switched on/off depending on the activity of a pointing device (mouse button). When a user clicks on an object, its color begins to brighten because of the new additional light. It is easy to arrange the scene graph so that the activation of one object (a switch) illuminates other object(s).

```
#VRML  V2.0  utf8
Group  {
    children  [
        Inline  {url  "blueball.wrl"}
        DEF  TOPLIGHT  DirectionalLight  {
                on                  FALSE       # initially no light
                ambientIntensity  1
                color               1  1  1     # white light
                direction           0.3  0  −0.8
        }
        DEF  CLICK  TouchSensor  { }
    ]
}
ROUTE  CLICK.isActive  TO  TOPLIGHT.on
#  end of Example 6
```

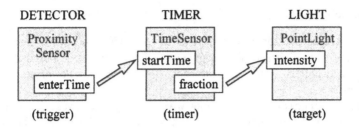

DETECTOR　　　　　　TIMER　　　　　　LIGHT

(trigger)　　　　　　(timer)　　　　　　(target)

**Fig. 5.** Event passing scheme from Example 7 – Proximity sensor

**Example 7.** Animated lighting using proximity sensor

Good timing can convince the user that the virtual world is like a real one. For this reason, TimeSensor nodes are parts of almost all dynamic actions. This example uses a time sensor for controlling illumination. When avatar comes near an object, a ProximitySensor sends an event to the timer. The TimeSensor then generates a sequence of events which increase the intensity of the light source. The duration of the action is 2 seconds.

```
#VRML  V2.0  utf8
Group  {
    children  [
        DEF  DETECTOR  ProximitySensor  {  size  2  2  2  }
        DEF  TIMER     TimeSensor  {
            cycleInterval      2          #  duration 2 sec.
            startTime          0          #  start immediately
        }
        DEF  LIGHT  PointLight  {
            intensity          0
            location           3  3  3
        }
        Inline  {  url  "object.wrl"  }  #  something to be lit
    ]
}
ROUTE  DETECTOR.enterTime       TO  TIMER.set_startTime
ROUTE  TIMER.fraction_changed   TO  LIGHT.set_intensity
#  end of Example 7
```

Another feature of VRML syntax is also shown in the above example. *Exposed fields* can be distinguished by the prefix set_ and the suffix _changed, whether they are used for receiving or sending events respectively. Although this fact is unambiguously done using ROUTE syntax, the possibility of adding a suffix and a prefix to the name of an exposed field improves readability of VRML files.

## 5.3   Interpolators

The group of interpolators represents a highly powerful concept for the animation of VRML worlds. With a given sequence of key values, a linear interpolation between each pair is performed. Interpolators generate events of all important data types.

ScalarInterpolator node works with floats, OrientationInterpolator node changes the parameters needed for rotation. A PositionInterpolator node outputs just one position in 3D space, while a CoordinateInterpolator node outputs a set of interpolated 3D vectors. Similarly, a NormalInterpolator node outputs a set of interpolated 3D normal vectors.

A ColorInterpolator node ensures the linear changes of color values.

**Example 8.** Opening a door using interpolation

The door in this example is animated using a sequence of three dynamic nodes. A touch sensor named CLICK waits for the door to be activated by a pointing device. Then a timer outputs events for a duration of 2 seconds. This time interval is subdivided into 2 equal parts inside the orientation interpolator (field key). In the first interval, the door is quickly opened by a rotation of $\frac{\pi}{2}$. The second interval is given to slowing down the opening process as would be the case in a real world. The door is rotated using a small angle, but again during one second which gives impression of slower motion.

```
#VRML   V2.0   utf8
Group   {
    children  [
        DEF   CLICK   TouchSensor  {  }
        DEF   TIMER   TimeSensor  {  cycleInterval  2  }
        DEF   ENGINE  OrientationInterpolator  {
                key       [0,  0.5,  1]
                keyValue  [0 1 0 0, 0 1 0 -1.57, 0 1 0 -1.8]
        }
        DEF  DOOR  Transform  {
                children    Transform {     #  nested because of
                translation 0.4  0.9  0     #  transf. order
                children                    #  door  geometry
                    Shape  { geometry Box { size 0.8 1.8 0.1} }
                }
        }
    ]
}
ROUTE  CLICK.touchTime         TO  TIMER.startTime
ROUTE  TIMER.fraction_changed  TO  ENGINE.set_fraction
ROUTE  ENGINE.value_changed    TO  DOOR.set_rotation
#  end of Example 8
```

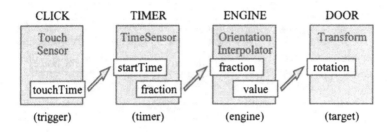

| CLICK | TIMER | ENGINE | DOOR |

Fig. 6. Event passing scheme from Example 8 – Opening door

## 5.4 Advanced Dynamics by Scripting

Should the users desire to do something which is beyond the scope of VRML, they can do this using a script – internal or external program. The connection to other programs arranges a Script node which is the most universal and powerful tool for dynamics in VRML.

The script node can improve all elements of an animation sequence. It can be a part of an engine controlling animation or a logic processing input conditions. Moreover, a script is able to manipulate a scene hierarchy – to add and/or remove child nodes. Advanced scripts allow communication across a network.

The syntax of a Script node is relatively similar to a PROTO statement. An unlimited number of user defined fields can be added to the definition of a Script node. Those fields are mostly of type eventIn and/or eventOut, because frequently the position of a Script node is "a crossing point" for a variety of events. For each eventIn field, a user function with the same name must be written. Each time a value is assigned to an eventOut field, an event is generated.

An obligatory field is called url. It contains either a program itself or a hyperlink to a program file. Two programming languages are currently supported – Java and JavaScript. Application Program Interfaces (APIs) between VRML and these languages are also standardized.

**Example 9.** Switching geometry using script

Example 6 showed the light being switched ON or OFF following user activity. The light source remained permanently OFF after user activation. In order to switch the light source back ON, the declaration and manipulation of a local variable is required. This cannot be easily achieved by common VRML nodes but is possible using a Script node.

This example shows, how to change and save the color of an object longer than mouse button is pressed. For simplification, two objects with the same shape but different colors are defined and arranged as children of Switch node. A local variable (a field) named "on" is declared in a Script node. It stores the current switch status. When a touch sensor named ONOFF is activated, an event is processed in a script node. Another event is then sent from a script node to a switch node.

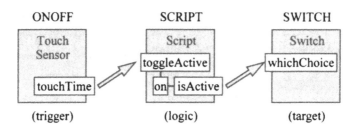

ONOFF        SCRIPT        SWITCH

(trigger)        (logic)        (target)

**Fig. 7.** Event passing scheme from Example 9 – Script

```
#VRML  V2.0  utf8
Transform  {
     children  [
          DEF  ONOFF   TouchSensor  {  }
          DEF  SWITCH  Switch  {
               whichChoice  0
               choice  [  Inline  {url  "redbox.wrl"   }
                          Inline  {url  "greenbox.wrl"  }
                       ]
          }
     ]
}
DEF  SCRIPT  Script  {
        eventIn    SFTime    toggleActive
        eventOut   SFInt32   isActive
        field      SFBool    on    FALSE
        url  "javascript:
               function  toggleActive(event_time)  {
                            if  (on)  on  =  0;
                               else  on  =  1;
                            isActive     =  on;
               }"
}
ROUTE  ONOFF.touchTime  TO  SCRIPT.toggleActive
ROUTE  SCRIPT.isActive  TO  SWITCH.set_whichChoice
#  end of Example 9
```

## 6  Other Approaches to VR on WWW

The base of current VRML specification has been selected from seven proposals
designed by leading companies in the computer graphics and virtual reality fields.
Most of the rejected proposals are no longer in current use, but a few of them
are still under development. For this reason another file formats for VR can be
found on the WWW. The main problem of their utilization is the computing

platform dependency. Most of those file formats have been designed solely for Microsoft Windows. This limitation prevents wider acceptance, although several advanced and useful features can be found in their specifications.

### SVR (Super VRML, Superscape VR)

is the name of file format developed by the Superscape company in UK. The format is binary encoded and is not publicly available. It is intended strictly for the use of Superscape software for authoring and browsing SVR files [5]. SVR format is used by people, which are developing non official presentation environment called *Virtual WWW* [6].

Superscape company is known by its incredibly fast VR browsers on PC platform. The good news for people interested in VR is that the next version of Superscape browser *Viscape* will support both SVR and VRML formats. The VRML advanced features have been probably recognized as good extension to current SVR specification.

### QuickTimeVR

comes from an Apple platform and represents a completely different approach [4]. The original format of *QuickTime* for audio and video compression has been taken as the base for virtual reality presentation. The main idea is thus based on the utilization of movie technology.

Virtual world is created by set of panoramic views. Each solid object is represented by spherical or cylindrical surface. All possible views on the object have to be mapped onto the surface at the time of object definition. When a user later manipulates the object, a browser shows an image corresponding with the current viewer position.

Initially, a user is placed inside a cylindrical or spherical area covered by panoramic images. This is similar to a Background node in VRML. The difference is that in QuickTime VR, a user's movement means zooming background images.

Since spatial objects are replaced by virtual spheres and cylinders, the visibility problem can be solved effectively. All panoramic images can be touch sensitive. It allows the user to "jump" from one room to next by activating a door on the image.

The creation of QuickTime VR files requires a relatively complex preparation of the data. Hundreds of views must be either snapped or rendered by a computer before the virtual world is ready to be presented. Also the size of QuickTime VR files is much larger than of VRML files.

To be complete, we should mention Microsoft's approach to VR called *ActiveVRML*. It represents one of the non-accepted proposals for VRML 2.0. The idea is to define a static virtual world using VRML 1.0 and to manipulate it using Microsoft's Active library. The library contains utilities for math and matrix computations, routines for interaction, animation, etc. This approach is again coupled with one computing platform only. Currently, Microsoft has stopped ActiveVRML development and supports the international standard VRML.

## 6.1 Other Modeling Systems

The VRML format is not intended to be a universal interchange format among systems for modeling and animation such as CAD systems or specialized animation packages. VRML is dedicated to real time presentation, not for the transportation of technological data or complex data structures. Large design systems can use VRML to present their designed products on the Web. Examples of this approach are software packages such as *3D Studio Max* from Autodesk or *TrueSpace* from Caligari. They are able to export either objects or entire scenes into VRML files.

A number of convertors from a variety of file formats to VRML can be found on the Web. Back conversion from VRML is not very interesting, because the advanced features of other systems remain unused. For instance, VRML does not contain any kind of complex surface (spline, NURBS). If a modelling system exports such shapes into VRML file, surfaces have to be replaced by sets of small patches and the original quality is lost.

## 7 Future of VRML

The acceptance of VRML as the International Standard is not the last step in VRML development. Several working groups preparing further extensions are organized by the VRML Consortium [8].

Binary coding and compression specialized to geometrical data is one direction of future VRML development. On the other hand, current ASCII format has high readability. All browsers accept "gzipped" VRML files. A request for special encoding is not too urgent.

Another area of VRML innovation is the support for a multi-user environment and interaction. Current VRML specification assumes that only one user (avatar) will be examining a virtual world. The world is defined in static file(s) without any possibility to change the content. An idea called *Living Worlds* (formerly VRML 3.0) may extend VRML in such a way that multiple avatars could meet in one virtual world and change the content of the world even if they leave it. This is similar to the approach used to voice chat on the Internet.

In spite of video conferencing, multi-user cooperation based on VRML will not require any special devices (i.e. cameras) or high speed Internet connections for real time video transfer. Real time dissemination of up-to-date information about positions and activities of all avatars can be be achieved even with low speed connections, however other problems such as avatar's interaction must be still solved. The development of a multi-user VRML extension is actively supported by Sony.

## 8 Conclusion

An introduction to the Virtual Reality Modeling Language has been presented. VRML nodes have been logically grouped. Typical VRML constructions have been shown step by step from static virtual worlds up to advanced dynamic ones.

VRML represents a new generation of languages oriented primarily toward the presentation of three-dimensional data and WWW utilization. VRML is considered a base for the "third generation" of GUI (Graphics User Interface). The reason for this claim comes from the evolution of user interfaces:

1st generation (1D) – command line
2nd generation (2D) – windowing systems
3rd generation (3D) – virtual reality ?

It is clear that all GUIs listed above will have their own users and applications. Successfully finished and widely accepted VRML standard will undoubtedly influence a number of applications and their user interfaces. With the help of VRML, the term *Virtual Reality* may overcome its original "game orientation" and become a natural tool for modern presentations, scientific visualization, data mining, simulation, and other fields in computer science. Moreover because VRML has become a standard part of Internet browsers, it is available for constantly growing community of Internet users working with non-specialized, commonly equipped computers.

# References

1. Hartman, J., Wernecke, J.: *The VRML 2.0 Handbook*. Addison-Wesley, 1996, ISBN 0-201-47944-3.
2. Wernecke, J.: *The Inventor Mentor*. Addison-Wesley,1994.
3. Open Inventor – http://www.sgi.com/Technology/Inventor/
4. QuickTimeVR – http://www.quicktimevr.apple.com/
5. SVR – http://www.superscape.com/products/viscape/
6. Virtual WWW – http://vwww.com/
7. The VRML Forum – http://vag.vrml.org/www-vrml/
8. The VRML Consortium – http://www.vrml.org/
9. VRML specifications – http://www.vrml.org/Specifications/

# Stepping Stones to an Information Society

Jiří Zlatuška

Masaryk University, Faculty of Informatics
Botanická 68a, 602 00 Brno, Czech Republic

**Abstract.** The information revolution is radically transforming lots of patterns along which society and enterprises have traditionally worked. These changes do not bring just minor technological improvements, but indeed a fundamental transformation of the industry-based society into an information-based one. The changes are most visible and documented within the business world, but the synergy between technological and social shifts does not stop there. In this paper we try to identify and summarize key trends and challenges which this development puts before us.

## 1 Introduction

The development of information processing and communication technologies based on the digital computer and digital information representation and manipulation has made the closing decades of the millennium also the beginning of the new "Age of Information". It is ushered by the information revolution caused by the rapid change of the technological base used by the developed society. Information storage, manipulation, and distribution have been part of the history of human civilization, and a fundamental element of societal development both in the spiritual and production sense. The development of writing from cave paintings within the span of c.20 000 BC to c.4000 BC introduced discrete representation into the way in which human memory was recorded or various forms of communication were utilized, followed closely by the use of linear representation (c.1500 BC) and alphabetic writing (shortly after 1000 BC). This coincided with early civilizations and the development of viable agriculture. When compared to block-print known to the Chinese already sometime BC, Gutenberg's invention of the movable type (1448) can be likened to a "universal machine" of the printed text (a difference similar to the difference between a simple calculator and a programmable universal computer). The movable printing press allowed to reassemble components needed for book printing into configuration of any sort, which made book publishing economically much more viable when compared to manual hand-copying by monks. This triggered an explosion of book production (some 20 million books are estimated to have been printed within the first 50 years, more than the number of books produced ever before that), led to general literacy and school system, and made it possible for the industrial revolution to have both sufficient number of knowledgeable inventors, as well as the uniformly educated workforce needed for the industrial era.

The development and massive utilization of digital computers (*electronic computers*, as we should stress in the year of the 100th anniversary of Thomson's discovery of the electron) and computer-based communication networks within the last few decades have been causing fundamental change in the character of our society. There are many aspects of this process which make it possible to identify the coming of an "information society" as a result of the computer revolution, in an analogous way to the "industrial society" which superseded the agricultural one as a result of the industrial revolution. Computers and telecommunication devices are becoming ubiquitous appliances used in nearly any area of human activity and they are the basis of the new infrastructure. The computer, once a huge technological monstrum dimming the street lights when having been switched on, has become a tiny appliance found everywhere from washing machines, car engines, telephones, or ordinary "things" of everyday use such as hotel doors or personal ID cards. We are not paying attention to them in many cases any longer, simply expecting that they would do the "information-enriched" job they are supposed to, similarly to having stopped paying attention to the way the electrical wiring is prepared in buildings so that our industrial devices can be powered up wherever we need them.

The technical advances make it possible to think realistically of pentaflop computers or terabit networks to appear very soon for affordable prices. Computer chips keep obeying the so-called *Moore's law*, which means doubling the performance each 18 months, or halving the price under constant power. The cost per communicated bit falls down much faster, allowing for forecasts for doubling the total network bandwidth within as short as 8 months [18]. With pentaflop computers and terabit networks coming soon (note that estimated 1M increase can be achieved without changing the already installed fiber-optic cable capacity), it is realistic to assume that computing becomes ubiquitous and the price of networking virtually disappears. Most popular software producers already take advantage of the fact that the worries to optimize for space or time belong to yesterday—and majority of today's customers seem quite willing to pay the price for it, even though the asymptotic benefits have not really materialized for them yet.

The story is however not just about computers as convenient appliances useful for embedding into various sorts of devices. On a more fundamental scale, it is about the increasingly important function of information as the main driving force of the post-industrial stage of the developed nations and deep changes in the character and organization of social and economic life induced by contemporary information technologies. The net effect of these changes is indeed revolutionary in its profound, pervasive, and permanent nature which transgresses just mere improvements of the way how to perform calculations (see also an earlier [21], or, e.g., [49]).

The nature of the changes has a variety of facets, which result from various rôles in which the use of information shifts the emphasis from producing and manipulating material goods to manipulating knowledge, from processing the "hard" to processing the "soft", or perhaps from creating the wealth to generating and manipulating knowledge. This concerns a range of issues from manipulating bits and digital information with all the consequences to human communication, perception, and understanding [39], economic activities, organizing and conducting business, and changes induced by networking [55], [24], the impact of new technological changes [12], to the overall pattern of changes which accompany the end of the industrial age [57] and the increased dominance of services on the expense of traditional manufacturing [7].

## 2 The pattern of change

The influence of the information revolution on businesses and industrial enterprises has manifested itself clearly enough over past several decades ([32] even argues for "organizational informatics"). None of these changes is limited to just "efficiency" of production, but rather to very far-reaching changes in the functional organization and structure, resulting in a real change of the "quality" as well. New organizations and new organizational forms exist in parallel with the old ones. The changes do not constitute any complete replacement. The new forms co-exist and compete with those inherited from the past. Various observers have put emphasis on slightly differing points when identifying the principal features of the change.

Alvin Toffler argues for the "third wave" constituting a development superseding the old "second wave" society of the industrial age [57]. The economies of scale, mass production, standardization, synchronization, concentration and centralization are the key features growing through the fabric which the society is contained in. The third wave society introduces a "practopian" future favoring individual difference, de-massification, variety of possible lifestyles, working and production patterns, and the value of individual skills and innovations. Toffler's third wave comes with the "prosumer" (merging production with consumption), with the change in the organizational structure from information-monopolizing hierarchies to information-spreading matrixes or to a host of other "flexible" organizational forms suiting individual business needs, and with the diminishing rôle of marketing and mediating agents. These changes have been enabled by modern information technology, but they in no means are constrained by it – this is, in fact, one of the reasons why Toffler would not say "computer" or "information" revolution, but rather the "third wave" as a concept with much wider applicability. Within the U.S., Toffler argues, the third wave arrived in the decade following 1955, within the time when the white-collar and service workforce outnumbered the blue-collar workers, followed soon by other developed countries.

Don Tapscott comes with a concept linked very tightly with the digital nature of information processing and computer-based communication [55], and draws very close analogies between information technologies (IT) based on digital technology, both hardware and software, and the transformation of the business, and of social, and political life. Among the themes Tapscott introduces, one can find "digitization", "virtualization", "molecularization", "internetworking", "convergence", or "prosumption", which relate very closely to a majority of concepts introduced by Toffler. Digital, or bit-based products are becoming the major product and material within the economy (this trend is also illustrated indirectly by the shift in spending proportion between packaged software and hardware, which changed from 7 cents for software per $1 spent on hardware to the current ratio of 1:1 according to [37]).

A related set of problems has been tackled by Stan Davis [10] who studied the change in the meaning of time, space, and mass, and introduced the term "mass customization" assuming a market where "customer's individualized needs are satisfied with mass-produced goods, and the customizing occurs at instantaneous speed in the matching-up process". This relates to Tapscott's "real-time economy" and Toffler's "just-in-time" or "on-demand" production with prosumer controlling the features of the resulting product.

The economy and means of production changes also bring about changes in organizational structure. This concerns the team-based organization [35], the matrix structure [57], the hypertext organization and the task-force use [40], the internetworked enterprise [55], or Quinn's infinitely flat, spider's web, or starburst, or inverted organizations and the focus on the functional "core competencies" of network organizations [44]. The resulting "virtual organization" [43] allows for flexible outsourcing of activities, encourages the "molecularization" [55] or "cottage industries" [57]. During the early nineties, several techniques have been introduced that enable organizational change and thus allow existing institutions to accommodate their organizational structure to their changing environment, even though their rate of success is rather debatable [14]. Although the use of information and communication technology is clearly a catalyst and a mediator of the change, some of the impacts are still not well enough understood, such as the "productivity paradox" (see [25] and [54] for more discussion on this).

The networking organizational paradigm extends beyond the individual organizations. The molecularization of the production units, subcontracting, and the use of internetworked organizational structures make the difference between individual company and a productive chain, or between a vendor and a consumer, a fuzzy one, by unifying these configurations into a new entity as it has been the case with blurred distinction between the producer and the consumer. The same applies to the virtual or niche markets where the community aspects of the network coalitions, including loyalty to the networks rather than to individual companies, help to create a very dynamic environment reinforcing the returns for everybody who participates in it[1]. The chaotic nature of these

---

[1] An unpublished report by Donald Hicks of University of Texas commissioned by the State of Texas discovered that the average half-life of businesses dropped by half during past 22 years, indicating that the economic churn may be one of the key factors in establishing long-term stability. Surprisingly enough, areas with the shortest life expectancy of businesses, such as Austin, are having the fastest-growing job base and the highest wages—high business-mortality rates are good for economic health. The rate of innovation helps to explain this. ([59])

processes, the continuing churn in structure, membership and functional linkages, the power of creative destruction within decentralized dynamic network-based systems generate a "net gain" from which every participant benefits [24]. Communication and networking becomes the key element of success here. The multiplicative effect of the network suddenly makes the contribution of an individual computer negligible. The history repeats itself: Had it not been for things other than just the ability to carry out numerical computations, the computer would have remained a scarce item in the post WW II world. Within the information-based world of today, it is the ability to network and to communicate information rather than just to process it, which makes computers the agents of the revolution we participate in.

The change in the working environment generates profound changes in other aspects of social life as well. They change the rôle of the individual within the information-age institutions, the rapid pace of change and makes it necessary for the individuals to adapt several times within their lifetime to change in their work and lifestyle, and to incorporate the element of continuous learning into the productive life. Informatics is emergeing as a new methodology of science and technology, accompanying the theoretical and experimental ones [22]. The way of how democratic societies function will surely transform. A democratization of information access and generation, enhancement of the individual's participation, and economic feasibility to effectively care about minority interests, establish themselves as the working modes of the society structure, eventually breaking the neck of the hierarchical bureaucracy as it has been the case within successful companies.

## 3 The Information Economy

Information becomes both an essential product and indispensable commodity for functioning of the modern society. There is a trade-off between the mass and the information in efficient functioning of companies in the information age which is similar to the trade-offs between space and time in the complexity theory. Communication links and virtual market communities allow to eliminate intermediaries, and to adjust the mismatch between production and consumption in terms of quantity, quality, and the time of delivery and/or production. The communication channels become a substitute for storage houses and store shelves which would otherwise be necessary. The instant knowledge of the volume of sales performed, the accessibility of information about the amounts of goods still available on the department store shelves (with built-in intelligence comparable by amount and size to that needed in an ordinary smart card), the possibility of combining the information gathered about the customer (leaving aside the privacy issues for the moment) with targeted marketing—each of these contribute to eliminating some portion of the production and/or distribution facilities, to lowering the volume of mass production of "ready-made" goods, and to substitute these by just-in-time production of products which are mass-customized to the needs of the individual customer.

Plentiful examples of this can be found almost everywhere. Motor vehicle factories can produce passenger cars in such a way that a desired setting of some 100 features can be chosen by the customer as in independent options (or in most cases independent). The potential number of variants, exponential in number of features, thus exceeds the real production possibilities, making the traditional production line very unpractical. Efficient handling of information concerning customers requirements, the knowledge immersed into the workflow organization, and a radical reorganization of the causal flow of actions makes it possible to produce a car as if custom-made for a particular customer, yet with the efficiency not lower than that of blind mass-production of the first half of the 20th century. The effective use of knowledge within the production process results internally in abolishing the beat of the production line as a central clock of every worker's work, and substituting it by automated work of robots fed on-line with the necessary data, and the workers performing individualized special tasks when needed. The resulting process effectively implements late-binding within the factory work organization, using a lazy-evaluation computational model as contrasted with the eager-evaluation strategy of the Taylorian production line model performing brute-force search, or product generation for storage from which the customer chooses later.

The "Levi's Personal Pair Jeans" has been a service established by a tiny network-based company called Custom Clothing Technology for Levi Strauss in order to link the retailers, cutters, stitchers, and washers in order to produce a pair of jeans made exactly to the measures taken from the customer and inserted into a networked computer. Within a matter of days the jeans made to perfect fit are delivered by Federal Express. The network becomes a device for mass customization and late-binding lazy-mode style production. No production for storage is needed but this is still not all. In the traditional mass-production style, 75% of cost are in distribution, not manufacture, and more than 70% of women who purchase jeans are unhappy with the fit [55]. These factors are almost completely eliminated as well as the seasonal sales of unsold stock.

Journals and other publications can be mass-customized for a particular subscriber or a special group of them. Major popular journals are published in mutations specialized for some geographical area or even a city region, such as the Time magazine differentiating in some details of itself even within various parts of Los Angeles, or the Farm Journal mixing individualized issues according to the reading profiles and individual interests in agriculture production or advertising for the customer to which it will be delivered [10]. McGraw-Hill offers its academic customers custom-made textbooks printed according to their individual requirements. Electronic web sites which major newspapers or journals run as an extra bonus, can not only feature up-to-date news flashes (the small guys need not necessarily be the slowest – the Czech electronic daily newspaper *Neviditelný pes* for example managed to report the take-off of U.S. bombers from Guam for the first strike in the Gulf war several hours before the news appeared on CNN, thanks to a virtual editorial board which spans the globe[2]), but also the latest news selection prepared so that it fits the pre-defined user's profile.

Federal Express, a package delivery company with 50% market share in the U.S., has been expanding its customer base in a major part because of the use of a sophisticated real-time electronic tracking which provides on-line a detailed information about the position of every single parcel, and even sharing this information with the customer. The company plans to use smart labels equipped with a single-use chip for active package identification, making the "information on the move" a part of the shipping business.

Clearly, the material aspect is no longer what makes a product interesting, but rather the information or knowledge contents in it. In fact, knowledge may be the only meaningful resource overshadowing the traditional production factors such as capital, labor, land, or raw materials [13]. The emphasis on the generation of material or financial wealth has been replaced by the emphasis on processing the knowledge. The number of various bit patterns which can be stored on a floppy is bigger than the number of elementary particles in the observable Universe. Each of them can turn out to be a product—maybe worthless, maybe a future killer application. The cost of reproduction of the digital product is close to nil, with no quality loss. What matters is the "inspiration", not some repetitive of reproducible labor ("perspiration" remains reserved for robots). Successful companies today do not depend on huge capital investments or production plants. It is rather the "knowledge worker" of theirs who matters most[3].

## 4  Building blocks of the new organizational structures

The organizational structure assumes new forms which emerge with dynamic and innovative organizations, either new ones or those ones who transform the way they function. These elements of the new structure are by no means a panacea. More than that, they represent a trend which can be observed more and more frequently, and which symbolizes the new structures enabled by new technologies and new ways of doing business. Some of them are to be listed in this section.

---

[2] The politicians are, however, sometimes very reluctant to accomodate to these changes. The advisor to the otherwise very market-favoring PM would publicly label the web daily newspaper as "unofficial" media, when explaining why the government claimed having had no knowledge of news which were already given enough publicity in it—incidentally just a few days after the number of readership hits of this newspaper exceeded 1 million.

[3] There are also negative aspects of this. Consider problems of financial guarantees of service-type companies towards their customers, and the amount of recoverable capital available in case of bancruptcy.

## Disintermediation

Disintermediation is a result of direct relationship between consumers and producers where intermediate steps or processes make it more difficult to use efficiently the information feedback which occurs when a close loop is engaged. This is an important element when production on demand is used. This trend is predominantly represented by companies selling CDs, concert tickets, or books. In these applications the network provides an environment in which the consumer visits a virtual store with functions that are similar to the activities one could be interested in when physically visiting a store (e.g., book browsing and book reading capabilities in the National Academy Press bookstore at http://www.nap.edu/. Some additional enhanced features such as new publication notices, and readers contents evaluations such as those available at Amazon.com, are usually added to cuch a functionality.

Areas will remain where the consumer avoids the intermediary because no added value, no added knowledge, or information enrichment occurs. This applies to direct air ticket sales, direct booking of hotels, or direct arrangements of vacations by virtually visiting the places of interest and arranging what needs to be done. Travel companies will soon cease to be just selling points for travel tickets, and will transform themselves into businesses which depend on selling knowledge-dependent products ("meta-knowledge") which offer services, expertise, and structure, which would otherwise be very difficult to acquire.

Newspapers and news journals won't disappear either. They may not be needed any longer as a privileged source of information, because that is more easily available on the Net, but they will actually sell their work on selecting relevant pieces of news and arranging them conveniently to the reader. The meta-knowledge and meta-information structure can easily be worth more than the object of it (e.g., the *TV Guide* which publishes programs of American TV channels, has been having greater market value than either of the TV networks it reports on [10]).

## Convergence

Convergence of key economic sectors as well as convergence of different technologies comes as a natural consequence of the fact that the product and/or mode of operation becomes digital. Services and products can merge one into another, creating new challenges and opportunities. Many of the professions of tomorrow, the products, and future industries do not exist today. Computing, communications, and contents traditionally used to be distinct areas of expertise, manipulation, and company activities. The technological base of each of them used to be clearly differentiable. There are plenty of examples of a trend to the reverse now.

Telephone, television set, and computer may shortly evolve into a single home appliance. There are products, either hardware or software, which already combine some of these functions.

Web sites of major newspapers add more and more interactive features and perpetual news of the day updating, so that they resemble TV or radio (when the network bandwidth allows this, it will be difficult to differentiate between those and customized personal TV channels). TV stations run informative web sites which can allow the visitors to play voice messages of other visitors, or leave one's own comments for others (examples of this may be ABCNEWS.com chat column or CNN.com voice message corner for specific key stories). Electronic newspapers run discussion forums for their readers allowing to discuss issues related to newspaper contents or any topic in general—besides of the function of an electronic newspaper this also allows to establish a virtual community of interested readers.

## Virtualization

Virtualization creates digital objects which substitute a real entity, causing the environment and the users to act as if the objects existed, with real effects and consequences. We can therefore introduce notions of, e.g., "virtual" institutions, staff, products, tools, services, markets, or appliances.

Virtual communities are formed by allowing sufficiently rich communication among the members of the community across the network. Physical proximity or temporal synchronization of communication by individual members is no longer an obstacle in forming an effectively working community. Groups with minority interests can more easily be formed as virtual ones, because most of the economic limitations of members' mutual interaction can be reduced next to nothing. The dynamics of virtual community formation depends on the interplay of attracting critical mass of members, user profiles and aggregation of vendors and advertisers giving rise to the amount of transactions and transaction categories [24]. In successful cases the combined impact amplifies the dynamics of increasing returns as a resonance-based effect within an otherwise chaotic system. The Amazon.com virtual bookstore has raised sales from $ 0.5 million in 1995 to $ 15.7 million in 1996, a 25% monthly grow, creating a virtual community base of 180,000 customers from 100 countries worldwide. It is still in its infancy, though, with net balance in red numbers – a sign of being still within the aggregating phase of critical mass accumulation. (The total volume of electronic commerce has been around $ 600 million in 1996, expected to grow to $ 66 billion by 2000, according to Forrester Research estimates.)

Virtual markets, virtual stores, virtual advertisement, and virtual jobs (100% telecommuting or teleworking) correspond to activities performed over the net, without any physical presence which would otherwise be necessary. Virtual immigrants can perform work in overseas companies without leaving their home country (such as Indian programmers for U.S. software companies), evading immigration laws, work permits, and often also paying taxes.

Virtual universities can bring together the best professors (no matter where their physical home and/or place of work is) and the brightest students in order to network top-quality resources. Virtual students may also be interested in continuing education or enhancement/refreshment/updating their current proficiency level. (Example of the latter is the Michigan Virtual Automotive College at http://www.mvac.org/.)

Virtual communities also present a sociological phenomenon, not just a technological one. If the members of the community are to co-operate effectively, it is necessary to build certain level of trust among its memebrs, and it is not enough to merely verify the communication channels or the other party's identity (compare [15] on the importance of the level of trust for the overall structure and functioning of economic structures in various countries). From all possible obstacles, this may be the single one most difficult to overcome in reality. In the digital world, it is becoming easier to falsify then to verify the truth[4]. Other interesting social contradictions related to the concept of virtual communities are raised in [45].

## Molecularization and decentralization

Molecularization and decentralization represents a trend towards establishing a network-based economic and social interactions between small units in contrast to large industrial-age corporations. Flexibility of information flow patterns enabled by information technology use have made huge bureaucracies which represented an organizational skeleton between the top management and the line worker disappear. The structure whose principal function was to organize information flow and information filtering has lost its function and has become an obstacle. The decision power has moved down, creating a flat structure instead of a hierarchy. More independence in decision making allows to commission more work on a contractual basis. This may lay a foundation to an "invisible economy" based in a larger extent on the prosumers working within the "electronic cottage" environment as very much autonomous economic agents [57][5].

The flexibility of digital information flow arrangement allows a liquid-crystal-like enterprise with elements of it able to act and reconfigure very flexibly, yet with a degree of structure where needed [55]. Similarly to that, "organizations without boundaries" [35] make flexible arrangement of the information flows which disregards the difference between the inside and

---

[4] See Web link http://www.chicago.tribune.com/news/current/schmich0601.htm for a recent amusing example of this where a Chicago Tribune column by Mary Schmich started to circulate the Internet as a commencement speech of Kurt Vonnegut at MIT. We may soon watch digitally fabricated news shots on TV news in the style of Forrest Gump. There are also tendencies going counter this: The news shots from the Boer War were actually made on Jersey island but the Gulf War was already converted by CNN into an on-line show which went for real.

[5] Although this trend is noticeable, the general tendency within the U.S. economy shows mixed patterns: According to 1993 U.S. Bureau of Census data, some 90% of the companies have fewer than 20 employees, but the total number of employees within them amounts to just 20% of the total employment and 17% of the net salaries payed. Companies with more than 500 employees represent 47% of total employed workforce, with net salaries 53% of the total.

outside of an organization and allows the molecules diffuse through its outer membrane. This is already happening with the use of extranets. (Forrester Research has found in a recent study that extranets represent service which is demanded most from the Fortune 1000 companies.) Interaction patterns based on the object-oriented approach are another manifestation of this.

## Outsourcing and subcontracting

Outsourcing and subcontracting together with the use of network relationships (internetworking), each of which are very closely related to molecularization, emerge as a new working paradigm. New products can be based on combining existing elements and fitting them together—a task much easier done within the digital world, in contrast to mechanical parts of industrial products. Within the educational sector the idea of virtual universities which provide "knowledge pointers" to sites which specialize in particular topics can provide the motivation for such an approach. A coherent outline of the use of the contracting paradigm applied to public education can be found in [26]. (For more cautious treatment of outsourcing in connection with development of software systems, see [54].)

## Globalization

Globalization comes out as a natural consequence of the way in which the information and communication technologies make the difference in time and space to disappear (or mostly disappear). The Age of Information coincides with the Era of Globalization not just by coincidence—it is indeed the digital paradigm which makes both of them work.

Telecommunications based on telephone or telegraph allowed to overcome distance in space in point-to-point communication between people, and changed the psychology of communication and human interaction in this respect. Some of the consequences have been rather surprising and seemingly without causal relationship—e.g., it was the telephone which made building of skyscrapers possible and usable. Digital packet switching networks as invented by Paul Baran [2] and brought into being as today's Internet ([23] or [48] provide interesting historical accounts of early Internet development) are also not just connecting the computational power of computers, but changing the behavior of the society, as can be illustrated on the disputes over free speech, privacy issues, or individual responsibility.

The ability to deliver information in various formats and amounts across a uniform network environment without distortion or quality degradation (in contrast to analog communications), allows the communication environment to stretch the understanding of "same place" and "same time" and to multiply human ability to interact. E-mail already changed the topology of time in multiple dialogues which the user is continually engaged in, and noone would assume any longer that, e.g., both parties in a discussion must be both awake at the same time, or that engagement in a discussion within one circle prevents taking part simultaneously in another one, on a different topic. The structure of the economic and social interactions in network-based virtual communities are similarly no longer bound by physical proximity. The notions of "niche economy" or "economy of scope" are based on these liberating trends. They have become essential ingrediences of success of the fast-growing knowledge companies.

## 5  Elements of transition

The general trends are better noticeable with new companies and enterprises but the existing one are undergoing a process of adaptation as well. Although there is no universally accepted explanation of the productivity paradox (based on U.S. data, productivity growth in non-farm business has slowed down from 3% p.a. to 1.1% p.a. during the 70's and 80's, with noticeable complete stagnation during 80-82), there seems to be prevailing consensus that a key ingredient of the explanation has to do with the inadequacy of how information technology tools are used in obsolete organizational contexts.

It is not surprising that the use of computers in manufacturing has generated huge productivity advances. When human workers are being replaced by machines or robots, this is, after all, the simplest situation in which computers can boost productivity as measured in terms of the volume of production.

It is not so straightforward, however, to apply computers in administration and services, and in the context of "knowledge work" in general. These applications deal with information and task flows within the company structure, yet their potential also allows changing the very same structure as well. If new technology is to be introduced with efficiency it promises to deliver, it is necessary to reorganize the information structure, to identify the core elements of the organization, to establish a clear vision of the future of the organization, and to try to identify new opportunities within the organization rather than to try solve emerging crises.

Existing strategies have focused on various priorities within the transformational task. The general drive towards smaller companies is reflected in *downsizing* in various disguises[6], ideally making the organization more efficient and more responsive to changes of its working environment. Making an organization work better under more potent technological conditions is the principal aim of *reengineeing* which aims at improving old processes (which mostly evolved due to history of events, not due to functional needs). *Top quality management* (TQM) is another approach focused on seeking perfectness in key areas related to quality of the function of the company.

---

[6] The Orwellian terminology used with very much the same meaning include: compressing, consolidating, demassing, dismantling, decomissioning, downshifting, rationalizing, reallocating, reassigning, rebalancing, redesigning, resizing, retrenching, redeploying, rightsizing, secondment, streamlining, or slimming down. (Most of these terms are taken from [14].)

There is really no theory behind any of these strategies, and there are few, if any, credible studies of the real effect of either of them [14], [30]. Reengineering seems to be the most relevant of these as far as information technologies are concerned. One of the classical cases of its use was a project at IBM where the number of steps in the process from customer orders to sales was reduced from seven-day six-step process to two-step process taking just four hours. Chase Manhattan bank has been reported recently to have automated e-mail processing by having a program which processes 94-95% of the e-mails sent automatically, and the rest is dispatched for manual processing. This allowed to cut the time needed to process one e-mail message from 8-15 minutes when done manually, to under 1 second after this change. (The investment costs should return within 6 months.) The transformation strategies may not be everything what counts in this respect, and also other dimensions which take into account functioning of the user-machine interfaces should be given more attention. A case is made in [34] for such a "user-centered" design as a software-related part of the solution of productivity paradox problems.

Besides of the inadequate organizational structure, the reason for productivity stagnation in jobs equipped by IT may also come from the fact that the investment into information technologies is actually very slow to return, and majority of companies are still under-equipped. One possible explanation of the current unusually long period of low inflation, low unemployment, and prolonged economic growth in the U.S. (productivity increase of 3.4% in 1995 and 3.8% in 1996 is unprecedented compared to 70's and 80's) is based on the idea that this is a result of 20 year investments into IT slowly beginning to generate returns. It is however too early to be able to say anything conclusive on this (see, e.g., [53] for very cautious arguments in this respect).

The pattern of investment into new technologies, the effect of learning curves, and the pace of technological adaptation can also be studied based on the experience of the industrial revolution as it is done in [19], even though the pace of change may be considerably accelerated. New technologies need substantial amount of capital investment which is essential for continuing growth. The returns on this investment take about two decades to start appearing. Dramatic differences are emerging between the opportunities available to qualified and unqualified workforce, high learning cost drives down productivity within timespan of about four decades, and the income differences between qualified and unqualified members of the society keep rising. The social consequences of these processes may be as dramatic as those which shake up company bureaucracies. Even though in the long run everybody will be better off, to arrive at this point will be as painful as it was the case before everybody gained from the industrial revolution: "Unskilled wages fall during initial stages of the Information Age. Twenty years elapse before this loss in unskilled wages is made up and about 50 go by before they cross their old path. Interestingly, during the early stages of the Information Age the stock market booms as it capitalizes the higher rates of return offered by the new investment opportunities. For many in the economy, though, waiting for the benefits of the technological miracles will be like watching grass grow; but grow it will. ... Thus, in the short run young, unskilled agents fare worst. In the long run the rising tide of technological change will lift everybody's boat." ([20]) The initial part of this scenario is very much consistent with the data on the current trends related to the effects of de-industrialization. Recent IMF study shows that the reduction of the share of civilian employment is declining worldwide (it started in late 60's in the industrial countries with Japan being the slowest to join this trend as late as in 73, and it started also in more advanced East Asian countries in late 80's) but this decline is accompanied by increased numbers of well-qualified jobs (information technologies, education, financial and legal services) [47].

When these effects are combined with the negative consequences of the majority rent-seeking efforts in the welfare states at the end of 20th century [4], the demographic trend in the developed countries, and the incompatibility of trade union bargaining mentality, with basic mechanisms of the information economy functioning [47], it is clear that the worst stirs, transformations, and shake-ups which the society faces will be much more dramatic than those faced by company bureaucracies. The increased emphasis on education, especially the continuing life-long education, may help to narrow the gap between those who get and who loose in the short time. Education will have to be run as an information-age business as well [63], and to create a virtual classroom [56] which everybody will be able to attend. There won't be just "learning organizations" but the society should become a "learning society" [17] as well. The multiplying effect, driven by the law of increasing returns, will form around virtual universities, and the synergy between academic research and the students transferring these state-of-the-art technologies into practical life may further catalyze future development and economic well-being. The conclusions of two major recent higher education system studies in the U. S. [5] and U. K. [38] support this claim. Information technologies also make it easier to accommodate for diverse educational opportunities compatible with the value of unique combination of knowledge and skills as opposed to uniform mass education of the industrial age. The acceleration of the perpetual change with the society and the economy makes it common for quite a few to change their jobs or carriers several times during their productive life. Requalification becomes easier if the educational opportunities at earlier stages really provide generic education, not just professional training tailored to a specific job (note that even the information-society motivation need not mean that "computers to every classroom" is really a good idea, given how incomplete our knowledge of the efficiency and effects of extensive computer actually is [42]). This may increase the value of foundational disciplines within the educational system.

# 6  Social consequences

The organizational elements of computing and communication systems provide a metaphor for organizational structures in the knowledge-based society and the intelligent enterprises (see also [50] on this), and Paul Baran's [2] ideas about decentralized, self-reconfigurable networks of autonomous agents come close to a blueprint for liberal democracy as opposed to heavily centralized structures of telecommunication monopolies of his time. Even though this does not mean that the society will become computer-controlled or that the enterprises will be run by robots, the parallels may serve as a mind opener which unleashes our creativity to find new solutions that match the future possibilities better, the principles of which have only been enabled by the new technological basis of the society.

There the trade-off between the value of hardware and software closely parallels the trade-off between manufacturing and agriculture and the knowledge-based sectors of modern society. The digital (or information) paradigm is the basis of these trends of both of these cases, and it is as possible for a society to be based on the intangible activities (which gain relevance once the basic material needs of humans are sufficiently covered by the existing technologies), as it is the case with computer systems. The cost of difference disappears, and the difference is actually becoming more a virtue than an obstacle (socially the uniform workforce is no longer needed in order to suit the basic material needs, and similarly the value of software components depends on the ideas, not some uniform components). The open society integrates for differences and facilitates to function with less rigidity of thought, less constraints for development and more interconnected world. The rationales for open software systems provide essentially the same motivations (the more parties can join in, the more benefit for everybody) as the liberal economy theories. Mobility, both physical and social, has contributed to these social tendencies, as well as networking has motivated some of the key efforts within the systems area. Ubiquitous computing and zero-cost high-bandwidth networking will contribute to either of these, as more and more social economic functions will be based on the virtual basis.

The implications for the political structures will have to surface up as well, even though the lessons of the twentieth century give strong reasons for caution. The hierarchical structure of representative democracy is based on obsolete communication channels, and some of its properties are clearly not able to respond in reasonable time to changes around, as well as enterprises with obsolete bureaucracies are unable to compete with flexible information-based companies running more efficient matrix, team-organized or ad-hocratic forms of organization. The shared/conflicting responsibility modes of function will find its way into social and political structures in order to synchronize better with the pace of the world ([55], [58], [12], or [46]). The naïve proposals of "electronic town hall" in Ross Perrot direct democracy style are unlikely to function, and some sophisticated way of representation, organization of intermediation, and reorganization of "virtual" voting districts in a way immune to gerrymandering will have to be designed [3]. Networks may help to overcome boundaries between groups which would otherwise have communication difficulties. Another aspect worth mentioning is the urge for politics to remove interference with technology and human creativity as much as possible—current regulatory frameworks often clash with inventing new things (e.g., digital convergence does not fit regulatory frameworks of separate licensing of telecommunication services).

## 7 The Internet and beyond

The success of the Internet is based on the success of a decentralized and distributed effort of many agents loosely coordinated by a loosely drafted set of open quasi-standards. A few years ago the "serious" business world would not pay much attention to a bunch of idealistic academics. Only at about the same time when NFSNET has stopped to function as an Internet backbone and Internet was "privatized" in a way (1994), the business world has begun to take it for real[7]. According to the host count performed by Network Wizards, www.nw.com, in July 1997 there have been 19,540,000 Internet hosts worldwide in 1,301,000 domains and 171 countries, making Internet *the* worldwide network (just a year ago, Internet was trailing UUCP which was then still 17 countries ahead). When combined with population figures (see Table 1 for OECD countries and July 97 host count), the diffusion of the Internet is highest in the Scandinavian countries, U.S., New Zealand, and Australia where the penetration figures are within 65.77 hosts per 1000 inhabitants for Finland, to 26.21 for Denmark (the U.S. ranks third with 44.97 hosts per 1000 inhabitants). Although the differences in penetration levels look quite huge for some countries (e.g., 14.98 for the U.K, 5.02 for France, or 4.75 for the Czech Republic), the absolute time difference is just a few years because of the exponential growth (Finland had penetration level of today's U.K. some 2.4 years ago, of France 4.2 years, and of the Czech Republic 4.3 years ago; the relative difference of the last two being just about 8 weeks)[8]. The WWW server count is estimated to 1,269,800 according to the statistics in [62] (see Figures 1 and 2 for the statistics of Internet host and WWW server counts from this reference), having grown from zero since 1993 introduction of http protocol by Tim Breners-Lee at CERN.

Some of the most ambitious undertakings are under way right now, whose realization should mean a considerable step forward along the vector of future trends. The Net will be the place where these effects are going to be visible most. Digitization of telecommunications and deregulatory efforts should put us closer to omnipresent bandwidth for negligible price. Several

---

[7] By that time, NFSNET Internet backbone traffic exceeded 10 Tbytes/month. Bill Gates would not think Internet deserved much space in his [16] until second edition appeared within less than a year. U.S. telecoms would similarly wait until 1996 to ask U.S. Congress to ban Internet telephony, about one year after it had become available (unsuccessfully, so in mid-1997, several major telecoms already announced they would offer Internet telephony as part of their services, in the *if you can't beat them, join them* style).

[8] Or taken the other way around, these figures show how fast the clock of change is running.

major large-scale Gigabps networks are currently being built, which present testbeds for future global-scale attempts. One of those is the Internet 2 consortium of the U.S. universities building a 2.5 Gbps network which should provide a successor of the Internet with qualitatively new functionality (see http://www.internet2.edu), and also provide new technology for the National Information Infrastructure [8]. Another unlikely competitor to these ambitious goals is the Multimedia Super Corridor project of the Malaysian government aiming at creating a 2.5 Gbps-based infrastructure in a 40km per 15km area around Kuala Lumpur, providing advanced infrastructure for for both the commercial institutions and by then a fully paper-less government before 2020 ([28], or see http://www.mdc.com.my/msc/).

The growth of the Net is definitely the most important technical and infrastructural discontinuity conceived by the information revolution. The exponential growth self-enforces its base along the law of increasing returns, and within the last few years we are witnessing the emergence of a (virtual) global community. Any society depends on interaction among its members, and hence it should not come as a surprise that it is not merely an "information economy" but to more extent even a "network economy" [33] which reinforces the foundations of the Information Society.

On a micro-level of social interactions, the Net and the advanced telecommunications bring new codes of behavior, aspiring to become a norm within the society segments larger than the initial set of network pioneers. Besides of benefits, unexpected friction appears such as the phenomenon of "flaming" (documented quite early after the e-mail communication started to be used [52], [29]), or etiquette problems with beeping cellular phones at theater performances or during religious services. MUDs, MOOs and other rôle-playing communication-based tools can be useful for unleashing the imagination and stimulating a creative environment free of real-world psychological obstacles [55]. Conceptualization of some of the consequences of electronic communications, teleworking tools, and the loss of emotional clues in inter-personal communications may, however, be very difficult to explain, as demonstrated, e.g., in [36].

## 8 Some of the problems ahead

The development of information and communication technology and the computer science advances in the architecture of software and network systems make it possible to fulfill the promise of Global Information Society. The remaining technological and scientific challenges cannot be underestimated, but the main problems ahead of us are related to the societal concerns and implications which are network-related.

The information revolution brings about another shake-up which is going to change the balance of power worldwide, and within the society, too. The speed with which it is happening tends to amplify differences, and the danger of creating an under-privileged class of information "have-nots" may be real.

Deregulation of telecommunications becomes a reality—digital technology has made the concept of natural monopoly obsolete, advances in convergence of communication technologies are impossible within the majority of the current regulatory frameworks, and last but not least, increased competition is driving communication costs down[9] (cf. [9]). The notion of "universal service", a level of telecommunication service for which efforts should be made so that it is accessible to every member of the society, is being scrutinized, and serious proposals already appeared advocating inclusion of e-mail into delivery mechanisms, working as universally as the delivery of paper mail [1].

The attempts to regulate Internet as if it were a broadcast medium threaten to undercut the innovative potential which has so far worked so well. Recent defeat of Communications Decency Act (part of the U.S. Telecommunications Act of 1996) by unanimous vote of the U.S. Supreme Court followed by the publication of the plan of the Clinton administration to establish a free zone for electronic commerce and to refrain from imposing additional regulation or taxes [61] (but compare [6], also reflected in [27], for arguments favoring introduction of a "bit tax" as an analogy to VAT, once sending bits is generally acknowledged as a common economic activity). Several governments have introduced Internet censorship laws ranging in their effect from establishing government-controlled access points, designating Internet a broadcast medium or controlling cryptography, to including censorship as legal measures of a more general scope. Legal problems which need to be solved in connection with the Internet-type means of communication include the problem of classification of this type of media, solving territoriality problems in some way, and to overcome national legislation attempts to tackle a problem which clearly transgresses national boundaries and national sovereignty. Until some agreement similar to the regulation of high seas or cosmic space have been reached, no legal framework will be completely adequate. Mankind will have to find some way of living in a world in which certain pieces of information are already public property, even though it would have been better it they never had existed (such as bomb construction, etc.)

Electronic commerce faces advances in availability of cryptography and identity verification systems in order to really start off the ground. It is likely that even the current Internet is still the safest place where to use your credit card (it is still much more difficult to monitor your electronic transactions and to extract card numbers, etc., than to break into your house and get it physically), but wider use of cryptography methods can make it even safer – e.g., the vendor who accepts a card payment need not be able to decipher even the card number which may be passed through him to the bank still in encrypted form. Digital money is another concept which is more natural than

---

[9] The general expectation is a fall in long-distance communication cost by 80% within 5 years after deregulation. The estimates of total annual benefits of the U.S. Telecommunications Act of 1996 are likely to exceed $50 billion, compared to $16-23 billion consumer and businesses annual benefits coming from Airline Deregulation Act of 1978, according to U.S. Commerce Committee staff estimates and [60].

traditional paper banknotes once computer becomes ubiquitous. After all, virtual banking will hardly be able to handle paper money. The dark side of the ubiquitous computer and the benefits of globalization will be the impossibility to shield any part of the world away from transfers of huge sums within fractions of second around the globe, potentially causing instabilities within the soft-economy-based world. New models of financing will have to appear which would be applicable to paying for information services over the nets such as per-use micropayments.

Within the interconnected world, the threads to individual privacy protection are paramount, and protection mechanisms should be built into key systems at very elementary level. In fact the privacy-protection principles may be a candidate for a set of laws which would be built into every single chip, and very likely with much more urgency than Asimov's laws of robotics. It may come as a paradox, but not only privacy-protection mechanisms are needed, but on the other hand, also reliable identity verification mechanisms are needed once more and more transactions and interactions will move into Cyberspace.

## 9 Conclusion

The Information Society concept is hardly a well-defined technical artifact, or an engineering blueprint for a technology-based future. The causes and effects are vaguely defined, and sometimes they fuse and melt together, as the "virtuous cycles" [51] of positive feedback based on the law of increasing returns keep generating exponential growth. No matter how loosely defined they may have been by the time of their dawn, the concepts of agricultural and industrial societies and the revolutionary changes which triggered them off are clearly visible as a sum of the technological and social changes happening on a large scale and generating irreversible change.

A clearer picture can only be seen from a distance. The metrics of our understanding requires to step outside of a set of bits and pieces, in order to be able to generate concepts which move us closer to each member of such a set. Some of the analogies we use when contemplating the changes which the digital and network revolution generates today, are most likely flawed to some extent, and some details of the large picture will perhaps have to be modified or adjusted. The experience of changes happening over past decades, however, do not leave much room for doubts about the picture at large.

The Information Society is also a metaphor for describing the synergy between technological and social changes. The mood associated with it at the end of the millennium is that of anxiety of change, infinite possibilities and new frontiers of the digital world, quite unlike the feeling that classical mechanics had been providing all the answers within the world defined by its physical/material side. As it is with every metaphor, neither this one can be taken verbatim. Sure, no matter how small chips we make, the theoretical limits based on mathematics won't be lifted away, and neither will be the effective barriers coming from the limitations which computability theory or complexity theory can prove. The technological metaphor does nonetheless play the rôle of an enabler, magnifier, multiplier, generator, or mind opener. It allows human creativity to assume new grounds and to move closer (perhaps asymptotically) to fulfilling the vision which is generated by it.

Given the speedup of technological advances we are witnessing today, the Information Society cannot possibly be *very* far away. In a way, it is sufficient to flow with the tidal waves of changes—and (one should not forget this) keep generating the very same waves by ourselves.

## References

1. Robert H. Anderson, Tora K. Bikson, Sally Ann Law, and Bridger M. Mitchell: Universal Access to E-Mail: Feasibility and Societal Implications. Center for Information evolution Analyses, RAND, Santa Monica, CA, 1995.
2. Paul Baran: On Distributed Communications, RAND Corporation, Santa Monica, CA, 1968.
3. John Seely Brown, Paul Duguid, and Susan Haviland: Towards Informed Participants: Six Scenarios in Search of Democracy in the Electronic Age, in: David Bollier, rapporteur: The Promise and Perils of Emerging Information Technologies, The Aspen Institute, Washington, D.C., 1993.
4. James M. Buchanan: Politics by Principle, Not Interest. Cambridge University Press, Cambridge, MA, 1997. (forthcoming)
5. Commission on National Investment on Higher Education: Breaking the Social Contract. The Fiscal Crisis in Higher Education. Council for Aid to Education, RAND Corporation, Santa Monica, CA, June 1997.
6. Arthur J. Cordell: Taxing the Internet: The Proposal for a Bit Tax, Journal of Internet Banking and Commerce, Vol 2, No. 2, March 1997, http://www.arraydev.com//commerce/JIBC/articles.htm.
7. Computer Science and Telecommunications Board: Information Technology in the Service Society. A Twenty-First Century Lever, National Academy Press, Washington, 1994.
8. Computer Science and Telecommunications Board: Realizing the Information Future. The Internet and Beyond. National Academy Press, Washington, 1994.
9. Computer Science and Telecommunications Board: The Changing Nature of Telecommunications/Information Infrastructure. National Academy Press, Washington, 1995.
10. Stan Davis and Bill Davidson: Vision 2020. Simon&Schuster, New York, 1991.
11. Stan Davis: Future Perfect. Tenth Anniversary Edition. Addison-Wesley, Reading, MA, 1996.
12. Michael Dertouzos: What Will Be, HarperEdge, New York, 1997.
13. Peter F. Drucker: Post Capitalist Society. HarperCollins, New York, 1993.
14. Daniel Druckman, Jerome E. Singer, and Harold Van Cott, eds.: Enhancing Organizational Performance, National Academy Press, Washington, 1997.
15. Francis Fukuyama: Trust. The Social Virtues and the Creation of Prosperity. The Free Press, New York, 1995.
16. Bill Gates, with Nathan Myhrvold and Peter Rinearson: The Road Ahead. Viking, New York, 1995.
17. Amy Korzick Garmer and Charles M. Firestone: Creating a Learning Society. Initiatives for Education and Technology. The Aspen Institute, Washington, D.C., 1996.

18. George Gilder: Telecosm. Simon&Schuster, to be published, chapters have been appearing in Forbes ASAP since December 7, 1992.
19. Jeremy Greenwood and Mehmed Yorukoglu: 1974. Working Paper No. 429, Rochester Center for Economic Research, University of Rochester, Rochester, NY, September 1996.
20. Jeremy Greenwood: The Third Industrial Revolution. Working Paper No. 435, Rochester Center for Economic Research, University of Rochester, Rochester, NY, October 1996.
21. Jozef Gruska, Ivan M. Havel, Juraj Wiedermann, and Jaroslav Zelený: Počítačová revolúcia ("Computer revolution", in Slovak and Czech), Proc. Sofsem'83, Ždiar, 1983, pp. 7–64.
22. Jozef Gruska and Roland Vollmar: Towards Adjusting Informatics Education to Information Era. FIMU-RS-97-03 technical report, Faculty of Informatics, Masaryk University, Brno, 1997, 33pp.
23. Katie Hafner and Matthew Lyon: Where Wizards Stay Up Late. The Origins of the Internet. Simon&Schuster, New York, N.Y., 1996.
24. John Hagel III and Arthur G. Armstrong: Net Gain. Expanding markets through virtual communities. Harvard Business School Press, Boston, MA, 1997.
25. Douglas E. Harris, ed.: Organizational Linkages. Understanding the Productivity Paradox, National Academy Press, Washington, 1994.
26. Paul T. Hill, Lawrence T. Pierce, and James W. Guthrie: Reinventing Public Education. How Contracting Can Transform America's Schools. A RAND Research Study. The University of Chicago Press, Chicago, 1997.
27. High-Level Expert Group: Building the European Information Society for us all. Final policy report. European Commission, April 1997.
28. Derrik Khoo: The Multimedia Super Corridor: An Island of Excellence. Business Times Annual 1997, Business Times Malaysia, 1997, pp. 64–67.
29. Sara Kiesler, Jane Siegel, and Timothy W. McGuire: Social Psychological Aspects of Computer-Mediated Communication, American Psychologist, 39 (October), 1984, pp. 1123–1134.
30. John Leslie King: Where Are the Payoffs from Computerization? Technology, Learning, and Organizational Change, in: [31], pp. 239–260.
31. Rob Kling, ed.: Computerization and Controversy. Value Conflicts and Social Choices. Academic Press, San Diego, CA, 2nd ed., 1996.
32. Rob Kling and Jonathan P. Allen: Can Computer Science Solve Organizational problems? The Case for Organizational Informatics, in: [31], pp. 261–276.
33. Kevin Kelly: New Rules for the New Economy, Wired, 5.09, September 1997, pp. 140–197.
34. Thomas K. Landauer: The Trouble With Computers. Usefulness, Usability, and Productivity. MIT Press, Cambridge, MA, 1996.
35. Don Mankin, Suzan G. Cohen, Tora K. Bikson: Teams and Technology. Fulfilling the Promise of New Organization, Harvard Business School Press, Boston, MA, 1996.
36. M. Lynne Markus: Finding a Happy Medium: Explaining the Negative Effects of Electronic Communication on Social Life at Work, ACM Transactions on Information Systems, 12(2), April 1994, pp. 119-149.
37. Nathan Associates: Building an Information Economy. Business Software Alliance, June 1997.
38. National Comittee of Inquiry into Higher Education: Higher Education in the Learning Society: Report of the National Comittee. Norwich, 1997.
39. Nicolas Negroponte: Being Digital, Alfred A. Knopf, New York, 1995.

40. Ikujiro Nonaka and Hirotaka Takeuchi: The Knowledge-Creating Company. Oxford University Press, Oxford, 1995.
41. OECD Committee for Information, Computer and Communications Policy: Global Information Infrastructure – Global Information Society (GII-GIS). Policy Requirements. OECD/GD(97)139 Report, OECD, Paris, 1997.
42. Tod Oppenheimer: The Computer Delusion, The Atlantic Monthly, July 1997.
43. W. W. Powell: Neither market nor hierarchy: Network forms of organizations. In: B. M. Starr and C. C. Cummings, eds.: Research in Organization Behavior, Vol. 12, JAI Press, Greenwich, CT, 1990.
44. James Brian Quinn: Intelligent Enterprise, A Knowledge and Service Based Paradigm for Industry, The Free Press, New York, 1992.
45. Howard Rheingold: The Virtual Community: Homesteading on the Electronic Frontier. Addison-Wesley, Reading, MA, 1993.
46. Christopher Rowe: People and Chips: The Human iImplications of Information Technology. Alfred Waller, Henley-on-Thames, 2nd ed., 1990.
47. Robert Rowthorn and Ramana Ramaswamy: Deindustrialization: Causes and Implications. IMF Working Paper WP/97/42, April 1997.
48. Peter H. Salus: Casting the Net. From ARPANET to INTERNET and beyond... Addison-Wesley, Reading, MA, 1995.
49. Stephen Saxby: The Age of Information. The Macmillian Press, London and Basingstoke, 1990.
50. Jorge Reina Schement and Terry Curtis: Tendencies and Tensions of the Information Age. The Production and Distribution of Information in the United States. Transaction Publishers, New Brunswick, N.J., 1995.
51. Peter Schwartz and Peter Leyden: The Long Boom: A History of the Future 1980–2020, Wired, 5.07, July 1997, pp. 115–173.
52. Norman Z. Shapiro and Robert H. Anderson: Toward an Ethics and Etiquette for Electronic Mail. RAND report R-3283-NSF/RC, Santa Monica, CA, 1995.
53. Daniel A. Sichel: The Computer Revolution. An Economic Perspective. Brookings Institution Press, Washington, D.C., 1997.
54. Paul A. Strassmann: The Squandered Computer. Evaluating the Business Alignment of Information Technologies. The Information Economics Press, New Canaan, CT, 1997.
55. Don Tapscott: The Digital Economy, Promise and Peril in the Age of Networked Intelligence, McGraw-Hill, New York, 1996.
56. John Tiffin and Lalita Rajasingham: In Search of the Virtual Class. Education in an Information Society. Routledge, London and New York, 1995.
57. Alvin Toffler: The Third Wave, William Morrow and Co., 1980.
58. Alvin Toffler and Heidi Toffler: Creating a New Civilization: The Politics of the Third Wave, Turner Publishing, Atlanta, VA, 1995.
59. Jerry Useem: Churn, Baby, Churn. Inc. magazine, State of Small Business issue, 1997, p. 25.
60. C. Winston: Economic Deregulation: Days of Reckoning for Microeconomists, Journal of Economic Literature, Vol. 31, No. 3, September 1993, pp. 1263-1289.
61. Working Group on Electronic Commerce: A Framework for Global Electronic Commerce, The White House, Washington, D.C., July 1997.
62. Robert H. Zakon: Hobbes' Internet Timeline. http://info.isoc.org/guest/zakon/ /Internet/History/HIT.html, September 1997.
63. Jiří Zlatuška: Education as an Information-Age Business. in: Proc. of the Rôle of the Universities in Future Information Society conference (RUFIS'97), ČVUT, Prague, September 25&26, 1997.

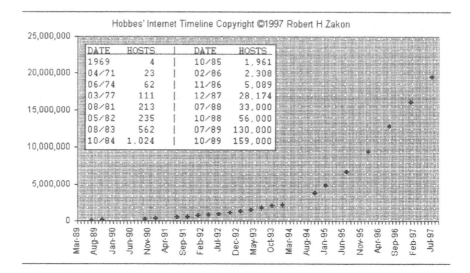

Figure 1: Internet host count
(http://info.isoc.org/guest/zakon/Internet/History/HIT.html)

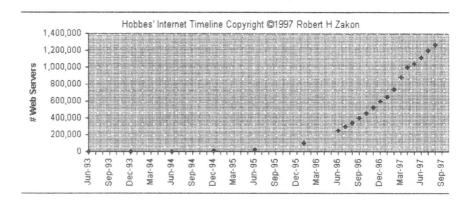

Figure 2: WWW server count
(http://info.isoc.org/guest/zakon/Internet/History/HIT.html)

| Country | July 91 | July 92 | July 93 | July 94 | July 95 | July 96 | July 97 |
|---|---|---|---|---|---|---|---|
| Finland | 1.74 | 3.12 | 5.34 | 9.75 | 21.90 | 54.27 | 65.77 |
| Iceland | 0.75 | 1.53 | 4.75 | 12.24 | 25.47 | 40.49 | 53.01 |
| Norway | 1.94 | 3.35 | 5.83 | 8.94 | 15.28 | 27.70 | 47.95 |
| USA | 1.69 | 2.87 | 4.87 | 7.84 | 16.23 | 31.26 | 44.97 |
| New Zealand | 0.35 | 0.53 | 0.91 | 4.21 | 12.25 | 21.76 | 43.48 |
| Australia | 1.26 | 2.78 | 4.65 | 7.15 | 11.49 | 22.02 | 33.44 |
| Sweden | 1.37 | 2.43 | 3.61 | 6.07 | 12.09 | 21.11 | 32.23 |
| Denmark | 0.30 | 0.53 | 1.19 | 2.33 | 7.07 | 14.72 | 26.21 |
| Canada | 0.69 | 1.37 | 2.45 | 4.36 | 8.87 | 14.33 | 23.32 |
| Netherlands | 0.49 | 1.39 | 2.33 | 3.88 | 8.76 | 13.89 | 22.10 |
| Switzerland | 1.46 | 2.50 | 4.42 | 6.78 | 9.01 | 14.50 | 20.90 |
| U.K. | 0.12 | 0.65 | 1.54 | 2.67 | 4.97 | 9.89 | 14.98 |
| Austria | 0.27 | 0.82 | 1.47 | 2.51 | 5.06 | 8.83 | 10.86 |
| Germany | 0.26 | 0.54 | 1.13 | 1.83 | 4.29 | 6.71 | 10.72 |
| Luxembourg | — | 0.21 | 0.47 | 1.06 | 3.67 | 6.97 | 9.33 |
| Ireland | 0.03 | 0.18 | 0.48 | 0.93 | 2.78 | 6.00 | 9.23 |
| Belgium | 0.03 | 0.15 | 0.43 | 1.20 | 2.34 | 4.27 | 8.49 |
| Japan | 0.05 | 0.13 | 0.29 | 0.58 | 1.28 | 3.96 | 7.63 |
| France | 0.16 | 0.33 | 0.69 | 1.24 | 1.96 | 3.26 | 5.02 |
| Czech Republic | — | 0.06 | 0.26 | 0.55 | 1.44 | 3.12 | 4.75 |
| Italy | 0.03 | 0.09 | 0.26 | 0.41 | 0.81 | 1.99 | 3.70 |
| Hungary | — | 0.00 | 0.14 | 0.53 | 1.10 | 2.45 | 3.30 |
| Spain | 0.03 | 0.09 | 0.22 | 0.54 | 1.02 | 1.59 | 3.11 |
| Korea | 0.00 | 0.10 | 0.12 | 0.30 | 0.50 | 1.10 | 3.00 |
| Greece | 0.02 | 0.06 | 0.13 | 0.28 | 0.53 | 1.21 | 1.88 |
| Portugal | — | 0.13 | 0.20 | 0.46 | 0.88 | 1.77 | 1.83 |
| Poland | — | 0.02 | 0.09 | 0.19 | 0.41 | 1.00 | 1.12 |
| Mexico | 0.00 | 0.00 | 0.02 | 0.06 | 0.09 | 0.22 | 0.39 |
| Turkey | — | — | 0.00 | 0.02 | 0.05 | 0.13 | 0.37 |

Table 1: Diffusion of Internet hosts in OECD countries per 1000 inhabitants

For 1996 and 1997, population figures from 1995 have been used; U.S. hosts include edu, com, gov, mil, org, net, and us.

Source: [41] and Network Wizards at http://www.nw.com/

# Contributed Papers

Arranged in alphabetic order of first author's family name

# Lower Bounds for the Virtual Path Layout Problem in ATM Networks *

Luca Becchetti[1] and Carlo Gaibisso[2]

[1] Istituto di Analisi dei Sistemi ed Informatica del CNR, Viale Manzoni 30, 00185 Roma, Italy, and Dipartimento di Informatica e Sistemistica, Università di Roma "La Sapienza", Via Salaria 113, 00184 Rome, Italy. E-mail: becchett@iasi.rm.cnr.it
[2] Istituto di Analisi dei Sistemi ed Informatica del CNR, Viale Manzoni 30, 00185 Roma, Italy. E-mail: gaibisso@iasi.rm.cnr.it

**Abstract.** In this paper we deal with the problem of designing virtual path layouts in ATM networks with given hop-count and minimum load, for which we prove a lower bound with respect to networks with arbitrary topology. The result is then applied to derive a tight lower bound for the one-to-many communication pattern in networks with arbitrary topology, and for the many-to-many communication pattern in planar graphs and graphs with bounded treewidth.

## 1 Introduction

Asynchronous Transfer Mode (ATM) ([1], [2], [3]) is the most common network protocol under consideration for the very fast Broadband Integrated Service Digital Networks of the future (B-ISDN). Due to this fact it has been studied intensively in the recent past.

In ATM, the transfer of data is based on packets of fixed length, called *cells*. In order to achieve the stringent transfer rate requirements, cell routing must be accomplished by dedicated hardware implementing very simple algorithms.

An ATM routing scheme is based on two types of predetermined routes in the network: *virtual paths* and *virtual channels*. Virtual paths are internal network entities, which, by bundling together several virtual channels sharing part of their routes, make network and connection management easier and more effective. Virtual channels connect, by sequences of virtual paths, pairs of end-users in the network.

In order to implement virtual paths and virtual channels, each cell header contains two routing labels: a *virtual path identifier* and a *virtual channel identifier*. Routing in ATM networks is hierarchical, in the sense that the virtual channel identifier of a cell is ignored by most of the switches it traverses, which route the cell just on the basis of its virtual path identifier.

A major problem in this framework is the one of defining the set of virtual paths in such a way that certain good properties are achieved:

---

* Work partially performed in the framework of Esprit Long Term Research Program project "ALCOM-IT" and of Italian MURST 40% project "Algoritmi, Modelli di Calcolo e Strutture Informative"

1. the number of virtual paths using the same physical link, denoted as *load* in what follows, should be kept as low as possible, so as not to exceed the capacity of the routing tables, which is fixed by the ATM protocol, and to allow a possibly fast recovery from faults;

2. the maximum number of virtual paths in a virtual channel, denoted as *hop count* in what follows, should be kept as low as possible, so as to guarantee low set up times for the virtual channels and high data transfer rates.

In its most general formulation, the Virtual Path Layout problem, VPL problem for short, is an optimization problem in which, given a certain communication demand between pairs of nodes and constrained load and hop count, it is required to design a system of virtual paths such that some given cost function is minimized ([4], [5], [6], [7]).

A second general approach to the VPL problem is the one typical of complexity theory, which aims at a precise characterization of the relations occurring between the involved quantities. This approach makes use of more specialized graph-theoretical techniques to derive upper and lower bounds both in the general case and in specific subproblems.

In this framework, interest is usually restricted to specific connection patterns, such as in the *one-to-many* virtual path layout problem, $VPL^{1-m}$ problem for short, in which a single node has to be connected to all other nodes in the network, and in the *many-to-many* virtual path layout problem, $VPL^{m-m}$ problem for short, in which a connection has to be established between all pairs of nodes.

Both these problems are of great practical relevance. In particular, the $VPL^{1-m}$ problem corresponds to the problem of setting up virtual path layouts in client-server networks, where data flows from a single source, the server, to multiple destinations, the clients, and viceversa ([8], [9]). In [10], [11], [12] and [13], upper and lower bounds are given for both the $VPL^{1-m}$ and $VPL^{m-m}$ problems with respect to different network topologies.

Finally, in [14], the problem of achieving a low load for a given value of the hop count, while preserving a good fault tolerance, is addressed with respect to complete graphs.

In [15] virtual path layouts have to be designed with fixed load and minimum hop count. The authors derive a lower bound for the hop count with respect to arbitrary network topologies and design asymptotically optimal virtual path layouts for mesh and chain networks.

In [16] and [17] the VPL problem has been studied by considering the node congestion, i.e. the number of virtual paths incident onto a node, as a cost measure.

In this paper, we deal with the problem of designing a virtual path layout with a given hop-count and minimum load. More in details we derive a lower bound for the load of any virtual path layout for any instance of the VPL problem, which is then very easily adapted to the one-to-many and many-to-many communication patterns. Such lower bounds are then instantiated to particular classes of graphs, such as planar graphs and graphs with bounded tree-width.

It is worth noting that the derived lower bounds also prove that it is possible to design asymptotically optimal one-to-many virtual path layouts for arbitrary network topologies ([11]) and many-to-many virtual path layouts for network topologies with bounded tree-width ([11], [12]).

## 2 The mathematical model

An instance of the virtual path layout problem can be represented by a triple $(G, U, h)$, where:

- $G = (V, E)$, is a connected graph describing the network topology; $V$, the set of nodes, models the set of network switches and end-users, and $E$, the set of edges, models the set of its physical links.
- $U \subseteq V \times V$ is a set of pairs of nodes , representing the set of connection requests;
- $h$ is a positive integer denoting the the maximum tolerable hop count.

The following definitions formalize the concept of virtual path layout:

**Definition 1.** A virtual path layout $\Phi$ for $(G, U, h)$ is a pair $(G_\Phi, I_\Phi)$, where: $G_\Phi$ is a graph $(V, E_\Phi)$ such that for each $(p, q) \in U$ there exists at least one simple path of length at most $h$ in $G_\Phi$ connecting $p$ and $q$, and $I_\Phi : E_\Phi \to S(G)$, where $S(G)$ is the set of simple paths in $G$, is an injective function assigning to each edge $(p, q) \in E_\Phi$ a simple path in $G$ connecting $p$ to $q$.

**Definition 2.** For any virtual path layout $\Phi$ for $(G, U, h)$, a simple path $\pi$ in $G$ is called a virtual path, if and only if an edge $e \in E_\Phi$ exists such that $\pi = I_\Phi(e)$.

**Definition 3.** For any virtual path layout $\Phi$ for $(G, U, h)$, a simple path $\gamma$ in $G_\Phi$ is called a virtual channel if and only if it connects any pair of nodes $p$ and $q$ such that $(p, q) \in U$.

Notice that any virtual channel connecting $p$ and $q$, can also be considered as a, not necessarily simple, path from $p$ to $q$ in $G$.

**Definition 4.** For any link $l \in E$, the load of $l$ with respect to any virtual path layout $\Phi$ for $(G, U, h)$ is defined as $L_\Phi(l) = |\{e \in E_\Phi | l \in I_\Phi(e)\}|$.

**Definition 5.** The load of any virtual path layout $\Phi$ for $(G, U, h)$ is defined as $L_\Phi = \max_{l \in E_\Phi} \{L_\Phi(l)\}$.

The we deal with is defined as follows: given any instance $(G, U, h)$ of the problem, design a virtual path layout $\Phi$ such that $L_\Phi$ is minimized.

In this general context, we will mainly concentrate on:

- the $VPL^{1-m}$ problem, in which a single node has to be connected to all other nodes in the graph, that is $U = \{p\} \times V$, for some $p \in V$. In what follows the generic instance of the VPL$^{1-m}$ problem will be simply denoted by $(G, \{v\}, h)$;

- the $VPL^{m-m}$ problem, in which all pairs of nodes in the graph have to be connected, that is $U = V \times V$. In what follows the generic instance of the $VPL^{m-m}$ problem will be simply denoted by $(G, h)$.

## 3  Lower bounds for general topologies

In this section we derive a lower bound for the load of any virtual path layout for any instance of the VPL problem. Such a lower bound will be then very easily adapted to the $VPL^{1-m}$ and $VPL^{m-m}$ problems.

In what follows, for any graph $G$, $\Delta(G)$ will denote the maximum degree of any of its nodes. In order to simplify the notation, whenever possible, $L$ and $\Delta$ will be respectively adopted in place of $L_\Phi$ and $\Delta(G)$.

For any instance $(G, U, h)$ of the VPL problem, let $C \subseteq E$ be an edge-cut which separates $V$ into two subsets $V_1$ and $V_2$, and $N_C$ be the number of pairs $(p, q) \in U$ such that $p \in V_1$ and $q \in V_2$.

**Lemma 6.** *For any $(G, U, h)$ and any virtual path layout $\Phi$ for $(G, U, h)$, at most $(h(\Delta L)^{h+1} + 1)/(\Delta L - 1)^2$ virtual channels of $\Phi$ can include the same virtual path.*

*Proof.* Let $\pi_j$ be a virtual path of $\Phi$. The maximum number of virtual channels including $\pi_j$ can be derived as follows:

1. for any $k$ and for any $i$, $k \leq h$, $1 \leq i \leq k$, $\pi_j$ appears as the $i$-th virtual path in at most $(\Delta L)^{i-1} \cdot (\Delta L)^{k-i} = (\Delta L)^{k-1}$ virtual channels of length $k$. In fact, let $p_j$ and $q_j$ be the endpoints of $\pi_j$. Node $p_j$ is the endpoint of at most $\Delta L$ virtual paths of $\Phi$ different from $\pi_j$. It is clear that, by iterating this consideration $i - 1$ times, at most $(\Delta L)^{i-1}$ different simple paths of length $i - 1$ exist in $G_\Phi$ rooted at $p_j$. In the same way it is possible to conclude that at most $(\Delta L)^{k-i}$ different simple paths of length $k - i$ exists in $G_\Phi$ rooted at $q_j$, thus proving the assumption;
2. consequently, the maximum number of virtual channels of length equal to $k$ including $\pi_j$ is given by $C_k \leq \sum_{i=1}^{k} (\Delta L)^{k-1} = k(\Delta L)^{k-1}$;
3. finally, the maximum number of virtual channels of length at most $h$ including $\pi_j$ is

$$\overline{C_h} \leq \sum_{k=1}^{h} C_k = \sum_{k=1}^{h} k(\Delta L)^{k-1} = \sum_{k=0}^{h} k(\Delta L)^{k-1} =$$

$$= \frac{d}{dx} \sum_{k=0}^{h} (\Delta L)^k < \frac{h(\Delta L)^{h+1} + 1}{(\Delta L - 1)^2}.$$

**Theorem 7.** *For any instance $(G, U, h)$ of the VPL problem and any virtual path layout $\Phi$ for $(G, U, h)$, if $C$ is an edge-cut that separates $N_C$ pairs in $U$, then $L > \frac{1}{\Delta} \sqrt[h]{\frac{\Delta}{4h} \left( \frac{N_C}{|C|} - \frac{4}{\Delta} \right)}$.*

*Proof.* Let us first derive the maximum number $\overline{C_h}(e)$ of virtual channels of length at most $h$ in $G_{\oplus}$ that include the same physical edge $e$ in $G$. By lemma 6 and since $e$ is included in at most $L$ virtual paths

$$\overline{C_h}(e) \leq L\overline{C_h} < L\frac{h(L\Delta)^{h+1}+1}{(L\Delta-1)^2} = \frac{L\Delta}{\Delta}\frac{h(L\Delta)^{h+1}+1}{(L\Delta-1)^2}.$$

It is now worth noting that $\Delta L \geq 2$, since we are obviously interested in graphs having at least one node with degree greater than 1 and in virtual path layouts with load greater than 0. As a consequence $\Delta L - 1 \geq \frac{\Delta L}{2}$, thus

$$\overline{C_h}(e) \leq \frac{4}{\Delta}\frac{h(L\Delta)^{h+1}+1}{L\Delta} < \frac{4}{\Delta}(h(L\Delta)^h+1) = 4h\Delta^{h-1}L^h + \frac{4}{\Delta}.$$

Given any edge-cut $C$ for $G$, the number of VCs which are disconnected by $C$ is then bounded by $\sum_{e \in C}\overline{C_h}(e) < |C|\left(4h\Delta^{h-1}L^h + \frac{4}{\Delta}\right)$.

Finally, since $N_C$ pairs in $U$ are separated by $C$, the following inequality must hold: $|C|\left(4h\Delta^{h-1}L^h + \frac{4}{\Delta}\right) > N_C$, which implies that for at least one edge $e$ of $G$, $L(e) > \frac{1}{\Delta}\sqrt[h]{\frac{\Delta}{4h}\left(\frac{N_C}{|C|} - \frac{4}{\Delta}\right)}$.

As an immediate corollary of theorem 7 it is possible to derive the following lower bound for the $VPL^{1-m}$ problem, which obviously applies to the $VPL^{m-m}$ problem too.

**Corollary 8.** *For any instance $(G, \{v\}, h)$ of the $VPL^{1-m}$ problem and for any virtual path layout for $(G, \{v\}, h)$, then $L > \frac{1}{\Delta}\sqrt[h]{\frac{|V|-5}{4h}}$.*

*Proof.* Just select $C$ as the set of edges incident to $v$. In such a case in fact $|C| \leq \Delta$ and $N_C = |V| - 1$.

A direct consequence of corollary 8, is that it is possible to the derive an asymptotically optimal virtual path layout for any instance $(G, \{v\}, h)$ of the $VPL^{1-m}$ problem, by simply considering any spanning tree $T_G$ for $G$ rooted at $v$ and applying the algorithm described in [11] for the solution of the $VPL^{1-m}$ to the instance $(T_G, \{v\}, h)$. The resulting virtual path layout has a load bounded by $h\sqrt[h]{n}$ and is consequently asymptotically optimal.

# 4 Lower bounds for particular classes of graphs

In this section, the lower bounds obtained for arbitrary network topologies will be instantiated to particular classes of graphs, such as planar graphs and graphs with bounded tree-width.

In deriving a lower bound for the load of many-to-many virtual paths layout for planar graphs, we rely on the following already known result ([18]):

**Theorem 9.** *For any planar graph $G = (V, E)$, $|V| = n$, $V$ can be partitioned into three subsets $A$, $B$ and $D$ such that: no edge connects a node of $A$ to a node of $B$, $|A| \leq \frac{n}{2}$, $|B| \leq \frac{n}{2}$ and $|D| \leq 3\sqrt{6n}$.*

By theorem 9 it is possible to prove that for any planar graph $G$ there exists at least one edge cut $C$, such that $|C| = O(\sqrt{n})$ and $\Omega(n^2)$ pairs of nodes in $G$ are separated by $C$. More in details it is possible to prove that:

**Lemma 10.** *For any planar graph $G = (V, E)$ of degree $\Delta$, there exists an edge-cut $C$, such that $|C| \leq 3\Delta\sqrt{6n}$ and $C$ separates at least $(n - 3\sqrt{6n})^2$ pairs of nodes in $G$, where $n = |V|$.*

*Proof.* By theorem 9, $V$ can be partitioned into three subsets $A$, $B$ and $D$, such that no edge connects a node of $A$ to a node of $B$ and the following set of inequalities are verified:

$$|A| + |B| + |D| = n$$
$$|D| < 3\sqrt{6n}$$
$$|A|, |B| < \frac{n}{2}.$$

The above inequalities imply that $|A|, |B| > \frac{n}{2} - 3\sqrt{6n}$. Let $C$ be the set of edges incident to nodes in $D$. Obviously $|C| = \Delta|D| < 3\Delta\sqrt{6n}$ and at least $\left(\frac{n}{2} - 3\sqrt{6n}\right)^2$ pairs of nodes in $G$ are separated by $C$, thus proving the theorem.

The next theorem is a consequence of lemma 10 and of theorem 7:

**Theorem 11.** *For any instance $(G, h)$ of the $VPL^{m-m}$ problem, where $G = (V, E)$ is a planar graph, and for any virtual path layout for $(G, h)$, $L = \Omega\left(\frac{1}{\Delta}\sqrt[2h]{n^3}\right)$, where $n = |V|$.*

The class of graphs with bounded tree-width includes many popular topologies, such as arrays, rings, trees and trees of rings, and is consequently of great theoretical and practical interest. Let us first recall few basic definitions ([19]):

**Definition 12.** A tree decomposition of a graph $G = (V_G, E_G)$ is a pair $(\mathcal{X}, T)$, where $\mathcal{X} = \{X_p | p \in V_T\}$ is a family of subsets of $V$, one for each node of the tree $T = (V_T, E_T)$, such that:

- $\cup_{p \in V_T} X_p = V_G$;
- for all edges $(u, w) \in E_G$, there exists a node $p \in V_T$ with $v, w \in X_p$;
- for all $p, q, r \in V_T$: if $q$ is on the path from $p$ to $r$ in $T$, then $X_p \cap X_r \subseteq X_q$.

**Definition 13.** Given a graph $G$ and a tree-decomposition $(\mathcal{X}, T)$ for $G$, the width of $(\mathcal{X}, T)$ is defined as $\max_{p \in V_T}\{|X_p| - 1\}$. The tree-width of $G$ is the minimum width over all tree-decompositions of $G$.

The following theorem holds for graphs with bounded tree-width ([19]):

**Theorem 14.** *For any graph $G = (V, E)$ with tree-width at most $k$, there exists a node separator $C$ for $G$, such that $|C| \leq k + 1$ and $C$ breaks $G$ into connected components whose size is not greater than $\frac{1}{2}(n - k)$, where $n = |V|$.*

It is now possible to state the following lemma:

**Lemma 15.** *For any graph $G = (V, E)$ with tree-width at most $k$ and degree $\Delta$, there exists an edge-cut $C$ for $G$, such that $|C| \leq (k + 1)\Delta$ and $C$ separates at least $\geq \frac{1}{12}(n - k)^2$ pairs of nodes in $G$, where $n = |V|$.*

*Proof.* By theorem 14 and since the tree-width of $G$ is bounded by $k$, there exists a node separator $F$ such that $|F| \leq k + 1$. Let us assume that $G$ is broken by $F$ into $l$ connected components $G_i = (V_i, E_i), i = 1, \cdots, l$. Let $m = \max_{i=1,\ldots,l}\{s \in \mathcal{N} \mid |V_i| \leq \frac{1}{s}(n - k)\}$. By theorem 14, $s \geq 2$. We consider two different cases:

1. $m = 2$, then for at least one $i \in \{1, \cdots, l\}$, $|V_i| > \frac{1}{3}(n - k)$. Choose $A = V_i$ and $B = V - A$, thus $\frac{1}{3}(n - k) < |A| \leq \frac{1}{2}(n - k)$ and therefore $\frac{1}{2}(n - k) \leq |B| \leq \frac{2}{3}(n-k)$. Consequently at least one edge cut $C$ will exist of size at most $\Delta|F| \leq (k+1)\Delta$, such that $C$ separates at least $\frac{1}{3}(n-k) \cdot \frac{1}{2}(n-k) = \frac{1}{6}(n-k)^2$ pairs $(p, q)$ in $G$ with $p \in A$ and $q \in B$.

2. $m \geq 3$, then choose $A$ as the smallest subset of $V$ obtained by the union of subsets in $\{V_1, \cdots, V_l\}$ and such that $|A| \geq \frac{1}{2}(n-k)$. Note that, since in this case $|V_i| \leq \frac{1}{3}(n - k), \forall\, 1 \leq i \leq l$, $|A| \leq \frac{1}{2}(n-k) + \frac{1}{3}(n-k) = \frac{5}{6}(n-k)$. Thus if we set $B = V - A$, $|A| \geq \frac{1}{2}(n-k)$ and $|B| \geq \frac{1}{6}(n-k)$ for all $m \geq 3$. By the same reasoning followed for the case $m = 2$, it is possible to conclude that there exists at least an edge-cut $C$ for $G$, such that $|C| = \Delta|F| \leq (k + 1)\Delta$ and $C$ separates at least $\frac{1}{12}(n - k)^2$ pairs $(p, q)$ in $G$ with $p \in A$ and $q \in B$, thus proving the lemma.

The next theorem is a consequence of lemma 15 and of theorem 7:

**Theorem 16.** *Given any instance $(G, h)$ of the $VPL^{m-m}$ problem, where $G = (V, E)$ is a graph with bounded tree-width at most $k$, and for any virtual path layout for $(G, h)$, $L = \Omega\left(\frac{1}{\Delta}\sqrt[h]{n^2}\right)$, where $n = |V|$.*

It is worth noting that theorem 16 also proves that the algorithm proposed in [11] and [12] to derive many-to-many virtual path layouts for network topologies with bounded tree-width is asymptotically optimal.

# References

1. McDysan, D.E., Spohn, D.L.: ATM: Theory and Applications. McGraw-Hill, New York, NY (1995)
2. de Prycker, M.: Asynchronous Transfer Mode: Solutions for Broadband ISDN. Ellis-Horwood, New York, NY (1993)
3. ATM Forum: ATM, User-Network Interface Specification. Version 3.0. Prentice-Hall (1993)

4. Eilam, T., Flammini, M., Zaks, S.: A Complete Characterization of the Path Layout Construction Problem for ATM Networks with given Hop Count and Load. Proc. of the 24th International Colloquium on Algorithms, Languages and Programming - ICALP 97, to appear (1997)

5. Gerstel, O., Zaks, S.: The Virtual Path Layout Problem in Fast Networks. Proc. of the 13th ACM Conf. on Principles of Distributed Computing (1994) 235–243

6. Dahl, G., Martin, A., Stoer, M.: Routing through Virtual Paths in Layered Telecommunication Networks. Tech. Rep. No. N-0316, University of Oslo, Blindern, Oslo, Norway (1996)

7. Chlamtac, I., Farago, A., Zhang, T.: Optimizing the System of Virtual Paths. IEEE Transactions on Networking (1994)

8. Lougher, P., Sheperd, D.: On the Complexity of the Disjoint-Paths Problem. Combinatorica 13 (1993)

9. Ramanathan, S., Rangan, P.V.: Feedback Techniques for Intra-Media Continuity and Inter-Media Synchronization in Distributed Multimedia Systems. The Computer Journal 36 (1) (1993)

10. Cidon, I., Gerstel, O., Zaks, S.: The Layout of Virtual Paths in ATM Networks. Tech. Rep. No. CS0831, Technion-Haifa, Israel (1994)

11. Cidon, I., Gerstel, O., Zaks, S.: A Scalable Approach to Routing in ATM Networks. Proc. of the 8th International Workshop on Distributed Algorithms (LNCS No. 857) (1994)

12. Gerstel, O.: Virtual Path Design in ATM Networks. PhD thesis, Technion-Haifa, Israel (1995)

13. Gerstel, O., Wool, A., Zaks, S.: Optimal Layouts on a Chain ATM Network. Proc. of the 3rd Annual European Symposium on Algorithms (1995)

14. Gasienic, L., Kranakis, E., Krizanc, D., Pelc, A.: Minimizing Congestion of Layouts for ATM Networks with Faulty Links. Tech. Rep., Carleton University, School of Computer Science, Ottawa, Canada (1995)

15. Kranakis, E., Krizanc, D., Pelc, A.: Hop-Congestion Tradeoffs for ATM Networks. Tech. Rep., Carleton University, School of Computer Science, Ottawa, Canada (1995)

16. Flammini, M., Nardelli, E., Proietti, G.: ATM Layouts with Bounded Hop Count and Congestion. Proc. of the 11th Workshop on Distributes Algorithms - WDAG 97, to appear (1997)

17. Gerstel, O., Cidon, I., Zaks, S.: Efficient Support for the Client-Server Paradigm over ATM Networks. Proc. of the IEEE Infocom Conference (1996) 1294–1301

18. Korte, B., Lovász, L., Prömel, H.J., Schrijver, A. Editors: Paths, Flows and VLSI-Layout. Springer-Verlag (1990)

19. Bodlaender, H.: A Partial k-Arboretum of Graphs with bounded Tree-Width. Tech. Rep. No. UU-CS-1996-02, Department of Computer Science, University of Utrecht, The Netherland (1997)

# Query Processing in Temporal Evidential Databases

Bingning Dai[1], David A Bell[1], and John G Hughes[2]

[1] School of Information and Software Engineering,
[2] Faculty of Informatics,
University of Ulster at Jordanstown, Newtownabbey, Co. Antrim BT37 0QB, UK
E-mail: {B.Dai, DA.Bell, JG.Hughes}@ulst.ac.uk

**Abstract.** This paper first presents the temporal evidential database (TED) model, which incorporates time into the evidential relational database model. Evidential databases support both uncertain and imprecise information using Dempster-Shafer's (D-S) theory of evidence (or evidence theory). The TED model also adopts an evidence theoretic approach to managing both relative and absolute times under uncertainty. The query language for the TED model is then presented. It is shown how to integrate the evaluations of both temporal and non-temporal predicates in query processing. This not only presents a solution in the TED model, but also points out this query processing problem is common in the database models that aim at managing both time and uncertainty. The query language also has a novel modal syntax with respect to evidential functions.

## 1  Introduction

Intensive research has been carried out on handling uncertainty in non-temporal databases (e.g. [1, 6, 14, 15]) or dealing with uncertain times in temporal databases (e.g. [7, 8, 17]). However, some applications, e.g. medical information systems, involve the management of both uncertain and temporal data. Though [17] claimed the uncertain temporal information could be treated in parallel with the traditional treatment towards incomplete non-temporal information, it did not show how to integrate the two in the query language and thus ignored a problem brought by the integration. This problem is described in the following paragraph.

When a query involves both uncertain non-temporal data and uncertain temporal data, two degrees of credibility are evaluated for a tuple to which the tuple satisfies the non-temporal predicate (or condition) in the query and the temporal one respectively. To decide whether this tuple should be presented in the answer to the query, these two degrees of credibility should be combined to show to what a degree the tuple satisfies the query's conditions as a whole. This problem should be explicitly addressed in the database models that aim at managing both time and uncertainty in the systems.

The temporal evidential database (TED) model incorporates time into the evidential relational database model [2, 14, 15, 16]. Both *transaction time* and *valid time* ([12]) are modelled. The transaction time of a database fact is the time when the fact is in the database and may be retrieved. The valid time of a fact is the time when the fact is true in the modelled reality. Evidential databases support both uncertain and imprecise information using Dempster-Shafer's (D-S) theory of evidence (or evidence theory [9, 10]). The TED model also adopts an evidence theoretic approach to managing time under uncertainty. The theoretic foundations for the management of time in TEDs have been addressed in [4], where the evidential temporal constraint networks (ETCNs) are proposed to represent both relative and absolute times.

Now, the TED query language has to deal with the integration of temporal and evidential representations. Informally, a query to a TED specifies both temporal and non-temporal conditions. The evaluation of the temporal ones employs the reasoning algorithms proposed in [4], while that of the non-temporal ones follows the similar approaches from [5]. After each evaluation returns a pair of belief and plausibility functions, the integration of them is accomplished according to the extension to evidence theory defined in [15]. The combined result then has to pass the credibility filter specified by the query, which has a novel modal syntax with respect to evidential functions.

The rest of the paper is organised as follows. The next section gives the basic definitions and an extension to D-S theory, which will be used in the TED model. Section 3 defines the TED model formally. Related issues are explained informally with an example. Section 4 focuses on query processing in TEDs. A query language is presented with its syntax and semantics interpreted informally. Some related work is discussed in the last section, which also concludes the paper and points out future work.

## 2   The Basics of and an Extension to Evidence Theory

In D-S theory, a *frame of discernment* $\Theta$ is a mutually exclusive and exhaustive collection of propositions in a domain, represented as a finite non-empty set.

**Definition 1 Mass function.** A mass function $m$ is:
  $m : 2^\Theta \to [0, 1]$, where $m(\emptyset) = 0$, $\sum_{X \subseteq \Theta} m(X) = 1$.
$X(\subseteq \Theta)$ is called a focal element of $m$ when $m(X) > 0$.          □

An uncertain variable taking values from the frame of discernment $\Theta$ then will be denoted as $\{X_1/m(X_1), X_2/m(X_2), \ldots, X_n/m(X_n)\}$, where $X_1, X_2, \ldots, X_n$ are the focal elements of the mass function that represents the evidence supporting the possible values of the variable.

**Definition 2 Evidential functions *bel* and *pls*.** Given a mass function $m$, a belief function *bel* is defined on $2^\Theta$, for all $A \subseteq \Theta$: $bel(A) = \sum_{X \subseteq A} m(X)$.
A plausibility function *pls* is defined as: $pls(A) = \sum_{X \cap A \neq \emptyset} m(X) = 1 - bel(\overline{A})$
for all $A \subseteq \Theta$.          □

Note that a mass function is a basic probability assignment to all the subsets $X$'s of $\Theta$. A belief function *bel* gathers all the support that a subset $A$ gets from such an assignment. It is not a probability distribution on $\Theta$.

The following proposition is proposed by [15].

**Proposition 3 Combination of *bel* and *pls* for independent events.** *For independent events $E_1$ and $E_2$ where $bel(E_1) = b_1$, $pls(E_1) = p_1$, and $bel(E_2) = b_2$, $pls(E_2) = p_2$, the pairs of bel and pls values of the conjunction and disjunction of the events $E_1$ and $E_2$ are defined as follows:*

$$\langle bel(E_1 \wedge E_2); pls(E_1 \wedge E_2) \rangle = \langle b_1 b_2; p_1 p_2 \rangle;$$
$$\langle bel(E_1 \vee E_2); pls(E_1 \vee E_2) \rangle = \langle 1 - (1 - b_1)(1 - b_2); 1 - (1 - p_1)(1 - p_2) \rangle.$$

*It follows naturally that this combination is associative.* □

## 3 The Temporal Evidential Database Model

This section first gives the formal definitions for the TED model. An example is then given to illustrate the definitions, which also serves as a run-through example in this paper. Some basic features of the TED model and query related issues are explained in this section too.

### 3.1 Formal Definitions

**Definition 4 Attribute.** Attributes are variable names. The set consisting of them is: $\mathcal{A} = \{A_1, A_2, \ldots, A_n\}$. □

**Definition 5 Domain.** The domain of an attribute $A_i$ is the set of values that the attribute can take, denoted by $D(A_i)$ or $D_i$. □

Note that every domain should be a finite set to enable an evidential representation.

**Definition 6 Evidential value.** An evidential value of an attribute $A_i$ is a mass function defined over $D_i \cup \{\perp\}$ where $\perp$ denotes an undefined value for $A_i$. The set of the evidential values for $A_i$ is denoted by $ED(A_i)$ or $ED_i$. □

Note the inclusion of $\perp$ is to represent different types of imperfect information in databases [14], which is not the focus of this paper thus not explained further.

**Definition 7 Temporal evidential tuple.** A temporal evidential tuple is represented by a function $t$:

$$t : TT \times VT \times K \rightarrow \prod_{\text{some } i\text{'s}, 1 \leq i \leq n} ED_i \times CL.$$ □

Briefly, $K$ is the key of the tuple. It is a special set of attributes, i.e. $K \subseteq \mathcal{A}$. $TT$ and $VT$ are two sets of system variables where the transaction time and valid time can take values from respectively. $CL$ is the *confidence level* of the tuple, which is represented by a pair $\langle bel, pls \rangle$ where $0 \leq bel, pls \leq 1$. Further explanation for this definition will be made in the following subsection.

Now, a *temporal evidential table* $T$ is defined as a set of temporal evidential tuples, a *temporal evidential database schema* is a set of temporal evidential tables, and a *temporal evidential database* is an instance of a temporal evidential database schema.

An example database is given below, which serves as a run-through example in the paper.

*Example 1 A TED.* There is a medical database. There are two attributes: *PatientNum* and *Disease* with *PatientNum* being the key. The domains are $D(PatientNum) = \{0, 1, \ldots, 9999\}$ and $D(Disease) = \{d_1, d_2, \ldots, d_{3000}\}$. The table *DiseaseRec* records patients' diseases. It contains 3 tuples as shown below:

| TT | VT | PatientNum | Disease | CL |
|----|----|----|----|----|
| $tt_1$ | $vt_1$ | 1 | $\{\{d_1\}/0.3, \{d_1, d_2\}/0.7\}$ | $\langle 1, 1 \rangle$ |
| $tt_2$ | $vt_2$ | 2 | $\{\{d_1\}/1\}$ | $\langle 1, 1 \rangle$ |
| $tt_3$ | $vt_3$ | 3 | $\{\{d_1\}/0.8, \{d_2, d_3\}/0.2\}$ | $\langle 1, 1 \rangle$ |

where $tt_1 = tt_2 = tt_3 = September$ *'96*, $vt_1 = August$ *'95*, $vt_2 = July$ *'95*, and $vt_3 = March$ *'96*. Note this table is just one of the tables in the database.  □

## 3.2  Managing of Evidence and Time in TEDs

Since managing either time or evidence alone in databases involves many complexities , there are even more intricacies in managing both of them. Here only the most basic features and those needed for the understanding of query processing in TEDs are explained. It should be noted that the TED model is a proper extension to the evidential database model based on [14, 15] and the evidential database model is a proper extension to the traditional relational database model.

**CL.** A *CL* value in Definition 7 is not derived from the attribute value uncertainties. It is an independent measure of the strength of the predicate represented by the tuple [2]. The interpretation of *CL* can vary in different applications. To simplify the discussion of query processing, all *CL* values are assumed to be $\langle 1, 1 \rangle$ in this paper. Since such a value indicates that the evidential information in the tuples should not be cast doubt on, it will not have any bearing on query evaluations. Note this simplification has been adopted in Example 1.

**Evidence and Evidential Relations.** It has been recognised that the comparison of values in databases is inevitable in query processing. This issue has been addressed in a general sense by [5]. Here the related definitions are formalised according to the TED model definitions, which will facilitate the discussion of query processing later on.

**Definition 8 Relation algebra.** A relation algebra for an attribute $A_i$ is a set of binary relations defined on $D_i$, denoted by $\Re(A_i)$ or $\Re_i$. The members of $\Re_i = \{r_{i1}, \ldots, r_{il_i}\}$ are exclusive and complete, i.e., they define a partition of the Cartesian product $D_i \times D_i$.  □

The above definition is for definite data. No uncertainty is involved. For example, in Example 1, $\Re(Disease) = \{=, \neq\}$.

**Definition 9 Evidential relation.** An evidential relation between two evidential values in $ED_i$ is a mass function defined over $\Re_i$.                              □

The Evidence Manager in a TED system accepts two evidential values and returns an evidential relation between them. Applying the algorithms proposed in [5], the evidential relation between Patient 1's and Patient 2's diseases (in Example 1) is $\{\{=\}/0.3, \{=, \neq\}/0.7\}$ (time is ignored here).

**Managing Time.** A TED system manages both transaction time and valid time via an evidential temporal constraint network (ETCN [4]) Manager. The function of the ETCN Manager in query processing is similar to the Evidence Manager's. It derives the evidential temporal relation between two evidential temporal elements. In fact, if $VT$ and $TT$ are taken as special domains, $\Re(VT)$ and $\Re(TT)$ will be the same as the set of temporal relationships defined in [4].

**Key.** The definition of *key* is essential in the investigations into database integrity issues such as redundancy and inconsistency. Note in Definition 7, both temporal dimensions are actually included in the primary key. $K$ is the key for the underlying non-temporal relational database. Two tuples with same $K$ values but different $TT$ values reflect different database states while two tuples with same $K$ values but different $VT$ values reflect a tuple's value history.

In the following discussion of query processing, data integrity is assumed to have been checked and the databases in question are consistent.

## 4   Query Processing in TEDs

**Definition 10 The query language.** The format of a query is as follows:

| (Keyword) | (Argument) | (Explanation) |
|-----------|------------|---------------|
| **select** | $t_1, \ldots, t_k$ | ;a list of the attributes that values will be returned to |
| **from** | $R_1, \ldots, R_l$ | ;a list of the tables where the attributes come from |
| **where** | $APred$ | ;a non-temporal predicate |
| **valid** | $VTPred$ | ;a valid time predicate |
| **when** | $TTPred$ | ;a transaction time predicate |
| **with** | $FilterPred$ | ;an evidential credibility filter |

There are six arguments for the query. The first two are common in any relational databases. They involve some other relational operations, such as *projection* and *join*, which are assumed to have been completed. This assumption has simplified the syntax of the term following **select** to a list of attribute names and the term following **from** to a list of table names. The focus now is on the other four arguments, which form the condition of the query. It is worth noting that it is not necessary that all four arguments should be present in a query.

In the condition, $VTPred$ and $TTPred$ concern the temporal dimensions of the databases while $APred$ deals with non-temporal restrictions. The syntax of these predicates are in accordance with predicate logic. The differences among them lie in their atomic terms.

In temporal predicates, the syntax of the atomic terms is a set of temporal relationships followed by a constant evidential temporal element or a variable temporal element. For example, $\{>,=\}$ *July '97* is an atomic term while $(\{>,=\}$ *July '97*$) \wedge (\{<,=\}$ *August '97*$)$ is a compound temporal predicate. Note the temporal elements before the set(s) of temporal relationships are omitted because it is always a tuple's $VT$ or $TT$ that is being compared.

In non-temporal predicates, the atomic terms are of the form $t.A_i \; r \; t'.A_i$, where the values of $t.A_i$ and $t'.A_i$ are $m, m' \in ED_i$ respectively, $r \subseteq \Re_i$. From the general discussion in [5], it can be seen that such predicates allow more possible relations to be considered than the simple scalar comparisons in common relational database queries.

The evaluations of $VTPred$ and $TTPred$ act upon the theoretic foundations in [4] (see also Section 3.2). The evaluation of the atom predicates in $APred$ follows the methods in [5] (see also Section 3.2). There is an equivalence between these methods and the fundamental method in evidence theory, which not only proves the correctness of the methods but also suggests they can be implemented in TEDs more efficiently. Each of these three evaluations yields a $CL$ value for the corresponding predicate. Note that although the transaction times of the tuples in a TED system are supposed to be generated by the system and thus always certainly known, $TTPred$ may involve an uncertain temporal element and thus becomes uncertain.

Now three parts of the query condition have been evaluated. It is time to combine the three $CL$ values and use the result to evaluate the evidential filter $FilterPred$. The syntax of the filter predicate is a scalar comparison operator $op$ from the set $\{>, \geq, =, \leq, <\}$, followed by a $(0 \leq)p(\leq 1)$, modified by **definitely** or **maybe**. It is evaluated according to the modifiers as shown below:

1. $op \; p$ **definitely:**
   (a) $op$ is $>$, $>=$, or $=$: $bel(APred \wedge VTPred \wedge TTPred) \; op \; p$;
   (b) $op$ is $<$ or $<=$: $pls(APred \wedge VTPred \wedge TTPred) \; op \; p$;
2. $op \; p$ **maybe,**
   (a) $op$ is $>$, $>=$, or $=$: $pls(APred \wedge VTPred \wedge TTPred) \; op \; p$;
   (b) $op$ is $<$ or $<=$: $bel(APred \wedge VTPred \wedge TTPred) \; op \; p$.

where $bel(APred \wedge VTPred \wedge TTPred) = bel(APred)bel(VTPred)bel(TTPred)$ and $pls(APred \wedge VTPred \wedge TTPred) = pls(APred)pls(VTPred)pls(TTPred)$ according to the extension to evidence theory in Proposition 3. Note this has assumed the independence of the non-temporal value uncertainty from the temporal uncertainty, which is the case in common applications. As for combining dependent evidential functions, [11] has proposed a solution for dichotomous belief functions. □

*Example 2.* This is a simplified query showing the functioning of the evidential filter in the query language.

> **select** *PatientNum*
> **from** *DiseaseRec*
> **valid** $\{<\}July$ '95
> **where** $t(PatientNum).Disease\{=\}\{d_i\}$
> **when** $\{=\}September$ '96
> **with** $\geq 0.5$ **definitely**

This query can be described in natural language as: find the numbers of the patients from the disease record, who suffered $d_i$ after *July '95*, as recorded in *September '96*, with a credibility degree definitely greater than or equal to 0.5. Using the values from Example 1, the answer is "3" with a credibility degree *(bel)* 0.8 while No. 1 is filtered out with 0.3 and No. 2 is filtered out with 0. Changing **definitely** to **maybe** in the filter will include "1" in the answer with a credibility degree *(pls)* 1 while No. 3's *pls* is 0.8 too.                                    □

## 5    Concluding Remarks

The temporal evidential database model that supports both time and uncertainty in databases has been presented. Previous research has been carried out on the theoretic foundations for this model. This paper shows how these works are integrated. The query language for TEDs is presented and illustrated. It has focused on solving the query evaluation problem brought by the integration.

The TED model has powerful expressiveness that encompasses those database models' that only deal with incomplete information, probabilistic information or temporal information. The query language also has a novel modal syntax. The models in [3, 13] can give modal *(possible* and *certain)* answers to a query involving uncertain information but only qualitative information is used there. [14, 15] only considered $bel >= p$ in the filter. Thus although evidential functions are employed, there is no modality in the answers. The most similar filter that uses quantitative information and considers modality is the one proposed by [1]. However, apart from the fact that [1]'s model is based on probability theory, only the "=" relation was considered for a certain domain, while the TED model's query language can handle any relations in a relation algebra defined on the domain ([5]).

As mentioned early in the paper, the TED model is an extended relational database model. It is necessary to re-define all the relational operations (besides *selection* that has been discussed in this paper), i.e., *union, intersection, difference, Cartesian product, projection, natural join, θ-join,* and *division.* Database modification and integrity maintenance should continue to be investigated, which is made complicated by the presence of both time and evidence in the databases. Issues related to efficient implementations will be looked into while the prototype of a TED system is under development.

# References

1. D. Barbara, H. Garcia-Molina, and D. Porter. The Management of Probabilistic Data. *IEEE Transactions on Knowledge and Data Engineering*, 4(5):487–502, October 1992.

2. D. A. Bell, J. W. Guan, and S. K. Lee. Generalized Union and Project Operations for Pooling Uncertain and Imprecise Information. *Data & Knowledge Engineering*, 18:89–117, 1996.

3. V. Brusoni, L. Console, P. Terenziani, and B. Pernici. Extending Temporal Relational Databases to Deal with Imprecise and Qualitative Temporal Information. In *Recent Advances in Temporal Databases, Proc. of the Int'l Workshop on Temporal Databases*, pp. 3–20, Zurich, Switzerland, 17–18 September 1995.

4. B. Dai, D. A. Bell, and J. G. Hughes. Evidential Temporal Representations and Reasoning. In *Proc. of the 6th Pacific Rim Int'l Conf. on Artificial Intelligence*, pp. 411–422, Cairns, Australia, August 1996.

5. B. Dai, D. A. Bell, and J. G. Hughes. Reasoning about Relations under Uncertainty. In *Proc. of the 10th Int'l Conf. on Industrial & Engineering Applications of Artificial Intelligence & Expert Systems*, pp. 93–98, Atlanta, Georgia, USA, June 1997.

6. D. Dey and S. Sarkar. A Probabilistic Relational Model and Algebra. *ACM Transactions on Database Systems*, 21(3):339–369, September 1996.

7. C. E. Dyreson and R. T. Snodgrass. Valid-time Indeterminacy. In *Proc. of the 9th Int'l Conf. on Data Engineering*, pp. 335–343, 1993.

8. A. Griffiths and B. Theodoulidis. SQL+i: Adding Temporal Indeterminacy to the Database Language SQL. In *Advances in Databases, Proc. of the 14th British Nat'l Conf. on Databases*, pp. 204–221, Edinburgh, UK, 1996.

9. J. Guan and D. A. Bell. *Evidence Theory and its Applications, Vol. 1*, North-Holland, 1991.

10. J. Guan and D. A. Bell. *Evidence Theory and its Applications, Vol. 2*, North-Holland, 1992.

11. H. Y. Hau and R. L. Kashyap. Belief Combination and Propagation in a Lattice-Structured Inference Network. *IEEE Transactions on Systems, Man, and Cybernetics*, 20(1):45–58, January/February 1990.

12. C. S. Jensen, J. Clifford, R. Elmasri, et al. A Consensus Glossary of Temporal Database Concepts. *SIGMOD RECORD*, 23(1):52–64, March 1994.

13. M. Koubarakis. Database Models for Infinite and Indefinite Temporal Information. *Information Systems*, 19(2):141–173, 1994.

14. S. K. Lee. Imprecise and Uncertain Information in Databases: An Evidential Approach. In *Proc. of the 8th Int'l Conf. on Data Engineering*, pp. 614–621, 1992.

15. S. K. Lee. An Extended Relational Database Model for Uncertain and Imprecise Information. In *Proc. of the 18th VLDB Conf.*, pp. 211–220, Vancouver, British Columbia, Canada, 1992.

16. E.-P. Lim. An Evidential Reasoning Approach to Attribute Value Conflict Resolution in Database Integration. *IEEE Transactions on Knowledge and Data Engineering*, 8(5):707–723, October 1996.

17. R. T. Snodgrass (ed.). *The TSQL2 Temporal Query Language*, Kluwer Academic Publishers, 1995.

# A First Approach to Temporal Predicate Locking for Concurrency Detection in Temporal Relational Databases Supporting Schema Versioning

Cristina De Castro

C.S.I.TE. - C.N.R.

University of Bologna, Viale Risorgimento 2, I-40136 Bologna, Italy

## Abstract

*Recently, a great deal of attention has been devoted to the problem of schema versioning. A possible solution [3, 7] maintains all the temporal intensional information in a unique "completed schemata" and all the underlying temporal data in a unique structure, here called "single-pool". As in [3, 7], the data can be accessed through any schema version, thus the problem arises of the concurrency control of the access to the single-pool. A simple solution is here proposed that detects conflicts checking first the intensional and then, if needed, the extensional level.*

**Keywords:**
*Temporal Data, Intensional Concurrency, Extensional Concurrency.*

## 1 Introduction

Two different and independent time dimensions are usually considered: *transaction-time* and *valid-time*. Transaction-time tells when an information is recorded or updated in a database, it is defined by the system and can only grow. Valid-time tells when an event occurs, occurred or is expected to occur in the real world, it is defined by the user and can concern present, past or future events. According to the temporal dimensions they support, temporal relations can be classified as *monotemporal* (transaction- or valid-time), *bitemporal* (both time dimensions) or *snapshot* (no time dimension) [6, 7].

When a schema change is applied in a non temporal environment, the old schema is substituted by a new one. In this way, a portion of intensional information may no longer be available and the corresponding extensional information could consequently be lost. In order to avoid information loss, the concept of *schema versioning* has been introduced. A rich survey on the subject can be found in [7]. A possible solution, consistent with the one adopted in [7, 3] for schema evolution and schema versioning is here adopted and briefly recalled. As far as the intensional level is concerned, a natural solution for the support of schema versioning is to define and manage the catalogues as temporal relations. As far as the extensional level is concerned, the data corresponding to all the schema versions of a relation are maintained in a single data repository, here named "single-pool". In the paper, the term *data pool* denotes a generic repository for extensional data. The solution is discussed only at a logical level. The single-pool consists of a repository where all the extensional data are stored according to a global schema (*completed schema* in [7]), that includes all the attributes introduced so far by successive schema changes. Data are thus maintained according to the largest schema so far defined and only the catalogues maintain the history of the changes. This approach can lead the same data to be concurrently accessed through different schema versions, thus the problem must be faced of the concurrent access to the single pool.

In this paper a simple method is proposed, based on controls on the involved temporal values specified by the concurrent transactions. In order to compare this approach with traditional serialization ones, we consider two possible solutions: first we apply traditional techniques for transaction serialization [1, 5]; second we use our method that, instead of directly applying a serialization, detects the situations of effective conflict on the basis of the temporal portions of data involved in the concurrent transactions. As a matter of fact, concurrent transactions can turn out to be in conflict or not on the basis of the temporal portions of data they are concerned with. For instance, if two transactions refer to disjoint portions of time, there is no need of further concurrency control. On the basis of the type of concurrency (write-write, read-read, read-write) and of the temporal intervals specified in the transactions, it is possible to detect those transactions or *portions of transactions* that are actually in conflict. The proposed approach can be as well used both to intensional and extensional temporal data and is in fact applied first at the intensional level and second at the extensional one. This method, thought for temporal data, is a particular simple case of the fascinating and difficult general problem of *Predicate Locking* [4].

## 1.1 Temporal Notation

In this paper we adopt the *Bitemporal Conceptual Data Model* (BCDM in [6]) for temporal representation. Time is represented by disjoint unions of discrete intervals, called *temporal elements*. A temporal element is composed of *chronons*, where a chronon is a non-decomposable interval of time, whose prefixed duration $p$ represents the chosen granularity of time (days, months, years, etc.) [6, 7]. Therefore, temporal data and transactions are assigned a time they refer to, in terms of temporal elements. In a temporal environment, the role of the traditional key, which usually identifies an object (entity or relationship), is assumed by a time-invariant identifier, such as a system defined *surrogate* [6, 7]. In this case, a time-invariant identifier identifies the collection of the versions of an object. The temporal attributes identify each version within a history, and thus they become part of the *temporal key* [6, 7].

In order to formally define a temporal relation, we need the following:

**Definition:** given a time-invariant identifier $k$, a set of non temporal attributes $r$, a temporal format $X$, where $X \in \{s, t, v, b\}$ (snapshot, transaction-time, valid-time, bitemporal), and a temporal element $\mathcal{T}_X$ of format $X$, a version along the temporal dimension $X$ can be defined as $r_X = (k, r, \mathcal{T}_X)$ and a relation of temporal format $X$ can thus be defined as a set $R_X$ of versions $r_X$.

The symbol 0 is used to denote the run-time of the first transaction and the minimum value of valid-time. $NOW$ denotes the current transaction time and the present valid time. The symbol $UC$ (*Until Changed*) refers to transaction-time and means that no further update has occurred yet. The symbol $FOREVER$ refers to valid-time and means that the end of data validity has not been specified. The full temporal domains of transaction-time, valid-time and the bitemporal one are thus:

$$\mathcal{U}_t = \{0 .. UC\} \quad \mathcal{U}_v = \{0 .. FOREVER\}$$
$$\mathcal{U}_b = \mathcal{U}_t \times \mathcal{U}_v = \{0 .. UC\} \times \{0 .. FOREVER\}$$

We say that: a transaction-time tuple $(k, r, \mathcal{T}_t)$ is *current* if $NOW \in \mathcal{T}_t$; a valid-time tuple $(k, r, \mathcal{T}_v)$ is *present* if $NOW \in \mathcal{T}_v$; a bitemporal tuple $(k, r, \mathcal{T}_b)$ is *current* if $\exists \, t \in \mathcal{U}_v \, : \, (NOW, t) \in \mathcal{T}_b$.

Depending on the represented time dimensions, the semantics of temporal relations vary deeply. For the sake of brevity, we address [2] for discussion and examples on the subject.

## 2 The Method for Concurrency Detection

### 2.1 Split of a Temporal Transaction

The transactions we consider are composed of a single read or write operation. Moreover, without loss of generality, we consider two concurrent transactions at a time.

Two transactions $T_1$ and $T_2$ are considered concurrent if they concern the same history within a temporal relation $R_X$, i.e. they refer to the same time-invariant identifier $k$ within $R_X$. This does not suffice to say that two temporal transactions are really in conflict. As a matter of fact, in a temporal environment we must consider the temporal pertinence of the transactions. In more details, two concurrent temporal transactions $T_1$ and $T_2$ can turn out to be in conflict depending on:

1. the type of the operations (write-write, read-read, read-write)
2. the temporal intervals $T_1.\mathcal{T}_X$ and $T_2.\mathcal{T}_X$ specified by $T_1$ and $T_2$

Therefore, we say that two temporal transactions are in *conflict* if at least one is a write operation and both share a non-empty portion of the temporal elements they refer to. More formally:

**Definition:** two temporal transactions $T_1$ and $T_2$ are in conflict iff:

$$T_1 = write \quad \text{or} \quad T_2 = write \quad \text{and} \quad T_1.\mathcal{T}_X \cap T_2.\mathcal{T}_X \neq \emptyset$$

The real conflict is only in the part where $T_1.\mathcal{T}_X$ and $T_2.\mathcal{T}_X$ overlap. The remaining parts which are not concerned with the intersection of the temporal elements of $T_1$ and $T_2$ can be confidently validated. Each temporal transaction can therefore be split in two sub-transactions, one validable and one where the conflict must be solved:

$$split(T_i) = (T_i^{validable}, T_i^{conflict}) \quad i \in \{1,2\}$$

The obtained sub-transactions can be executed separately.

## 2.2 Concurrency on Transaction-Time Data

As follows from the background, in case of transaction-time relations, read operations can be performed along the whole transaction-time domain $\{0..UC\}$, whereas write operations can only be effected on the temporal element $\{NOW..UC\}$. In other words, the read-area of a transaction-time history is the whole transaction-time axis and the write area is the semiaxis $\{NOW..UC\}$. Moreover, since transaction-time is defined by the system, we argue that *write operations on transaction-time can not be split* but only serialized. This safe approach will be re-discussed in section 3.4. Therefore, two concurrent transactions $T_1$ and $T_2$ can be in the following concurrent situations:

1. $T_1$ **writes** on $\{NOW..UC\}$, $T_2$ **writes** on $\{NOW..UC\}$
2. $T_1$ **reads** on $T_1.\mathcal{T}_t$, $T_2$ **reads** on $T_2.\mathcal{T}_t$
3. $T_1$ **reads** on $T_1.\mathcal{T}_1$, $T_2$ **writes** on $\{NOW..UC\}$

Let us describe two possible concurrency managements: a semi-traditional serialization method and our proposal, that, in case of transaction-time data, is based on the split of *read* operations in case of a read-write conflict.

In both solutions, the first and the second case are managed in the same way: the write-write conflict needs serialization, the read-read operations can be validated. The difference is in the read-write conflict.

### ...with a Semi-Traditional Serialization Method

In this case, as far as the read-write conflict is concerned, if the read operation refers to archived data, both transactions can be validated, since the temporal elements of $T_1$ and $T_2$ do not overlap and there is no actual conflict.

A read transaction $T_1$ refers to the archived portion of a history iff it does not involve the current version of such history. The condition $NOW \notin T_1.\mathcal{T}_t$ does not assure this condition. As a matter of fact, a current record can have been recorded before the current instant, (which is maintained such by the system!). For instance, the temporal element of the tuple (Ann, Manager 4.000, $\{93..UC\}$) in Tab. ?? contains $NOW$, and it is thus current, but was recorded in 1993. A read query on Ann could refer to $\{80..95\}$ and thus involve the current version. Let us now detail the condition that assures a read operation $T_1$ to refer only to the archived portion:

If we indicate by $r_t^{current}$ the current version to be updated by $T_2$ and by $r_t^{current}.\mathcal{T}_t$ its temporal element, such condition can be formalized as follows:

$$max(T_1.\mathcal{T}_t) < min(r_t^{current}.\mathcal{T}_t) \tag{1}$$

If ( 1) is verified, $T_1$ and $T_2$ can be validated; otherwise, they must be serialized. This approach uses both a check on the temporal pertinence of data and a traditional serialization method. It is thus a *semi-direct* application of traditional serialization methods, since a check on temporal intervals is first made for the detection of those transactions which do not overlap any other one.

Note that this method never splits a transaction.

### ...with Temporal Transaction Split (of Read Operations)

In this case, as far as the read-write conflict (3) is concerned, the read transaction $T_1$ is split as follows:

$$split(T_1) = (T_1^{validable}, T_1^{conflict})$$

where the temporal elements of the sub-transactions are respectively:

$$T_1^{validable}.\mathcal{T}_t = T_1.\mathcal{T}_t \cap \{0..NOW\} \quad , \quad T_1^{conflict}.\mathcal{T}_t = T_1.\mathcal{T}_t \cap \{NOW..UC\}$$

In this case, $T_1^{validable}$ is validated, whereas $T_1^{conflict}$ and $T_2$ must be serialized.

It is easy to check that this method leads to the same result produced by the serialization and to generalize this result to $n$ concurrent transactions.

### 2.3  Concurrency on Valid-Time Data

As follows from the background, valid-time data are all current and there are not read areas or write areas delimited *a priori*: data can be read or written along the whole valid-time axis.

Let us consider two concurrent transactions $T_1$ and $T_2$, whose valid-time intervals are respectively $T_1.\mathcal{T}_v, T_2.\mathcal{T}_v$. $T_1$ and $T_2$ can be in the following concurrent situations:

1. $T_1$ writes on $T_1.\mathcal{T}_v$, $T_2$ writes on $T_2.\mathcal{T}_v$
2. $T_1$ reads on $T_1.\mathcal{T}_v$, $T_2$ reads on $T_2.\mathcal{T}_v$
3. $T_1$ reads on $T_1.\mathcal{T}_v$, $T_2$ writes on $T_2.\mathcal{T}_v$

Let us now describe the two possible concurrency managents. In both cases, the second situation, i.e. the read-read operations can be validated. Conflict must be managed in the write-write and in the read-write cases.

### ...with a Semi-Traditional Serialization Method

As far as two temporal transactions $T_1$ and $T_2$ on valid-time data are concerned, there is no difference in the management of the write-write and the read-write conflict. $T_1$ and $T_2$ can be validated iff:

$$T_1.\mathcal{T}_v \cap T_2.\mathcal{T}_v = \emptyset \tag{2}$$

and they must be serialized otherwise. As a matter of fact, 2 assures that $T_1$ and $T_2$ are concerned with two non-overlapping valid-time regions, and thus there is no real conflict.

Again, this approach is a semi-direct application of traditional serialization methods, since a check on temporal intervals is first made for the detection of real conflicts and no transaction is ever split.

## ...with Temporal Transaction Split

In case of valid-time data, this approach can be formalized as follows. Each transaction $T_i$, $i \in \{1, 2\}$ is split in a validable portion $T_i^{validable}$ and in a conflict portion $T_i^{conflict}$, whose valid-time temporal pertinence is:

$$T_1^{validable}.\mathcal{T}_v = T_1.\mathcal{T}_v - T_2.\mathcal{T}_v \quad , \quad T_2^{validable}.\mathcal{T}_v = T_2.\mathcal{T}_v - T_1.\mathcal{T}_v \qquad (3)$$

$$T_1^{conflict}.\mathcal{T}_v = T_2^{conflict}.\mathcal{T}_v = T_1.\mathcal{T}_v \cap T_2.\mathcal{T}_v \qquad (4)$$

i.e. the conflict must be managed on that portion of $T_j.\mathcal{T}_v$ where the other transaction is concerned.

Again, we could check that this method leads to the same result produced by the serialization.

## 3  Concurrency Detection in Schema Versioning

Let us now apply the approach above to the considered environment supporting schema versioning. The query structure we are interested in is defined as follows:

$$Q_i =^{def} Q_i(\mathcal{T}^{int}, \mathcal{T}^{ext}) \qquad (5)$$

where $\mathcal{T}^{int}$ denotes the temporal portion of the schema specified by the query and $\mathcal{T}^{ext}$ denotes the temporal portion of the extensional data specified by the query.

As two temporal versions of the same instance are identified by a time-invariant surrogate $k$ and the extensional temporal element, two temporal versions of the schema of the same table are identified by a surrogate $s$ and the intensional temporal element. Therefore, let us review the conflict conditions in the presence of schema versioning. Two concurrent temporal transactions $Q_i$ and $Q_j$ referring to the same schema surrogate $s$ can turn out to be in conflict depending on:

- **Intensional Level:**
    1. the type of the intensional operations (write-write, read-read, read-write) to be performed on the schema.
    2. the intensional temporal intervals $Q_i.\mathcal{T}^{int}$ and $Q_j.\mathcal{T}^{int}$ specified by $Q_i$ and $Q_j$.

    These conditions state that $Q_i$ and $Q_j$ refer to overlapping schema versions.
- **Extensional Level:**
    1. the type of the extensional operations (write-write, read-read, read-write)
    2. the extensional temporal intervals $Q_i.\mathcal{T}^{ext}$ and $Q_j.\mathcal{T}^{ext}$ specified by $Q_i$ and $Q_j$.

    These conditions state that $Q_i$ and $Q_j$ want to access the same portion of extensional data.

Therefore, we get the following conditions:

**Intensional Conflict:** two temporal transactions $Q_i$ and $Q_j$ are in intensional conflict iff:

$$Q_i = write \quad \text{or} \quad Q_j = write \quad \text{and} \quad Q_i.\mathcal{T}^{int} \cap Q_j.\mathcal{T}^{int} \neq \emptyset$$

**Extensional Conflict:** two temporal transactions $Q_i$ and $Q_j$ are in extensional conflict iff:

$$Q_i = write \quad \text{or} \quad Q_j = write \quad \text{and} \quad Q_i.\mathcal{T}^{ext} \cap Q_j.\mathcal{T}^{ext} \neq \emptyset$$

The real conflict is only in the part where there is an intensional or extensional overlap. The remaining parts that are not concerned with the intersection of the temporal intervals or non temporal attributes of $Q_i$ and $Q_j$ can be confidently validated. Each temporal transaction can therefore be split in two sub-transactions, one validable and one where the conflict must be solved:

$$split(Q_h) = (Q_h^{validable}, Q_h^{conflict}) \quad h \in \{i, j\}$$

Note that this method must obey the following schedule:

1. at the intensional level, the split method must be applied to the temporal intervals of intensional data
2. at the extensional level, the split method must be applied to the temporal intervals of extensional data

Consider two queries $Q_i$ and $Q_j$ and suppose they respectively concern the schema versions $SV_l$ and $SV_m$. $Q_i$ and $Q_j$ are first checked at the intensional level and, as follows from 2, they are in effective conflict iff:

$$SV_l.\mathcal{T}_X \cap SV_m.\mathcal{T}_X \neq \emptyset \tag{6}$$

If the temporal elements of $SV_l$ and $SV_m$ do not overlap, the queries $Q_i$ and $Q_j$ can be validated at the intensional level and the concurrency control skips at the extensional level. As a matter of fact, even if the schema versions $SV_l$ and $SV_m$ are disjoint, $Q_i$ and $Q_j$ can still try to access the same portion of extensional data.

If the temporal elements of $SV_l$ and $SV_m$ overlap, as follows from the method in 2, two cases must be distinguished:

1. in the non-overlap portions $SV_l.\mathcal{T}_X - SV_m.\mathcal{T}_X \quad SV_m.\mathcal{T}_X - SV_l.\mathcal{T}_X$ the queries $Q_i$ and $Q_j$ are not in effective intensional conflict and the concurrency must now be detected at the extensional level, just as in the previous case.
2. In the overlap portion $SV_l.\mathcal{T}_X \cap SV_m.\mathcal{T}_X$ $Q_i$ and $Q_j$ must be serialized.

# 4 Conclusions

In this paper the problem of concurrency detection in schema versioning was discussed. The considered solution for schema versioning used a "completed schema" for intensional data and a "single-pool" for extensional data. The approach to concurrency consisted of two steps: detection of concurrency at the intensional level and succesive detection of concurrency at the extensional level. This method detects real conflicts and splits a temporal transaction in a validable part and in a conflict part, where concurrency must be solved. The method returns the same results as a traditional serialization method does and, provided that in distributed and concurrent systems multiprocess environments can be used, the proposed method can optimize the overall execution time of operations by delay reduction. The price to pay is to revise the concept itself of transaction atomicity, by considering it from a temporal point of view. Further work will be devoted to deepen the general problem of temporal predicate locking, on which the proposed method was based and to the consequent definition of suitable temporal synchronization protocols.

# References

1. S. Ceri, G. Pelagatti, *Distributed Databases: Principles and Systems*, Mc Graw-Hill, 1985.
2. C. De Castro, "Temporal Conversion Functions for Multitemporal Relational Databases", Proc. of *23rd Seminar on Current Trends in Theory and Practice of Informatics (SOFSEM '96)*, Brno Milovy, Czech Republic, 1996.
3. C. De Castro, F. Grandi, M.R. Scalas, "On Schema Versioning in Temporal Databases", Proc. of International Workshop on Temporal Databases, Zürich, Switzerland, 1995.
4. J. Gray, A. Reuter, Transaction Processing: Concepts and Technologies. *Morgan Kaufmann Publishers, San Francisco, California, 1993.*
5. M. Tamer Özsu, P. Valduriez, Principles of Distributed Database Systems, *Prentice Hall International Inc., 1991.*
6. A. Tansel, J. Clifford, V. Gadia, A. Segev, R.T. Snodgrass (eds), Temporal Databases: Theory, Design and Implementation. *The Benjamin/Cummings Publishing Company, Redwood City, California, 1993.*
7. R.T. Snodgrass (ed.) The TSQL2 Language Design Committee, The TSQL2 Temporal Query Language. *Kluwer Ac.Pub., 1995.*

# Efficient Insertion of Approximately Sorted Sequences of Items into a Dictionary

Carlo Gaibisso[1] and Guido Proietti[2]

[1] Istituto di Analisi dei Sistemi ed Informatica del CNR, Viale Manzoni 30, 00185 Roma, Italy. E-mail: gaibisso@iasi.rm.cnr.it

[2] Dipartimento di Matematica Pura ed Applicata, Università di L'Aquila, Via Vetoio, 67010 L'Aquila, Italy. E-mail: proietti@univaq.it

**Abstract.** In this paper a new data structure is introduced for the efficient on-line insertion of a sequence $S$ of keys into a dictionary. The data structure takes advantage by the presence in $S$ of subsets of adjacent keys which maintain their adjacency in the dictionary too, supporting the insertion of $t$ keys in $O(p \log s + (t - p))$ worst case time and $O(s)$ worst case space, where $s$ is the size of the dictionary and $p$ is the number of such subsets. Some applications, as specified in the paper, specifically require for this property.

## 1 Introduction

The dictionary problem has been largely studied and many efficient data structures and algorithms have been designed for its solution. The problem consists in maintaining a subset $D$ of a totally ordered set of keys interested by the following operations: *Insert(x)*, add a new key $x$ to $D$, *Delete(x)*, delete $x$ from $D$ and *Search(x)*, search for $x$ into $D$.

In this paper, we restrict our attention to the class of comparisons based solutions and rely to a quite natural abstract model of computation in this context, the *pointer machine* ([8]). See [6], for an exhaustive treatment of the subject.

Classical data structures, like AVL-trees ([1]) or RB-trees ([5]) and sophisticated approaches, like the one introduced by Brown and Tarjan ([2]), can be adopted in order to deal with the on-line insertion of a sequence of keys into a dictionary.

In this paper a new data structure is introduced, that takes advantage by the presence of sets of adjacent keys in the sequence which maintain their adjacency in the dictionary too: more formally, let $S = k_1, k_2, \ldots, k_t$ be any sequence of keys to be inserted into $D$. $S$ can be intended as composed by $m$, for some $m \geq 1$, sorted subsequences of keys, $S_1, S_2, \ldots, S_m$, such that $k_1, k_2 \in S_1$ and two keys $k_j$ and $k_{j+1}$, $j = 2, 3, \ldots, t - 1$, belong to two different subsequences if and only if $k_{j-1}$ and $k_j$ both belong to the same subsequence and either $k_{j-1} > k_j$ and $k_j < k_{j+1}$ or $k_{j-1} < k_j$ and $k_j > k_{j+1}$. Let $k$ be any key in $S_j$, $1 \leq j \leq m$. If $S_j$ is an increasing (resp. decreasing) subsequence, then $k$ is said to be an *interrupting key* if and only if either $k$ is the first key in $S_j$ or, in $D$,

$k$ immediately follows (resp. precedes) some other key which does not belong to $S_j$. By the data structure here introduced it is possible to insert $S$ into the dictionary in $O(p \log s + (t - p))$ worst case time and $O(s)$ worst case space, where $p$ is the number of interrupting keys in $S$ and $s$ is the final size of the dictionary.

Motivations for the proposed extension have been suggested from our experience on spatial data applications. Usually, spatial data structures are *hierarchical*, that is, they are based on the principle of recursive decomposition of the space. Typical representatives of hierarchical spatial data structures are *quadtrees* [7]. For example, the *region quadtree* is defined to represent two-dimensional spatial regions and is based on recursive decomposition of the space into four quadrants of equal size, until homogeneous quadrants are obtained. There exist a number of linear representations of region quadtrees, which are generally thought for secondary memory applications but can be applied in main memory too. Such linear representations have the common property of assigning to each underlying element of the workspace, a *locational key* [3] belonging to a totally ordered universe and describing both the decomposition level and the spatial position of a given quadrant. The crucial point is that when a refinement process is required for a given subregion of the workspace, a large amount of increasing locational keys has to be inserted. Such keys intersect only a small subset of the preexisting keys, that is those associated to refined quadrants. As a consequence, the number of interrupting keys is quite small.

## 2 The data structure

The data structure is logically organized on two levels. The lower level is made up by a forest $F$ of *Upward Binomial Trees* and *Downward Binomial Trees*, *UBTs* and *DBTs* for short, defined as follows:

**Definition 1.** A binary search tree $T$ with root $r$ is a $h - UBT$ (resp., $h - DBT$), $h \geq 1$, if and only if the subtree rooted at the left (resp., right) son of $r$ is a complete $(h - 1) - UBT$ (resp., $(h - 1) - DBT$), and the subtree rooted at the right (resp., left) son of $r$, if any, is a $h^* - UBT$ (resp., $h^* - DBT$), where $h^* < h$. A single node is a $1 - UBT$ (resp., $1 - DBT$).

Each node in $F$ stores a key. Let $K_T$ denote the set of keys stored by any tree $T$ of $F$, and let $l_T$ and $u_T$ respectively denote the minimum and the maximum keys in $K_T$. For any given couple of trees $T$ and $T^*$ in $F$, if $l_T < l_{T^*}$ then $u_T < l_{T^*}$. The upper level of the data structure is an index implemented by a balanced binary tree $U$ storing the set of keys $\{u_T | T \in F\}$. In addition to $u_T$, each node of $U$ stores $l_T$, a bit indicating whether $T$ is a *UBT* or a *DBT* and a pointer to $T$.

In what follows we will restrict our attention to *UBTs*. The considerations we will introduce and the properties we will prove will apply to *DBTs* too.

Given any *UBT T* and any of its nodes $n$, let *Key(n)* denote the key stored by $n$. Analogously given any key $k \in K_T$, by *Node(k)* we will denote the node of $T$ storing $k$.

Let $n = Node(k)$, together with $k$, $n$ maintains two pointers, *Right(n)* and *Left(n)*, respectively to its right and left son in $T$, and *Height(n)*, the height of $n$ in $T$, defined as 1, if $n$ is a leaf, and as *Height(l) + 1*, if $l$ is the left son of $n$. Given any *UBT T* and any of its nodes $n$, $n$ will be referred to as an *incomplete* node if and only if *Height(n) > 1* and at least one between *Right(n)* and *Left(n)* is null, and as an *out of level* node, if and only if *Height(p) − Height(n) > 1*, where $p$ is the parent of $n$. The following observations are a direct consequence of the above definitions:

1. any *UBT* contains at most 1 incomplete node $n$, and for such a node *Right(n) = null* and *Left(n) ≠ null*;
2. all the out of levels nodes and the incomplete node of any *UBT* appear on its rightmost path;
3. given any *UBT T* and any of its nodes $n$, *Height(n)* is given by the number of nodes on the longest leftmost path downward $T$ starting from $n$;
4. the height of any $h - UBT$ is exactly $h$.

Together with any *UBT T*, we maintain some additional information: a pointer to each out of level and to the only, if any, incomplete node of $T$. All pointers to out of level nodes are organized in a stack in which they appear, from top to bottom, in the same order in which the pointed nodes appear, from right to left, on the rightmost path of $T$. Let us now prove some fundamental properties of *UBT*s:

**Lemma 2.** *Let $T$ be any $h-$UBT, then $2^{h-1} \leq |K_T| \leq 2^h - 1$.*

*Proof.* If $h = 1$ then $|K_T| = 1$ and the lemma is trivially true. Let us assume the lemma true for any $1 \leq k \leq h-1$ and let $T_{min}$ denote the smallest legal $h- UBT$. By definition, $T_{min}$ is composed by its root and a complete $(h - 1)- UBT$. Consequently, $|K_{T_{min}}| = 1 + (2^{h-1} - 1) = 2^{h-1}$. Analogously, let $T_{max}$ denote the largest legal $h- UBT$. $T_{max}$ is composed by its root and 2 complete $(h-1)- UBT$s, then, by the inductive hypothesis, $|K_{T_{max}}| = 1 + 2 \cdot (2^{h-1} - 1) = 2^h - 1$.

**Theorem 3.** *The height of any UBT $T$ is $O(\log |K_T|)$ in the worst case.*

*Proof.* The proof is a direct consequence of observation 4 and of lemma 2.

**Lemma 4.** $O(\log |K_T|)$ *worst case time is required to associate the required additional information to any UBT.*

*Proof.* The proof is a direct consequence of theorem 3 and observation 2

# 3   How to insert, delete or search for a single key

In this section we will prove that inserting a single key in the data structure requires $O(\log s)$ worst case time and $O(s)$ worst case space, where $s$ is the size of the dictionary. In a quite similar way it is possible to prove that both the *Delete()* and the *Search()* operations can be implemented with the same time and space complexity. Such proofs are omitted from the paper. Most of the considerations we will introduce in this section will be of great aid in efficiently processing sequences of keys.

**Lemma 5.** *Given any $h$–UBT $T$ and any node $n$ appearing on its rightmost path, then $O(\log|K_T|)$ worst case time is required to split $T$ in at most $4$ UBTs, $T_1$, $T_2$, $T_3$ and $T_4$ such that: $K_{T_1} = \{k|k \in K_T \wedge k \leq \mathrm{Key}(p)\}$, where $p$ is the parent of $n$; $K_{T_2} = \{k|k \in K_T \wedge \mathrm{Key}(p) < k < \mathrm{Key}(n)\}$; $K_{T_3} = \{\mathrm{Key}(n)\}$; $K_{T_4} = \{k|k \in K_T \wedge k > \mathrm{Key}(n)\}$.*

*Proof.* Figure 1 shows how, once $p$ has been identified, which by theorem 3 requires $O(\log|K_T|)$ worst case time, $T_1$, $T_2$, $T_3$ and $T_4$ can be generated from $T$. Futhermore, by observation 3, $n$ is the only node for which the value of the *Height* field must be updated and set to 1. Finally, associating the required additional information to $T_1$, $T_2$, $T_3$ and $T_4$, requires, by lemma 4, $O(\log|K_T|)$ in the worst case, thus proving the theorem.

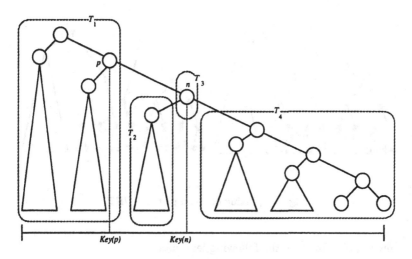

**Fig. 1.** How to obtain $T_1$, $T_2$, $T_3$ and $T_4$ from $T$.

**Lemma 6.** *Let $T$ be any complete $h$–UBT and let $n$ be any node of $T$, then $O(\log|K_T|)$ worst case time is required to split $T$ in a $h_U$–UBT, $T_U$, and a $h_D$–DBT, $T_D$, such that $h_U, h_D \leq h$, $K_{T_U} = \{k|k \in K_T, k \leq \mathrm{Key}(n)\}$ and $K_{T_D} = \{k|k \in K_T, k > \mathrm{Key}(n)\}$.*

*Proof.* Without loss of generality let us assume $n$ is not a leaf and let $P = n_1, n_2, \ldots, n_l$, $l \leq h$, be the sequence of nodes appearing on the path from the root of $T$ to the right son of $n$, as shown in figure 2. $T$ is broken into $l$ pieces by $l - 1$ steps. Step $i$, $i = 1, 2, \ldots, l - 1$, processes $T_i$, the subtree of $T$ rooted at $n_i$, and generates either a $(h - i + 1) - UBT$ or a $(h - i + 1) - DBT$ as follows: if $n_{i+1}$ is the right (resp., left) son of $n_i$, then $T_i$ is split in the subtree of $T_i$ rooted at $n_{i+1}$ and the $(h - i + 1) - UBT$ (resp., $(h - i + 1) - DBT$) made up by $n_i$ and the subtree rooted at its left (resp., right) son. In such a way $T$ is split in at most $h$ components some lying to the right of $P$, $UBTs$, some to the left, $DBTs$, plus one complete tree, the one rooted at the right son of $n$, which we will be considered as a $DBT$.

Let $U = \{U_{i_1}, U_{i_2}, \ldots, U_{i_q}\}$, $q \leq h - 1$, $i_1 \leq i_2, \ldots, i_q$, be the set of the so generated $UBTs$, where $U_{i_j}$, $1 \leq j \leq q$, is generated by processing $T_{i_j}$. $T_U$ can be easily obtained by connecting the root of $U_{i_j}$ to the root of $U_{i_{j+1}}$, for $j = 1, 2, \ldots, q - 1$. It is not difficult to realize that $T_U$ is a $UBT$ and that for each key $k$ stored at any of its node, $k \leq Key(n)$. Furthermore, since we always connect a $h - UBT$ to the right of a $h' - UBT$, with $h' > h$, $Height(n)$ is properly set for each node $n$ of $T_U$.

As far as the time complexity is concerned, by theorem 3, $O(\log |K_{T_U}|)$ worst case time is required in order to split $T$, to generate $T_U$ and, by lemma 4, to produce the required additional information.

$T_D$ can be obtained in an analogous way, thus proving the theorem.

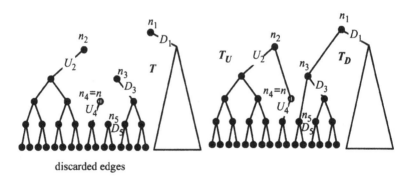

**Fig. 2.** Splitting of a complete $h - UBT$

It is now possible to state the following lemmata.

**Lemma 7.** *Any* Insert$(x)$ *operation can be accomplished in* $O(\log |F| + \log |K_{T_{max}}|)$ *worst case time, where* $T_{max}$ *is the tree of $F$ with maximum size and $|F|$ denotes the number of trees in $F$.*

*Proof.* Given any key $x$, $O(\log |F|)$ worst case time is required to determine either the unique couple of trees $T$ and $T'$ of $F$ s.t. $u_T < x < l_{T'}$, or the unique tree $T$ of $F$ such that $l_T < x < u_T$.

In the first case a new *UBT* with just one node storing $x$ is added to $F$ and $U$ is correspondingly updated, which requires $O(\log|F|)$.

Otherwise, let us assume that $T$ is a *UBT*. Then, in $O(\log|K_T|)$ worst case time it is possible to identify the unique node $n$ of $T$ storing the maximum key smaller than $x$. If $n$ lies on the rightmost path of $T$ then, by separating the subtree rooted at the right son of $n$ from $T$, it is possible to split $T$ into two *UBTs*, $T_1$ and $T_2$, such that $K_{T_1} = \{k|k \in K_T \wedge k < x\}$ and $K_{T_2} = \{k|k \in K_T \wedge k > x\}$.

By lemma 4, $O(\log|K_T|)$ worst case time is spent in order to associate to both $T_1$ and $T_2$ the required additional information. Finally, a new *UBT* with just one node storing $x$ is added to $F$, and, since a constant number of new data structures has been introduced, $O(\log|F|)$ worst case time is spent in order to update the content of $U$.

If $n$ does not ly onto the rightmost path of $T$ then it lies inside a complete subtree $T_n$ of $T$, whose root is the left son of some node $n^*$ on the rightmost path of $T$. By lemma 5, in $O(\log|K_T|)$ worst case time, $T$ can be split into four *UBTs*, $T_1$, $T_2 = T_n$, $T_3$ and $T_4$ such that: $K_{T_1} = \{k|k \in K_T \wedge k \leq Key(p)\}$, where $p$ is the parent of $n^*$; $K_{T_2} = \{k|k \in K_T \wedge Key(p) < k < Key(n^*)\}$; $K_{T_3} = \{Key(n^*)\}$; $K_{T_4} = \{k|k \in K_T \wedge k > Key(n^*)\}$.

Successively, by lemma 6, $T_2$ can be split in $O(\log|K_T|)$ worst case time, in a *UBT*, $T_U$, and a *DBT*, $T_D$, in such a way that, $K_{T_U} = \{k|k \in K_{T_2} \wedge k < x\}$ and $K_{T_D} = \{k|k \in K_{T_2} \wedge k > x\}$.

Finally a new *UBT* with just one node storing $x$ is added to $F$, and since a constant number of new data structures has been generated, then $U$ can be updated in $O(\log|F|)$ worst case time. Since $|T| \leq |T_{max}|$ the lemma is proved.

**Theorem 8.** *By the introduced data structure it is possible to implement any* Insert($x$) *operation in* $O(\log s)$ *worst case time and* $O(s)$ *worst case space, where* $s$ *is the number of keys currently maintained in the dictionary.*

*Proof.* The theorem follows from lemmata 7 and from the fact that $|K_{T_{max}}| = |F| = O(s)$.

## 4  How to efficiently insert a sequence of keys

Let us first of all prove the following theorem, which is crucial in efficiently processing non-interrupting keys:

**Theorem 9.** *For any* UBT *T and for any key* $k > u_T$, *k can be added to T in* $O(1)$ *worst case time.*

*Proof.* Let us consider the following procedure of insertion, which has already been introduced in a quite different context ([4]). A new node $n_k$ for $k$ is added to $T$ in the following way:

1. if any incomplete node $n$ exists in $T$, then $Height(n_k)$ is set to 1 and $n_k$ is inserted as the right son of $n$. If $Height(n) - Height(n_k) > 1$ then a pointer to $n_k$ is pushed onto the top of the stack of pointers associated to $T$;

2. if $T$ does not contain either incomplete or out of level nodes, let $r$ be the root of $T$. Then $n_k$ is inserted as the the new root of $T$, $r$ becomes the left son of $n_k$, $Height(n_k)$ is set to $Height(r) + 1$ and a pointer to $n_k$ is maintained as the only incomplete node of $T$;

3. otherwise, let $l$ be the rightmost out of level node in $T$ and $l'$ be its parent, then $n_k$ is inserted as the right son of $l'$, $l$ becomes the left son of $n_k$, $Height(n_k)$ is set to $Height(l) + 1$ and a pointer to $n$ is maintained as the only incomplete node of $T$. Finally, the pointer to $l$ is popped from the top of the stack associated to $T$ and, if $Height(l') - Height(n_k) > 1$, a pointer to $n_k$ is pushed onto the top of the same stack.

Let us prove that such procedure correctly works with respect to *UBT*s. If $T$ contains an incomplete node $n$, this must be the rightmost node in the tree. If $T$ does not contain either out of level or incomplete nodes, $T$ is a complete tree. If $T$ does not contain an incomplete node but at least one out of level node, then the subtree rooted at the leftmost of such nodes will be a complete tree, since it will not contain either incomplete or out of level nodes. In any case, $n_k$ will be correctly positioned inside $T$, the resulting data structure will still be a *UBT* and the additional information will be correctly updated. In addition, by observation 2, $Height(n)$ is properly set for each node $n$ of $T$. Obviously the above introduced procedure requires $O(1)$ worst case time to be implemented.

**Theorem 10.** *Any on-line sequence of keys $S = k_1, k_2, \ldots, k_t$, $t \geq 1$, can be inserted into the dictionary in $O(p \log s + (t - p))$ worst case time, where $p$ is the number of interrupting keys in $S$ and $s$ is the final size of the dictionary.*

*Proof.* We will prove that any $Insert(k_i)$ operation, $1 \leq i \leq t$, can be accomplished in $O(\log s)$ worst case time if $k_i$ is an interrupting key, in $O(1)$ worst case time otherwise.

Let us first of all assume that the following facts are true before inserting $k_i$, for some $2 \leq i \leq t$:

1. it is known whether $k_{i-1}$ is an interrupting key;
2. if $k_{i-1}$ is not an interrupting key, it is known whether the subsequence to which it belongs is increasing or decreasing;
3. let $T_{i-1}$ be the tree of $F$ storing $Node(k_{i-1})$, then a pointer to each of the records in $U$ for $T_{i-1}$, $T_{i-1}^+$ and $T_{i-1}^-$ is available, where $T_{i-1}^+$ and $T_{i-1}^-$ are the trees of $F$ storing the minimum key greater than $u_{T_{i-1}}$ and the maximum key smaller than $l_{T_{i-1}}$, respectively;
4. $k_{i-1}$ is the maximum key stored by $T_{i-1}$, if $T_{i-1}$ is a *UBT*, the minimum one otherwise.

We will prove that the same facts, which are obviously true for $k_1$, will still be true with respect to $k_i$, once it has been added to the dictionary.

If $k_{i-1}$ is an interrupting key then $k_i$ is surely a non-interrupting key. Otherwise if $k_{i-1}$ belongs to an increasing (resp., decreasing) sequence then $k_i$ is an interrupting key if and only if either $k_{i-1} > k_i$ (resp., $k_{i-1} < k_i$) or $k_i > l_{T_{i-1}^+}$

(resp., $k_i < l_{T_{i-1}^-}$). Consequently if the above listed information is available, then it is possible in constant time to decide whether $k_i$ is an interrupting key or not.

If $k_i$ is an interrupting key then, by quite the same procedure introduced by lemma 7, in $O(\log s)$ worst case time a new tree is added to $F$ with just one node storing $k_i$ and $U$ is correspondingly updtated. The only difference with respect to such procedure is that, at the time a second key will be added to the tree, it will be specified whether $T_i$ is a $UBT$ or a $DBT$. This will not create any kind of problem since a tree made by just one node is both a $UBT$ and a $DBT$. Moreover, with respect to $k_i$, fact 1 is true. As far as fact 3 is concerned, a pointer for the record in $U$ for $T_i$ is clearly available, while $O(\log s)$ worst case additional time is required to obtain the pointers to those for $T_i^+$ and $T_i^-$. Fact 4 is obviously true.

Let us now consider the case $k_i$ is not an interrupting key.

If $k_{i-1}$ is an interrupting key, then $T_{k_{i-1}}$ will contain exactly one node. Consequently if $k_i > k_{i-1}$ then $T_i$ will be identified as a $DBT$, as a $UBT$ otherwise. Furthermore by theorem 9 and by the available information, $k_i$ will be added to $T_{k_{i-1}}$ and $U$ will be correspondingly updated in $O(1)$ worst case time. Finally fact 2 is clearly true. Fact 3 is true since the already available pointers have not to be changed and fact 4 is obviously true.

The case $k_i$ and $k_{i-1}$ are both not-interrupting keys is analogously processed, with the main difference in the fact that at the time $k_i$ has to be added to $T_{k_{i-1}}$, it has already been identified either as a $UBT$ or as a $DBT$.

The theorem is consequently proved.

# References

1. Adelson-Velskii, G.M., Landis, E.M.: An Algorithm for the Organization of the Information. Dokl. Akad. Nauk SSSR 146 (1962) 263–266
2. Brown, M.R., Tarjan, R.E.: A Fast Merging Algorithm. J. ACM 26 (2) (1979) 211–226
3. Gargantini, I.: An Effective Way to Represent Quadtrees. Communication of the ACM 25 (12) (1982) 905–910
4. Gaibisso, C., Gambosi, G., Talamo, M.: A Partially Persistent Data Structure for the Set-Union Problem. Theoretical Informatics and Applications 24 (2) (1990) 189–202
5. Guibas, L.J., Sedgewick, R.: A Dichromatic Framework for Balanced Trees Proc. of the 10th Symp. on Foundations of Computer Science. Washington, DC, (1978) 8–21
6. Knuth, D.E.: The Art of Computer Programming, Volume 3: Sorting and Searching. Addison-Wesley, Reading, MA, (1973)
7. Samet, H.: The Quadtree and Related Hierarchical Data Structures. Computing Surveys 16 (2) (1984) 187–260
8. Tarjan, R.E. : Reference Machines Require Non-Linear Time to Maintain Disjoint Sets. Proc. of the 9th ACM Symp. on Theory of Computing, Boulder, Colorado (1977) 19–29

# High Availability Support in CORBA Environments*

Pablo Galdámez, Francesc D. Muñoz-Escoí and José M. Bernabéu-Aubán

Universitat Politècnica de València,
Camí de Vera, s/n,
46071 València, Spain

**Abstract.** Distributed systems are a good basis to support highly available applications. In a distributed system there are multiple nodes which have independent behavior when failures arise, i.e., the failure of one node does not mean the failure of the others. So, if some support is given by the underlying system, applications can be made highly available decomposing them in components, and placing replicas of those components into independent nodes. This paper describes Hidra, a CORBA-based architecture, where object oriented distributed applications can increase their availability placing a number of object replicas in different domains or different nodes. The system provides failure detection and failure recovery mechanisms to maintain the applications' state consistent.

## 1 Introduction

CORBA [14] has been taken as the basis to develop a big amount of object oriented distributed applications, using an *object request broker* as the system component which deals with object intercommunication in distributed environments. Moreover, there are a set of additional services [15] which can be used to extend the functionality of the ORB, providing support for object names, persistence, event notification, transactions, concurrency control, etc. However, in the current CORBA specification there is no support for object replication, which is the basis to develop highly available applications.

Our Hidra architecture [8] extends the basic ORB functionality with integrated support for several types of object replication and with some ORB related components needed to develop highly available applications following the *coordinator-cohort* replication model [5].

When object replication is considered, several models can be distinguished. The *passive replication* approach [7] maintains a primary replica which receives all the client requests, updating its state accordingly and later checkpointing its state to a set of secondary or backup replicas. This technique has been used in several systems [6]. Its main drawback is that the primary replica may become a bottleneck if the requests being served require a lot of computing effort. On

---

* This work was partially supported by the CICYT (Comisión Interministerial de Ciencia y Tecnología) under project TIC96-0729.

the other hand, its communication requirements are low, at least when protocol complexity is considered. In the *active replication* model [16] all object replicas receive the client requests and simultaneously update their local states which remain consistent due to the communication protocols (as those used in group communication toolkits [1, 3, 17]) used to guarantee the appropriate delivery order when the client request is multicast. This approach was used in [11], also integrated in an ORB. Its main advantage is that failure recovery does not require the election of a new distinguished entity (as is the case of the primary replica if passive replication is used and the primary fails), but this technique requires additional efforts to ensure that all object replicas receive the same requests in the same order.

An intermediate approach is the *coordinator-cohort replication* model. It is a variation of the passive replication approach where the primary replica may vary for different requests. In this case, the replica which receives the request is its coordinator, and all other replicas act as the secondaries for that request. They are the cohorts. So, if this model is used, the problems of the passive replication technique are avoided; i.e., the load is balanced among all the object replicas and multiple requests can be served concurrently if some care is taken to guarantee that the same state of the object is not modified simultaneously by two different coordinators. Other examples of replication approaches that are not strictly passive or active can be found in other systems, such as [10].

In Hidra, some support for several kinds of replication has been integrated into the ORB, which provides different types of object references to manage them. Additionally, some other components have been provided to deal with the special requirements of the coordinator-cohort model, such as concurrency control mechanisms and support for light weight transactions which ensure that all replica changes are consistent.

The rest of the paper is structured in two additional sections. Section 2 describes the Hidra architecture components, focusing on the Object Request Broker, transactions and concurrency control, terminating our description in Sect. 3 where we present some conclusions about this work.

## 2  Architecture

As shown in Fig. 1, Hidra is composed by several layers, each one making use of the lower ones and providing services to the upper ones. The lowest level is a datagram transport layer that provides unreliable message transfer among all the nodes in the system.

On top of this layer a group membership protocol [13] continuously monitorizes the nodes in the system to keep distributed agreement on the set of nodes that are considered to be up and running. This protocol, while providing the fail-stop failure model used in Hidra, serves for two other basic purposes. First, when a node is computed out of the membership, every ORB in the system is notified to reconfigure itself as required. Reconfiguration is done synchronously by the ORB's of all the surviving nodes in successive steps of reconfiguration

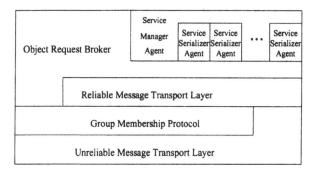

**Fig. 1.** The Hidra core architecture

triggered by the protocol. And second, the protocol also notifies the upper transport layer of node failures to help it implement reliable message transfers. Each time the system reconfigures, a new incarnation number is set to the surviving nodes as the current incarnation number. This way, orphan messages and invalid resources that belong to failed nodes can be promptly detected.

The reliable message transport layer uses the group membership protocol to detect node failures in order to abandon communications in course with the failed nodes. Message communications are retried until they complete or until the destination node is computed out of the membership set. However, messages are not expected to reach their destinations in the same order as they were sent.

On top of the reliable transport layer, an ORB based on the one described in [2], provides object invocation services, object reference counting and support for different types of replicated objects (using several types of object references). Object invocation uses the reliable message passing implemented by the transport layer, returning specific exceptions when the destination node fails. These exceptions may be either propagated to the client domain or intercepted by the ORB itself to retry the invocation on other object replicas if the destination object was replicated. The second main function of the ORB, is to count how many references point to each object implementation. When the count of references on a given object drops to zero, its implementation receives an *unreferenced* notification which means that its operations can not be requested by any other object in the system and that all its resources have to be released.

Two other distributed components complete the Hidra architecture: the *service manager* and the *service serializer*. They provide additional support to allow a coordinator-cohort replication model. The *service manager* (SM) provides all the administrative service operations which include the creation and destruction of services. It also knows where each service replica is located and which is their replica class, being involved thus, on replica registration, replica de-registration, promotion or degradation of replicas, ... participating also in the protocols run to recover from replica failures. Second, there exists a *service serializer* (SS) entity for each service installed in Hidra. This component implements a concurrency control mechanism to serialize the requests made to a service.

The service manager is itself a replicated service. To replicate it, there exists at each node a *service manager agent* (SMA) with the same interface as the SM. Each SMA acts as the local representative of the SM at each node and also performs any service management operations that could be resolved locally inside a node. For instance, if two primary replicas are placed into the same node, failures that affect one of these replicas do not require the SM intervention, sufficing the local knowledge maintained by the SMA to reconfigure the service.

At each Hidra node with primary replicas for a particular service, there exists a *service serializer agent* (SSA) to manage the concurrency control needs of its local service primaries. In addition, one of the service SSA's also acts as the *service serializer* (SS), that provides concurrency control decisions that cannot be locally resolved. To guarantee SS availability, in the case its node fails, a distributed protocol is run by the surviving SSA's to elect a new SS and to reconstruct its state from the state stored at each SSA.

## 2.1 The Object Request Broker

The Hidra ORB follows a similar design to the ORB described in [2]. Every domain in a Hidra node has access to the ORB services, where a special, trusted domain, also hosting the complete Hidra components, translates object references when inter-domain or inter-node invocations are requested. The rest of the domains, being untrusted, contain a reduced ORB structure, requiring the trusted domain intervention to access objects placed outside their address space. To increase system availability, we place the Hidra core at the operating system kernel which will be also the trusted domain. This results in the accomplishment of some other interesting properties. For instance, any operating system service such as device drivers or file systems, may access at low cost the object services and the high availability support provided by Hidra.

The ORB is composed by two layers, the handler layer and the reference layer. The handler level holds client and server handlers. Each server handler is attached to one object implementation while client handlers are connected to proxy representations of the server object. Handlers are responsible for marshaling input arguments, sending them to the object implementation, waiting for a reply, and unmarshaling the output arguments. To provide object implementors with different marshaling, unmarshaling or exception handling procedures, a number of handlers exist in our ORB. For instance, for objects whose marshaled representation includes a big stream of data, there exists a specialized handler providing marshaling and unmarshaling methods that avoids making unnecessary copies of the marshal stream.

Each handler is connected to an xdoor from the reference layer. Server handlers are connected to server xdoors while client handlers to client xdoors. Xdoors serve as the communication endpoints for object invocation. The state of client xdoors contain the location of the server xdoor they stand for, redirecting invocations to them when invocations over client handlers are requested. Xdoors are included in the trusted domain, while in the rest of the domains, an implementation-reduced version of them, called gated-xdoors, provide the attachment from the object handlers to the xdoors residing at the kernel level. Similarly to handlers, a number of xdoor classes provide different marshaling and unmarshaling procedures. The most common xdoor type is the standard xdoor, which running a distributed protocol, counts the object references sent to remote nodes. The reference counting protocol is responsible for delivering an unreferenced notification to the object handler, when no external references point to the object implementation.

To give support for object replication as the basic construction to replicate services, we make use of the extensibility mechanism provided by this ORB. We implement a hierarchy of new classes of xdoors and handlers that implement new marshaling, unmarshaling and invocation processing procedures for replicated objects. The new base handler class is called the replicated object handler (ROhandler). When this handler receives the standard CORBA exception raised when the invocation destination fails, the invocation is simply retried[1]. The new base class of xdoor, called the replicated object xdoor (ROxdoor), is responsible for choosing an adequate target node or set of nodes for each invocation it starts. The ROxdoor closely interacts with the SMA to access the current service configuration, allowing a quick redirection of invocations to living object replicas in case of failures. The close relationship among ROxdoors and the SMA also allows to block and resume all the object replica activity when processing any of the protocols that modify the service configuration.

Several subclasses of these ROhandler and ROxdoor entities can be developed. For instance, CC-ROhandlers and CC-ROxdoors provide the support needed to deal with invocations in the coordinator-cohort replication model. To this end, the client CC-ROxdoor and the SMA provide methods to change easily the reference of the coordinator replica. There are also Chk-ROhandlers and Chk-ROxdoors that manage the invocations needed to do checkpoints in both the passive and coordinator-cohort models. In this case, the client Chk-ROxdoor is used to invoke all the checkpoint objects in the service except the one assigned to the caller replica.

## 2.2 Concurrency Control

In Hidra, highly available services use the coordinator-cohort replication model. When this approach is followed, a concurrency control mechanism is needed to guarantee that different requests being served by different coordinators do not modify the same part of the service's state. Traditionally, two-phase locking has been applied as the general mechanism to preserve consistency. Even within the CORBA proposal, concurrency control mechanisms have been described for distributed systems [15], which also support nested transactions [12]. In Hidra, exploiting the object oriented model offered to applications, we propose a different, lower cost and mostly asynchronous concurrency control mechanism which

---

[1] As in Spring [9], we retry failed invocations at the handler level. Replication in Spring is made at the subcontracts level which is similar to our handlers level.

borrows some of the concepts presented in [4]. In Hidra, an IDL-like syntax allows service programmers to specify which service object operations may proceed concurrently and which others must proceed sequentially. When installing a service, an object built from this specification is provided as an argument to the ORB replica registration operation. This operation will store at the SS the *service concurrency specification* (SCS) that will be used to serialize invocations made over the service objects.

When a request arrives to the server ROhandler, this object uses the services provided by the SSA to find out if it can proceed. The SSA blocks the incoming request until all the currently running requests which can not be executed concurrently with this one have been completed. The SSA's and the SS maintain the set of operations that are being served and, using the SCS data, they can guarantee that all the incoming requests are served in an order that respects the consistency of the replicated objects.

To reduce the network traffic required to serialize requests, the SCS can be dynamically distributed, placing partitions of it into different SSA's. The SS will always know where do these partitions reside, redirecting serializing requests to them when required. This scheme benefits of the usual service client behavior, which is expected to invoke several highly related operations over the same service replica. A high level policy may direct the SS to place partitions of the concurrency specification into the nodes that most use them.

## 2.3 Light Weight Transactions

If the passive or coordinator-cohort replication models are used, as it is the Hidra case, some support is needed to guarantee the state consistency in all the replicas. This can be achieved if each request is made as a transaction, which is either successfully completed or it does not modify the state of any replica. CORBA services [15] propose transactions as the construction to modify the system state atomically, providing a set of interfaces that allow the construction of a general purpose transaction support. Our approach in Hidra is to use a kind of transactions, that we call *light weight transactions* (LWT), attached to every request made to a replicated service; i.e., our support is only provided for requests on replicated services and it can not be given to requests on other kinds of objects. Our approach while being less general than the CORBA one, is notably less expensive, since it only requires a minimal set of objects to support our transactions, and suffices to provide an environment for the development of highly available applications.

LWT's are initiated by client ROhandlers, that create a small object, called a TID to identify it during the transaction lifetime. The server ROhandler of the request coordinator, after contacting the SSA to serialize the request and once it is allowed to proceed, will expedite the invocation to the replica implementation. During the operation execution, the implementation may require to checkpoint the state modification made by the request to the rest of the replicas. To do so, it invokes the service checkpoint object. This invocation produces that every other replica will receive the checkpoint invocation. Piggybacked into the first checkpoint invocation message, the TID will be sent to the request cohorts, that will learn about the transaction when they receive the message. If the coordinator fails before replying to the client, the appropriate exception will be caught by the client ROhandler, that will automatically reissue the invocation to a different replica, using the same TID as the transaction identifier. The newly elected request coordinator will be able to continue with the operation once it realizes that it corresponds to a previously started request, replying to the client with the operation results.

## 3 Conclusions

CORBA technology aims to provide a complete framework for the development of distributed applications. While the central component is an Object Request Broker, a number of extra services [15] have been proposed and many others will be specified in the future, to address aspects like transactions, replication, persistence, concurrency control, etc. that altogether will provide support for highly available application development. The work we have presented, inspired in this emerging technology, proposes alternative mechanisms for transactions and concurrency control, whose architecture details can be found in [8]. While our system components serve the same mission as the CORBA specified ones, our concern has been to achieve system availability, focusing on object invocation reliability and efficiency.

On the other hand, the development of highly available services over Hidra, just requires as additional tasks, compared to non highly available application development, the design and implementation of checkpoint interfaces and the specification of the level of concurrency allowable inside the service. Moreover, clients of Hidra replicated services are programmed as if those services where non-replicated, being completely unaware of the underlying transactions attached to each invocation they request.

## References

1. Ö. Babaoğlu, R. Davoli, L. A. Giachini, and M. Baker. RELACS: A communications infrastructure for constructing reliable applications in large-scale distributed systems. Technical report, UBLCS-94-15, Dept. of Comp. Sc., Univ. of Bologna, Italy, June 1994.

2. J. Bernabéu, V. Matena, and Y. Khalidi. Extending a traditional OS using object-oriented techniques. In USENIX Assoc., editor, *2nd Conf. on Object-Oriented Technologies & Systems, Toronto, Canada*, pages 53–63, Berkeley, CA, USA, June 1996. USENIX.

3. K. P. Birman. Replication and fault-tolerance in the ISIS system. In *Proc. of the 10th ACM Symp. on Operating System Principles, Orcas Island, Washington*, pages 79–86, Dec. 1985.

4. K. P. Birman, T. Joseph, and T. Raeuchle. Concurrency control in resilient objects. Technical report, TR 84-622, Dept. of Computer Science, Cornell Univ., Ithaca, NY, July 1984.

5. K. P. Birman, T. Joseph, T. Raeuchle, and A. El-Abbadi. Implementing fault-tolerant distributed objects. *IEEE Trans. on SW Eng.*, 11(6):502–508, June 1985.
6. A. Borg, W. Blau, W. Graetsch, F. Herrmann, and W. Oberle. Fault tolerance under UNIX. *ACM Transactions on Computer Systems*, 7(1):1–24, Feb. 1989.
7. N. Budhiraja, K. Marzullo, F. B. Schneider, and S. Toueg. The primary-backup approach. In S. J. Mullender, editor, *Distributed Systems (2nd edition)*, pages 199–216. Addison-Wesley, Wokingham, England, 1993.
8. P. Galdámez, F. D. Muñoz-Escoí, and J. M. Bernabéu-Aubán. HIDRA: Architecture and high availability support. Technical report, DSIC-II/14/97, Univ. Politècnica de València, Spain, May 1997.
9. G. Hamilton, M. L. Powell, and J. J. Mitchell. Subcontract: A flexible base for distributed programming. In B. Liskov, editor, *Proc. of the 14th Symp. on Operating Systems Principles*, pages 69–79, New York, NY, USA, Dec. 1993. ACM Press.
10. R. Ladin, B. Liskov, L. Shrira, and S. Ghemawat. Providing high availability using lazy replication. *ACM Trans. on Comp. Sys.*, 10(4):360–391, Nov. 1992.
11. S. Maffeis. *Run-Time Support for Object-Oriented Distributed Programming.* PhD thesis, Dept. of Computer Science, University of Zurich, Febr. 1995.
12. J. E. Moss. Nested transactions: An approach to reliable distributed computing. Technical report, MIT/LCS/TR-260, MIT Laboratory for Computer Science, 1981.
13. F. D. Muñoz-Escoí, V. Matena, J. M. Bernabéu-Aubán, and P. Galdámez. A membership protocol for multi-computer clusters. Technical report, DSIC-II/20/97, Univ. Politècnica de València, Spain, May 1997.
14. OMG. *The Common Object Request Broker: Architecture and Specification.* Object Management Group, July 1995. Revision 2.0.
15. OMG. *CORBAservices: Common Object Services Specification.* Object Management Group, Nov. 1995. Revised Edition.
16. F. B. Schneider. Replication management using the state-machine approach. In S. J. Mullender, editor, *Distributed Systems (2nd edition)*, pages 166–197. Addison-Wesley, Wokingham, England, 1993.
17. R. van Renesse, K. P. Birman, B. Glade, K. Guo, M. Hayden, T. M. Hickey, D. Malki, A. Vaysburd, and W. Vogels. Horus: A flexible group communications system. Technical report, TR95-1500, Dept. of Comp. Sc., Cornell Univ., NY, March 1995.

# On f–Sparse Sets in NP − P

Vladimír Glasnák[1]

*Abstract.* In this paper a new upward separation technique is developed. One of its consequences solves an open question of Hartmanis [6]: There is an $O\left(n^{\log n}\right)$–sparse set in $NP - P$ iff

$$\bigcup_{c>1} NTIME\left(2^{c\sqrt{n}}\right) \neq \bigcup_{c>1} DTIME\left(2^{c\sqrt{n}}\right).$$

The technique is similar to that ones proving that any sparse set is conjunctively truth–table reducible to a tally set (by Buhrman, Longpré and Spaan [4], Saluja [9]) or that any sparse set is randomly m–reducible to a tally set (by Schöning [10]). Our technique applies perfect hashing functions with small descriptions.

## 1 Introduction

The most important questions in structural complexity are the questions of the collapse of classes ($NP \overset{?}{=} P$, $NEXT \overset{?}{=} DEXT$, etc ... ). It seems very difficult to prove the collapse or a separation of complexity classes. Therefore many distinct characteristics of the collapses are given in literature. The aim of this paper is to characterize a collapse in term of the existence of special sets of a distinct complexity.

The first result of this type was proved by Book [3]: $NEXT \neq DEXT$ iff $NP-P$ contains a tally set. This result was strenghtened by the upward separation technique due to Hartmanis [6]. He proved that a separation of the exponential classes is equivalent to the existence of a polynomially sparse set in $NP - P$. The paper [6] also studied the existence of a set of nonpolynomial density in $NP - P$. It is natural that the existence of a set in $NP - P$ with density greater than polynomial should imply a separation of classes with complexity less than exponential and, on the other side, the existence of a set in $NP - P$ with density less than polynomial should imply a separation of classes with complexity greater than exponential. The technique of [6] is not so strong to prove such results.

Sets with small density were studied by Allender [1]. He showed that the method of [6] cannot prove a result of type "if $NP - P$ contains a set of very

---

[1]Charles University, Faculty of Mathematics and Physics, Department of Theoretical Computer Science, Prague, Czech Republic; e–mail: glasnak@kti.ms.mff.cuni.cz

low density, then there are classes above exponential ones which do not collapse." He proved such result under an additional assumption of low generalized Kolmogorov complexity of the very sparse set.

A result on sets of greater than polynomial density was proved by Hartmanis, Immerman and Sewelson [7]. They proved that $NP - P$ contains an $n^{O(\log n)}$-sparse uniformly distributed set iff $NTIME(2^{O(\sqrt{n})}) \neq DTIME(2^{O(\sqrt{n})})$. Analogously as for very sparse sets, it might seem that the assumption of the uniform distribution (or something similar) is necessary.

The main result of this paper shows that this assumption can be omitted. For a set of functions $\mathcal{F}$ and a function $g$ between $n + 1$ and $2^n$ fulfilling some suitable conditions we prove:

If $NTIME(\mathcal{F}) - DTIME(\mathcal{F})$ contains an $g$–sparse set, then $NTIME(\mathcal{G}) \neq DTIME(\mathcal{G})$, where $\mathcal{G} = \{f \circ g^{-1}(2^n); f \in \mathcal{F}\}$.

One of its corollaries solves the open question of Hartmanis [6].

**Corollary 1.** There is an $O\left(2^{\lceil \log n \rceil}\right)$–sparse set in $NP - P$ iff

$$\bigcup_{c>1} NTIME\left(2^{c\lceil \sqrt{n} \rceil^2}\right) \neq \bigcup_{c>1} DTIME\left(2^{c\lceil \sqrt{n} \rceil}\right).$$

The main result uses a new method of the upward separation appropriate for sets of the density between $n$ and $2^n$. The method is close to the techniques of Buhrman, Longpré, Spaan [4], Saluja [9] and Schöning [10] for proofs that every sparse set is conjunctively truth–table reducible to a tally set [4], [9], and that every sparse set is randomly m–reducible to a tally set [10].

Our method exploits hashing functions with small descriptions. The existence of such functions was independently proved by Fredman, Komlós, Szemerédi [5] and Mehlhorn [8].

**Theorem 2** [5], [8]. Let $S \subseteq \{0, 1, \ldots N - 1\}$ be a set of the cardinality $|S| = m$. Then there exists a prime $p \in O(m^2 \log N)$ such that the hashing function $h(x) = x \bmod p$ is perfect for $S$ (i.e. $h(x) \neq h(y)$ for every two different $x, y \in S$).

## 2 Preliminaries

Our computation model is multi-tape Turing machine with an alphabet $\{0, 1\}$. We say that a function $f$ is time constructible if there is a Turing machine $M$ working in time $O(f)$ such that $M$ on input $1^n$ produces output $1^{f(n)}$. The class of sets recognized deterministically (or nondeterministically) in time $O(f)$ is denoted by $DTIME(f)$ (or $NTIME(f)$). Let us define (see [2]):

$P = \bigcup_i DTIME(n^i)$, $NP = \bigcup_i NTIME(n^i)$,
$DEXT = \bigcup_i DTIME(2^{in})$, $NEXT = \bigcup_i NTIME(2^{in})$,

Length of a string $x$ is denoted by $|x|$. The binary expansion of a natural number $n$ is denoted by $\mathrm{bin}(n)$. For a string $x \in 1\{0, 1\}^*$, we denote $\mathrm{nn}(x)$ the natural number with $\mathrm{bin}(\mathrm{nn}(x)) = x$. For a set $A$ define $A' = \{0, 1\}^* - A$, $A_{\leq n} = \{x \in A; |x| \leq n\}$ and $A_{=n} = \{x \in A; |x| = n\}$.

Let $h : \{0, 1, \#\}^* \to \{0, 1\}^*$ be a homomorphism such that $h(0) = 00$, $h(1) = 11$ and $h(\#) = 10$. Let us define

$$\langle x_1, x_2, \ldots x_n \rangle = \begin{cases} x_1, & \text{if } n = 1, \\ h(x_1 \# x_2 \# \ldots \# x_n), & \text{if } n > 1. \end{cases}$$

For $n > 1$ we have $|\langle x_1, x_2, \ldots x_n \rangle| = 2|x_1| + 2|x_2| + \cdots + 2|x_n| + 2n - 2$.

Let $f : \{0, 1\}^* \to \mathbb{N}$ be a partial function. We define

$$[x_1, x_2 \ldots x_n]_f = \langle x_1, x_2, \ldots x_n \rangle 01^{f(x_1) - |\langle x_1, x_2, \ldots x_n \rangle| - 1},$$

if $f(x_1)$ is defined and $f(x_1) - |\langle x_1, \ldots x_n \rangle| \geq 1$ and

$$[x_1, x_2, \ldots x_n]_f = \langle x_1, x_2, \ldots x_n \rangle 0, \quad \text{otherwise.}$$

Thus if $f$ grows quickly and $x_1$ has a sufficient length (with respect to $\langle x_1, \ldots, x_n \rangle$), then $|[x_1, \ldots x_n]_f| = f(x_1)$.

For a function $f : \mathbb{N} \to \mathbb{N}$ and a set $A$ the set $A_f$ is defined by $A_f = \{[x]_g; x \in A\}$, where $g(x) = f(|x|)$.

We work with nondecreasing functions on natural numbers. For a nondecreasing unbounded function $f$ define $f^{-1}(n) = \min\{t; f(t) \geq n\}$. Observe that $f^{-1}$ coincides with standard inverse function of an increasing function $f$. The notion "$^{-1}$" is used here in this sense only.

**Lemma 3.** Let $f : \mathbb{N} \to \mathbb{N}$ be a nondecreasing unbounded function. Then $f^{-1}$ is a nondecreasing unbounded function and

$$f^{-1}(f(n)) \leq n < f^{-1}(f(n) + 1), \quad n \leq f(f^{-1}(n)), \quad \text{for all } n.$$

## 3 The base classes

The complexity classes *NP* and *P* are defined by the sequence of polynomials $n, n^2, n^3 \ldots$ Similarly, we can define many other complexity classes by sequences of functions. A sequence determines some important properties of the corresponding classes. Our result depends on these properties. The following definition introduces necessary notions.

**Definition 4.** Let $\mathcal{F} = \{f_i\}$ be a sequence of nondecreasing time constructible functions such that $n \leq f_i(n) \leq f_{i+1}(n)$ for all natural numbers $i, n$. The classes $D(\mathcal{F})$, $N(\mathcal{F})$ are defined by:

$$D(\mathcal{F}) = \bigcup_i DTIME(f_i), \quad N(\mathcal{F}) = \bigcup_i NTIME(f_i).$$

We say that $\mathcal{F}$ is a *base* of the classes $D(\mathcal{F})$, $N(\mathcal{F})$ and these classes are called *base classes* with the base $\mathcal{F}$.

Let $\mathcal{F} = \{f_i\}$, $\mathcal{G} = \{g_i\}$ be bases. We say that $\mathcal{F}$ is

(1) *closed under compositions*, if for all natural numbers $i, j$ there is a natural number $k$ such that $f_i \circ f_j \in O(f_k)$,

(2) *closed under products*, if for all natural numbers $i, j$ there is a natural number $k$ such that $f_i \cdot f_j \in O(f_k)$,

(3) *closed under right compositions with $\mathcal{G}$*, if for all natural numbers $i, j$ there is a natural number $k$ such that $f_i \circ g_j \in O(f_k)$,

Let $\mathcal{F} = \{f_i\}$ and $\mathcal{G} = \{g_i\}$ be bases. The *composition* of the bases $\mathcal{F}$ and $\mathcal{G}$ is the base $\mathcal{F} \circ \mathcal{G} = \{f_i \circ g_i\}$. Analogously, given a nondecreasing time constructible function $f$ and a base $\mathcal{G} = \{g_i\}$, let $\mathcal{G} \circ f$ denote the base $\{g_i \circ f\}$.

We say that a Turing machine works in $D(\mathcal{F})$-time if it works deterministically in time $O(f)$ for some $f \in \mathcal{F}$. Analogously, we use the notion of $N(\mathcal{F})$-time, $D(\mathcal{F})$-algorithm, etc ...

Let $f : \mathbb{N} \to \mathbb{N}$ be a function. We say that $S$ is $f$-*sparse*, if $|S_{\leq n}| \in O(f)$.

Let $\mathcal{F}$ be a base. A set $S$ is $\mathcal{F}$-*sparse*, if there exists a function $f \in \mathcal{F}$ such that $S$ is $f$-sparse.

A base $\mathcal{F}$ *collapses*, if $N(\mathcal{F}) = D(\mathcal{F})$.

Finally, we give a few examples of base classes. We define some standard bases of classes.

**Definition 5.** Let $k > 0$. Let us define

$$\mathcal{P} = \{n^i\}_{i>0}, \quad \mathcal{EXT} = \{2^{in}\}_{i>0}, \quad \mathcal{EL}_k = \left\{2^{i \lceil \log n \rceil^k}\right\}_{i>0}.$$

The classes with the bases $\mathcal{P}$, $\mathcal{EXT}$, are *NP, P, NEXT, DEXT*. The corresponding classes to the base $\mathcal{EL}_k$ are denoted by $NEL_k$, $EL_k$ (for $k > 0$). It is easy to prove that the base $\mathcal{EL}_k$ is closed under products.

## 4 Upward separation technique for sets of great density

This paragraph shows a relation between the collapse of a base and the existence of $f$-sparse sets between classes of an another base. These results generalize and extend some of the results of [6], [7] and [1].

The first method of a separation called the downward separation is well-known. The existence of a special set in $N(\mathcal{F}) - D(\mathcal{F})$ is equivalent to a separation of a base depending on $\mathcal{F}$ and $g$.

**Theorem 6** (*downward separation*). *Let* $\mathcal{F} = \{f_i\}$ *be a base and let* $g_1$ *be an unbounded nondecreasing function with* $g_1(n) < 2^n$ *such that the function* $h(n) = g_1^{-1}(2^n)$ *is time constructible. Let* $g_2$ *be a function such that* $g_2(n) \leq 2^n$ *for all* $n$ *and let* $A$ *be a* $g_2$-*sparse set. Then* $A_h \in N(\mathcal{F}) - D(\mathcal{F})$ *if and only if* $A \in N(\mathcal{F} \circ h) - D(\mathcal{F} \circ h)$.

*Moreover, the set* $A_h$ *is* $g_2(\lfloor \log g_1 \rfloor)$-*sparse.*

*Proof sketch.* By padding arguments. $\square$

An importance of Theorem 6 consists in the implication from right to left, therefore it is called a downward separation. The implication from left to right is, in fact, a weak upward separation because the strong assumption — the existence of a special set $A_h \in N(\mathcal{F}) - D(\mathcal{F})$ — is necessary. Hartmanis [6] developed the stronger upward separation technique. This technique was generalized in [7] and [1], but the results of both of the papers cannot be used without additional assumptions (uniform distribution) for example if there is an $\mathcal{EL}_2$-sparse set in $NP - P$. Later, new techniques of coding sparse sets to tally sets were developed (see [4], [9] and [10]), but none of them was appropriate for sets of great densities. This paper develops an upward separation technique which can be used in such cases. The technique is similar to that one of [10], but there is a key difference — a sparse set is hashed and then a perfect hashing function is found (by Theorem 2) therefore we can reconstruct the sparse set.

We begin with a technical lemma showing a construction of $\lfloor \log g(n) \rfloor$ if $g^{-1}(2^n)$ is time constructible.

**Lemma 7.** *Let $g$ be a nondecreasing function with $1 \leq g(0)$ such that $h(n) = g^{-1}(2^n)$ is time constructible. Then there exists an algorithm working in time $O(n^2)$, which for an input $1^n$ provides the output $1^{\lfloor \log g(n) \rfloor}$.*

*Proof sketch.* The algorithm computes $t$ with $g^{-1}\left(2^t\right) \leq n < g^{-1}\left(2^{t+1}\right)$. Such $t$ is equal to $\lfloor \log g(n) \rfloor$ by Lemma 2. $\square$

One important step in the upward separation technique of [6] is to guess nondeterministically the members of a sparse set $S$. The problem of the method consists in a great census of the sparse set. For example, if $S$ is $\mathcal{EL}_2$-sparse, it is impossible to guess all words of $S_{=n}$ in nondeterministic polynomial time with respect to $n$. The upward separation technique of [6] is based on this step, so it cannot be simply applied to sets of greater than polynomial density. We develop a method with guessing only one word of $S$. It is based on perfect hashing. Let us suppose that $h$ is a perfect hashing function for $S_{=n}$ and we want to determine whether a word $x$ belongs to $S_{=n}$. Then it suffices to guess a word $y \in S_{=n}$ with $h(x) = h(y)$ and to verify whether $x = y$. The crucial point here is the existence and finding of a perfect hashing function. Let us define $h_a(x) = \mathrm{nn}(1x) \bmod a$, for a natural number $a > 1$. The function $h_a$ can be described by the string $\mathrm{bin}(a)$ because a computation of $h_a(x)$ for a given $x$ depends only on $a$. Theorem 2 gives us the existence of a perfect hashing function $h_p$ with the description of length $O(\log(m^2 \log N))$. In our case $m$ equals to $|S_{=n}|$ and $\log N = n + 1$, thus the representation of the perfect hashing function is small enough.

**Theorem 8.** *Let $\mathcal{F} = \{f_i\}$ be a base closed under products and right compositions with $\mathcal{P}$. Let $g$ be a nondecreasing function such that $n + 1 \leq g(n) \leq 2^n$ and let $h(n) = g^{-1}(2^n)$ be time constructible. If there is a constant $\alpha \in \mathbb{N}$ such that $g^{-1}(n^2) \leq (g^{-1}(n))^{\alpha}$ for all $n$ and $N(\mathcal{F}) - D(\mathcal{F})$ contains a $g$-sparse set, then the base $\mathcal{G} = \mathcal{F} \circ h$ does not collapse.*

*Proof.* Let us assume that $N(\mathcal{G}) = D(\mathcal{G})$ and that $S \in N(\mathcal{F})$ is a $g$-sparse set. We will show that $S \in D(\mathcal{F})$.

According to Theorem 2, there exists a constant $c_0$ such that for every $n$ and $M \subseteq \{0,1\}^n$ of cardinality $g(n)$ there exists a prime $p_n$ satisfying $p_n \leq c_0(n+1)g^2(n)$ and $h_{p_n}$ is a perfect hashing function for $M$.

Let $n_0$ be an integer such that $c_0(n+1)g^2(n)$ has the binary expansion less than $4\lfloor \log g(n) \rfloor$ and $\lfloor \log g(n) \rfloor > 14$ for all $n \geq n_0$. Set $q_n = 2^{4\lfloor \log(g(n)) \rfloor}$ for $n \geq n_0$. Then $1 < p_n < q_n$, i.e. the set $\{2, 3, \ldots, q_n - 1\}$ contains a number determining a perfect hashing function for $S_{=n}$. Note that every number in the set $\{2, 3, \ldots, q_n - 1\}$ has the binary expansion of length at most $4\lfloor \log g(n) \rfloor$ (for $n \geq n_0$).

Let us define $f : 1\{0,1\}^* \to \mathbb{N}$ by $f(x) = 21\lfloor\log(g(\text{nn}(x)))\rfloor$.

In the sequel, three sets $A, B, C$ are defined and we prove that they belong to $N(\mathcal{G})$. By the assumption $N(\mathcal{G}) = D(\mathcal{G})$, they are used to create a $D(\mathcal{F})$-algorithm recognizing $S$.

The set $A$ characterizes the numbers $a$ such that the function $h_a$ (defined by $h_a(x) = \text{nn}(1x) \bmod a$) is not perfect for $S_{=n}$. Let us define

$$A = \{[\text{bin}(n), \text{bin}(a)]_f; \; n_0 \leq n \; \& \; 1 < a < q_n \; \&$$
$$\text{there exist two distinct } x, y \in S_{=n} \text{ with } h_a(x) = h_a(y)\}.$$

Note that if $[\text{bin}(n), \text{bin}(a)]_f \in A$, then

$$|\langle\text{bin}(n), \text{bin}(a)\rangle| \leq 2\lfloor\log n\rfloor + 2 + 8\lfloor\log g(n)\rfloor + 2 \leq 10\lfloor\log g(n)\rfloor + 4 < f(\text{bin}(n)).$$

Therefore $[\text{bin}(n), \text{bin}(a)]_f \in A$ implies $|[\text{bin}(n), \text{bin}(a)]_f| = f(\text{bin}(n))$.

A nondeterministic algorithm for $A$ tests whether an input is a correct code of a tuple and $a < q_n$. Then it guesses $x, y$ and it verifies whether they belong to $S_{=n}$ and whether $h_a(x) = h_a(y)$.

Since $S \in N(\mathcal{F})$, the algorithm works in $N(\mathcal{F})$-time with respect to $n$. Since

$$n \leq g^{-1}\left(2^{\lfloor\log g(n)\rfloor+1}\right) \leq g^{-1}\left(2^{2\lfloor\log g(n)\rfloor}\right),$$

the algorithm works in $N(\mathcal{G})$-time. Therefore $A \in N(\mathcal{G}) = D(\mathcal{G})$. Since every deterministic class is closed under complements, $A' \in D(\mathcal{G})$.

Let us define

$$B = \{[\text{bin}(n), \text{bin}(a), \text{bin}(b)]_f; \; n_0 \leq n \; \& \; 1 \leq a < b < q_n \; \&$$
$$\text{there exists } k, \; a < k \leq b \text{ such that } [\text{bin}(n), \text{bin}(k)]_f \notin A\}.$$

Note that if $[\text{bin}(n), \text{bin}(a), \text{bin}(b)]_f \in B$, then

$$|\langle\text{bin}(n), \text{bin}(a), \text{bin}(b)\rangle| \leq 2\lfloor\log n\rfloor + 2 + 16\lfloor\log g(n)\rfloor + 4 \leq$$
$$18\lfloor\log g(n)\rfloor + 6 < f(\text{bin}(n)).$$

Thus $[\text{bin}(n), \text{bin}(a), \text{bin}(b)]_f \in B$ implies $|[\text{bin}(n), \text{bin}(a), \text{bin}(b)]_f| = f(\text{bin}(n))$.

A simple nondeterministic algorithm decides the membership of an input in $B$. It tests the correctness of the input, guesses $k$ between $a$ and $b$ and then verifies the condition $[\text{bin}(n), \text{bin}(k)]_f \notin A$. The last step (membership in $A'$) is done in $D(\mathcal{G})$-time. Hence, $B \in N(\mathcal{G}) = D(\mathcal{G})$. The membership of $B$ in $N(\mathcal{G})$ can be proved only under assumption $N(\mathcal{G}) = D(\mathcal{G})$.

The third set is defined as follows.

$$C = \{[\text{bin}(n), \text{bin}(t), \text{bin}(a), \text{bin}(i), d]_f; \; n_0 \leq n \; \& \; t < a < q_n \; \&$$
$$1 \leq i \leq n \; \& \; \text{there exists } x \in S_{=n} \text{ such that } h_a(x) = t \; \&$$
$$i\text{-th digit of } x \text{ is } d\}.$$

Note that if $[\text{bin}(n), \text{bin}(t), \text{bin}(a), \text{bin}(i), d]_f \in C$, then

$$|\langle \text{bin}(n), \text{bin}(t), \text{bin}(a), \text{bin}(i), d \rangle| \leq 4\lfloor \log n \rfloor + 4 + 16\lfloor \log g(n) \rfloor + 2 + 8$$
$$\leq 20\lfloor \log g(n) \rfloor + 14 < f(\text{bin}(n)).$$

Therefore $[\text{bin}(n), \ldots]_f \in C$ implies that its length is $f(\text{bin}(n))$.

A nondeterministic algorithm for $C$ tests the correctness of an input, guesses $x$ and verifies whether $x \in S_{=n}$, then it computes $h_a(x)$ and compares it with $t$ and at the end it verifies whether the $i$-th digit of $x$ is $d$. Similarly to the previous algorithms, this algorithm works in $N(\mathcal{G}) = D(\mathcal{G})\text{–time}^2$.

Having a perfect hashing function $h_a$ for $S_{=n}$, it is possible to verify the membership of $x$ in $S_{=n}$ by the set $C$. The binary searching and the set $B$ enable us to find $h_a$ quickly. Using $D(\mathcal{G})$–algorithms for $B$ and $C$ we construct a $D(\mathcal{F})$–algorithm deciding the membership of an input $x$ in $S$. The algorithm:

(1) By the binary searching find a perfect hashing function $h_a$ for $S_{=n}$. Start with the interval $\langle 2, q_n - 1 \rangle$ and by the set $B$ choose the half of the current interval containing a perfect hashing function.

(2) Compute $h_a(x)$ and use the set $C$ to generate a member $y$ of $S_{=n}$ with $h_a(y) = h_a(x)$.

(3) If such $y$ exists and $x = y$ then *ACCEPT*, else *REJECT*

The step (1) always finds a perfect hashing function for $S_{=n}$, because the interval $\langle 2, q_n - 1 \rangle$ contains at least one, hence the set $B$ always navigates the computation to the interval containing a perfect hashing function. If an input $x$ is in $S$, then $x = y$ for the generated $y$, because the function $h_a$ is perfect. If an input $x$ is not in $S$, then either there is an $y \in S$ with $h_a(x) = h_a(y)$ (and so $x \neq y$) or there is no $y \in S$ with $h_a(x) = h_a(y)$. Therefore, the algorithm is correct.

We compute the complexity of the algorithm.

Step (1). The binary searching consist of $O(\log q_n) = O(\log g(n))$ substeps. Each substep works in $D(\mathcal{G})$–time with respect to $21\lfloor \log g(n) \rfloor$. Thus there is an $i$ such that every substep works in time $f_i(g^{-1}(2^{21\lfloor \log g(n) \rfloor}))$. By the assumptions

$$g^{-1}\left(2^{21\lfloor \log g(n) \rfloor}\right) \leq g^{-1}\left(2^{2^{\lfloor \log 21 \rfloor}\lfloor \log g(n) \rfloor}\right) \leq (g^{-1}(g(n)))^{\alpha^{\lfloor \log c_2 \rfloor}} \leq n^{\alpha^{\lfloor \log 21 \rfloor}}.$$

Hence every substep works in time $f_i\left(n^{\alpha^{\lfloor \log 21 \rfloor}}\right)$. Together with the closeness of $\mathcal{F}$ under right compositions with $\mathcal{P}$ we obtain that the step runs in $D(\mathcal{F})$–time.

Step (2). There are $2n$ calls of $D(\mathcal{G})$–algorithm deciding $C$. Again, it is $D(\mathcal{F})$–time by the same computation.

Since step (3) works in $O(n)$.

The given $D(\mathcal{F})$–algorithm decides the membership in $S$ for $|x| \geq n_0$. Now we add to the algorithm a table containing all members of $S_{<n_0}$ and we obtain a correct $D(\mathcal{F})$–algorithm deciding the membership in $S$. $\square$

Since our method extends the method of Hartmanis, we obtain his result as a corollary.

---

[2]This is a place, where the method of [6] does not work for sets with a great census. A similarly constructed set $C_1$ in [6] has to guess all members of $S$, hence we do not know to show that it is in $N(\mathcal{G})$. Therefore, the paper [7] has the assumption that $S$ is uniformly distributed.

**Corollary 9 [6].** *There is a sparse set in NP − P iff NEXT ≠ DEXT.*

The open question of [6] is solved by Corollary 1.

Theorems 6 and 8 can be used to transfer the existence of a set of certain density in $N(\mathcal{F}) - D(\mathcal{F})$ to the existence of a set of another density in $N(\mathcal{G}) - D(\mathcal{G})$.

**Corollary 10.** *Let c be a natural number and let f be a function satisfying these conditions:*

(1) $n + 1 \leq f(n) \leq 2^{2^{\lceil \sqrt[c]{\log n} \rceil - 1}}$,

(2) $h(n) = 2^{\lceil \sqrt[c]{\log f^{-1}(2^n)} \rceil}$ *is time constructible,*

(3) *there exists a such that* $f^{-1}(n^2) \leq (f^{-1}(n))^a$.

*Then the following conditions are equivalent.*

a) *NP − P contains an f-sparse set,*

b) *$NEL_c - EL_c$ contains an $f\left(2^{\lceil \log n \rceil^c}\right)$-sparse set,*

c) *the base $\{(f^{-1}(2^n))^i\}_{i \in \mathbb{N}}$ does not collapse.*

# References

[1] E. Allender, *Limitations of the upward separation technique*, Mathematical Systems Theory **24** (1991), 53–67.

[2] J. L. Balcázar, J. Díaz, J. Gabarró, *Structural complexity I*, Springer Verlag, Berlin, 1988.

[3] R. Book, *Tally languages and complexity classes*, Information and Control **26** (1974), 186–193.

[4] H. Buhrman, L. Longpré, E. Spaan, *SPARSE reduces conjunctively to TALLY*, Proc. 8th IEEE Conf. on Structure in Complexity Theory (1993), 208–214.

[5] M. L. Fredman, J. Komlós, E. Szemerédi, *Storing a sparse table with O(1) worst case access time*, 23rd Annual Symposium on Foundations of Computer Science (1982), 165–169.

[6] J. Hartmanis, *On sparse sets in NP−P*, Information Processing Letters **16** (1983), 55–60.

[7] J. Hartmanis, N. Immerman, V. Sewelson, *Sparse sets in NP−P: EXPTIME versus NEXPTIME*, Proc. 15th ACM STOC (1983), 382–391.

[8] K. Mehlhorn, *On the program size of perfect and universal hash functions*, 23rd Annual Symposium on Foundations of Computer Science (1982), 170–175.

[9] Saluja, *Relativized limitations of left set technique and closure classes of sparse sets*, Proc. 8th IEEE Conf. on Structure in Complexity Theory (1993), 215–222.

[10] U. Schöning, *On random reductions from sparse sets to tally sets*, Information Processing Letters **46** (1993), 239–241.

# Zero-Overhead Exception Handling
# Using Metaprogramming

Markus Hof, Hanspeter Mössenböck. Peter Pirkelbauer
Johannes Kepler University Linz. A-4040 Linz
{hof, moessenboeck}@ssw.uni-linz.ac.at

We present a novel approach to exception handling which is based on metaprogramming. Our mechanism does not require language support, imposes no run time overhead to error-free programs, and is easy to implement. Exception handlers are implemented as ordinary procedures. When an exception occurs, the corresponding handler is searched dynamically using the type of the exception as a search criterion. Our implementation was done in the Oberon System but it could be ported to most other systems that support metaprogramming.

## 1. Motivation

Exception handling is the ability to separate the reaction to a program failure (i.e., to an exception) from the place where the failure occured. This keeps algorithms free of error flags and error checking code code and allows a programmer to implement the reaction to different occurrences of the same exception in a single place.

Exception handling was suggested in the seventies [Good75] and is part of many modern languages such as *Java* [ArGo96] , *C++* [Stro86], *Eiffel* [Meye92], *Modula-3* [Nels91] or *Smalltalk* [GoRo83]. The exception handling facilities of these languages differ from each other in the syntactical notation they use, in the way how they allow a program to continue after an exception, how they check that all possible exceptions will be handled, and how exception handling is implemented.

Besides supporting exception handling in a programming language it can also be supported by library functions (e.g., [Mill88]). This has the advantage of keeping the language small although it might not be as readable as with language support.

In this paper we present a novel approach to exception handling which is based on metaprogramming. Our mechanism does not require language support, imposes no run time overhead to error free programs, and is easy to implement. We implemented exception handling for the Oberon System [WiGu92] but it could have been done for most other systems that support metaprogramming.

The rest of the paper is organized as follows: In Section 2 we give an overview of existing exception handling notations and implementations. In Section 3 we explain the metaprogramming facilities of Oberon, which are then used in Section 4 to introduce our approach to exception handling. In Section 5 we compare our technique with mechanisms used in other languages.

## 2. Common Exception Handling Mechanisms

Exception handling is part of many modern programming languages. In this section we give an overview of the notations that are used in these languages and sketch some common implementation techniques. In Section 4 we introduce our own notation and implementation.

**General principles.** Exception handling is built around three concepts: a *block* of statements that is protected against exceptions; one or more *exception handlers* that are specified for the protected block; and a mechanism to *raise exceptions*. If an exception is raised during the excution of the protected block, the corresponding handler is executed. Some exceptions can also be raised by the system because of illegal operations (e.g. division by zero). After the exception was handled the program can be continued in one of three ways:

- *Terminate semantics*: The protected block is terminated and the program continues with the first statement after the block.
- *Resume semantics*: The execution of the protected block is resumed after the point where the exception was raised.
- *Retry semantics*: The protected block is re-executed from the beginning (after the exception handler has repaired some conditions which caused the block to fail in the first try).

Many languages support only terminate semantics. For a discussion of the pros and contras of the various semantics see [Stro94].

**C++ and Java.** The programmer can protect a statement sequence against exceptions by enclosing them in a *try* statement. If the execution of such a statement sequence raises an exception by means of a *throw* statement an appropriate handler takes control. Handlers are appended to the *try* statement as *catch* clauses. After the execution of the *catch* clause the program continues with the first statement after the *try* statement. Exceptions are objects which are thrown (i.e. raised) in a *throw* statement and caught in a *catch* statement. The *catch* clause specifies the class of the exception object that is to be caught. Both C++ and Java support only terminate semantics. For details see [ElSt90] and [ArGo96].

```
try {
    ... some calculations ...
    if (...) throw Overflow();
    Foo(); // may also throw exceptions
}
catch (Overflow& ovfl) {... handle overflow ...}
catch (Underflow& ufl) {... handle underflow...}
```

**Eiffel.** The Eiffel exception handling mechanism is essentially based on the principle of contracts. Classes and methods establish contracts with their clients by specifying preconditions, postconditions and class invariants. If one of these conditions fail, an exception is raised, which can be handled in a *rescue* clause of some currently executing method. In addition, the user may also raise user-defined exceptions. Exceptions are denoted by integer codes. The rescue clause has to analyze these codes with an if statement. A rescue clause usually repairs the failure (if possible) and retries the method. If no retry is specified the method fails, propagating the exception to the rescue clause of its caller. Thus Eiffel supports retry semantics. For details see [Meye92].

```
Foo (...) is
    require ... precondition ...
    do
        Foo1();   -- may raise exceptions
        Foo2();
    ensure ... postcondition ...
```

```
rescue
  if exception = ... then
    ... repair the failure ...
    retry;
  end
end;
```

**Smalltalk.** Smalltalk provides the most powerful mechanism. Exception handling is applied to blocks; exceptions are represented by classes. An exception is raised by sending a *signal* message to an exception class. The corresponding handler is itself specified as a block. After the exception was handled, the program can terminate the block that raised the exception, retry it or resume it after the point where the exception was raised. The user may specify an *ensure* block which is always executed after the protected block, no matter if an exception occurred or not. For details see [GoRo83].

[*... some action ...* MyException **signal**. *... some action...*]
**on**: MyException **do**: [:theException | *...handle exception...* theException **return**]
**on**: MyException2 **do**: [:theException | *...handle exception...* theException **resume**]
**on**: MyException3 **do**: [:theException | *...handle exception...* theException **retry**]
**ensure**: [*...local cleanup...*]

**Implementations.** Current implementations are either based on C's *setjmp / longjmp* mechanism or on tables generated by the compiler. We will shortly sketch both variants. A more extensive description can be found in [KöSt90].

*Setjmp/longjmp.* The function *setjmp(s)* saves the current machine state in a buffer *s* and returns 0. The function *longjmp(s)* restores the machine state from the buffer *s* (including the program counter). As a result, the execution will continue in the *setjmp* routine where this state was saved. After a *longjmp*, however, *setjmp* will return 1. This makes the following implementation of

**try** {... block ...} **catch** {... handler ...}

possible:

```
IF setjmp(s) = 0 THEN Push(s); ... block ... Pop(s)
ELSE ... handler ... (*execution continues here after a longjmp*)
END
```

If an exception is raised in block, the following code is executed:

```
Pop(s); longjmp(s);
```

If the handler cannot handle the exception, it re-raises it, so that a *longjmp* to the previous *setjmp* is executed, and so on. This implementation is straightforward but it leads to some run-time overhead (*setjmp, Push, Pop*) even in the case of error-free programs.

*Range tables.* The compiler generates a table in which the range of every *try* block (start address, end address) as well as the addresses of the corresponding handlers are recorded. If an exception occurs, the current program counter is looked up in the table (if it does not fall into any of the ranges, the program counter of the caller is tried). When the appropriate range is found, the stack is unwound and the corresponding handler is called. This imposes no run-time overhead in the case of error-free programs but requires the compiler and loader to set up the table.

# 3. Metaprogramming

Our exception handling technique makes use of *metaprogramming*, so we will shortly explain this term and show how it can be used in the Oberon System.

Metaprogramming means the ability to treat programs as data, for example to get information about the names and the structure of their variables, types and procedures. If a program can acquire information also about itself, this is called *reflection*. Metaprogramming and reflection were pioneered by *Lisp* [McCa60] and *Smalltalk* [GoRo83]. Today this feature is available in many modern languages such as *Java* [ArGo96], *CLOS* [Atta89] or *Beta* [MMN93]. An implementation for *Oberon* is described in [Temp94] and [StMö96].

In Oberon, information about programs is organized in sequences and can be accessed with iterators, which are called *Riders*. A Rider can enumerate the procedures, types or variables of a module as well as the activation frames of the currently active procedures on the stack. When a Rider is positioned on an element of such a sequence (e.g., on a variable), its fields contain information about this element, for example its name and its data type. If an element is itself structured (e.g., a record variable), one can "zoom in" and enumerate the elements of the inner structure.

The following example shows how to iterate over the currently active procedures and print their names as well as the names of the modules in which the procedures are declared.

```
VAR r: Ref.Rider;

Ref.OpenStack(NIL, r);            (*place r on the topmost frame of the stack*)
WHILE r.mode # Ref.End DO
    Out.String(r.name); Out.Ln;   (*print the name of the corresponding procedure*)
    Out.String(r.mod); Out.Ln;    (*print the module declaring this procedure*)
    r.Next                        (*proceed to the next frame*)
END
```

A second example shows how to iterate over the procedures of a module "M" and look for a procedure that has a reference parameter of type "T" as its first parameter. The list of parameters is obtained by zooming into the procedure.

```
VAR r, r1: Ref.Rider; type: Types.Type;

Ref.OpenProcs("M", r);            (*r is placed on the first proc. of module M*)
WHILE r.mode # Ref.End DO
    r.Zoom(r1);                   (*r1 is placed on the first param. of the proc.*)
    IF r1.mode = Ref.VarPar THEN  (*if it is a reference parameter*)
        type := r1.Type();        (*get its type*)
        IF type.name = "T" THEN ...found...; RETURN END;
    END;
    r.Next                        (*proceed to the next procedure*)
END
```

# 4. Exception Handling with Metaprogramming

In this section we introduce our exception handling technique. It needs no special language constructs and does not require compiler support. Error free programs are not slowed down. Overhead occurs only in the case of exceptions. We describe our implementation for the Oberon system, but the same technique could also be applied in any other system that supports metaprogramming.

**Exception objects.** Exceptions are objects of an exception class, which is a subclass of *Exception*. There are *system exceptions* and *user exceptions*. System exceptions (e.g., division by zero) are triggered automatically while user exceptions are raised by the user program using the library call *Exceptions.Raise(exception)*.

**Exception handlers.** Exceptions are *caught* (i.e. handled) by an *exception handler* which is an ordinary procedure *H* with the following characteristics:

- *H* is declared local to some currently executing procedure *P*, i.e., to one with an activation frame on the procedure stack.
- *H* has a single reference parameter of type *E*, which is the type of the exception to be caught or a superclass thereof.
- Both *H* and *P* have the same return type or none.

The following example shows a procedure *Foo* that calls a procedure *Read* in order to read from a file. If the end of the file is reached, *Read* raises an exception of type *EofException* (a subclass of *Exception*), which is caught by the handler *HandleEof* in *Foo*:

```
PROCEDURE Foo (): INTEGER;
  VAR f: File; ch: CHAR;

  PROCEDURE HandleEof (VAR eof: EofException): INTEGER;  (*the handler*)
  BEGIN
    Close(f);
    RETURN 1 (*error code for eof*)
  END H;

BEGIN (*Foo*)
  ... Read(f, ch); ...
  RETURN 0 (*no error*)
END Foo;

PROCEDURE Read (f: File; VAR ch: CHAR);
  VAR eof: EofException
BEGIN
  IF ...end of file ... THEN Exceptions.Raise(eof) ELSE ... END
END Read;
```

Raising an exception (e.g. *Exceptions.Raise(eof)*) leads to a call of the appropriate handler (e.g. *HandleEof*). When the handler returns, execution continues after the call of the procedure to which the handler is local. In the above example control returns to the caller of *Foo*.

A procedure like *Foo* may contain multiple handlers for different kinds of exceptions (i.e., multiple local procedures with parameters of different exception types). If a handler for *EofException* is not found in *Foo*, the search continues in the caller of *Foo*, then in its caller and so on, until a matching handler is found or the topmost procedure in the call chain is reached. If no matching handler is found, a standard error message is produced and the program terminates.

A handler may again raise an exception, in particular, it may re-raise the same exception that it is currently handling. In this case the search for a new handler starts in the caller of the procedure that contains the current handler (in the above example, the search starts in the caller of *Foo*).

Note that a handler has access to the local variables of the enclosing procedure. *HandleEof* can, for example, close the file *f* declared in *Foo* .

**Exception handling semantics.** The above example makes use of *terminate semantics*: After the handler was executed, the containing procedure $P$ is terminated and the program continues after the call of $P$. Instead, we can also make use of *resume semantics*. That means, that the handler returns to the point where the exception was raised, and execution continues with the instruction after the *Raise*. Resume semantics can be requested for user-raised exceptions with the library call *Exceptions.Resume* like in the following example:

```
PROCEDURE Handler (VAR e: SomeException);
BEGIN
    IF ... the failure can be repaired ... THEN
        ...Repair it...;
        Exceptions.Resume  (*return using resume semantics*)
    END;
    (*return using terminate semantics*)
END Handler;
```

**Implementation.** When an exception occurs, the system has to look for an appropriate handler. It uses metaprogramming facilities to search the procedure stack for a procedure with a local procedure that can be used as a handler. The following pseudo code shows an outline of this algorithm.

```
PROCEDURE Raise (VAR e: Exception);
    E := dynamic type of e;
    FOR all frames on the stack in reverse order DO
        P := procedure that created this frame;
        FOR all local procedures H of P DO
            IF (H has a single reference parameter of type E or a supertype of E)
            & (H has the same return type as P) THEN
                Execute H;
                Return to the caller of P or to the point where the exception was raised,
                    depending on the chosen exception handling semantics
            END
        END
    END;
    Terminate the program with a standard error message
END Raise;
```

The metaprogramming facilities necessary for this implementation are described in the examples of Section 3. *Raise* uses a rider to iterate over all stack frames. For every frame it gets the procedure $P$ and the module $M$ to which the frame belongs. It then uses another rider to iterate over all procedures of module $M$ looking for a procedure $P.H$ (a procedure $H$ local to procedure $P$) that has the required characteristics.

Assume that we found the handler $H$. What remains to be done is to call the handler and to return, depending on the chosen exception handling semantics. When an exception is raised, the procedure stack looks as in Fig. 1.

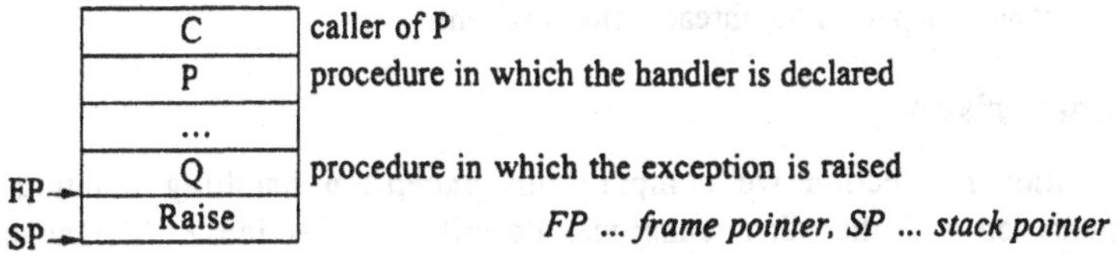

**Fig. 1.** Stack of procedure frames at the time when an exception occurs in procedure Q

For system exceptions, which are triggered automatically, the system will generate a call to the procedure *Raise* as if they were user exceptions. This will lead to the same picture as in Fig. 1. In both cases *Raise* will search for the handler H and its enclosing procedure P, as described above, and then execute the following code:

```
Push e;                        (*exception parameter*)
Push frame pointer of P;       (*static link*)
Call H;                        (*this point is only reached under terminate semantics*)
FP := beginning of C's frame;
SP := end of C's frame;
PC := return address of P      (*return to the caller of P*)
```

When the handler H is executing, the stack looks as in Fig. 2a. The static link of H allows accessing the local variables of P. When H returns, *Raise* modifies the registers FP (frame pointer), SP (stack pointer) and PC (program counter) so that the stack is cleaned up and execution continues after the call of P (Fig. 2b).

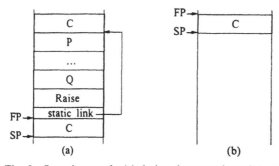

**Fig. 2.** Procedure stack: (a) during the execution of the handler H and (b) after the control was transferred back to the caller of P

If H calls *Resume* (which is only possible for user-raised exceptions), this procedure returns control in the following way:

```
PROCEDURE Resume;
    FP := beginning of Q's frame;    (*Q is the procedure that raised the exception*)
    SP := end of Q's frame;
    PC := return address of Raise    (*return to Q*)
END Resume;
```

If an exception handler H raises an exception itself the same mechanism starts again: The end of the stack will contain another frame of *Raise* as well as a frame of some new handler H1. Note, that the search for H1 must start with the frame of C (the caller of P in Fig. 1) in order to avoid cycles. Therefore, every invocation of *Raise* must remember where to continue the search if an exception is raised in the handler. This information is stored in some global exception context before the handler is invoked. Since in a multi-threaded environment *Raise* must be reentrant, the exception context is part of the thread's environment.

## 5. Comparison

In the following section we compare our exception handling notation and implementation with the other techniques described in Section 2. A summary is shown in the following table.

|  | C++ / Java | Eiffel | Smalltalk | Oberon |
|---|---|---|---|---|
| *Notation* | lang. based | lang. based | library based | library based |
| *Exceptions* | objects | numbers | objects | objects |
| *Granularity* | block level | procedure level | block level | procedure level |
| *Resumption semantics* | terminate | retry | terminate resume retry | terminate resume (retry) |

**Notation.** Our exception handling technique does not require special language support. Protected blocks are procedure bodies, handlers are local procedures with special parameters, and exceptions are raised with a library call. The main advantage of this approach is that we did not have to change an existing language (in our case Oberon). Our technique can be combined with any language as long as the environment supports metaprogramming.

A possible disadvantage of our notation is that exception handling does not stand out as clearly as with special keywords for protected blocks and handlers. It is also not easy to check statically which exceptions a procedure may raise, since we do not require the unhandled exceptions to be specified in the signature of a procedure, as it is the case for example in Java.

In our approach, as in most other implementations, exceptions are objects. This has the advantage that exceptions can be subclassed to carry arbitrary information from the exception point to the handler. In Eiffel this is not possible because exceptions are just numbers.

**Granularity.** We support exception handling on the procedure level and not on the block level. Due to our experience this is sufficient. If a finer granularity is needed, any statement sequence can be turned into a procedure.

**Program resumption.** Our implementation currently supports terminate semantics and resume semantics. In principle we could also support retry semantics but this was not implemented so far because it was not considered necessary. Java and C++ support only terminate semantics and Eiffel supports just retry semantics. Only Smalltalk is as flexible as our approach, supporting all three variants of program resumption.

**Efficiency.** Our implementation does not impose any run time overhead on error free programs. Only in the case of an exception, the system has to search for a handler, which takes about 1 ms in a typical case (measured on a Power Macintosh with 66 MHz). In contrast to that, the *setjmp/longjmp* mechanism described in Section 2 leads to run time costs for every protected block, even if no exception is raised. The *range table* technique (also described in Section 2) does not slow down error free programs, but it requires storage overhead (the range tables) and compiler support, because the tables have to be generated by the compiler. This is not the case with our implementation. The meta-information that we need is already there in the Oberon system so that it does not impose any additional overhead.

**Heap cleanup**. If blocks are terminated due to an exception, there may be data on the heap for which the deallocate statements or destructor invocations were skipped. This is not a problem in most of the mentioned languages (including Oberon) since they rely on automatic garbage collection. In C++, however, proper deallocation of objects is an issue and complicates exception handling considerably [KöSt90].

## 5. Conclusions

We suggested a zero-overhead exception handling technique based on metaprogramming. It was implemented without extending the programming language or the compiler. Exception free programs are not slowed down at all. A slight run time penalty has to be paid only if an exception occurs. Our implementation does not need special data structures at run time, except for meta information about programs, which is available anyway in many languages (e.g. in Java or Smalltalk). Our technique supports various program resumption semantics and allows the programmer to declare custom exception classes with exception-specific information. We implemented our technique for Oberon but it could have been done for any other language that supports metaprogramming.

We found that our technique is powerful, efficient and easy to implement. It can be the method of choice if one does not want to change the programming language or the compiler.

### References

[ArGo96]  Arnold K., Gosling J.: The Java Programming Language. Addison Wesley 1996.

[Atta89]  Attardi G. et al.: Metalevel Programming in CLOS. Proc. European Conference on Object-Oriented Programming (ECOOP'89). Cambridge University Press, 1989.

[ElSt90]  Ellis M.A., Stroustrup B.: The Annotated C++ Reference Manual. Addison-Wesley 1990.

[Good75]  Goodenough J. B.: Exception Handling: Issues and a Proposed Notation. Communications of the ACM, vol.18, no.12, December 1975.

[GoRo83]  Goldberg A., Robson D.: Smalltalk-80, the Language and its Implementation. Addison Wesley 1983.

[McCa60]  McCarthy J.: Recursive Functions of Symbolic Expressions and their Computation by a Machine. Comm. of the ACM, vol.3, no.4, 1960.

[KöSt90]  König A., Stroustrup B.: Exception Handling for C++ (revised), Proc. USENIX C++ Conference, 149-176, (1990)

[Meye92]  Meyer B.: Eiffel — The Language. Prentice Hall 1992.

[Mill88]  Miller W. M.: Exception Handling without Language Extensions. Proc. USENIX C++ Conference. Denver CO., October 1988.

[MMN93]  Lehrmann-Madsen O., Moller-Pedersen B., Nygaard K.: Object-Oriented Programming in the BETA Programming Language. Addison-Wesley 1993.

[Nels91]  Nelson G. (ed.): Systems Programming with Modula-3. Addison-Wesley 1991.

[StMö96]  Steindl C., Mössenböck H.: Metaprogramming Facilities in Oberon for Windows and Power Macintosh. Technical Report 4, Institute of Computer Science, University of Linz, 1996.

[Stro86]  Stroustrup B.: The C++Programming Language. Addison Wesley 1986.

[Stro94]  Stroustrup B.: The Design and Evolution of C++. Addison Wesley 1994.

[Temp94]  Templ J.: Metaprogramming in Oberon. Dissertation, ETH Zurich, 1994.

[WiGu92]  Wirth N., Gutknecht J.: Project Oberon—TheDesign of an Operating System and Compiler. Addison-Wesley 1992.

# The Output–Store Formal Translator Directed by *LR* Parsing

Jan Janoušek and Bořivoj Melichar

Department of Computer Science and Engineering
Czech Technical University, Karlovo nám. 13, 121 35, Prague, Czech Republic
e–mail: {janousej|melichar}@cs.felk.cvut.cz

**Abstract.** This paper presents a one–pass formal translator that can be constructed for each translation grammar with an $LR(k)$ input grammar. The formal translator is the conventional $LR$ parser whose operations are extended by actions performing both an output and a temporary storing of output symbols. The temporary storing of some symbols before their appending to the output string is implemented by using one synthesized attribute.

## 1    Introduction

There exists a class of formal translations that can be described by context–free translation grammars with $LR(k)$ input grammars. One–pass implementations of such formal translations are mostly directed by $LR$ parsing process of the input string. Formal translators creating the output string during the $LR$ parsing process have been described for certain classes of the translation grammars: first, for grammars with a postfix property (output symbols may appear at the ends of grammar rules, see [AU71], [PB80]), and later for the class of $Kernel(R)$ translation grammars (output symbols may also appear in front of certain input symbols in grammar rules, see [Me92], [Me95]). This class can be further increased by some transformations of translation grammars (see [Me95], [MB96]). In spite of various transformations the resultant class does not include all the translation grammars with $LR(k)$ input grammars.

The only known way of performing the one–pass translations for the class of translation grammars with $LR(k)$ input grammars is a method evaluating the attributes (see [ASU86]). However, a feature of this method is the availability of the output string only when the parsing process finishes.

The intention of this paper is to introduce a new one–pass formal translator that can be constructed for each translation grammar with an $LR(k)$ input grammar. The formal translator is the conventional $LR$ parser whose operations are extended by actions performing not only an output but also a temporary storing of output symbols. The temporary storing of some symbols before their appending to the output string is implemented by a non–conventional use of one synthesized attribute. In comparison with the method evaluating the attributes the presented translation technique has these two advantages: First, it decreases the size of memory used. Second, the output string is being created during the parsing process of the input string.

# 2 Basic notions and notations

A *context-free translation grammar* (CFTG) is a 5-tuple $TG = (N, T, D, R, S)$, where $N, T$ and $D$ are finite sets of *nonterminal, input* and *output symbols*, respectively. $R$ is a finite set of *rules* $A \to \alpha$, where $A \in N$, $\alpha \in (N \cup T \cup D)^*$, $S \in N$ is the *start symbol*.

The *augmented context-free translation grammar* (aCFTG) $TG'$ derived from a CFTG $TG$ is defined as $TG' = (N \cup S', T, D, R \cup \{S' \to S\}, S')$.

Relation $\Rightarrow_{TG}$ is called *derivation*: if $\alpha A \gamma \Rightarrow_{TG} \alpha \beta \gamma$, $A \in N$ and $\alpha$, $\beta$, $\gamma \in (N \cup T \cup D)^*$, then rule $A \to \beta \in R$. The *rightmost derivation* $\Rightarrow_{rm}$ is relation $\alpha A \gamma \Rightarrow \alpha \beta \gamma$, where $\gamma \in (T \cup D)^*$.

The *input homomorphism* $h_i^{TG}$ and the *strong output homomorphism* $h_{so}^{TG}$ from $(N \cup T \cup D)^*$ to $(N \cup T \cup D)^*$ are defined as follows:

$$h_i^{TG}(a) = \begin{cases} a \text{ for } a \in N \cup T \\ \varepsilon \text{ for } a \in D \end{cases} \qquad h_{so}^{TG}(a) = \begin{cases} a \text{ for } a \in D \\ \varepsilon \text{ for } a \in N \cup T \end{cases}$$

The *formal translation* defined by a translation grammar $TG$ is the set

$$Z(TG) = \{(h_i^{TG}(w), h_{so}^{TG}(w)) : S \Rightarrow_{TG}^* w, w \in (T \cup D)^*\}.$$

The *input context-free grammar* of a CFTG $TG$ is the 4-tuple $G_i = (N, T, R_i, S)$, where $R_i = \{A \to h_i^{TG}(\alpha) : A \to \alpha \in R\}$.

The subscripts $_{TG}$ and the superscripts $^{TG}$ will be omitted whenever no confusion arises.

Given an input string $x \in T^*$, the $LR$ parser constructed for the input grammar $G_i$ of an aCFTG $TG'$ reads the input string from left to right and follows the rightmost derivation $S' \Rightarrow_{rm}^* w$, where $h_i(w) = x$, in reverse. The parser is described in two tables called the *parsing* and the *goto table*.

The parsing table $PT$ is a collection of rows with columns for all elements of the set $T^{*k}$. The values in the table are either *shift, reduce j, error*, or *accept*, where $j \geq 1$. The goto table $GT$ has the same rows as the parsing table, its columns are created for all elements of the set $(N \cup T)$. The values in the table are either $M_i$ or *error*, where $M_i$ is the name of a row of the parsing table.

$LR$ parser is implemented by a deterministic pushdown automaton. Its configuration is a pair $(\alpha, \tilde{x})$, where $\alpha$ is the contents of the pushdown store and $\tilde{x}$ is the unread part of the input string. Given the input string $x$ the initial configuration is pair $(\#, x)$, the accepting configuration is a pair $(\# M_i, \varepsilon)$, where $M_i$ is a symbol at the top of the pushdown store and $PT(M_i, \varepsilon) = accept$.

To parse the input string the $LR$ parser performs a sequence of *shift* and *reduce* operations finished by the *accept* operation:

The shift operation of a symbol $a$ is to be performed in configuration $(\alpha M_i, av\tilde{x})$, where $a \in T$, $av \in T^{*k}$, $\tilde{x} \in T^*$ and $M_i$ is the top symbol of the pushdown store, if $PT(M_i, av) = shift$. The shift makes move $(\alpha M_i, av\tilde{x}) \vdash (\alpha M_i a', v\tilde{x})$, where $a' = GT(M_i, a)$.

The reduce operation by a rule $A \to \beta$ is to be performed in configuration $(\alpha M_i, u\tilde{x})$, where $u \in T^{*k}$, $\tilde{x} \in T^*$, and $M_i$ is a symbol at the top of the pushdown store, if $PT(M_i, u) = reduce\ j$, where $A \to \beta$ is the $j$-th rule in $R_i$. The reduction consists of these two parts: The *first part of the reduction* makes

move $(\alpha M_i = \delta\zeta, u\tilde{x}) \vdash (\delta, u\tilde{x})$, where $|\zeta| = |\beta|$. The *second part of the reduction* makes move $(\delta = \delta' M_j, u\tilde{x}) \vdash (\delta A', u\tilde{x})$, where $A' = GT(M_j, A)$. Thus, the whole reduction makes the move $(\alpha M_i, u\tilde{x}) \vdash (\delta A', u\tilde{x})$.

Input grammars for which $LR$ parsers can be constructed are called $LR(k)$ *grammars*. See [AU71] for further information on $LR$ parsing.

Attributed grammars represent a formalism for specifying the semantics. Values of attributes are defined by attribute rules. These rules are associated with the grammar rules and specify how to evaluate the values of certain attribute instances as a function of other attribute instances. In this paper an *attributed translation grammar* will be written as a triple $ATG = (TG, AD, F)$, where

a) $TG = (N, T, D, R, S)$ is the underlying context–free translation grammar.

b) $AD = (A, I, S, TYPE)$ is a description of *attributes*. $A$ is the set of all attributes. $A$ is partitioned into two disjoint subsets $I$ and $S$ of *inherited* and *synthesized* attributes, respectively. For $X \in (N \cup T \cup D)$, there is a fixed set $A(X)$ of attributes associated with $X$. For $a \in A$, $TYPE(a)$ is the set of possible values of $a$. An attribute $a$ of $X$ is denoted $X.a$.

c) $F(p)$ is a finite set of attribute evaluation rules associated with rule $p \in R$.

See [AM91] for more details and [DJ90] for related concepts.

An *S-atributted translation grammar* is an attributed translation grammar with synthesized attributes only. In such a case the attribute evaluation during $LR$ parsing process is performed within the reduce operations only (see [AM91]). *Note.* In this paper the only case of an attributed grammar used is with $AD = (\{s\}, \emptyset, \{s\}, TYPE(s) = D^*)$, i.e. with one synthesized attribute $s$ whose values are strings of output symbols.

## 3 The output–store formal translator

This section deals with the behaviour of the output–store formal translator. The translation process is directed by $LR$ parsing input strings. The conventional $LR$ parser is extended by a facility to perform two kinds of fixed actions within the parsing operations:

• Actions, called *output actions*, appending output symbols to the output string.

• Actions, called *store actions*, evaluting one synthesized attribute $s$ whose values are strings of output symbols. These actions store instances of attribute $s$ onto the pushdown store and can be performed within the reductions only.

The strings stored by the store actions (i.e. values of the stored instances of attribute $s$) are appended to the output string later, via the output actions. Attribute $s$ is always associated with nonterminal symbols from the set $NOut$:

**Definition 1.** (Set $NOut$)

Let $TG = (N, T, D, R, S)$ be a CFTG. Then the set $NOut$ of *output–generating nonterminal symbols* is defined as $NOut = \{A \in N : A \Rightarrow^+ w_1 y w_2, y \in D\}$. □

A formal translator is described in two tables called the *translation parsing table* and the *translation goto table*:

- A translation parsing table $TPT$ is a collection of rows. Its columns are for all elements of the set $T^{*k}$. The values in $TPT$ are either $shift(z_0, X_1.s, z_1, \ldots, X_m.s, z_m)$, $reduce$ $j[\Psi](z_0, X_1.s, z_1, \ldots, X_m.s, z_m)$, $error$, or $accept$, where $j \geq 1$, $m \geq 0$, $X_1, \ldots, X_m \in NOut$, $z_0, \ldots, z_m \in D^*$, $\Psi \in \{\bot, A.s : A \in NOut\}$.

- A translation goto table $TGT$ has the same rows as the table $TPT$. Its columns are for all elements of the set $(N \cup T)$. The values in $TGT$ are either $M_i$ (for columns $a \in T$), $M_i(z_0, X_1.s, z_1, \ldots, X_m.s, z_m)$ (for col. $A \in N$) or $error$, where $M_i$ is the name of a row of $TPT$, $m \geq 0$, $X_1, \ldots, X_m \in NOut$, $z_0, \ldots, z_m \in D^*$.

The following lines give an informal description of the operations of the formal translator (the formal description is given by Algorithm 3):

The front parts of the values in the tables – $shift$, $reduce$ $j$, $accept$, $M_i$, and $error$ – describe parsing operations and have the same meanings as in the conventional parsing and goto tables. The meanings of the parts $(z_0, X_1.s, z_1, \ldots, X_m.s, z_m)$ and $[\Psi]$ are:

1. The part $(z_0, X_1.s, z_1, \ldots, X_m.s, z_m)$ defines output actions. Within a parsing operation the output action performs the output of a string which is created as the concatenation of parameters $z_0, x_1, z_1, \ldots, x_m, z_m$, where $x_1, \ldots, x_m \in D^*$ are values of the attribute instances $X_1.s, \ldots, X_m.s$, respectively, stored in the pushdown store. After that the attribute instances $X_1.s, \ldots, X_m.s$ are removed from the pushdown store. Output actions can be performed within the following three kinds of parsing operations:

a) For $TPT(M_i, av) = shift(z_0, X_1.s, z_1, \ldots, X_m.s, z_m)$,
   within the shift $(\alpha M_i, av\tilde{x}) \vdash (\alpha M_i a', v\tilde{x})$.
b) For $TPT(M_i, u) = reduce$ $j[\Psi](z_0, X_1.s, z_1, \ldots, X_m.s, z_m)$,
   within the reduce first part $(\alpha M_i = \delta\zeta, u\tilde{x}) \vdash (\delta, u\tilde{x})$, where $|\zeta| = |h_i(\beta_1)|$ and $A \rightarrow \beta_1$ is the $j$-th rule in $R$.
c) For $TGT(M_i, A) = A'(z_0, X_1.s, z_1, \ldots, X_m.s, z_m)$,
   within the reduce second part $(\delta = \delta' M_i, u\tilde{x}) \vdash (\delta A', u\tilde{x})$.
   The output is performed after the possible storing of the attr. instance $A.s$.

2. The part $[\Psi]$ defines whether store actions are to be performed or not.

a) For $TPT(M_i, u) = reduce$ $j[A.s](z_0, X_1.s, z_1, \ldots, X_m.s, z_m)$,
   where $A \in NOut$ and $A \rightarrow \beta_1$ is the $j$-th rule in $R$,
   the attribute instance $A.s$ is evaluated and stored onto the pushdown store within the corresponding reduction $(\alpha M_i, u\tilde{x}) \vdash (\delta A'(A.s), u\tilde{x})$. The value of $A.s$ is evaluated according to the attribute rule associated with the grammar rule $A \rightarrow \beta_1$. All attribute instances necessary for the evaluation are already present in the pushdown store.
b) For $TPT(M_i, u) = reduce$ $j[\bot](z_0, X_1.s, z_1, \ldots, X_m.s, z_m)$,
   where $A \in N$ and $A \rightarrow \beta_1$ is the $j$-th rule in $R$,
   no attribute instance is evaluated and stored onto the pushdown store within the corresponding reduction $(\alpha M_i, u\tilde{x}) \vdash (\delta A', u\tilde{x})$.

Two output–store formal translators are presented in Examples 2 and 4. Both of them are directed by $LR(1)$ parser.

**Example 2.** Consider an $S$-attributed translation grammar $ATG = (TG', AD, F)$, where $TG' = (\{S', S, A, B\}, \{a, b\}, D, R, S')$, $D = \{x, y, w, z\}$, $AD = (\{s\}, \emptyset, \{s\}, TYPE(s) = D^*)$, $R$ and $F$ are given by these rules:

| | | | | |
|---|---|---|---|---|
| (0) | $S'$ | $\rightarrow$ | $S$ | $S'.s = S.s$ |
| (1) | $S$ | $\rightarrow$ | $A$ | $S.s = A.s$ |
| (2) | $S$ | $\rightarrow$ | $B$ | $S.s = B.s$ |
| (3) | $A^0$ | $\rightarrow$ | $xaA^1$ | $A^0.s = concat('x', A^1.s)$ |
| (4) | $A$ | $\rightarrow$ | $y$ | $A.s = 'y'$ |
| (5) | $B^0$ | $\rightarrow$ | $aB^1w$ | $B^0.s = concat(B^1.s, 'w')$ |
| (6) | $B$ | $\rightarrow$ | $zb$ | $B.s = 'z'$ |

Translation grammar $TG'$ defines translation $Z(TG') = \{(a^n, x^n y), (a^n b, zw^n) : n \geq 0\}$. Attribute $s$, some instances of which are being evaluated and stored onto the pushdown store as results of store actions, is associated with nonterminal symbols from set $NOut = \{S', S, A, B\}$. The attributed version of the translation grammar is always systematically derived from the translation grammar (the algorithm of this construction is postponed to the next section).

The following table includes both $TPT$ and $TGT$ tables describing the formal translator constructed for $ATG$.

| | $a$ | $b$ | $\varepsilon$ | $S$ | $A$ | $B$ | $a$ | $b$ |
|---|---|---|---|---|---|---|---|---|
| # | shift($\varepsilon$) | shift('z') | reduce $4[\bot]('y')$ | $S(\varepsilon)$ | $A_1(\varepsilon)$ | $B_1(\varepsilon)$ | $a_1$ | $b$ |
| $S$ | | | accept | | | | | |
| $A_1$ | | | reduce $1[\bot](\varepsilon)$ | | | | | |
| $A_2$ | | | reduce $3[\bot](\varepsilon)$ | | | | | |
| $A_3$ | | | reduce $3[A.s](\varepsilon)$ | | | | | |
| $B_1$ | | | reduce $2[\bot](\varepsilon)$ | | | | | |
| $B_2$ | | | reduce $5[\bot]('w')$ | | | | | |
| $a_1$ | shift($\varepsilon$) | shift('z') | reduce $4[A.s](\varepsilon)$ | | $A_2('x', A.s)$ | $B_2(\varepsilon)$ | $a_2$ | $b$ |
| $a_2$ | shift($\varepsilon$) | shift('z') | reduce $4[A.s](\varepsilon)$ | | $A_3(\varepsilon)$ | $B_2(\varepsilon)$ | $a_2$ | $b$ |
| $b$ | | | reduce $6[\bot](\varepsilon)$ | | | | | |

In the above tables there are defined the following output actions:

1. The output of symbol $y$ within the first part of the reduction by rule (4) $A \rightarrow y$ in configuration $(\#, \varepsilon)$ of the $LR$ parser.
2. The output of symbol $w$ within the first part of the reduction by rule (5) $B \rightarrow aBw$ in configuration $(\alpha B_2, \varepsilon)$ of the $LR$ parser.
3. The output of symbol $z$ within the shifts of symbol $b$.
4. The output of symbol $x$ and the value of attribute instance $A.s$ within the second part $(\#a_1, \varepsilon) \vdash (\#a_1 A_2, \varepsilon)$ of reductions.
5. The output actions within the other parsing operations perform the output of an empty string.

The tables define the following store actions:

1. Attribute instance $A.s$ is stored onto the pushdown store within the reductions by rule (4) $A \to y$ in configurations $(\#a_1, \varepsilon)$ and $(\alpha a_2, \varepsilon)$.
2. Attribute instance $A.s$ is stored onto the pushdown store within the reduction by rule (3) $A \to xaA$ in configuration $(\alpha A_3, \varepsilon)$.

No store actions are performed within the other reductions. $\square$

A configuration of the formal translator includes an output string that has been created – it is a triple $(\alpha, \tilde{x}, \tilde{y})$, where $\alpha$ is the contents of the pushdown store, $\tilde{x}$ is the unread part of the input string, and $\tilde{y}$ is the created part of the output string. Given an input string $x \in T^*$ to be translated, the initial configuration is triple $(\#, x, \varepsilon)$, the accepting configuration is a triple $(\#M_i, \varepsilon, y)$, where $(x, y) \in Z(TG)$, $M_i$ is a symbol at the top of the pushdown store and $TPT(M_i, \varepsilon) = accept$.

The translation process is formally described by the following algorithm.

**Algorithm 3.**    Formal translation directed by $LR$ parsing.

**Input:** Input string $x \in T^*$, $k \geq 0$, translation parsing table $TPT$ and translation goto table $TGT$ constructed for $S$-attributed translation grammar $ATG = (TG', AD, F)$, where $TG' = (N, T, D, R, S')$.

**Output:** Output string $y$ where $(x, y) \in Z(TG')$. Otherwise an error indication.

**Method:** Symbol $\#$ is the initial symbol on the pushdown store. Repeat steps 1 and 2 until accept or error appears. Let $M$ denote a symbol at the top of the pushdown store.

1. Fix string $r$ of first $k$ symbols from the unread part of the input string.
2. (a) If $TPT(M, r) = shift(z_0, X_1.s, z_1, \ldots, X_m.s, z_m)$, then read one symbol $a$ from the input string and push symbol $a' = TGT(M, a)$ onto the pushdown store.

    Call procedure OUTPUT$(z_0, X_1.s, z_1, \ldots, X_m.s, z_m)$ (see step 3).

   (b) If $TPT(M, r) = reduce\ j[\Psi](z_0, X_1.s, z_1, \ldots, X_m.s, z_m)$, where the $j$-th rule in $R$ has the form $(j)\ A \to \beta$, then:

    Call procedure OUTPUT$(z_0, X_1.s, z_1, \ldots, X_m.s, z_m)$.

    If $\Psi = A.s$ then according to the attribute rule associated with the rule $A \to \beta$ evalute attribute instance $A.s$ and store it in $t_s$.

    Pop from the pushdown store $|h_i(\beta)|$ symbols and attribute instances.

    Let $TGT(M, A) = A'(y_0, Y_1.s, y_1, \ldots, Y_l.s, y_l)$.

    Push $A'(t_s)$ (if $\Psi = A.s$) or $A'$ (if $\Psi = \perp$) onto the pushdown store.

    Call procedure OUTPUT$(y_0, Y_1.s, y_1, \ldots, Y_l.s, y_l)$.

   (c) If $TPT(M, r) = accept$, then finish the translation. The created output string is declared to be the formal translation of the input string.

   (d) If $TPT(M, r) = error$, finish the translation with an error indication.

3. Procedure OUTPUT$(x_0, Z_1.s, x_1, \ldots, Z_n.s, x_n)$:

   $d_s := concat(x_0, z_1, x_1, \ldots, z_n, x_n)$, where $z_1, \ldots, z_n$ are values of the attr. instances $Z_1.s, \ldots, Z_n.s$, respectively, stored in the pushdown store. Remove the attribute instances $Z_1.s, \ldots, Z_n.s$ from the pushdown store. Append $d_s$ to the output string. $\square$

**Example 2., contd.** The formal translator in question performs the following sequences of moves for pairs $(aaa, xxxy)$ and $(aaab, zwww) \in Z(TG')$:

$(\# \quad ,aaa ,\varepsilon) \quad \vdash (\#a_1 \quad\quad ,aa ,\varepsilon) \quad \vdash (\#a_1a_2 \quad\quad ,a ,\varepsilon) \vdash$
$(\#a_1a_2a_2 ,\varepsilon \quad ,\varepsilon) \quad \vdash (\#a_1a_2a_2A_3(y) ,\varepsilon \quad ,\varepsilon) \quad \vdash (\#a_1a_2A_3(xy) ,\varepsilon ,\varepsilon) \vdash$
$(\#a_1A_2 \quad ,\varepsilon \quad ,xxxy) \vdash (\#A_1 \quad\quad ,\varepsilon ,xxxy) \vdash$
$(\#S \quad\quad ,\varepsilon \quad ,xxxy) \vdash accept$

$(\# \quad\quad ,aaab ,\varepsilon) \quad \vdash (\#a_1 \quad ,aab ,\varepsilon) \quad \vdash (\#a_1a_2 \quad ,ab ,\varepsilon) \vdash$
$(\#a_1a_2a_2 ,b \quad ,\varepsilon) \quad \vdash (\#a_1a_2a_2b ,\varepsilon \quad ,z) \quad \vdash (\#a_1a_2a_2B_2 ,\varepsilon ,z) \vdash$
$(\#a_1a_2B_2 ,\varepsilon \quad ,zw) \quad \vdash (\#a_1B_2 \quad ,\varepsilon ,zww) \vdash$
$(\#B_1 \quad\quad ,\varepsilon \quad ,zwww) \vdash (\#S \quad\quad ,\varepsilon ,zwww) \vdash accept \qquad \square$

Another example, in which the translation grammar contains an output symbol in front of a left–recursive nonterminal symbol, is presented below:

**Example 4.** Consider an $S$-attributed translation grammar $ATG = (TG', AD, F)$, where $TG' = (\{S', S, A\}, \{a, b\}, D, R, S')$, $D = \{x, y, z\}$, $AD = (\{s\}, \emptyset, \{s\}, TYPE(s) = D^*)$, $R$ and $F$ are given by these rules:

| (0) | $S'$ | $\to$ | $S$ | | $S'.s$ | $= S.s$ |
|-----|------|-------|-----|--|--------|---------|
| (1) | $S$ | $\to$ | $Azb$ | | $S.s$ | $= concat(A.s, 'z')$ |
| (2) | $A^0$ | $\to$ | $xA^1a$ | | $A^0.s$ | $= concat('x', A^1.s)$ |
| (3) | $A$ | $\to$ | $yb$ | | $A.s$ | $= 'y'$ |

Translation grammar $TG'$ defines translation $Z(TG') = \{(ba^nb, x^nyz): n \geq 0\}$. The following table includes both $TPT$ and $TGT$ tables describing the formal translator constructed for $ATG$.

| | $a$ | $b$ | $\varepsilon$ | $S$ | $A$ | $a$ | $b$ |
|---|-----|-----|---------------|-----|-----|-----|-----|
| $\#$ | | shift($\varepsilon$) | | $S(\varepsilon)$ | $A(\varepsilon)$ | | $b_1$ |
| $S$ | | | accept | | | | |
| $A$ | shift($\varepsilon$) | shift('$z$') | | | | $a$ | $b_2$ |
| $a$ | reduce $2[A.s](\varepsilon)$ | reduce $2[\bot]('x', A.s)$ | | | | | |
| $b_1$ | reduce $3[A.s](\varepsilon)$ | reduce $3[\bot]('y')$ | | | | | |
| $b_2$ | | | reduce $1[\bot](\varepsilon)$ | | | | |

The formal translator performs the following sequences of moves for pairs $(bb, yz)$ and $(baab, xxyz) \in Z(TG')$:

$(\# \quad ,bb ,\varepsilon) \vdash (\#b_1 ,b ,\varepsilon) \vdash (\#A \quad ,b ,y) \vdash$
$(\#Ab_2 ,\varepsilon \quad ,yz) \vdash (\#S ,\varepsilon ,yz) \vdash accept$

$(\# \quad\quad ,baab ,\varepsilon) \quad \vdash (\#b_1 \quad\quad ,aab ,\varepsilon) \quad \vdash (\#A(y) \quad ,aab ,\varepsilon) \quad \vdash$
$(\#A(y)a ,ab \quad ,\varepsilon) \quad \vdash (\#A(xy) ,ab ,\varepsilon) \quad \vdash (\#A(xy)a ,b ,\varepsilon) \quad \vdash$
$(\#A \quad ,b \quad ,xxy) \vdash (\#Ab_2 ,\varepsilon \quad ,xxyz) \vdash (\#S \quad\quad ,\varepsilon ,xxyz) \vdash acc. \quad \square$

# 4 Attributed grammar for the formal translator

As it is shown in the previous section, an output–store formal translator is always constructed for an $S$-attributed translation grammar $ATG$ whose attribute rules define values of one synthesized attribute $s$. This attributed version of the underlying translation grammar is always derived from the translation grammar by the following algorithm.

**Algorithm 5.** Construction of $S$-attributed translation grammar $ATG$.

**Input:** Augmented context–free translation grammar $TG' = (N, T, D, R, S')$.
**Output:** $S$-attributed translation grammar $ATG = (TG', AD, F)$.
**Method:** Create $S$-attributed translation grammar $ATG = (TG', AD, F)$, where $AD = (\{s\}, \emptyset, \{s\}, TYPE(s) = D^*)$. The sets $F$ and $A(X)$, where $X \in N$, are constructed in the following way:
  1. Construct set $NOut = \{A \in N : A \Rightarrow^+ w_1 y w_2, y \in D\}$.
  2. Associate attribute $s$ with nonterminal symbols as follows:
     $A(B) = \{s\}$, $A(C) = \emptyset$, where $B \in NOut$ and $C \in (N - NOut)$.
  3. With each rule $A \to q_0 B_1 q_1 B_2 \dots B_n q_n \in R$, where $B_1, \dots, B_n \in NOut$, $A \in NOut$, $q_0, q_1, \dots, q_n \in ((N - NOut) \cup T \cup D)^*$, $n \geq 0$, associate attribute rule $A.s = concat(h_{so}(q_0), B_1.s, h_{so}(q_1), \dots, B_n.s, h_{so}(q_n))$. $\square$

The $S$-attributed translation grammar $ATG = (TG', AD, F)$ constructed by Algorithm 5 has the following property: Given a derivation $A \Rightarrow^*_{TG'} \alpha$, where $A \in NOut$, $\alpha \in (T \cup D)^*$, the corresponding attribute instance $A.s = h_{so}(\alpha)$.

# 5 Conclusion

The output–store formal translator has been described. The way of its constructing for a given translation grammar with an $LR(k)$ input grammar and further information can be found in [Ja97].

# References

[AM91]   Alblas, H., Melichar, B. (Eds.) *Attribute Grammars, Applications and Systems*. LNCS, vol 545, Springer–Verlag, Berlin, 1991.

[ASU86]  Aho, A.V., Sethi, R., Ullman, J.D. *Compilers – Principles, Techniques and Tools*. Addison–Wesley, Reading, Mass., 1986.

[AU71]   Aho, A.V., Ullman, J.D. *The Theory of Parsing, Translation and Compiling*. Vol.1: Parsing, Vol.2: Compiling, Prentice–Hall, New York, 1971, 1972.

[DJ90]   Deransart, P., Jourdan, M. (Eds.) *Attribute Grammars and their Applications*. LNCS, vol 461, Springer–Verlag, Berlin, 1990.

[Ja97]   Janoušek, J. *Formal Translations Described by Translation Grammars with LR(k) Input Grammars*. Postgraduate Study Report, Department of Computer Science and Engineering, CTU, Prague, September 1997.

[Me92]   Melichar, B. *Syntax Directed Translation with LR Parsing*. In: LNCS, vol 641(*Compiler Construction*), Springer–Verlag, Berlin, pp. 30-36, 1992.

[Me95]   Melichar, B. *LR Parsing and Formal Translation*. Doctoral dissertation thesis, Department of Computer Science and Engineering, CTU, Prague, 1995.

[MB96]   Melichar, B., Bac, N.V. *Transformations of Grammars and Translation Directed by LR Parsing*. Research report DC-96-02, Department of Computer Science and Engineering, CTU, Prague, March 1996.

[PB80]   Purdom, P., Brown, C.A. *Semantic Routines and LR(k) Parsers*. Acta Informatica, Vol. 14, No. 4, pp. 229-315, 1980.

# Parallel Processing on Alphas Under MATLAB 5

Jiří Kadlec

Institute of Information Theory and Automation
of the Academy of Sciences of the Czech Republic*
P.O. Box 18, 182 08 Prague, Czech Republic

**Abstract.** Presented software toolbox connects the PC MS W95 based MATLAB 5 with a network of AD66 high performance boards. Each board includes one DEC AXP 64 bit 21066 and 21164 RISC processor working 233MHz-500MHz with additional circuits supporting T4/T8 transputer links, C40 comports or T9000 link standards. The ParaMat starts and controls parallel applications from within MATLAB 5. Users can program the parallel Alpha applications, which may use MATLAB and SIMULINK as a programmable high level graphical interface. This significantly reduces the development time and costs of the final parallel application and brings the high-end workstations performance to the standard, PC based, Windows platforms.

## 1 Introduction

MATLAB is an open shell system, providing means for coding interfaces to other software in the form of (dynamically linked) MEX - functions in C language. The design of MEX functions is straightforward and the resulting software enjoys broad range of standardized system services.

This paper describes an ParaMat toolbox for MATLAB ver 5, which combines the connectivity of transputers with the computing power of DEC Alpha processors. In compute-intensive applications, a single Alpha chip is able to replace a network of approximately 70-100 T800 transputers!

ParaMat accelerates Matlab applications and moves the performance of PC based MATLAB MEX-functions from the performance limited by, say, 200 MHz Pentium-Pro to the performance of the high-end workstations clusters.

The ParaMat toolbox [7] operates as a network of MATLAB-like (reduced) interpreters running on each of the AD66 boards [4] under control of MATLAB 5. The boards and the Alpha AXP architecture is described in next section.

### 1.1 Alpha AXP and AD66 Boards

The AD66 boards [4] used by the ParaMat are intended for compute-intensive applications, such as DSP, image processing, molecular simulation, fluid dynamics, weather forecasting and financial modeling. Boards serve as building-blocks

* This work was partialy supported by grants GAČR 102/95/0926, and GAČR 102/97/0118 of the Grant Agency of the Czech Republic

of a high-performance parallel processing network under the control from MAT-LAB. See Figure 1 and Figure 2.

The DEC 21066 (21164) Alpha processor is fully pipelined, dual-issue (quadruple) 64-bit advanced RISC processor. Running at 233MHz (500 MHz), it is capable of a maximum 466 mips (1000 mips). It includes an integral memory-management unit and cache controller, IEEE and VAX-compatible floating-point and a PCI I/O controller. See [4] for details.

Boards include an ISA or PCI interface to PC/AT or PCI bus. This interface is supported by the 32 bit Windows 95 0 server ws32 developed by 3L company. See the User Guide [4] for the board options supporting the C40 comports and T9000 link standards. For simplicity, this paper concentrates on the networks of the AD66 boards equipped with four transputer links or PCI implementation of links.

The key software development tool is the 3L Parallel C [2], and the DEC ANSI C compiler [3]. 3L Parallel C/AXP is source code compatible with 3L Parallel C for transputers. The DEC compiler runs in the Alpha processor, supported by PC via axp server. A minimal subset of the DEC OSF operating system is emulated for the compiler. Therefore, the DEC 21066 (21164) RISC processor is programmed by the standard, UNIX like optimizing DEC ANSI C. In addition, the standard 3L Parallel C functions known from the transputer world can be used.

From the programmer's point of view, the twin T425/DEC 21066 can be programmed and configured into transputer networks like a standard transputer with 4 Inmos compatible links. The 3L Parallel C/AXP ver. 1.1.1 [3] supports true multi-tasking, dynamic process (thread) creation, synchronized message passing and semaphores.

The AD66 Alpha boards can be used as a parallel numerical accelerators programmable from MATLAB. Individual SIMULINK blocks can be implemented as parallel Alpha MEX functions execute on a statically configuratd network of AD66 boards or as a data parallel "farm" with a master omplemented as a M-function or MEX-function converted to a new "data-parallel" S-function and corresponding S-block of SIMULINK.

Alternatively, the AD66-ISA board can be viewed as the master processor of a larger network of Alphas, transputers and transputer-link compatible I/Os. In such applications, MATLAB acts mainly as a graphical server for the initialization, user interfaces and the presentation of results under Windows.

## 2 ParaMat

ParaMat is a software environment for design, verification and implementation of parallel transputer and DEC Alpha applications controlled directly from MAT-LAB 5 under Windows 95.

The ParaMat offers the possibility of running parallel applications on multiple AD66 boards directly from MATLAB.

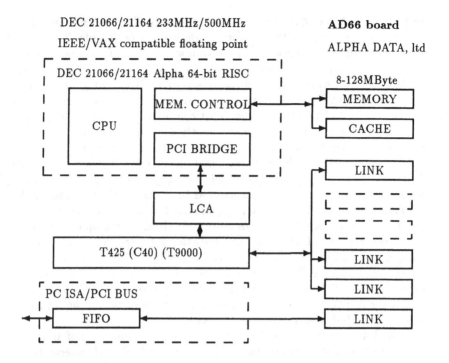

**Fig. 1.** AD66 High Performance Parallel Processor Board

A reduced, Matlab-like interpreters runs on each AD66 Alpha board. The interpreters have their own ability to work with separate matrix environments located in the memory of the DEC Alpha processors. Users can control the communication between the MATLAB environment and each of AD66 Alpha boards.

It is possible to define sequences of calls to MEX-functions for each local interpreter. These MEX-functions are source code compatible with the MATLAB MEX-functions, but executed by the Alpha AXP processor and operate directly on the matrices located in the memory of AD66 Alpha boards.

If convenient, a MEX-function, executed by the Alpha processor, can be called with the input/output matrix arguments from the original MATLAB workspace.

ParaMat automates the communication between the MATLAB and the network of Alpha processors. All the MATLAB's External Interface Library functions are emulated on each board. Therefore, the original PC MATLAB MEX functions can run (without change of C code) on AD66 Alpha boards.

This includes the callbacks from Alpha boards back to MATLAB to execute a MEX-function or a M-function with input/output matrix parameters located in the transputer or Alpha workspace. Callbacks are used for graphics, user dialogs, or for calls to MATLAB interpreted scripts and toolboxes. See Figure 2.

Figure 2. presents a PC platform with four AD66 boards. There are four independent interpreters of MATLAB compatible MEX-functions. The parallel algorithm is defined and controlled from the top MATLAB interpreter level.

MATLAB controls the switch process running in the first Alpha processor on the AD66-ISA board. The switch process selects which board is able to communicate with MATLAB. All subsequent commands are applied to the selected board. All boards can operate concurrently and the independent interpreters inform the switch block about the progress of the computation.

Naturally, this default configuration of one AD66-ISA board and additional AD66 boards can be extended. Parallel, statically configured networks of Alpha boards and transputers can be constructed by the use of the remaining links and 3L Parallel C/AXP functions for synchronized message passing. Each of such parallel functions is controlled from one of Alpha or transputer interpreters. Therefore, the MATLAB script and ParaMat software can operate with such parallel function identically as with the sequential functions running on other interpreters. The parallel interactions are hidden. The spare links can be used for connection of the special hardware interfaces too.

See the AD-66 User Guide [4] for the detailed description of the Alpha boards with the C40 link and T9000 link support.

MATLAB interpreter can read the progress of computation of all connected interpreters by a service **switchip**. Therefore, the MATLAB interpreter script can decide what MEX-function or sequence of MEX-functions to start on what Alpha board.

## 3   Elementary ParaMat Services

Using the ParaMat toolbox, the user operates with two related matrix environments: The MATLAB shell and a network of Alpha interpreters running on the AD66 boards. Both environments communicate using a set of PC service functions compiled as special MEX-functions of MATLAB. The service functions can communicate with the Alpha and transputer networks via a B004 compatible link interface.

This section provides a list of elementary services supported by the ParaMat [7].

### 3.1   Copy PC Matrices to Alpha

This operation is solved by the functions **boot** and **boottp**. The first sends permanent (named) MATLAB matrices to the Alpha board, eg.: **boot(A,B,C)** copies the matrices **A**, **B** and **C** to the Alpha interpreter workspace under the same name. The second service allows the copy of results of MATLAB expressions to the selected Alpha board, eg.: **boottp(A*B,'AB',i,'i')** sends the (noname) product to Alpha matrix **AB** and the imaginary unit **i** to the Alpha under the same name **i**.

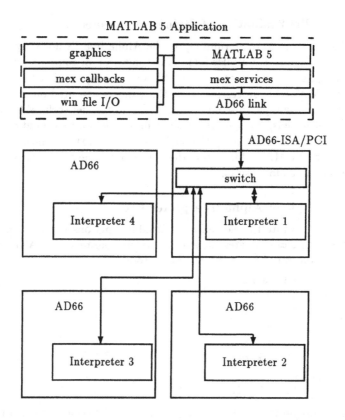

MATLAB 5 Application

**Fig. 2.** MATLAB with 4 Interpreters on AD66 Boards

## 3.2 Copy Alpha Matrices to PC

The function bootpc carries out this operation, eg.: [A, B] = bootpc('TPA', 'TPB') creates MATLAB matrices A and B as copies of the Alpha matrices TPA and TPB.

## 3.3 Allocate Matrices in Alpha Workspace

This task is solved by the function alloctp, eg.: alloctp('TPM', 10, 20) creates the Alpha 10-by-20 matrix TMP.

## 3.4 Display and Clear of Alpha Matrices

The functions whotp, whostp and clearip are functionally similar to the who, whos and clear MATLAB functions.

## 3.5 Parallel Evaluation of String Commands

The ParaMat interpreter running on the Alpha boards not only maintains the matrix environment, but also allows interpretation of command lines. The function **evalpar** and **evalwait** are available for this purpose. The function **evalwait** suspends MATLAB processing while the string is executed on the Alpha board. The function **evalpar** disconnects the PC and the network. MATLAB and the network of Alphas/transputers continue in parallel up to a point of synchronization which is implemented by the service function **waitip**. The service **switchip** can be used by MATLAB to test the status of the network and to decide when and where to go for results.

## 3.6 Example of Alpha Interpreter Use

Let us demonstrate the elementary ParaMat functions using a simple example. The function **sum** evaluates the sum of matrix columns. In MATLAB, the standard solution would be to call the MEX code by:

```
»A = magic(3)
A =
8    1    6
3    5    7
4    9    2
»U=sum(A)
U =
15   15   15
```

The same problem can be solved by a MEX-function on Alpha board by:

```
»boot(A)
»evalwait('U=sum(A)')
»U=bootpc('U')
U =
15   15   15
```

The repetitive interpretation of commands is both inconvenient and time consuming. Therefore the ParaMat toolbox introduces a concept of programming of Alpha boards as described in next section.

## 3.7 Evaluation of Program on Alpha

Programs consist of sequentially numbered lines which are introduced and recorded in an Alpha interpreter by a function **pgmline** and listed by **pgmlist**. Let us for example introduce the following program:

```
»pgmline('U=sum(A)');
»pgmline('V=sum(U)')
»pgmlist
Alpha board program
1   U=sum(A)
2   V=sum(U)
```

The Alpha interpreter is instructed to start execution of the current program by services `callpar` or `callwait`. These services differ in the time schedule. After `callwait` MATLAB waits for the results, connected to the board. After `callpar` MATLAB continues in interpretation of it's script and the interpreter on the Alpha board executes the program without connection to MATLAB. The above listed example can continue like:

```
»callpar
...
»waitip
»V = bootpc('V')
V =
45
```

## 3.8 Multiple Alpha Boards/Interpreters

On systems with more then one Alpha board, the service `switchip` is used to select the board for communication and to test the progress of computation of all boards from MATLAB interpreter level.

Multiple interpreters can run in parallel even on a system with only one AD66 board. This can be used for development of the final parallel application on a one-processor development system. Service `switchip` is used to select one of the parallel running interpreters in this case.

The service `switchip` communicates with the process `switchip` running on the AD66-ISA board connected to MATLAB.

`S = switchip(I)` selects the interpreter I for the subsequent services and returns the (1,M) vector S containing the status of all M interpreters.

`[S, A] = switchip(I)` works as above and returns in addition the scalar A indicating the number of the selected interpreter (A=I).

`[S, A, M] = switchip(I)` works as above and returns in addition the scalar M with the total number interpreters connected to the switch. The elements of the status vector S can have these values:

S(j)=0 Interpreter j is ready.
S(j)=2 Interpreter j finished a parallel call.
S(j)=3 Interpreter j works in parallel mode.

MATLAB scripts can monitor the progress of functions running in parallel on the network of Alphas by testing the status returned from `switchip` service.

## 3.9 Alpha MEX Function Design

MATLAB 4 has established a standard for accesses to matrices. This standard was considerably chabged in Matlab 5. Standard 4.2 Matrices are manipulated by a library of functions used by MEX-functions for MATLAB. These routines are grouped into three classes distinguished by prefixes: (1) `mx` for access to matrices in the MATLAB workspace; (2) **mex** for access to other interpreted .m scripts or M-functions and to the compiled MEX-functions;(3) **mat** for access to MATLAB data files in .**MAT** format.

Present versoion of ParaMat toolbox emulates only Matlab 4.2 routines for MEX-functions running on Alpha boards.

## 4 Conclusion

The paper describes concept of the ParaMat toolbox environment for parallel programming under MATLAB 5.

The ParaMat concept of parallel interpreters running on the network of Alpha boards provides new cost/performance attractive alternative for MATLAB and SIMULINK users as well as for developers of real-time DEC Alpha parallel applications.

## References

1. Matlab Reference Guide, The MathWorks, Inc. Natick, Mass 01760, (1992).
2. Parallel C ver 2.2.2 Reference Manual, 3L Edinburgh. UK., (1991).
3. Parallel C/AXP V1.1.1. Parallel C, User Guide, Digital Equipment Corporation Alpha/AXP, 3L Ltd., Edinburgh. UK (1995).
4. AD-66 Parallel Processor Boards. User Guide, ALPHA DATA, parallel systems ltd, 86 Causewayside, Edinburgh EH9 1PY,Scotland, (1995).
5. Kadlec J., P. Nedoma. Matlab-Transputer-Bridge 1.0, User Guide. Institute of the Information Theory and Automation of The Academy of Science of The Czech Republic, Prague (1995).
6. J. Kadlec Direct Software Bridge Matlab–Transputer Boards Preprints of the European Signal Processing Conference EUSIPCO'94, C.F.N. Covan editor, Edinburgh, Scotland (1994) 1601–1604.
7. MATLAB Bridge to Parallel Alpha, User Guide, ALPHA DATA, *parallel systems* ltd. (1996).

# PRAM Lower Bound for Element Distinctness Revisited

Petr Kolman

Department of Applied Mathematics, Faculty of Mathematics and Physics, Charles University, Prague, Czech Republic
kolman@kam.ms.mff.cuni.cz

**Abstract.** This paper considers the problem of element distinctness on CRCW PRAMs with unbounded memory. A complete proof of the optimal lower bound of $\Omega\left(n \log n / \left(p \log(\frac{n}{p} \log n + 1)\right)\right)$ steps for $n$ elements problem on $p$ processor COMMON PRAM is presented. This lower bound has been previously correctly stated by Boppana, but his proof was not complete. Its correction requires some additional observations and some not straightforward changes.

## 1 Introduction

This paper considers the PRAM model with unbounded memory. A *PRAM* consists of $p$ processors $P_1, \cdots, P_p$ and shared memory with cells indexed by integers $N$. Initially the contents of all shared memory cells are 0 and processor $P_i$, $1 \leq i \leq n$, has the input value $x_i$ in it's local memory. One step of computation consists of three phases: local computation, global read, global write. On the PRIORITY model the minimal index processor (among those attempting to write to the same cell) succeeds in writing to a memory cell if a write conflict appears. On the COMMON model all processors writing to the same cell have to write the same value. Assume also that the result of the computation is known to processor $P_1$.

The problem of *Element Distinctness* is defined as follows: Given $n$ integers $x_1, \cdots, x_n$ determine whether they are pairwise distinct. This problem has been studied for some time as a tool for separating different variants of PRAM. On $n$ processor PRIORITY PRAM it is solvable in $O(1)$ steps [11]. For COMMON PRAM several lower bounds were proved implying the separation. Fich, Meyer auf der Heide, and Wigderson [7] showed a lower bound of $\Omega(\log \log \log n)$, Ragde, Steiger, Szemeredi, and Wigderson [11] improved it to $\Omega(\sqrt{\log n})$. Finally Boppana [1] presented a proof of the optimal lower bound $\Omega(\frac{\log n}{\log \log n})$ (resp. $\Omega\left(\frac{n \log n}{p \log(\frac{n}{p} \log n + 1)}\right)$) for the case with $n$ elements and $p$ processors). Unfortunately there is an error in the proof of his theorem and in fact the assertion is not proved (see Appendix A). All these results require an infinite input domain size. Later on Edmonds [3, 4] proved (using a new technique) the same lower bound even for the case of much smaller input domain size (his proof requires only $2^{2^{\Omega(n)}}$ ). The proof is very complicated and the published version [3] gives mainly an overview of the proof, the key ideas, not the details.

The present paper gives the complete proof of the lower bound for infinite input domain size. The basic idea is the same as in the proof of Boppana – to measure the number of pairs of variables that were learned to be different by entropies of associated random variables. The difference is that the unjustly omited pairs in the previous proof are considered now. The other difference is that now the lower bound is proved directly for the COMMON PRAM.

## 1.1 Definitions

A pair $(x, y)$ of vectors of variables is said to *cover* a pair $(x_i, x_j)$ of variables, if $x$ and $y$ agree on all coordinates but one, and on that coordinate $x$ is equal to $x_i$ and $y$ to $x_j$ (e.g. $((x_1, x_2), (x_1, x_3))$ covers $(x_2, x_3)$).

Let $l$ be a mapping of a set of $n$ variables $V = \{x_1, \cdots, x_n\}$ into some totally ordered set. We say that a vector $a = (a_1, \cdots, a_n) \in N^n$ is *consistent with $l$* if for every $1 \le i, j \le n$, $l(x_i) < l(x_j) \Rightarrow a_i < a_j$. Let $(S^n)_l$ denotes the set $(S^n)_l = \{a \in S^n \mid a \text{ is consistent with } l\}$. Because $l$ corresponds to some partial order on $V$ we will sometimes speak about $l$ as about an partial order. For $l$ and subsets $A, B \subseteq V$ define the *comparison graph* $C_l(A, B) = (V, E)$, where $E = \{(x_i, x_j) \mid x_i \in A, \ x_j \in B, \ i \ne j, \ l(x_i) = l(x_j)\}$. The number of edges in $C_l(A, B)$ will be denoted by $|C_l(A, B)|$.

A *D-function* is a function $f$ whose domain is of the form $D^V$ for $D \subseteq N$ and for some finite set $V$. A function $f|_I$, where $f|_I$ denotes the restriction of $f$ to $I \subseteq D^V$, is said to *depend on the variable $x_i$* if there are $a = (a_1, \cdots, a_i, \cdots, a_n) \in I$ and $\hat{a} = (a_1, \cdots, \hat{a}_i, \cdots, a_n) \in I$ such that $f(a) \ne f(\hat{a})$ ($a$ and $\hat{a}$ agree on all coordinates but the $i^{th}$).

Let $f, g$ be $S$-functions and let $I \subseteq (S^n)_l$ be such a set that the set of variables which $f|_I$ (resp. $g|_I$) depends on is lineary ordered in $l$. Suppose $f|_I$ and $g|_I$ are indentical when seen as functions just of those variables they depend on and when these variables are ordered according to $l$. Then we say that $f|_I$ and $g|_I$ are *weakly indentical* (e.g. if $l(x_1) = 1$, $l(x_2) = 2$, $l(x_3) = 2$, $f(x_1, x_2, x_3) = x_1 + x_2^3$, $g(x_1, x_2, x_3) = x_1 + x_3^3$ and $I \subseteq (S^3)_l$, then we would say that $f|_I$ and $g|_I$ are weakly indentical).

Given a random variable $X$ with range $\{c_1, \cdots, c_m\}$, its *entropy* is $H(X) = \sum_{j=1}^m -Pr[X = c_j] \log_2 Pr[X = c_j]$. The *entropy of a function $f$* with a finite domain is $H(f) = H(f(v))$, where $v$ is a random variable with uniform distribution on the domain of $f$. Given two random variables $X$ and $Y$, the notation $H(X, Y)$ will be used for the entropy of a random variable $(X, Y)$. The most useful property of entropy for us is subadditivity: $H(X, Y) \le H(X) + H(Y)$.

Finally let $L$ denotes the function $L(z) = (z + 1) \log_2(z + 1) - z \log_2 z$.

## 2 Proof

In this section a proof of the lower bound will be given. It is based on those of Boppana [1] and Ragde, Steiger, Szemeredi, Wigderson [11]. The proof makes crucial use of Ramsey theory. The following theorem states the results we shall need.

**Theorem 1.** *Let $F$ be a finite set of S-functions of $n$ variables $\{x_1, x_2, \cdots, x_n\}$, for infinite $S \subseteq N$, let $l$ be a mapping of $\{x_1, x_2, \cdots, x_n\}$ into some totally ordered set and let $I \subseteq (S^n)_l$ be such a set that for every $f \in F$ the set of variables which $f|_I$ depends on is lineary ordered in $l$. Then there exists an infinite $\bar{S} \subseteq S$ such that for $I' = I \cap \bar{S}^n$*

$\forall f \in F \; : \; f|_{I'}$ *is one-to-one on variables it depends on,*

$\forall f, g \in F \; : \; f|_{I'}$ *and $g|_{I'}$ either have disjoint ranges or are weakly identical.*

*Proof.* The theorem follows from Lemma 1 [10], which is based on Erdös-Rado Theorem [5].

**Theorem 2.** *Every COMMON PRAM with $p$ processors that solves the problem of element distinctness of size $n, n \leq p$, requires $\Omega\left(\frac{n \log n}{p \log(\frac{n}{p} \log n + 1)}\right)$ steps.*

*Proof.* The given algorithm defines for each processor $P_i$ and for each time step $t$ two *addressing functions* $R_t^i, W_t^i \; : \; N^n \to N$, where $R_t^i(x)$ specifies the number of a cell of the global memory, that the processor $P_i$ at time step $t$ for input $x$ is reading from. Similarly with $W_t^i$ for writing. By doubling the amount of steps of the algorithm we may assume that $R_t^i = W_t^i$ for each $i$ and $t$.

Suppose, by contradiction, there is a COMMON PRAM, that solves the problem in less than $\Omega\left(\frac{n \log n}{p \log(\frac{n}{p} \log n + 1)}\right)$ steps, say in $T$ steps. We will show there are two different inputs, one element distinct and the other not, that are indistinguishable to the given PRAM at the beginning of step $T$. To reach the desired goal, a sequence of input sets $I_t \subseteq N^n$ will be inductively constructed, such that for every $t$, $1 \leq t \leq T$, there are two inputs in $I_t$, one element distinct and the other not, that are indistinguishable to the PRAM at the beginning of step $t$. The set $I_t$ is described by a set $S_t$, $S_t \subseteq N$, by a mapping $l_t$ of the set $V = \{x_1, \cdots, x_n\}$ of input variables into some totally ordered set and by a graph $DG_t$ on the set $V$, in the following way:

$$I_t = \{ \; a \mid a \in (S_t^n)_{l_t} \; \& \; a_i = a_j \Rightarrow (x_i, x_j) \notin DG_t \; \&$$
$$\text{there is at most one pair } a_i = a_j, i \neq j \}.$$

Sets $V_t^i$, $1 \leq t \leq T$, $1 \leq i \leq p$, will help us to define the objects. The meaning of the objects is roughly as follows. If for each $x \in I_{t-1}$ the PRAM is able to recognize during the first $t - 1$ steps the relative order of $x_i$ and $x_j$, then $l_t(x_i) \neq l_t(x_j)$. If it is only able to recognize wheather $x_i = x_j$ or $x_i \neq x_j$, then $(x_i, x_j) \in DG_t$. If the values for all $x_j \in V_t^i$ are fixed then no matter how the other values are changed, the state of $P_i$ at the beginning of step $t$ will be the same, provided $x \in I_t$ (we say that only variables in $V_t^i$ may *affect* the state of processor $P_i$ at the beginning of step $t$). The restriction of inputs to a subset of $(S_t^n)_{l_t}$ ensures that the interactions among processors via the shared memory appear only in a very regular way. This enables us to bound the amount of information about the input values the algorithm is able to learn in a single step.

All the mentioned objects will be constructed inductively and will satisfy the following properties for $m = 8 \log p$:

1. $S_t \subseteq S_{t-1}$ for $t \geq 1$,
2. $l_{t-1}(x_i) < l_{t-1}(x_j) \Rightarrow l_t(x_i) < l_t(x_j)$ for $t \geq 1$, and for $t \geq 0$ hold:
   (a) $H(l_t) \leq \left(L(\frac{pm}{n}) + 3\right) t$,
   (b) $|C_{l_t}(V_t^i, V_t^j)| \leq m$ for $1 \leq i < j \leq p$,
3. $DG_t \supseteq DG_{t-1}$ for $t \geq 1$,
4. $l_t$ is a linear order on $V_t^i$ (i.e. $l_t$ is one-to-one on $V_t^i$) for $t \geq 0$, $\forall i, 1 \leq i \leq p$,
5. $V_t^i$ contains all the variables that may affect the state of processor $P_i$ at the beginning of step $t$, for all inputs in $I_t$.

This will make sure, besides others, that $I_t \subseteq I_{t-1}$.

According to the definition of PRAM we put $V_0^i = \{x_i\}$ for $1 \leq i \leq n$, $V_0^i = \emptyset$ for $n < i \leq p$. Let $S_0 = N$, $l_0$ be a constant mapping and $DG_0$ be an empty graph (i.e. with no edges). All desired properties are satisfied for $t = 0$.

Now the inductive step can be described. Let

$$F = \{W_{t'}^i \mid 1 \leq i \leq p, 1 \leq t' \leq t\}.$$

By applying the Theorem 1 to the set $F$ of functions, mapping $l_t$ and sets $I_t$ and $S_t$ (we have $I_t \subseteq (S_t)_{l_t}^n$) we obtain $\bar{S} \subseteq S_t$. We put $S_{t+1} = \bar{S}$ and $I_t' = I_t \cap S_{t+1}^n$. Then the set $F$ can be divided into several disjoint subsets $F_1, \cdots, F_{m_t}$ such that

$$\forall f, g \in F_i : f|_{I_t'} \text{ is weakly identical with } g|_{I_t'},$$
$$\forall f \in F_i, \ \forall g \in F_j, i \neq j : f|_{I_t'} \text{ and } g|_{I_t'} \text{ have disjoint ranges }.$$

Let's choose the indices of $F_i$ in such a way that $F_i$ from the previous step is a subset of $F_i$ from the current step, for $1 \leq i \leq m_{t-1}$. Consider processors $P_j$ and $P_k$. If there is no pair of addressing functions $f$ of $P_j$ and $g$ of $P_k$ (used up to step $t$) such that $f, g \in F_i$ for some $i$, then processors $P_j$ and $P_k$ cannot exchange directly (i.e. without help of any other processor) any information up to step $t$, for inputs in $I_t'$.

Now we are ready to construct the objects $l_{t+1}$, $DG_{t+1}$ and $V_{t+1}^i$. We have to ensure that properties 2, 3, 4 and 5 will hold for $t + 1$. For $f \in F$ let $\mathbf{k}_f$ denote the vector of variables that $f|_{I_t'}$ depends on, ordered according to $l_t$. Consider one by one each $F_i$, $1 \leq i \leq m_t$. Let's define

$$C_t^i = \{\mathbf{k}_f \mid f \in F_i \ \& \ f = W_t^j \text{ for some } j\}.$$

At this point, in the construction of $DG_{t+1}$ and $V_{t+1}^i$, the fact that a COMMON PRAM is considered will be used. An edge $(x_k, x_l)$ is added to $DG_{t+1}$ if there exists $\mathbf{a} \in C_{t'}^i$ for some $t' < t$ and $\mathbf{b} \in C_t^i$, such that $(\mathbf{a}, \mathbf{b})$ covers $(x_k, x_l)$ and $\mathbf{b} \notin A_{t''} \ \forall t'', t' \leq t'' < t$. This will make sure that the communication among processors via the shared memory in step $t$ will be the same for all inputs in $I_{t+1}$. Let $R = \{P_j \mid W_t^j \in F_i\}$. For each $P_j \in R$ let $t'$ be the last time $\tau < t$ such that there exists processor $P_k$ with $W_\tau^k \in F_i$ and $\mathbf{k}_{W_\tau^k} = \mathbf{k}_{W_t^j}$, and let $W_j$ be a set of all such $P_k$ (i.e. $W_j = \{P_k \mid W_{t'}^k \in F_i \ \& \ \mathbf{k}_{W_{t'}^k} = \mathbf{k}_{W_t^j}\}$), let $W_j = \emptyset$ if no such $\tau$

exists. $W_j$ is the set of all processors that wrote at the last time to the memory cell that $P_j$ is reading from in step $t$, for inputs in $I_{t+1}$. The important thing for us is that $W_j$ is the same for all inputs in $I_{t+1}$. Thus the contents of the memory cell $P_j$ is reading from in step $t$ is determined by the same set of processors, for all inputs in $I_t'$. Therefore if we put $U_{t+1}^j = \bigcap_{P_i \in W_j} V_t^i$ then $V_{t+1}^j = V_t^j \cup U_{t+1}^j$ has the desired property 5. After doing this for each $P_j \in R$ proceed to next $F_i$.

By invariant 2(b) in step $t$ the graph $G = \bigcup_{i=1}^p C_{l_t}(V_t^i, U_{t+1}^i)$ has at most $pm$ edges. Thus by Lemma 3.4 [1] there is a coloring $\chi$ of graph $G$ with entropy at most $L(\frac{pm}{n})$. Order the colors of $\chi$ arbitrarily and define $l' = (l_t, \chi)$, ordering its range in lexicographic order. Notice that for any $l_{t+1}$ that is a refinement of $l'$ the invariant 4 follows.

Next we shall show there is a function $f : V \to \{1, ..., 8\}$ such that for $l_{t+1} = (l', f)$ with lexicographicaly ordered range the properties 2(a)(b) hold. Choose a random function $f : V \to \{1, ..., 8\}$ (we assume a uniform distribution of the random functions from $V$ to $\{1, ..., 8\}$, then $H(f) = 3$). Since for each $i, j$ the graph $C_{l'}(V_{t+1}^i, V_{t+1}^j)$ is a matching, the events "$f(a) = f(b)$" (i.e. the edge $(a, b)$ remains in $C_{l_{t+1}}(V_{t+1}^i, V_{t+1}^j)$ ) over all edges $(a, b)$ in this graph are mutually independent. The probability of each such event is $\frac{1}{8}$. By Chernoff bound [2] we get ($C = |C_{l_{t+1}}(V_{t+1}^i, V_{t+1}^j)|$ is a random variable with a binomial distribution, $E(C) \leq \frac{m}{2}$):

$$ Pr[C \geq m] \leq e^{-\frac{1}{2}\left(\frac{m}{E(C)}-1\right)^2 E(C)} \leq e^{-\frac{m}{4}} = \frac{1}{p^2}. $$

Thus

$$ Pr[\exists i < j, |C_{l_{t+1}}(V_{t+1}^i, V_{t+1}^j)| \geq m] \leq \binom{p}{2}\frac{1}{p^2} < 1, $$

i.e. there exists $f$ such that for $l_{t+1} = (l', f)$ the invariant 2(b) holds for time $t + 1$. Invariant 2(a) follows for time $t + 1$ from the subadditivity property of the entropy: $H(l_{t+1}) \leq H(l_t) + H(\chi) + H(f) \leq (t+1)\left(L\left(\frac{pm}{n}\right) + 3\right)$. This is the end of the inductive step.

It remains to show there are the two inputs in $I_T$, one element distinct and the other not, that are indistinguishable to the given PRAM up to step $T$. Suppose there is at least one pair of variables $(x_j, x_k)$ such that (i) $(x_j, x_k)$ is not an edge in $DG_T$ graph and (ii) $l_T(x_j) = l_T(x_k)$. Then due to invariants 4 and 5 for every $a \in I_T$ at most one of variables $x_j, x_k$ can affect the state of $P_1$ at the beginning of step $T$. Suppose that $x_k$ does not affect the state of $P_1$. Choose $a \in I_T$ with all coordinates $a_i$ distinct and let $\bar{a}$ be such that $\bar{a}_i = a_i$ for $i \neq k$ and $\bar{a}_k = a_j$. Surely $\bar{a} \in I_T$. Furthermore, since the state of $P_1$ is not affected at the beginning of step $T$ by the variable $x_k$, the PRAM cannot distinguish the two inputs $\bar{a}$ and $a$ up to step $T$. So it suffices to show that there exists a pair of variables $(x_j, x_k)$ such that (i) and (ii) hold for $DG_T$ and $l_T$.

Let us make some definitions first. For a function $f$ with domain $A \subseteq V$, let $K(f)$ be a graph $(V, E)$ such that $E = \{(a, b) \mid a, b \in A : f(a) \neq f(b)\}$. For a set $F$ of such functions we define $K(F) = \bigcup_{f \in F} K(f)$. Let's define next

$$cost(f) = \frac{|dom(f)|}{|V|} H(f), \quad cost(F) = \sum_{f \in F} cost(f).$$

For a graph $G$ on $V$ define

$$content(G) = \min_{K(F) \supseteq G} cost(F).$$

The subadivity of entropy implies subadivity of content.

Let $CG = K(l_T)$. Clearly if $CG \cup DG_T$ is not complete graph $K_n$ then the desired pair of variables $(x_j, x_k)$ exists. A useful result by Fredman and Komlós [6] says that content of the complete graph $K_n$ is equal to $\log_2 n$. Thus to show that $CG \cup DG_T \neq K_n$ it suffices to show that $content(CG) + content(DG_T) < \log_2 n$. By the definition of $CG$ and by invariant 2(a) for step $T$ we have $content(CG) \leq H(l_T) \leq \left(L\left(\frac{8p \log p}{n}\right) + 3\right) T$. Since $content(CG) > 0.1 \log n$ would imply $T = \Omega\left(\frac{n \log n}{p \log(\frac{n}{p} \log n + 1)}\right)$ let us assume that $content(CG) \leq 0.1 \log n$.

Bounding $content(DG_T)$ will be more complicated. We will describe $DG_T$ in other way and then divide it into two parts $DG_1$ and $DG_2$ for which it is easier to bound their content. Let's define $A_t^i = \bigcup_{i=t+1}^T C_t^i$, for $1 \leq i \leq n_T$, $1 \leq t \leq T$. Put

$$TS = \{(A_i^t - C_i^t, C_i^t) \mid 1 \leq i \leq n_T, \ 1 \leq t \leq T - 1\}.$$

Let $E = \{(x_i, x_j) \mid \exists (A, C) \in TS \ \exists a \in A \ \exists c \in C \ : \ (a, c) \text{ covers } (x_i, x_j)\}$. Then for graph $DG' = (V, E)$ it follows $DG_T \subseteq DG'$, and $content(DG_T) \leq content(DG')$.

We may assume that $|V_T^i| \leq n^{0.1}$ for $1 \leq i \leq p$ since the other case would imply $T = \Omega\left(\frac{n \log n}{p \log(\frac{n}{p} \log n + 1)}\right)$. Lemma 4.1 [4], when applied to our set $TS$, says that there is a set $\Phi = \{(B_i, D_i) \mid 1 \leq i \leq r\}$ such that

1) $B_i \cup D_i = V$, $B_i \cap D_i = \emptyset$,
2) $|E - E_1| \leq n^{0.1} \sqrt{n} \log(|TS|)|TS| + \frac{|TS|^2}{\sqrt{n}}$
   for $E_1 = \{(x_k, x_l) \mid \exists (B_i, D_i) \in \Phi \ : \ x_k \in B_i, x_l \in D_i\}$.

where $|TS| = \sum_{i,t}(|A_i^t - C_i^t| + |C_i^t|)$. In other words it is possible to divide the graph $DG'$ conveniently in two parts $DG_1 = (V, E_1)$ and $DG_2 = (V, E - E_1)$ such that $DG_1 \cup DG_2 = DG'$.

By Lemma 3.4 [1] every graph with $n$ vertices and $|E - E_1|$ edges has a coloring $\chi$ with entropy $H(\chi) \leq \log_2\left(\left(\frac{|E-E_1|}{n} + 1\right) e\right)$. Since $DG_2 \subseteq K(\chi)$, it follows that $content(DG_2) \leq cost(\chi) = H(\chi)$. For $p > n \log n$ the lower bound of the main theorem is trivial, so assume $p \leq n \log n$. Remember also that we assume $T < \log n$. Then because $|TS| \leq 2pT^2$ we have

$$|E - E_1| \leq n^{0.1} \sqrt{n} \log(2pT^2) 2pT^2 + \frac{4p^2 T^4}{\sqrt{n}} \leq n^{1.6}.$$

Thus $H(\chi) \leq 0.7 \log n$ and $content(DG_2) \leq 0.7 \log n$.

For each $(B_i, D_i) \in \Phi$ define function $h_i : V \to \{0, 1\}$ in the following way: for $x \in B_i$ let $h_i(x) = 0$, for $x \in D_i$ let $h_i(x) = 1$. We put $H = \bigcup_{i=1}^{r} h_i$. Then $DG_1 \subseteq K(H)$. By definition $cost(H) = \sum_{i=1}^{r} \frac{|dom(h_i)|}{n} H(h_i)$. With the help of Lemma 4.2 [4] we have

$$cost(H) \leq \frac{2pT^2}{n} \frac{2 \log T}{T} = 4\frac{p}{n} T \log T .$$

Let us assume $cost(H) \leq 0.1 \log_2 n$ since the other case would imply $T = \Omega\left(\frac{n \log n}{p \log(\frac{n}{p} \log n + 1)}\right)$. It follows that $content(DG_1) \leq 0.1 \log_2 n$ and $cost(DG') \leq 0.8 \log_2 n$.

Putting together the bounds on $content(CG)$ and $content(DG')$ we get the desired $content(CG) + content(DG) < \log_2 n$ which implies the existence of a pair of variables $(x_j, x_k)$ that are neither an edge in $DG_T$ neither comparable in $l_T$. Thus, as described earlier, there are two different inputs, one element distinct and the other not that are indistinguishable to the given PRAM through step $T$ which is the desired contradiction.

# References

1. Boppana R.V.: Optimal Separations Between Concurrent-Write Parallel Machines, Proc. 21th ACM Symp. on Theory of Computing (1989) 320-326
2. Chernoff H.: A Measure of Asymptotic Efficiency for Tests of a Hypothesis Based on the Sum of Observations, Annals of Math. Stat. **23** (1952) 493-509
3. Edmonds J.: Lower Bounds with Smaller Domain Size, Structures in Complexity Theory (1991) 322-331
4. Edmonds J.: Removing Ramsey Theory: Lower Bounds with Smaller Domain Size, Journal of Theoretic Computer Science, submited, available as Postscript file at http://www.cs.yorku.ca/~jeff/research/index.html
5. Erdös P. and Rado R.: A Combinatorial Theorem, J. London Math. Soc. **25** (1950) 376-382
6. Fredman M., Komlós J.: On the Size of Separating Systems and Families of Perfect Hash Functions, SIAM J. Alg. Disc. Meth. **5** (1984) 61-68
7. Fich F. E., Meyer Auf Der Heide F., Wigderson A.: Lower Bounds for Parallel Random-Access Machines with Unbounded Shared Memory, Advances in Computing Research **4** (1986) 1-15
8. Fich, F . E., Ragde, P. L., Wigderson, A.: Relations Between Concurrent-Write Models of Parallel Computation, SIAM J. Comp. **17** (1988) 606-627
9. Kolman, P.: Complexity of Some Problems on COMMON PRAM, Masters Thesis, Charles University, Prague (1995), in Czech
10. Meyer Auf Der Heide F., Wigderson A.: The Comlexity of Parallel Sorting, 26th FOCS (1985) 532-540
11. Ragde P. L., Steiger W., Szemeredi E., Wigderson A.: The Parallel Complexity of Element Distinctness is $\Omega(\sqrt{\log n})$, SIAM J. Disc. Math. **1** (1988) 399-410

# A   Counter Example

In this part a brief description of a counter example to Lemma 4.3 [1] together with a short discussion of the problem will be given.

Let's consider a COMMON PRAM that consists of two processors $P_1$ and $P_2$ and unbounded memory. Let $D$ be the infinite set of possible input values. The computation proceeds as follows: In the first step $P_i$, $1 \le i \le 2$, reads the contents of cell $x_i + 3$ and writes 19 to the same cell. In the second step $P_1$ reads the contents of cell $x_1 + 3$ and then writes 1 there, while $P_2$ reads the contents of cell 2 and writes 7 there. In the third and last step $P_1$ reads the contents of cell 1 and writes 9 there, $P_2$ reads from cell $x_2 + 3$ and writes 17 there. If $P_2$ reads 1 in this step the input numbers are not element distinct, otherwise they are (the written values have no special meaning, some of them only need to be different). Let us notice that the described PRAM fulfills the requirements of the COMMON model and that it indeed solves the element distinctness problem.

The read-index and write-index functions of the processors are easy to derive. They satisfy without any additional adjustment the assumption on identity of the read-index and write-index functions of each processor in each step as required in the proof of Lemma 4.3 [1]. For the set $F$ of write-index functions it holds that for $C = D$, $F|_C$ is reducible to the one-to-one set $G = \{g_1(y) = y+3, g_2() = 1, g_3() = 2\}$, i.e. the set $F|_C$ is canonical.

A merging and distinctness-checking machine constructed according to the proof of Lemma 4.3 [1] has a possible path of the computation (in fact there is only one path, there is no branching in the computational tree) that ends with the pair $\{x_1, x_2\}$ neither comparable in the partial order $\Gamma$, nor an edge in the distinctness graph $DG$, i.e. it does *not* solve the problem whereas the original COMMON PRAM *does*. For lack of space we omit detail description of the construction, it is given in [9].

The problem is that according to the definition of merging and distinctness-checking machine every access pair may be choosen at most one time. This fact is used in the the proof of the lower bound for the merging and distinctness-checking machine in Lemma 4.6 [1]. The consequence of this limitation is that the simulating merging and distinctness-checking machine is unable to compare some pairs of variables that the original PRAM was able. It is possible to show [9] that, to make the simulation possible, it is necessary either to impose no limitation on the nuber of times an access pairs may be choosen by the merging and distinctness-checking machine, or to allow it to be choosen at least $\Omega\left(\frac{\log n}{\log \log n}\right)$ times. However, in either way it would weaken the lower bound of Lemma 4.6, if no other changes would be done in the proof of it.

# Optimal Trees for Searching in Codebook

Ivan Kopeček

Faculty of Informatics, Masaryk University
Botanická 68a, 60200 Brno, Czech Republic
e-mail: kopecek@fi.muni.cz
http://www.fi.muni.cz/~kopecek/

**Abstract.** Finding nearest neighbour of a given vector in a codebook leads to the following model of searching. In a metric space V, a vector x and a finite subset of vectors S (representing a codebook) are given. We have to find an element of S which is „nearest" to the element x. In what follows, the problem is formulated more exactly and a characterization of optimal search trees for this model of searching is given. It turns out, that balanced quasi-ternary trees are optimal search trees for the discussed problem. The result enables to speed up finding a codebook representation vector of a given acoustic vector, which is important for applications in speech recognition and synthesis.

## 1 Introduction

Vector quantization, creation of a codebook and tree search in the codebook belong to standard techniques used in speech synthesis and recognition (see, e.g., [2], [3], [8]). In what follows, we will formulate the problem of tree searching in a codebook in a slightly generalized way and characterize optimal search trees.

Let $V = (M, d)$ be a metric space, S a finite subset of M and $x \in X \subseteq M$ (restriction to X can avoid cases, when search tree does not exist. Our problem is to minimize $d(x, y)$ (with respect to the variable y) in S (a codebook). Let us remark, that the problem can arise in some other applications (see e.g.[1]).

We suppose that we have constructed a rooted tree R which nodes are elements of M and leaves are exactly all elements of the set S. The search algorithm is performed in the following way.

(1) We start at the root r of the tree.

(2) In a node s we choose such next node, which is a successor of s and minimizes $d(x, z)$ on the set of all successors of the elements s.

(3) We finish when the actual node is a leaf.

Such a rooted tree is called search tree for our above formulated problem, if there is, for every $x \in X$, only one path (described by (1), (2), (3)) from root to a leaf and if the leaf minimizes the function $d(x, y)$ on the set S. Construction of the suitable search tree (even for a restricted set X) can be non-trivial problem, which is, however, not discussed here (of course, a search tree does not always exist for a given search problem ).

When performing search algorithm, in each inner node s on the path (described by (1), (2), (3)) from the root to a leaf we have to perform computations of the metrics $d(s, z)$ for each successor z of the node s. We take one computation of the metrics d(s, z) as a unit of work, which is to be minimized. Of course, in each node s we must compute d(s, z) for all successors of s. In such a way we can evaluate the work which is needed to „find a leaf". Naturally, the function which characterizes the optimality of a search tree will be defined as the sum of all works needed to find any leaf. This leads to Definition 3.1. Consequently, Definition 3.3. formalizes optimality of a rooted tree (called W-optimality). In section 3, basic lemmas are formulated. In section 4 we discuss basic properties of balanced quasi-ternary trees with respect to the optimality. Section 5 summarizes main results.

## 2 Notation and terminology

In what follows all sets are finite (card(M) denotes the number of elements of a set M). Rooted trees (see e. g. [6]) are denoted by ordered pairs $R = (V, A)$ of vertices (nodes) and arcs (edges). We use also notation $V(R)$ for R and $A(R)$ for A. A root is denoted by r. We denote the set of all successors of a node x by $q(x)$ and the set of all antecedents by $p(x)$. $p(x)$ can contain one element or it can be empty. To make notation more simple, we identify the element which $p(x)$ contains (if $p(x)$ is not empty) with $p(x)$. We denote by $L(R)$ the set of all leaves of R.

Let $R = (V, A)$ be a rooted tree, $x, y \in V$, and suppose that there exists an $(n+1)$-tuple $(x_0, x_1, ..., x_n)$ such that $(x_{i-1}, x_i) \in A$, $x_0 = x$, $x_n = y$. The set $\{x_0, x_1, ..., x_n\}$ will be denoted $P(x,y)$ and we use notation $y < x$. If $x_0 = x = r$ then n is called level of y and is denoted $lvl(y)$. Further, $lvl(R)$ will denote the number $\max\{lvl(x); x \in V(R)\}$. If $R = (V, A)$ is a rooted tree and x is its arbitrary node, we denote by $R(x)$ the subtree generated by the node x, i.e. $R(x) = (V', A')$ where $V' = \{y \in V; y < x\} \cup \{x\}$ and $A' = \{(y, z) \in A(R); y, z \in V'\}$;

## 3 W-optimality and W-optimal trees

**3.1. Definition.** Let $R = (V, A)$ be a rooted tree, $x, y \in V$, card(V) > 1 and $y < x$. We put $w(x,y) = \sum_{z \in P(x, y)-\{y\}} card(q(z))$ and $W(R) = \sum_{z \in L(R)} w(r, z)$;

The function W means optimality measure (it corresponds to the computational costs by searching, see Introduction). In what follows, we shall consider only rooted trees with at least three nodes; the other case is trivial.

**3.2. Lemma.** Let $R = (V, A)$ be a rooted tree, $x, y, z \in V$ and $z < y < x$. Then

$$w(x, z) = w(x, y) + w(y, z).$$

**3.3 Definition.** We shall say that a rooted tree $R = (V, A)$ is W-optimal, if for each rooted tree R' satisfying $L(R) = L(R')$ the condition $W(R) \leq W(R')$ holds;

**3.4. Lemma.** If a rooted tree $R = (V, A)$ is W-optimal, then for every $x \in V$ exactly one of the following two cases occurs.

$\quad\quad$ (1) $card(q(x)) > 1$, $\quad\quad$ (2) $x \in L(R)$.

$\quad$ *Proof:* Suppose $x \notin L(R)$, $card(q(x)) = 1$ and there is no antecedent of x. Then $x = r$. Let us denote by y the only element of the set $q(x)$. If $y \in L(R)$, the case is trivial. Hence, suppose $y \notin L(R)$. If we substitute the root r by y, we get a rooted tree R' with the same number of leaves but with the property $W(R') < W(R)$. This contradicts the W-optimality of R.

$\quad$ Suppose now that the antecedent of the element x exists and let us denote $z = p(x)$. Let y be the successor of x. Consider the rooted tree $R' = (V', A')$ defined in the following way: $V' = V - \{x\}$, $A' = A - \{(p(x), x), (x, y)\} \cup \{(p(x), y)\}$. Hence, $L(R) = L(R')$ and $W(R') < W(R)$ This contradicts the W-optimality of R.

**3.5. Definition.** Suppose R and R' are rooted trees with no common nodes and $x \in V(R)$. $R(x, R')$ will denote the rooted tree that arises from R by substitution of the subtree $R(x)$ by R', i.e.: $R(x, R') = (V'', A'')$, where $V'' = \{V(R) - V(R(x))\} \cup V(R')$ and $A'' = (A(R) - A(R(x)) - \{(p(x), x)\}) \cup (A(R') \cup \{(p(x), r')\})$ (r' is the root of R').

**3.6. Lemma.** Let $R=(V, A)$ be a W-optimal rooted tree, $y \in V$. Then $R(y)$ is W-optimal.

$\quad$ *Proof:* Suppose R is W-optimal and $y \in V$. According to Definition 3.1. we obtain $W(R) = \sum_{x \in L(R)} w(r, x) = \sum_{x \in L(R(y))} w(r, x) + \sum_{x \in L(R)-L(R(y))} w(r, x)$. Applying Lemma 3.2.

we get $\quad W(R) = \sum_{x \in L(R(y))} w(r, x) + c = n \cdot w(r, y) + W(R(y)) + c \quad\quad (*)$

where $\quad c = \sum_{x \in L(R)-L(R(y))} w(r, x)$ and $n = card(L(R(y)))$.

Suppose there exists a rooted tree R' such that $L(R') = L(R(y))$ and $W(R') < W(R(y))$ (without loss of generality we can suppose that R' and $R(y)$ are disjoint). From (*) it follows $W(R(y, R')) = n \cdot w(r, y) + W(R') + c < n \cdot w(r, y) + W(R(y)) + c = W(R)$ contradicting W-optimality of R. Hence, $R(y)$ is W-optimal.

**3.7. Definition.** For $n > 2$ we denote by $R[n] = (V, A)$ the rooted tree defined in the following way: $V = \{r, x_1, ..., x_n\}$ and $A = \{(r, x_1), ..., (r, x_n)\}$.

**3.8. Lemma.** $n > 4$ implies that $R[n]$ is not W-optimal.

$\quad$ *Proof:* Consider the rooted tree $R_n = (V_n, A_n)$ defined in the following way:

$\quad V_n = \{r, x_1, ..., x_{n-1}, x_{11}, x_{12}\}, \quad\quad A_n = \{(r, x_1), ..., (r, x_{n-1}), (x_1, x_{11}), (x_1, x_{12})\}$.

By simple calculation we get $W(R[n]) = n^2$ and $W(R_n) = n^2 - n + 4$. For $n > 4$ we get $W(R_n) < W(R[n])$ whereby $L(R_n) = L(R[n])$, contradicting the W-optimality of $R[n]$.

**3.9. Lemma.** Let $R^*$ be the rooted tree derived from a rooted tree $R = (V, A)$ in the following way: $R^* = (V^*, A^*)$, $V^* = \{r\} \cup q(r)$, $A^* = \{(r, x); x \in q(r)\}$. Then the W-optimality of $R$ implies the W-optimality of $R^*$.

*Proof:* Suppose that $R$ is W-optimal and that $R^*$ is not W-optimal. Then there exists a rooted tree $R' = (V', A')$ such that $L(R') = L(R^*)$ and $W(R') < W(R^*)$. Consider the rooted tree $R'' = (V'', A'')$ defined by: $V'' = (V - V^*) \cup V'$, $A'' = (A - A^*) \cup A'$. Let $q(r) = \{y_1, \ldots, y_k\}$. For the rooted tree $R$ we have $w(r, y_i) = k$ for $i = 1, \ldots, k$, $W(R) = k \cdot (n_1 + \ldots + n_k) + W(R(y_1)) + \ldots + W(R(y_k))$, where $n_1 = \text{card}(L(R(y_1))), \ldots, n_k = \text{card}(L(R(y_k)))$. Without loss of generality we can suppose

$$n_1 \geq n_2 \geq \ldots \geq n_k \quad (*) \quad \text{and} \quad c_1 \leq c_2 \leq \ldots \leq c_k \quad (**)$$

where for the rooted tree $R''$ we obtain $c_1 = w(r, y_1), \ldots, c_k = w(r, y_k)$. Hence $W(R'') = c_1 n_1 + \ldots + c_k n_k + W(R(y_1)) + \ldots + W(R(y_k))$. We get $W(R) - W(R'') = (k - c_1)n_1 + \ldots + (k - c_k)n_k$. From the assumption $W(R') < W(R^*)$ it follows $c_1 + \ldots + c_k < k^2 = W(R^*)$. Denoting $g_i = k - c_i$ we get $g_1 + \ldots + g_k > 0$. However, from $(*)$ and $(**)$ it follows $W(R) - W(R'') = g_1 n_1 + \ldots + g_k n_k > 0$, i.e. $W(R) - W(R'') > 0$ contradicting the W-optimality of $R$.

**3.10. Lemma.** Let $R = (V, A)$ be a W-optimal rooted tree satisfying $L(R) > 2$. Then $x \notin L(R)$ implies $1 < \text{card}(q(x)) < 5$.

*Proof:* follows immediately from 3.4., 3.6, 3.8. and 3.9.

**3.11. Lemma.** If $n > 2$ then there exists a rooted tree $R = (V, A)$ such that $R$ is W-optimal, $\text{card}(L(R)) = n$, and $\text{card}(q(r)) = 3$.

*Proof:* We will prove the assertion by induction with respect to the number $n$. Obviously, it holds for $n = 3$. Suppose, that the assertion holds for every $n$ satisfying $3 \leq n \leq k$. We have to prove it for $n = k + 1$. Let $R = (V, A)$ be a W-optimal rooted tree and suppose $L(R) = k + 1$. According to the previous lemma the following cases can occur:

(1) $\text{card}(q(r)) = 2$;      (2) $\text{card}(q(r)) = 3$;      (3) $\text{card}(q(r)) = 4$.

If case (2) occurs, the assertion of the lemma holds. Case (3) can be transformed to case (1) as follows. We put $q(r) = \{y_1, y_2, y_3, y_4\}$ and consider the rooted tree $R' = (V', A')$ defined in the following way:

$$V' = \{r, x_1, x_2\} \cup q(r), \quad A' = \{(r, x_1), (r, x_2), (x_1, y_1), (x_1, y_2), (x_2, y_3), (x_2, y_4)\}.$$

Further, consider the rooted tree $R'' = (V'', A'')$ defined by:

$$V'' = (V - q(r)) \cup V', \quad A'' = (A - \{(r, y_1), (r, y_2), (r, y_3), (r, y_4)\}) \cup S A'.$$

For $R''$ $w(r, y_1) = \ldots = w(r, y_4)$. Obviously $W(R'') = W(R)$.

From our consideration we get that it suffices to prove the inductive hypothesis for case (1). Let us denote $q(r) = \{y_1, y_2\}$. We will distinguish the possible cases a), b), c), d).

a) Let $card(L(R(y_1))) = 2$ and $card(L(R(y_2))) = 2$ hold.

Let us define the rooted tree $R' = (V', A')$ in the following way

$$V' = \{r, x_1, x_2, x_3, x_{11}, x_{12}\}, \quad A' = \{(r, x_1), (r, x_2), (r, x_3), (x_1, x_{11}), (x_1, x_{12})\}.$$

Then the rooted tree $R'$ is W-optimal and $card(q(r)) = 3$;

b) Let $card(L(R(y_1))) > 2$ and $card(L(R(y_2))) = 2$ holds.

W-optimality of the rooted tree $R$ implies $card(q(y_2)) = 2$. Let $q(y_1) = \{z_1, z_2, z_3\}$, $q(y_2) = \{z_4, z_5\}$ hold and consider the rooted tree $R' = (V', A')$ defined in the following way: $V' = \{r, x_1, x_2, z_1, z_2, z_3, z_4, z_5\}$, $A' = \{(r, x_1), (r, x_2), (r, z_5), (x_1, z_1), (x_1, z_2), (x_2, z_3), (x_2, z_4)\}$. Consider the rooted tree $R'' = (V'', A'')$ defined by: $V'' = (V - \{y_1, y_2\}) \cup V'$, $A'' = (A - A^*) \cup A'$, $A^* = \{(r, y_1), (r, y_2), (y_1, z_1), (y_1, z_2), (y_1, z_3), (y_2, z_4), (y_2, z_5)\}$. For both $R$ and $R''$ $w(r, z_i) = 5$ holds (for $i = 1, 2, 3$). For $R$ we obtain $w(r, z_4) = 4$, $w(r, z_5) = 4$, and for $R''$ $w(r, z_4) = 5$, $w(r, z_3) = 3$. This implies $W(R) = W(R'')$, whence $card(q(r)) = 3$ for $R''$.

c) Suppose $card(L(R(y_1))) > 2$ and $card(L(R(y_2))) > 2$. From the inductive hypothesis and from Lemma 3.6. it follows that we can suppose without loss of generality that $card(q(y_1)) = 3$ and $card(q(y_3)) = 3$. Let $q(y_1) = \{z_1, z_2, z_3\}$ and $q(y_2) = \{z_4, z_5, z_6\}$. Consider the rooted tree $R' = (V', A')$ defined by: $V' = \{r, y_1, y_2, y_3, z_1, z_2, z_3, z_4, z_5, z_6\}$, $A' = \{(r, y_1), (r, y_2), (r, y_3), (y_1, z_1), (y_1, z_2), (y_2, z_3), (y_2, z_4), (y_3, z_5), (y_3, z_6)\}$. Further, consider the rooted tree $R'' = (V'', A'')$ defined by: $V'' = (V - \{y_1, y_2\}) \cup V'$, $A'' = (A - A^*) \cup A'$, $A^* = \{(r, y_1), (r, y_2), (y_1, z_1), (y_1, z_2), (y_1, z_3), (y_2, z_4), (y_2, z_5), (y_2, z_6)\}$. For both $R$ and $R''$ $w(r, z_i) = 5$ (for $i = 1, ..., 6$) and hence we have $card(q(r)) = 3$ for $R''$, whence $W(R) = W(R'')$.

d) Let $card(L(R(y_1))) = 2$ and $card(L(R(y_2))) > 2$. This case is analogous to b).

For all cases a) - d) we have found W-optimal trees satisfying $q(r) = 3$. Hence, inductive implication holds and the assertion is proved.

# 4 L₃ trees and balanced quasi-ternary trees

**4.1. Definition.** A rooted tree $R = (V, A)$ will be called $L_3$ tree if it satisfies:

(1) $card(q(x)) < 4$ holds for every $x \in V$,

(2) $card(q(x)) < 3$ implies $q(x) \subseteq L(R)$.

**4.2. Definition.** Balanced quasi-ternary tree is an $L_3$ tree which satisfies:

(1) $lvl(x) < lvl(R) - 1$ implies $card(q(x)) = 3$;

(2) $lvl(x) = lvl(y) = lvl(R) - 1$ implies $|card(q(x)) - card(q(y))| < 3$;

(3) $card(q(x)) \neq 1$ for every $x \in V$.

*An example of a balanced quasi-ternary tree*

**4.3. Lemma.** Let $R = (V, A)$ be an $L_3$ tree, $x, y \in (V - L(R))$, $lvl(x) > lvl(y)$. Then $w(r, x) = w(r, y) + 3(lvl(x) - lvl(y))$.

**4.4. Lemma.** Suppose $R = (V, A)$ is a W-optimal $L_3$ tree and $\{x, y\} \subseteq L(R)$. Then $|lvl(x) - lvl(y)| < 2$.

*Proof.* Suppose that $R = (V, A)$ is a W-optimal $L_3$ tree, $x, y \in L(R)$ and $lvl(x) > lvl(y) + 1$ (1). Let $x = x_1$, $q(z) = \{x_1, ..., x_k\}$ ($1 < k < 4$), (the case k=1 is eliminated by Lemma 3.4.) $z \in q(u)$, $y \in q(v)$. According to condition (1) and Lemma 4.3. we obtain $w(r, u) \geq w(r, v) + 3$ (2). Consider the rooted tree $R' = (V', A')$ defined in the following way: $V' = V$, $A' = A - \{(z, x)\} \cup \{(v,x)\}$. Obviously, $L(R) = L(R')$ holds. Now, let us estimate $W(R) - W(R')$. Because in $R'$ the set $q(v)$ contains maximally four elements, we obtain (in $R'$) $w(r, x) \leq w(r, v) + 4$ (3).

In R $w(r, x) = w(r, u) + 3 + k$ holds and according to (2) we get (for R): $w(r, x) \geq w(r, v) + 6 + k$ (4). From (3) and (4) it follows that the difference between the amount of the work needed to find x is in the tree R and the amount of the work needed to find x in the tree R' is at least $(w(r, v) + 6 + k) - (w(r, v) + 4) = k + 2$ (5). Because there exist at most three elements of the set $q(v)$ that differ from x, we obtain $W(R) - W(R') \geq (k + 2) - 3 \geq 1$, which contradicts the W-optimality of R.

**4.5. Lemma.** Let $R = (V, A)$ be a W-optimal $L_3$ tree. Then $card(q(z)) = 3$ for every node $z \in V$ which satisfies $lvl(z) < lvl(R) - 1$.

*Proof.* Suppose that $lvl(z) < lvl(R) - 1$ for $z \in V$. From Lemma 4.4. we get $card(q(z)) > 0$. Lemma 3.4. implies $card(q(z)) > 1$. Suppose $card(q(z)) = 2$.

Let $x \in L(R)$, $lvl(x) = lvl(R)$ and $x \in q(y)$. Consider the rooted tree $R' = (V', A')$ defined in the following way: $V' = V$, $A' = (A - \{(y, x)\}) \cup \{(z, x)\}$. For R we obtain $w(r, y) \geq w(r, z) + 3$, which implies $W(R') \leq W(R) - 3 + 2$, i.e. $W(R') < W(R)$, contradicting the W-optimality of R.

**4.6. Lemma.** Let $R = (V, A)$ be a W-optimal $L_3$ tree and x, y nodes satisfying the condition $lvl(x) = lvl(y) < lvl(R)$. Then $w(r, x) = w(r, y)$ ( $= 3 \cdot lvl(x)$ ).

# 5 Main theorems

**5.1. Theorem.** Each W-optimal $L_3$ tree is a balanced quasi-ternary tree.

*Proof.* If R is a W-optimal $L_3$ tree, then the property (1) of the definition of balanced quasi-ternary tree (4.2.) follows from Lemma 4.5. and property (3) follows from Lemma 3.4. We have to prove that R possesses the property (2). Let $R = (V, A)$ be a W-optimal $L_3$ tree. Suppose that there exist nodes x, y satisfying

$$lvl(x) = lvl(y) = lvl(R) - 1 \text{ and } |card(q(x)) - card(q(y))| = 3.$$

Without loss of generality we can suppose that $card(q(x)) = 3$ and $card(q(y)) = 0$. Let $y \in q(z)$ hold (according to Lemma 4.5. we obtain $card(q(z)) = 3$) and let $q(x) = \{x_1, x_2, x_3\}$ be satisfied. Consider the rooted tree $R' = (V', A')$ defined in the following way: $V' = V \cup \{y'\}$, $A' = A - \{(x, x_3), (z, y)\} \cup \{(z, y'), (y', y), (y', x_3)\}$, where $y'$ is an added node, i.e. $y' \notin V$. Obviously $L(R) = L(R')$. From Lemma 4.6. it follows that $w(r, x) = w(r, y)$ holds for the tree R. Analogously, $w(r, x) = w(r, y')$ for the tree R'.

Hence, for $R'$ (in comparison with R) the amount of the work needed for finding all leaves decreased from nine to four for the elements of the set $q(x)$ and increased for the elements of the set $q(y')$ (+4). We get $W(R') = W(R) - 5 + 4$, i.e. $W(R') < W(R)$, contradicting the W-optimality of R.

**5.2. Theorem.** For every $n > 3$ there exist a W-optimal balanced quasi-ternary tree satisfying $card(L(R)) = n$.

*Proof.* According to Lemma 3.11. there exists a rooted tree $R = (V, A)$ such that $L(R) = n$, R is W-optimal and $card(q(r)) = 3$. From Lemma 3.6. it follows that $R(x)$ is W-optimal for each node $x \in V(R)$. According to Lemma 3.11. we can suppose without loss of generality that in the rooted tree $R(x)$ $card(q(x)) = 3$ holds for every $x \in q(r)$. This consideration can be done for each node $y \in q(r)$, etc. Assumption $card(\{z; z \in R(u)\}) < 3$ implies $q(u) \subseteq L(R)$. Thus R is an $L_3$ tree. By the previous theorem R must be a balanced quasi-ternary tree.

**5.3. Corollary.** Balanced quasi-ternary trees are W-optimal.

*Proof.* It is easy to see that if $R_1$ and $R_2$ are balanced quasi-ternary trees that satisfy $L(R_1) = L(R_2)$, then $W(R_1) = W(R_2)$. The assertion now follows from the previous theorem.

**5.4. Corollary.** A rooted tree R is W-optimal if there exists a balanced quasi-ternary tree R' satisfying $L(R) = L(R')$ and $W(R) = W(R')$.

## Conclusions

The characterization of optimal trees enables to speed up algorithms for searching in codebook by constructing optimal search trees. This is important especially for real-time recognizing, where saving time by speeding up routines plays important role. Algorithms for optimal tree searching in codebook were implemented in software products described in [4].

## Acknowledgements

The author is very grateful to Miroslav Novotný and Vítězslav Veselý for carefully reading a draft of this paper, for valuable comments, and helpful ideas.

## References

1. M. Berka, I. Kopeček. *Optimality of Decision Structures of Information Systems.* 16<sup>th</sup> IFIP Conference on System Modelling and Optimization,1993, pages 653-657.
2. D.Y. Cheng, A. Gerscho. *A Fast Codebook Search Algorithm for Nearest-Neighbour Pattern Matching.* Proc ICASSP 86, Tokyo 1986, pages 265-268.
3. R.M. Gray. *Vector Quantization.* IEEE ASSP Magazine, 1, 1984, pages 4-29.
4. I. Kopeček. *Speech Synthesis of Czech Language in Time Domain and Applications for Visually Impaired.* Proceedings of the 2-nd SQEL Workshop, Plzeň 1997, pages 141-144.
5. D. E. Knuth. *The Art of Computer Programming.* Volume 3 - Sorting and Searching. Addison -Wesley, 1973.
6. W. Mayeda. *Graph Theory.* New York, Wiley 1972.
7. J. Wiedermann. *Searching.* SNTL, Praha 1991 (in Czech).
8. T.P. Yunck. *A Technique to Identify Nearest Neighbours.* IEEE Trans. Syst. 6(76), pages 678-683.

# Time Optimal Self-Stabilizing Algorithms[*]

Rastislav Kráľovič

Department of Computer Science
Faculty of Mathematics and Physics
Comenius University, Bratislava, Slovakia
kralovic@dcs.fmph.uniba.sk

**Abstract.** In this paper we present lower bounds on the stabilization time for a number of graph theoretic problems, as leader election, spanning tree construction, computing the diameter, the number of nodes, the connectivity or orientation on tori, rings, hypercubes and CCC. These bounds are of the form $\Omega(D)$, where $D$ is the diameter of the network. Moreover, time-optimal self-stabilizing algorithms for computing the orientation on tori, rings, hypercubes and CCC are presented. This gives an answer to the problem 15.4 for tori stated in [Tel94b].

## 1 Introduction and Model

The study of self-stabilizing agorithms was initiated by Dijkstra [Dij74], but it did not attract much attention until the late eighties. Since that time a number of efficient self-stabilizing algorithms was proposed [AK93], [AV91].

Self-stabilizing algorithms are algorithms on distributed point-to-point networks that can always reach a configuration where a particular stable property holds, regardless of the initial configuration. Self-stabilizing algorithms are fault-tolerant in the following sense. If local states of network processors are allowed to be changed infrequently as a result of a failure, then by definition the algorithm recovers from that failure by reaching a configuration at which the desired stable property is once again valid. The complexity measure for self-stabilizing algorithms is the stabilization time.

Recently much effort is devoted to the study of oriented networks [Tel94a], [Tel95b], [FMS94]. The reason is to study the impact of the structural information on the complexity of distributed algorithms. The main issues in this research are the efficiency of computation on oriented/unoriented networks [Tel95a], [DR97] and the computational complexity of network orientation [Tel95b].

The aim of this paper is to present time-optimal self-stabilizing algorithm for computing tori orientation in the stabilization time $T_s = O(D)$, where $D$ is the diameter of the network. Using a similar approach algorithms for other topologies (rings, hypercubes, cube-connected-cycles) are also given. In Section 2 a lower bound on the stabilization time of the form $\Omega(D)$ is given for a number of problems on arbitrary topology. As a corollary we obtain lower bounds for

---

[*] This research has been partially supported by VEGA project 1/4315/97.

computing the number of nodes, the maximal degree of a node, the diameter or connectivity of the network, leader election and spanning tree construction. Lower bounds on the stabilization time have been previously mentioned only in [AK93]. In Section 3 we present a lower bound on the stabilization time of the form $\Omega(D)$ for computing orientation on a wide class of networks. As a corollary lower bounds for computing the orientation on rings, tori, hypercubes and CCC are given together with time-optimal algorithms.

## 1.1  Definitions

A point-to-point network is modelled by a graph where vertices are processors and undirected edges are communication links. Every processor has its unique identifier and is able to distinguish its links. Processors communicate via exchanging messages. The behaviour of a distributed system is modelled by a transition system. We shall use the formal definition from [Tel94b].

**Definition 1.** A transition system is a pair $\mathcal{S} = (\mathcal{C}, \rightarrow)$, where $\mathcal{C}$ is a set of configurations and $\rightarrow$ is a binary transition relation on $\mathcal{C}$. An execution of $\mathcal{S}$ is a maximal sequence $E = (\gamma_0, \gamma_1, \ldots)$ such that for all $i \geq 0$, $\gamma_i \rightarrow \gamma_{i+1}$.

A configuration of the distributed system is given by the states of the processors and the messages on links. The transition relation $\rightarrow$ of system $\mathcal{S}$ covers the transition relations $\rightarrow_v$ of every processor $v$. A local transition relation of processor $v$ covers the actions of sending and recieving messages and inner transitions. A more detailed definition can be found in [Tel94b].

The correct behaviour of the system is described by a specification. A specification $S$ is a predicate defined on sequences of configurations.

**Definition 2.** A system $\mathcal{S}$ stabilizes to a specification $S$ if there exists a subset $\mathcal{L} \subseteq \mathcal{C}$ of legitimate configurations with the following properties:

- **Correctness.** Every execution starting in a configuration in $\mathcal{L}$ satisfies $P$.
- **Convergence.** Every execution contains a configuration of $\mathcal{L}$.

Let $\gamma$ be a configuration, then $\gamma(v)$ denotes the evaluation of variables in a vertex $v$. By $\mathcal{G}$ we denote a class of graphs. We consider algorithms, the specification of which can be expressed by a specification constraint $\mathcal{P}$. A constraint $\mathcal{P}$ is a predicate on configurations such that for each execution $E = (\gamma_0, \gamma_1, \ldots)$ it holds $S(E)$ iff $\mathcal{P}(\gamma_i)$ for all $i$. Moreover we assume the closeness of $\mathcal{P}$ (i.e., for configurations $\gamma, \delta$ satisfying $\gamma \rightarrow \delta$ and $\mathcal{P}(\gamma)$ it holds $\mathcal{P}(\delta)$) and satisfiability of $\mathcal{P}$ (for $G \in \mathcal{G}$ there is a configuration $\gamma$ such that $\mathcal{P}(\gamma)$ ).
$G_d(v)$ denotes a $d$-neighbourhood of vertex $v$, i.e. subgraph of $G$ induced by vertices $\{w \in G \mid d(v, w) \leq d\}$, where $d(v, w)$ is the distance from $v$ to $w$. The constraint $\mathcal{P}$ holds on $X \subseteq G$ (denoted as $\mathcal{P}(\gamma|X)$ ) iff there exists $X' \in \mathcal{G}$ obtained from $X$ by adding edges of the form $(u, v)$ where $u, v \in X - Int(X)^2$

---

[2] $Int(X)$ denotes the interior of $X$, i.e. $Int(X) = \{v \in X \mid \forall w : d(v, w) = 1 \Rightarrow w \in X\}$.

and a configuration $\gamma'$ on $X'$ such that $\mathcal{P}(\gamma')$ holds and for all $w \in Int(X)$ it holds $\gamma(w) = \gamma'(w)$.

To estimate the stabilization time we assume that one unit of time is needed to deliver a message via a communication line and inner computation in processors can be neglacted. Stabilization time is the time spent by the algorithm to stabilize in the worst case. We use the definition of orientation from [Tel94a], [FMS94] and the composition rule from [PS95].

## 2 General Lower Bound on Stabilization Time

**Lemma 3.** *Consider a problem $P$ (with a constraint $\mathcal{P}$) on the class $\mathcal{G}$ of graphs such that for a graph $G \in \mathcal{G}$ there exist vertices $u$, $v$ and subgraphs $X \subseteq G$, $Y \subseteq G$, $X, Y \in \mathcal{G}$ such that the following conditions hold:*

- *$Int(X) \cap Int(Y) = \emptyset$*
- *$u \in Int(X)$, $v \in Int(Y)$*
- *for an arbitrary configuration $\gamma$: if there exists a configuration $\delta$ such that $\mathcal{P}(\delta|X)$ in $\mathcal{G}$, $\mathcal{P}(\delta|Y)$ in $\mathcal{G}$, $\delta(u) = \gamma(u)$ and $\delta(v) = \gamma(v)$, then $\neg\mathcal{P}(\gamma)$.*

*Suppose there exists $d > 0$ such that $G_d(u) \subseteq X$ and $G_d(v) \subseteq Y$. Then the stabilization time of the problem $P$ is $T_s = \Omega(d)$.*

*Proof.* Assume a starting configuration $\gamma$ such that $\mathcal{P}(\gamma|X)$ and $\mathcal{P}(\gamma|Y)$. Such a configuration exists because $Int(X) \cap Int(Y) = \emptyset$ and by the definition there are configurations satisfying $\mathcal{P}$ on both $X$ and $Y$.

It holds $\neg\mathcal{P}(\gamma)$ on $G$. Consider an arbitrary execution on $G$ starting from $\gamma$. We show that at least $d$ steps are needed to reach a configuration that satisfies $\mathcal{P}$. Denote the configuration at the time $0 \leq t < d$ as $\gamma^{(t)}$.

Let $\beta_X^{(0)}$ be a configuration on $X'$ such that $\mathcal{P}(\beta_X^{(0)})$ holds on $X'$ and for all $w \in Int(X)$ it holds $\beta_X^{(0)}(w) = \gamma(w)$ (see that such a configuration exists). We show that for each $0 \leq t < d$ there is a configuration $\beta_X^{(t)}$ on $X'$ such that there exists an execution $\beta_X^{(0)} \rightarrow^* \beta_X^{(t)}$ on $X'$ and for all $w \in G_{d-t}(u)$ it holds $\beta_X^{(t)}(w) = \gamma^{(t)}(w)$.

The proof is by induction on $t$. For $t = 0$ the claim is true, as $\gamma^{(0)} = \gamma$. Now suppose the claim holds for $t < d$. We show that all actions performed in the vertices from $G_{d-t-1}(u)$ in time $t+1$ during the execution on $G$ can be also performed during the execution on $X'$. As for all $w \in G_{d-t-1}(u)$ it holds $\beta_X^{(t)}(w) = \gamma^{(t)}(w)$ (i.e. during both executions they are in the same state), all inner transitions or sending a message actions in the execution on $G$ can be performed also during the execution on $X'$. In the case of receiving a message action by a vertex from $G_{d-t-1}(u)$, this message has to be generated in a vertex from $G_{d-t}(u)$ in time at most $t$. Thus, also this action can be performed during the execution on $X'$. As $\mathcal{P}(\beta_X^{(0)})$ holds on $X'$, then due to closeness of $\mathcal{P}$ also $\mathcal{P}(\beta_X^{(t)})$ is satisfied on $X'$ for all $0 \leq t < d$. Moreover $\beta_X^{(t)}(u) = \gamma^{(t)}(u)$. Analogicaly we can obtain configurations $\beta_Y^{(t)}$ such that for all $0 \le t < d$ it holds $\mathcal{P}(\beta_Y^{(t)})$ on $Y'$ and $\beta_Y^{(t)}(v) = \gamma^{(t)}(v)$. As $Int(X) \cap Int(Y) = \emptyset$, for all $t < d$ we can construct a configuration $\delta^{(t)}$ such that for all $w \in Int(X)$ it holds $\delta^{(t)}(w) = \beta_X^{(t)}(w)$ and for all $w \in Int(Y)$ it holds $\delta^{(t)}(w) = \beta_Y^{(t)}(w)$. It holds $\mathcal{P}(\delta^{(t)}|X)$ and $\mathcal{P}(\delta^{(t)}|Y)$ on $G$ and $\delta^{(t)}(u) = \gamma^{(t)}(u)$ and $\delta^{(t)}(v) = \gamma^{(t)}(v)$. Therefore we get $\neg\mathcal{P}(\gamma^{(t)})$ on $G$. □

Following Lemma 3 we obtain lower bounds for some particular problems.

**Corollary 4.** *Given an arbitrary graph $G$ with the diameter $D$, leader election (LE), computing the number of vertices (NV) and computing the diameter of a graph (DG) have the stabilization time $T_s = \Omega(D)$.*

*Proof.* The solution of LE is a configuration in which every vertex $v$ contains the variable $Leader_v$ and exactly one vertex $w$ has the value $Leader_w = true$.

For a given LE problem, a constraint $\mathcal{P}$ and a graph $G$, choose $u$ and $v$ such that $d(u, v) = D$. Let $d = \lfloor \frac{D}{2} \rfloor - 1$, $X = G_d(u)$ and $Y = G_d(v)$. Following Lemma 3 we obtain $T_s(LE) = \Omega(d) = \Omega(D)$.

The proof of NV and DG is similar to the proof of LE. □

**Corollary 5.** *Given an arbitrary graph $G$ with the diameter $D$, constructing a spanning tree (ST) has stabilization time $T_s = \Omega(D)$.*

*Proof.* The solution of ST is a configuration in which every vertex $v$ in its variable $Parent_v$ contains a pointer to one of its neighbours or to itself and these pointers form the spanning tree of the graph $G$.

In every configuration satisfying $\mathcal{P}$ on $G$ there exists exactly one vertex $u$ for which $Parent_u = u$. So if there would be a self-stabilizing algorithm $\mathcal{A}$ solving ST in time $T_s = o(D)$, we could use this algorithm also for solving LE in time $o(D)$ by adding $Leader_v := true$ iff $Parent_v = v$. □

**Corollary 6.** *Given an arbitrary graph $G$ with the diameter $D$, the problems of computing the maximal degree of a node and the vertex (edge) connectivity of the graph has the stabilization time $T_s = \Omega(D)$.*

*Proof.* The first part follows from Lemma 3 using a graph consisting of two cliques $K_{\Delta_1}$, $K_{\Delta_2}$, $\Delta_1 \ne \Delta_2$ connected with a path of length $3d$, $d > 3$. The second part follows from Lemma 3 using a graph consisting of two rings of length $2d$ with one common vertex $w$ (connected by one edge). □

## 3 Orientation Problem

**Definition 7.** Let $\mathcal{G}$ be a class of graphs with a constriant $\mathcal{P}$, which defines the orientation by assigning some values to variables $Label_v[l]$ for a vertex $v$ and its link $l$. The specification is given by the costraint $\mathcal{P}$ and by the additional condition that the values of $Label_v[l]$ may not change.

The following lemma gives a lower bound on the stabilization time for computing orientation.

**Lemma 8.** *Let $\mathcal{G}$ be a class of graphs with a constriant $\mathcal{P}$, which defines the orientation such that following conditions hold:*

1. *for $G \in \mathcal{G}$ there are at least two distinct configurations $\gamma$, $\gamma'$ such that $\mathcal{P}(\gamma)$, $\mathcal{P}(\gamma')$,*
2. *for each $\gamma \neq \gamma'$ such that $\mathcal{P}(\gamma)$ and $\mathcal{P}(\gamma')$ it holds $\forall v \in G : \gamma(Label_v) \neq \gamma'(Label_v)$, where $\gamma(Label_v)$ means the evaluation of variables $Label_v$ in configuration $\gamma$.*

*Then the computation of orientation on $\mathcal{G}$ has the stabilization time $T_s = \Omega(D)$, where $D$ is the diameter of the graph.*

*Proof.* Take two vertices $u$, $v$ such that $d(u, v) = D$. Denote $X = G_d(u)$, $Y = G_d(v)$, where $d = \lfloor D/2 \rfloor$. Obviously $X \cap Y = \emptyset$. Following the condition 1 there are two distinct configurations $\gamma_1 \neq \gamma_2$ such that $\mathcal{P}(\gamma_i)$. We construct a configuration $\alpha$ such that for all $v \in X$ it holds $\alpha(v) = \gamma_1(v)$ and for all $v \in G - X$ it holds $\alpha(v) = \gamma_2(v)$. Following the condition 2 it holds $\neg\mathcal{P}(\alpha)$.

Now consider an arbitrary stabilizing execution $\mathcal{A}$ on $G$ starting from $\alpha$. In analogy to the proof of Lemma 3 we find an execution $\mathcal{B}$ on $G$ starting from $\gamma_1$ which is in time $t < d$ identical to $\mathcal{A}$ on $G_{d-t}(u)$. Because $\mathcal{B}$ satisfies the specification, the values of variables $Label_u$ do not change. Similarly, the values of variables $Label_v$ do not change. Hence following condition 2 the execution $\mathcal{A}$ is not in a configuration satisfying $\mathcal{P}$. □

## 3.1 Time Optimal Self-stabilizing Algorithm for Tori Orientation

**Definition 9.** The $m \times n$ torus is a graph consisting of $N = mn$ vertices, where each vertex has an unique name from the set $\mathbb{Z}_m \times \mathbb{Z}_n$ such that the vertex $(i, j)$ is connected to four vertices $(i, j \oplus 1)$, $(i, j \ominus 1)$, $(i \ominus 1, j)$, $(i \oplus 1, j)$ [3].

A labeling of the torus is an assignment of different labels from the set $\{U, D, L, R\}$ in each vertex $v$ to each edge outgoing from $v$.

The tori orientation is a labeling where each vertex has an unique name $\mathcal{N}(v) = (i, j)$ from the set $\mathbb{Z}_m \times \mathbb{Z}_n$ such that the edge $(v, w)$ in vertex $v$ is labeled $U$ (resp. $D, L, R$) iff $\mathcal{N}(w) = (i \ominus 1, j)$ (resp.$(i \oplus 1, j)$, $(i, j \ominus 1)$, $(i, j \oplus 1)$).

**Corollary 10.** *The tori orientation problem on an $m \times n$ torus, $m, n > 10$, has the stabilization time $T_s = \Omega(D)$, where $D$ is the diameter of the torus.*

*Proof.* Obviously there are only four distinct orientations, thus the assumptions of Lemma 8 hold giving $T_s = \Omega(d) = \Omega(n) = \Omega(D)$. □

---

[3] $\oplus$ (resp. $\ominus$) is addition (subtraction) in corresponding modulus.

**Claim 11.** ([AK93]) *There exists a self-stabilizing spanning tree algorithm on an arbitrary graph with the stabilization time $T_s = O(D)$, where $D$ is the diameter of the graph. Moreover the specification of this algorithm satisfies the following condition: the identity of the root never changes and for a vertex $v$ its distance from the root in the tree is the same as in the whole graph.*

**Algorithm 12.** *Self-stabilizing algorithm for tori orientation.*

Assume that each vertex $v$ knows its identity $ID_v$ and it has unique identifiers to distinguish its links $L_v = \{l_0, \ldots, l_3\}$. The algorithm consists of five parts joined together with the composition rule. The resulting orientation will be stored in variables $Label_v[l]$.

In the first part, the algorithm from Claim 11 is used to build a spanning tree. This part sets the variables $Distance_v$ and $Parent_v$. In the second part, the preorientation is computed by an algorithm which is a stabilizing version of the algorithm from [Tel94a]. Preorientation in each vertex $v$ assigns to each outgoing edge $e$ a label from the set $\{H, V\}$ (in the variable $PreLabel_v[l]$) such that it holds $PreLabel_v[l] = PreLabel_v[l']$ iff there exists exactly one path of length 2 between vertices of these edges different from $v$. Next, the vertices are assigned coordinates with which the orientation is easy to compute. Every vertex has variables $[x_v, y_v]$ containing its name from $\mathbb{Z}_m \times \mathbb{Z}_n$. In the third part, the leader (the vertex with $Distance_v = 0$) chooses the orientation and sends messages $\langle D, 0 \rangle$ and $\langle R, 0 \rangle$. They will propagate along the preorientation and set the coordinates of vertices $(i, 0)$ and $(0, j)$. In the next part, every vertex with more than one neighbour at the smaller distance computes its coordinates by "inheriting" from its "horizontal" neighbour the first coordinate and from its "vertical" neighbour the second coordinate. There are variables $NDist_v[l]$ containig the value $Distance$ of the corresponding neighbour and messages of type INFO. The algorithm is given in Fig. 1. Finally, every vertex sends its coordinates to neighbours by messages of type MYCOORD. Using this information every vertex is able to compute the orientation.

**Claim 13.** *After time of $O(1)$, a vertex $v$ contains in its variable $NDist_v[l]$ the value $Distance$ of its corresponding neighbour (via the link $l$).*

**Lemma 14.** *After stabilization of the spanning tree algorithm, a vertex at the distance $d$ from the leader computes correct coordinates with additional time $O(d)$ and its prelabelings correspond to the global "horizontal" and "vertical" directions.*

*Proof.* By induction on $d$. Every $P$ units of time the leader sets its coordinates to $[0, 0]$. Moreover, from the uniqueness of the link identifiers it follows that the leader does not change the labels of its links.

Let vertices at the distance $d$ satisfy the assumption of Lemma 14. Now consider a vertex at the distance $d + 1$. If it has exactly one neighbour at the distance $d$ it receives a message $\langle D, d \rangle$ (resp. $\langle R, d \rangle$) and sets correct coordinates and orientation. If it has at least two such neighbours, they have correct orientation and so $v$ gets messages INFO from both directions $H$ and $V$ and sets correct coordinates and orientation. □

**Theorem 15.** *Algorithm 12 is a time optimal self-stabilizing algorithm for computing tori orientation.*

*Proof.* Spanning tree algorithm stabilizes at the time $O(D)$. According to Lemma 14 a vertex at the distance $d$ from the leader has correct coordinates at the time $O(D)$. With correct coordinates of all vertices, the stabilization time is clearly $O(1)$. Hence the Algorithm 12 has stabilization time $O(D)$ and it is time optimal. □

## 3.2 Time Optimal Self-Stabilizing Algorithms for Computing Orientation on other Topologies

In this subsection we conclude results from [Kra97] concerning orientation on rings, hypercubes, cube-connected-cycles and butterflies.

**Theorem 16.** *The ring (hypercube, CCC) orientation problem has stabilization time $T_s = \Theta(D)$, where $D$ is the diameter of the graph.*

The orientation on a ring assigns labels $L$, $R$ to each link such that global "left" and "right" directions are set. The orientation on a hypercube assigns to each link the dimension of the bit in which the names of vertices (in the standard labeling) differ. Vertices of a CCC are labeled with bit strings with a cursor pointing to some bit. Edges correspond to operations left (right) cyclic shifting cursor and negating the bit on cursor. The orientation on a CCC assigns labels $L$, $R$, $N$ to each link to determine the operation corresponding to a link. Precise definition of hypercubes and CCC can be found in [Lei92].

The lower bounds follow from Lemma 8. The algorithms are similar to tori orientation algorithm. For the case of ring and hypercube no preorientation is needed. Local names in hypercube are computed from the names of all neighbours with smaller distance from the leader. The preorientation in CCC distinguishes between "shift" (left/right) links and "negate" links. It exploits the fact that (for dimension other than 8) a "negate" link is contained in exactly two circles of length 8 while a "shift" link is in exactly one. Additional information of the length $O(\log|V|)$ is carried in the further messages to distinguish "left" and "right" edges.

**Theorem 17.** *The butterfly orientation problem has stabilization time $T_s = \Theta(1)$.*

A butterfly has vertices arranged in layers. Each layer has vertices labeled with binary strings. Edges connect vertices in neighbouring layers iff they have the same label or their labels differ in the position corresponding to the number of layer. Precise definition can be found in [Lei92]. The orientation on a butterfly assigns labels to distinguish between layers and between "shift" and "negate" links. The orientaion exploits the circles of length four which can be oriented independently.

# 4 Conclusion

We presented lower bounds on the stabilization time of the form $T_s = \Omega(D)$ for some fundamental graph theoretic problems on an arbitrary topology, where $D$ is the diameter of the graph. These lower bounds are asymptotically optimal for leader election problem, spanning tree construction, computing the number of vertices, computing the diameter of the graph. Next, we presented a general lower bound on the stabilization time of the form $T_s = \Omega(D)$ for computing orientation on a large class of topologies and as a corollary the lower bounds for tori, ring, hypercube and CCC topologies. Moreover we presented asymptotically time optimal algorithms for computing orientation on these topologies with the stabilization time $T_s = O(D)$. In the case of tori we solved the problem from [Tel94b].

Further study is oriented on computing orientation for other structures (Shuffle-Exchange, Star) and on considering communication complexity.

# References

[AK93]    Aggarwal, S., Kutten, S.: Time Optimal Self-Stabilizing Spanning Tree Algorithms. In Proc. of the Conference on Foundations of Software Technology and Theoretical Computer Science, LNCS 652 1993, pp. 400—410.

[AV91]    Awerbuch, B., Varghese, G.: Distributed Program Checking: a Paradigm for Building Self-stabilizing Distributed Protocols. In 32nd FOCS, October 1991, pp. 258—267.

[Dij74]   Dijkstra, E. W.: Self stabilization in spite of distributed control. In Comm. ACM 17, 1974 pp. 643–644.

[DR97]    Dobrev, S., Ružička, P.: Linear broadcasting and $N \log \log N$ election in unoriented hypercubes. In SIROCCO, 1997, to appear.

[FMS94]   Flocchini, P., Mans, B., Santoro, N.: Sense of Direction: Formal Definitions and Properties. In SIROCCO, 1994, pp. 9—34.

[Kra97]   Královič, R.: Time Optimal Self-Stabilizing Algorithms. Technical Report, Dept. of Computer Science, Comenius University, Bratislava, 1997. Submitted for publication.

[Lei92]   Leighton, F. T.: Introduction to parallel algorithms and architectures: arrays, trees, hypercubes. Morgan Kaufmann Publishers, Inc., San Mateo, USA, 1992.

[PS95]    Prasetya, I.S.W.B., Swierstra, S.D.: Formal Design of Self-stabilizing Programs. Technical report UU-CS-1995-07, Dept. of Computer Science, Utrecht University, 1995.

[Tel94a]  Tel, G.: Network Orientation. International Journal of Foundations of Computer Science 5, 1994, pp. 23—57.

[Tel94b]  Tel, G.: Introduction to Distributed Algorithms. Cambridge University Press, Cambridge, UK, 1994.

[Tel95a]  Tel, G.: Linear election in oriented hypercubes. Parallel Processing Letters 5, 1995, pp. 357—366.

[Tel95b]  Tel, G.: Sense of Direction in Processor Networks. In SOFSEM, LNCS 1012, 1995, pp. 50—82.

```
(* COMPUTING PREORIENTATION *)
   every P units of time
      for all l ∈ L_v do
         Neigh := ⋃_{l'≠l} Id_v[l]
         send ⟨MSGA, Neigh, ID_v⟩ via  l

Ok(v, l, l') ≡ PreLabel_v[l] = PreLabel_v[l'] ∧ (∀k ≠ l, l' : PreLabel_v[k] ≠ PreLabel_v[l])

      when received ⟨MSGA, Set, id⟩ via  l
         Bag_v[l] := Set, Id_v[l] := id
         if ∃l, l' ∈ L_v : (Bag_v[l] ∩ Bag_v[l'] = ∅ ∧ ¬Ok(v, l, l')) then
            for all k ∈ L_v do
               if k ∈ {l, l'} then  PreLabel_v[k] := H
               else  PreLabel_v[k] := V

(* INITIALIZATION *)
   every P units of time
      if Distance_v = 0 then
      x_v := y_v := 0
      send ⟨R, 0⟩ via  min{l ∈ L_v | PreLabel_v[l] = H}
      send ⟨D, 0⟩ via  min{l ∈ L_v | PreLabel_v[l] = V}

(* ORIENTATION PROPAGATION *)

   when received ⟨D, i⟩ via  l
      if Distance_v > 0 ∧ PreLabel_v[l] = PreLabel_v[Parent_v] then
      x_v := i, y_v := 0
      if PreLabel_v[l] ≠ V then
         {Swap H and V in preorientation}
      l' ≠ l, PreLabel_v[l'] = V
      send ⟨D, i + 1⟩ via  l'
   (* Similar for messages ⟨R, i⟩ *)

(* COORDINATES TO ALL VERTICES *)
   every P units of time
      for every l ∈ L_v send  ⟨INFO, Distance_v, x_v, y_v, PreLabel_v[l]⟩ via  l

   when received ⟨INFO, D, x, y, P⟩ via  l
      NDist_v[l] := D
      if PreLabel_v[l] ≠ P then
         {Swap H and V in preorientation}
      if |{l ∈ L_v; NDist_v[l] = Distance_v − 1}| > 1 ∧ D = Distance_v − 1 then
         if P = H then  x_v := x
         else  y_v := y
```

**Fig. 1.** Tori orientation

# Requirements Specification Iteratively Combined with Reverse Engineering

Kroha,P., Strauß, M.

Technical University of Chemnitz,
Department of Informatics, 09107 Chemnitz, Germany

**Abstract.** A new concept of elicitation of requirements specification supported by an implemented CASE tool TESSI will be described in this paper. It concerns the problem of understanding, eliciting, and describing user requirements. The presented method consists of three steps. First, a semi-automatic transformation of the text of requirements into an object-oriented model will be supported. Second, a corresponding textual description will be automatically generated based on the just identified OO model. The generated text represents the analyst's understanding of the user's requirements. Third, the generated text will be read by the user and the domain experts in order to correct/complement it. A new version of requirements in the textual form will be created. This process will be iteratively repeated until there are no doubts about the analyst's good understanding of the problem. We argue that this approach which combines analysis and reengineering improves the quality of requirements specification.

## 1  Introduction

At the very beginning of a software development there should be a document containing a detailed textual description of requirements. The software analyst writes it in a cooperation with the user, i.e. with the group consisting of the customer, of the end users, and of the domain experts. He writes it in a natural language because this is the only form of requirements specification which the user really understands. Charts which are usually a part of this document should make the understanding deeper but they should not contain some facts not mentioned in the text of the requirements specification. The description of requirements specification in a natural language is informal, imprecise and vague. Every sentence, every word of it could be understood in a different way by the user and by the analyst.

It is very well known that the most serious mistakes and failures in software products have their origin in analyst's bad understanding of user's needs. The analyst describes how he/she understands the problem but he/she doesn't really understand the problem deep enough at this time, i.e. at the very beginning of the project. The analyst's understanding cannot be easily confronted with the user's understanding because these groups of specialists have no common formalism how to describe systems.

In this paper a method will be described (including an implemented CASE tool TESSI) which offers a pragmatic solution. This method and the derived CASE tool support following steps:

- A semi-automatic construction of main features of the object-oriented model of the proposed software system during an object-oriented analysis will be provided based on the text of requirements specification. In this step the tool creates also links between entities of requirements and entities of the OO model.
- Automatic generation of a textual description of the just derived OO model. This reversely generated textual description represents the analyst's understanding of the problem.
- Comparison of analyst's understanding and user's understanding. The generated text will be read by the user. It will be compared with the original text in order to approve it or to correct/complement it. The corrected original text will be then analysed again by using the same method. This process will be iteratively repeated until the user doesn't see any lacks and suspicious formulations in the generated text of requirements.

This idea represents the new approach described in this contribution. The rest of the paper is organised as follows. In section 2, we discuss the related work. In section 3, we describe the presented method together with its implementation. We discuss the search for both static and dynamic structures. Section 4 describes the phase of automatic synthesis of text based on the identified OO model. In section 5 we briefly mention the importance of this concept for other phases of the software development, especially for adaptive maintenance. The last section discusses some details of the implementation and the experiences obtained.

## 2 Related work

The process of aiming to understand an unstructured collection of data in a textual form has been studied since a very long time. There are many papers and books [1] coming from software design and discussing the method of a grammatical inspection of a text. Nouns will be held for candidates for object or classes, adjectives will be held for candidates for attributes, and verbs will be held for candidates for methods. The Object Behaviour Analysis [6] belongs to this group of methods, too. The important difference is that it doesn't work with an unstructured text but it builds an structured description using templates from the very beginning. Nevertheless, we had not any opportunity to prove a CASE tool which supports these methods.

Other papers coming from artificial intelligence and from the construction of expert systems discuss similar problems as a process of knowledge acquisition which should generate knowledge structures from the data produced by an expert and transform them into machine executable code. Programs doing that are called protocol editors [5]. Some their authors dream to automate this process completely which is sure very difficult. Mostly, some predefined frameworks and templates for tasks, behaviour, domain, etc. are used, e.g. in [5].

Although the iterative nature of the software development process is very well known we didn't find any related work to the problem of the automatic synthesis of a textual description corresponding to the identified OO model having as its goal to reach an agreement in formulation of requirements between the software analyst and the user.

# 3 From text to classes

The process of requirements specification is based on discussions between the software analyst and the user. During these discussions the user tries to explain what he/she want and the analyst tries to understand it, and to notice the relevant facts into the first version of the requirements specification.

After the first version of textual requirements specification has been written (by the analyst) it will be analysed (supported by our CASE tool TESSI - TExtual aSSIstent). The first phase of the method transforms an unstructured and uninterpretable set of "raw data" (represented as a sequence of sentences in a natural language) into structured data in a framework of an object-oriented model. Its consistency can be checked by a C++ and POET compiler.

This process can be divided in:

- identification of static features of the system,
- identification of dynamic features of the system.

## 3.1 Identification of static features

In the first step, the analyst tries (supported and forced by TESSI) to decide about the role of every word of requirements in in the OO model.

A word can be:

- irrelevant for the OO model,
- candidate for an object,
- candidate for a class (problem class, presentation class, control class, container class, persistent class),
- candidate for an attribute,
- candidate for a method.

After the analyst decides about the role of the investigated word (a choice from a menu of TESSI), the CASE tool TESSI marks the investigated word using different kinds of brackets and modifies its internal structures including a dictionary of marked entities. All occurrences of the investigated word will be then found in the text of requirements, the corresponding sentences will be shown and the analyst should decide whether his/her decision about the role of the investigated word is compatible with the context of other sentences of requirements. Being forced by TESSI he writes his definition of the semantics of this word as a notice in plain text into the dictionary.

In the next step TESSI forces the analyst to look for relations between identified entities, i.e. for:

- inheritance structure between classes (Is-A relation),
- types of attribute in classes (aggregation, Is-Part-Of relation),
- tupels of objects (association).

The identified relations will be again inserted into the dictionary including their definitions. The process of identification of static structures, i.e. methodical hints for the analyst, is based on a slightly modified version of [1] described in [3].

## 3.2  Identification of dynamic features

The process of identification of dynamic features is based on Object Behaviour Analysis [6]. Such words will be searched and marked in the text of requirements specification which are :

- irrelevant,
- candidates for events as initiators of actions,
- candidates for messages being sent, i.e. messages the reacting objects sends as a reaction at the event,
- candidates for objects whom the sent messages are addressed,
- describing how objects react on some event changing their internal states and sending messages to other objects.

These dynamic features will be modelled as a sequential automaton using transition diagrams. The tool TESSI forces the analyst to build the automaton according to the offered menu. More complex mechanisms for description of dynamic features like Petri nets are the topic of the ongoing project. The semantics of actions in a sequential automaton will be described in plain text, stored in the dictionary and later as comments in bodies of methods of the OO model.

After the static and dynamic structures have been identified, the problem of synonyms should be solved. One entity should be denoted by just one word in textual descriptions of requirements. To recognise it is a serious problem usually partially solved by external dictionaries of the natural language. A complete solution is not always possible in this way because the requirements specifications contain often special, application specific words not contained in common dictionaries. The only support by TESSI is that the dictionary will be discussed by analyst and user with the goal of looking for synonyms.

After this step, an object-oriented model represented by structures in C++ and POET (persistent extension of C++) will be automatically generated using description which has been written into internal tables of TESSI. The OO model is not complete, of course. It is, better to say, a skeleton of the model. For example, constraints and bodies of methods describing the most essential part of semantics of actions cannot be described precisely enough using this method and therefore they are only as comments there.

# 4 From classes to textual description of requirements

After the analyst means he identified both static and dynamic structures of the proposed software system he/she starts the reengineering phase.

The CASE tool TESSI generates texts of requirements which will be derived from the stored C++ and POET structures. For each used programming structure there is a template in a natural language. Using these templates the tool generates a new textual version of requirements specification. This new, generated version describes how the analyst understands the problem. The old, original version will be compared with the generated new version to decide whether the software analyst's understanding has an acceptable level.

The differences found will be discussed and a corrected version of requirements specification will be written anew in cooperation between the customer, the end user, the domain expert, and the software analyst. The whole process will be repeated until no important differences can be found. The expression possibilities of templates for generating a natural language texts are, of course, limited. Literature scientists and novel writers would not like the style of the generated language.

Example:
A template for the dynamic structure:
Objects of the class: " ... " can reach the following states: " ... ".
The start state is: " ... " and will be reached by the event: " ... ".
The behaviour of the system starting from the start state: " ... " can be described like this:
From the state: " ... " through the event " ... " will be reached the state " .... ".
#

The generated text sounds unusually but it contains all important features, precisely said, all features which the software analyst holds for important. All words which were not marked in the original text will be held for irrelevant and omitted in the generated text. New irrelevant words have come from the used templates.

The original version of the text contains descriptions of relevant objects often in a chaotic order which doesn't support the overview. The generated text is hierarchically ordered which makes its investigation more easy.

# 5 Advantages for the software development

Using the described method supported by the CASE tool TESSI we can get advantages not only during the analysis (requirements will be better understood

) and during the design (main structures will be created during the analysis and are directly bound to entities of the analysis) but also during other phases of the life-cycle of the software system, especially during:

- estimation of resources,
- maintenance,
- reuse.

## 5.1 Advantages for estimation of resources

The semi-automatic transformation of textual requirements into OO model (in C++ and POET) brings as a side effect the possibility of using metrics at this phase of the development. Any object-oriented metric can be embedded in the module which generates the OO model. Currently, the most simple object-oriented metric has been used in TESSI. All identified structures of the static model will be counted. Corresponding to every version of the requirements specification, the analyst knows how many classes, methods, attributes, relationships etc. will be probably used in the description of the system.

The costs of the product development, its time schedule, the volume of tests, personal resources, etc. can derived from this knowledge. The condition is that experiences in this field has been collected during similar projects. Anyway, the possible impact of every change in the requirements specification on resources can be followed during the discussion with the user.

## 5.2 Advantages for the maintenance

The CASE tool TESSI builds and maintains some important relationships, e.g.:

- bidirectional links between the identified entities in sentences of the original text and the entities of the OO model, i.e. C++ and POET structures,
- links between the entities of the OO model and the sentences of the generated text,
- links between the identified entities in sentences of the original text and the related sentences of the generated text.

The links representing semantic dependencies between entities in sentences of requirements specification and structures of the OO model will be followed during an adaptive maintenance. This helps to hold requirements and programs consistent and supports the concept of software development in which every change in the software system should follow the life-cycle of development. The location of changes in the program structures of the OO model will be directly derived using the links for every change in the requirement specifications and vice versa. The impact of the proposed changes can be much better estimated.

## 5.3 Advantages of the dictionary for reuse

A library of predefined, reusable classes together with a dictionary containing their names and templates of their textual description can be used for development of related software products. The analyst uses it like follows. He finds e.g. a noun in the original requirements, identifies it as candidate for a class and asks the dictionary whether this found noun is contained in it. If it is so, he will get a generated text describing the predefined, reusable class which contains all static features and an important part of dynamic features of the reused class. The user has to decide the problem of homonyms, i.e. whether the meaning of the reused class from the library is the same as the one mentioned in the original text.

## 6 Implementation of TESSI

The first version used the German language for the analysis and synthesis of the requirements specification. The CASE tool TESSI has been written in Borland Pascal for Windows on PC [7]. The target language of the generated object-oriented model is C++ (Version Borland 3.1) and the system POET (Version 2.1) which has been used for storing persistent objects.

Some linguistic problems occured concerning the complexity of the German language , e.g. irregular plurals, suffixes of adjectives, articles, etc. The language dependent parts have been programmed in separated units. There is also a new derived version available which is based on English. The generated language sounds a little funny but it is good understandable. The linguistic part of this project is out of the scope of this contribution.

## 7 Conclusions and Future work

The activity of eliciting, interpreting, and organising the information gained from user is a complex and painful process. This process should be supported by CASE tools.

Without having some exact measured quantitative data from a big project we argue that the described concept will essentially improve the quality of

- requirements specification because the possibility of bad understanding of the problem by the software analyst will be reduced,
- estimations of time and costs of the software development because it will be provided in parallel with the refinement of requirements,
- efficiency of the adaptive maintenance because of the bidirectional links between entities of requirements and entities of design.

Except of simple examples the CASE tool TESSI has been used for processing requirements specification of its own which contained more hundreds of sentences.

Using the tool TESSI on simple examples in practice we proved the very known fact that the most important parts of the problem's semantics are usually not included in the first version of the requirements specification. The reason is that the software analyst doesn't know about their existence and the customer holds them for so much self-evident that he/she doesn't mention them at all.

Currently, we are trying to identify subsystems in the origin text to reduce the complexity of the generated text and to produce groups of classes, to use this tool in projects and to compare the effort necessary with the effort when using manual methods, to rewrite TESSI into Java for making possible the remote access to processing of requirements.

## References

1. Coad, P., Yourdon,E.: Object-Oriented Analysis. Yourdon Press - Prentice Hall, 1991.
2. Emmerich, W., Kroha, P., Schäfer, W.: Object-Oriented Database Management Systems for Construction of CASE Environments. In: Marik, V. et al (Eds.): Proceedings of the 4th International Conference DEXA'93, Lecture Notes in Computer Sciences, No. 720, Springer, 1993.
3. Kroha, P.: Objects and Databases. McGraw-Hill, 1993.
4. Kroha, P.: Softwaretechnologie. Prentice Hall, 1997 (in German).
5. Motta, E., Rajan, T., Eisenstadt, M.: Knowledge acquisition as aprocess of model refinement. Knowledge Acquisition 2 (1), pp. 21-49, 1990.
6. Rubin,K., Goldberg,A.: Object Behaviour Analysis. Communication of ACM, Vol. 35, No.9, pp. 48-62, September 1992.
7. Strauß, M.: Development and Implementation of a Prototype of a CASE Tool for Object- Oriented Analysis and Design with Aspects of Reverse Engineering. Master Thesis, Technische Universitt Chemnitz, Faculty of Informatics, 1996 (in German).

# On Finite Representations of Infinite-State Behaviours

Antonín Kučera[*]

e-mail: tony@fi.muni.cz

Faculty of Informatics, Masaryk University
Botanická 68a, 60200 Brno
Czech Republic

**Abstract.** We examine the problem of finite-state representability of infinite-state processes w.r.t. certain behavioural equivalences. We show that the classical notion of regularity becomes insufficient in case of all equivalences of van Glabbeek's hierarchy except bisimilarity, and we design and justify a generalization in the form of strong regularity and finite characterizations. We show that the condition of strong regularity guarantees an existence of finite characterization in case of all equivalences of van Glabbeek's hierarchy, and we also demonstrate that there are behaviours which are regular but not strongly regular w.r.t. all equivalences of the mentioned hierarchy except bisimilarity.

## 1  Introduction

The problem whether a given infinite-state behaviour (process) can be equivalently represented by a finite-state one has recently attracted a lot of attention. A similar problem has been actually known from the theory of formal languages for a long time—given a grammar $G$, one can ask whether there is an equivalent *regular* grammar $G'$. The grammar $G'$ can be seen as a 'finite-state representation' of $G$ because of the associated finite-state automaton. However, it is folklore that the mentioned problem is *undecidable* even for context-free grammars.

The situation is more complicated within the framework of concurrency theory. Transition systems are widely accepted as structures which can exactly define semantics of concurrent process; however, there are many *behavioural* equivalences over the class of transition systems which try to formally express 'sameness' of two concurrent systems. Rob van Glabbeek presented in [vG90] a hierarchy of equivalences, relating them w.r.t. their *coarseness* (see Figure 1).

The problem whether for a given process there is an equivalent finite-state one has been intensively studied w.r.t. *bisimulation equivalence (bisimilarity)*; it is also known as the "regularity problem". Regularity has been proved to be decidable for BPA processes [MM94, BG96, BCS96], labelled Petri nets (and thus also BPP processes) [JE96], normed PA processes [Kuč96], and one-counter processes [Jan97]. Those results are also interesting from the practical point of view—verification of infinite-state systems is generally difficult, but if we replace an infinite-state system with some equivalent finite-state one, the procedure can be much easier. Moreover, decidability

---

[*] Supported by GA ČR, grant number 201/97/0456

of regularity can simplify various considerations about infinite-state behaviours (see e.g., [ČKK97, Kuč97]).

In this paper we examine a general question what properties should have a finite-state transition system if it is to be used as a 'reliable' description of some infite-state one. We argue that in case of all equivalences of van Glabbeek's hierarchy except bisimilarity the notion of regularity becomes insufficient, as it does not characterize reachable states (see the first paragraph of Section 3). We design and justify a new notion of finite *characterization* and we examine its basic properties. We prove that the condition of *strong regularity* guarantees an existence of a finite characterization w.r.t. all equivalences of van Glabbeek's hierarchy. As regularity and strong regularity *coincide* in case of bisimilarity, the condition of strong regularity can be seen as a 'proper' predicate expressing the feature of finite representability. We also prove that regularity and strong regularity do *not* coincide in case of all equivalences of van Glabbeek's hierarchy except bisimilarity, i.e., strong regularity is really a 'stronger' condition than regularity. We conclude with some remarks on future work.

## 2 Definitions

**Definition 1.** A transition system $T$ is a tuple $(S, Act, \rightarrow, r)$ where $S$ is a set of *states*, $Act$ is a set of *labels*, $\rightarrow \subseteq S \times Act \times S$ is a *transition relation* and $r \in S$ is a distinguished state called *root*. The class of all transition systems is denoted by $\mathcal{T}$.

As usual, we write $s \xrightarrow{a} t$ instead of $(s, a, t) \in \rightarrow$ and we extend this notation to elements of $Act^*$ in an obvious way (we sometimes write $s \rightarrow^* t$ instead of $s \xrightarrow{w} t$ if $w \in Act^*$ is irrelevant). A state $t$ is said to be *reachable* from a state $s$ if $s \rightarrow^* t$. The states which are reachable from the root are said to be *reachable*.

Various *behavioural equivalences* over the class of transition systems were proposed in the literature—each of them tries to express a certain level of 'sameness' which is proper in certain situations. Rob van Glabbeek presented in [vG90] a hierarchy of behavioural equivalences, relating them w.r.t. their *coarseness*, i.e., how many identifications they make. The resulting lattice is presented in Figure 1.

**Definition 2.** Let $T$ be a transition system and let $\leftrightarrow$ be an equivalence over $\mathcal{T}$. The system $T$ is *regular* w.r.t. $\leftrightarrow$ if there is a finite-state transition system $F$ such that $T \leftrightarrow F$. Such a system $F$ is called a finite *representation* of $T$.

## 3 Finite Characterizations

The notion of finite representation can be used for any equivalence of van Glabbeek's hierarchy. It is extremely useful in case of bisimilarity—and we argue this is due to the following fact: if we take bisimilar transition systems $T$ and $F$ such that $F$ has finitely many states, then for each reachable state $t$ of $T$ there is a bisimilar reachable state $f$ of $F$. In other words, $F$ gives a complete characterization of *all* reachable states of $T$. This is no more true for the other equivalences of van Glabbeek's hierarchy; if we take e.g., trace equivalence (see Definition 12) and two transition systems $T$ and $F$ such that

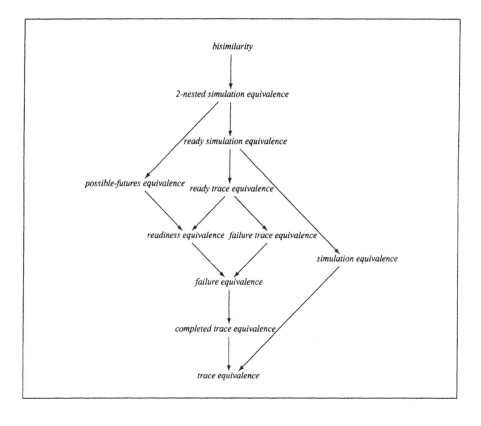

**Fig. 1.** van Glabbeek's hierarchy of behavioural equivalences

$T$ and $F$ are trace equivalent and $F$ has finitely many states, then the only thing we can say about $T$ and $F$ is that their *roots* have the same sets of traces—but if we take a reachable state $t$ of $T$, it need not be trace equivalent to any reachable state of $F$. If we want to check some temporal property of $T$ (e.g., something bad never happens), then we are usually interested in *all* reachable states of $T$. It is thus sensible to ask whether there is a finite-state transition system $F'$ such that *each* reachable state of $T$ is equivalent to some state of $F'$. If so, we can examine features of $F'$ instead of $T$ and as $F'$ has only finitely many states, it should be easier. This is the basic idea which leads to the notion of finite *characterization*.

**Definition 3.** Let $T$ be a transition system and let $\leftrightarrow$ be an equivalence over $\mathcal{T}$. A finite-state transition system $F$ is a finite *characterization* of $T$ w.r.t. $\leftrightarrow$ if all the following conditions are true: $T \leftrightarrow F$, states of $F$ are pairwise nonequivalent w.r.t. $\leftrightarrow$, and for each reachable state $t$ of $T$ there is a reachable state $f$ of $F$ such that $t \leftrightarrow f$.

Now we examine the question when finite characterizations exist and what is their relationship with finite representations. First we need to introduce further notions.

**Definition 4.** Let $\leftrightarrow$ be an equivalence over $\mathcal{T}$. For each transition system $T = (S, Act, \rightarrow, r)$ we define the transition system $T/\!\!\leftrightarrow = (S', Act, \rightarrow', r')$ in the following way:

- $S'$ contains equivalence classes of $S/\!\!\leftrightarrow$ (the equivalence class containing $s \in S$ is denoted by $[s]$).
- The relation $\rightarrow'$ is determined by the rule $s \xrightarrow{a} t \implies [s] \xrightarrow{a}' [t]$.
- $r' = [r]$

The equivalence $\leftrightarrow$ is said to *have quotients* if for any $T \in \mathcal{T}$ the natural projection $p : T \longrightarrow T/\!\!\leftrightarrow$, assigning to each state $s$ of $T$ the state $[s]$ of $T/\!\!\leftrightarrow$, is a part of $\leftrightarrow$ (i.e., $s \leftrightarrow [s]$ for each state $s$ of $T$).

The notion of finite characterization is naturally motivated. Now we can ask what features of a transition system $T$ guarantee an existence of a finite characterization of $T$. This is the aim of the following definition:

**Definition 5.** Let $\leftrightarrow$ be an equivalence over $\mathcal{T}$. A transition system $T$ is *strongly regular* w.r.t. $\leftrightarrow$ if $T$ can reach only finitely many states up to $\leftrightarrow$.

The next lemma says when the condition of strong regularity guarantees an existence of a finite characterization.

**Lemma 6.** *Let $\leftrightarrow$ be an equivalence over $\mathcal{T}$ which has quotients. Then $T$ has a finite characterization w.r.t. $\leftrightarrow$ iff $T$ is strongly regular w.r.t. $\leftrightarrow$.*

*Proof.*
"$\Rightarrow$" Obvious.
"$\Leftarrow$" As $T$ is strongly regular w.r.t. $\leftrightarrow$ and $\leftrightarrow$ has quotients, the transition system $T/\!\!\leftrightarrow$ is a finite characterization of $T$. $\qquad\square$

Now we prove that the requirement of "having quotients" from the previous lemma is not too restrictive in fact—all equivalences of van Glabbeek's hierarchy have this property. Due to the lack of space we cannot give a separate proof for each of them; instead we present just two full proofs which "cover" the whole hierarchy in the sense that all remaining proofs can be obtained by slight modifications of one of the two indicated approaches.

**Definition 7.** Let $T = (S, Act, \rightarrow, r)$ be a transition system. For each state $s \in S$ we define the set $I(s) = \{a \in Act \mid \exists t \in S \text{ such that } s \xrightarrow{a} t\}$. A pair $(w, \Phi) \in Act^* \times \mathcal{P}(Act)$ is a *failure pair* of $T$, if there is a state $s \in S$ such that $r \xrightarrow{w} s$ and $I(s) \cap \Phi = \emptyset$. Let $F(T)$ denote the set of all failure pairs of $T$. Transition systems $T_1, T_2$ are *failure equivalent*, written $T_1 =_f T_2$, if $F(T_1) = F(T_2)$

**Lemma 8.** *Failure equivalence has quotients.*

*Proof.* Let $T = (S, Act, \rightarrow, r)$ be a transition system. We show that $F(p) = F([p])$ for each state $p \in S$.

"$\subseteq$": Let $(w, \Phi) \in Act^* \times \mathcal{P}(Act)$ be a failure pair of $p$. By definition, there is a state

$p' \in S$ such that $p \xrightarrow{w} p'$ and $I(p') \cap \Phi = \emptyset$. But then also $[p] \xrightarrow{w} [p']$. The set $I([p'])$ is the union of all $I(q)$ such that $q \in [p']$. As $u =_f v$ implies $I(u) = I(v)$, we can conclude that $I([p']) = I(p')$, hence $I([p']) \cap \Phi = \emptyset$, thus $(w, \Phi) \in F([p])$.

"$\supseteq$": Let $(w, \Phi) \in Act^* \times \mathcal{P}(Act)$ be a failure pair of $[p]$ and let $w = a_k \ldots a_1$. By definition, there is a sequence of transitions $[p_k] \xrightarrow{a_k} [p_{k-1}] \xrightarrow{a_{k-1}} \ldots \xrightarrow{a_1} [p_0]$ in $T/=_f$ such that $p \in [p_k]$ and $I([p_0]) \cap \Phi = \emptyset$. We show that for each state $q$ of $T$ such that $q \in [p_i]$, where $i \in \{0, \ldots, k\}$, the pair $(a_i \ldots a_1, \Phi)$ belongs to $F(q)$. We proceed by induction on $i$:

- **$i = 0$** : as $I(q) = I([p_0])$, we have $(\epsilon, \Phi) \in F(q)$.
- **induction step:** as $[p_i] \xrightarrow{a_i} [p_{i-1}]$, there are states $u, v$ of $T$ such that $u \xrightarrow{a_i} v$, $u \in [p_i]$ and $v \in [p_{i-1}]$. By induction hypothesis we have $(a_{i-1} \ldots a_1, \Phi) \in F(v)$, hence $(a_i \ldots a_1, \Phi) \in F(u)$. As $q =_f u$, the pair $(a_i \ldots a_1, \Phi)$ belongs to $F(q)$. $\qquad\square$

The same technique can be also applied to trace equivalence, completed trace equivalence, readiness equivalence, failure trace equivalence, ready trace equivalence and possible-futures equivalence.

**Definition 9.** Let $T_1 = (S_1, Act_1, \rightarrow_1, r_1)$ and $T_2 = (S_2, Act_2, \rightarrow_2, r_2)$ be transition systems. A relation $R \subseteq S_1 \times S_2$ is a *simulation* if whenever $(s, t) \in R$ then

$$\forall a \in Act_1 : s \xrightarrow{a}_1 s' \implies \exists t' : t \xrightarrow{a}_2 t' \wedge (s', t') \in R$$

Transition system $T_1$ is *simulated* by $T_2$, written $T_1 \sqsubseteq_s T_2$, if there is a simulation $R$ with $(r_1, r_2) \in R$. It is easy to see that $\sqsubseteq_s$ is a preorder. Transition systems $T_1, T_2$ are *simulation equivalent*, written $T_1 =_s T_2$, if $T_1 \sqsubseteq_s T_2$ and $T_2 \sqsubseteq_s T_1$.

**Lemma 10.** *Simulation equivalence has quotients.*

*Proof.* Let $T = (S, Act, \rightarrow, r)$ be a transition system. We show that $t =_s [t]$ for each state $t \in S$. By definition, we must show an existence of two simulations $P, R$ such that $(t, [t]) \in P$ and $([t], t) \in R$. The simulation $P$ is exactly the natural projection $p : T \rightarrow T/=_s$, i.e., $P = \{(u, [u]) : u \in S\}$. It is easy to check that $P$ is a simulation. The simulation $R$ is defined as follows:

$$([u], v) \in R \overset{def}{\Longleftrightarrow} \exists p \in [u] : p \sqsubseteq_s v$$

We prove that $R$ is indeed a simulation. Suppose $[u] \xrightarrow{a} [u']$. By definition of $T/=_s$, there are $q, q' \in S$ such that $q \xrightarrow{a} q'$, $u =_s q$, and $u' =_s q'$. Moreover, by definition of $R$ there is $p \in S$ with $p =_s q$ and $p \sqsubseteq_s v$. As $q \sqsubseteq_s p \sqsubseteq_s v$, we also have $q \sqsubseteq_s v$ by transitivity of $\sqsubseteq_s$. Hence $v \xrightarrow{a} v'$ for some $v' \in S$ with $q' \sqsubseteq_s v'$. As $q' \in [u']$, the pair $([u'], v')$ belongs to $R$ and the proof is finished. $\qquad\square$

This method also works for ready simulation equivalence and 2-nested simulation equivalence. As bisimilarity has quotients (this is obvious), we can now state the following theorem:

**Theorem 11.** *Each equivalence in van Glabbeek's hierarchy has quotients.*

There are also other well-known equivalences which have quotients, e.g., weak bisimilarity (see [Mil89]) or branching bisimilarity (see [vGW89]). But this property is naturally not general—there are also equivalences which do not have quotients. To present a concrete example, we first need several definitions.

**Definition 12.** Let $T = (S, Act, \rightarrow, r)$ be transition system. A *trace* of $T$ is any sequence $w \in \Sigma^+$ such that $r \overset{w}{\rightarrow} s$ for some $s \in S$. A trace $w$ of $T$ is *completed* if $r \overset{w}{\rightarrow} s$ for some $s \in S$ which does not have any successors. Transition systems $T_1, T_2$ are

- *trace equivalent* if they have the same sets of traces.
- *completed trace equivalent* if they have the same sets of traces and the same sets of completed traces.
- *language equivalent*, written $T_1 =_L T_2$, if they have the same sets of completed traces.

Language equivalence is well-known from the theory of formal languages and automata. Note that it is incomparable even with trace equivalence.

**Theorem 13.** *Language equivalence does not have quotients.*

*Proof.* A simple counterexample looks as follows:

Clearly $r \neq_L [r]$ because the set of completed traces of $r$ is $\{a\}$ while the set of completed traces of $[r]$ is empty. □

We have seen that if we restrict our attention to behavioural equivalences which have quotients, then the condition of strong regularity becomes necessary and sufficient for an existence of a finite characterization. An interesting question is, what is the exact relationship between conditions of regularity and strong regularity. First, we already know that there are equivalences for which these two conditions coincide (e.g., bisimilarity, weak bisimilarity or branching bisimilarity). But there are also equivalences for which conditions of regularity and strong regularity express different properties.

**Theorem 14.** *Let $\leftrightarrow$ be an equivalence of van Glabbeek's hierarchy which lies under bisimilarity. Then there is a transition system $T$ such that $T$ is regular w.r.t. $\leftrightarrow$ and $T$ is not strongly regular w.r.t. $\leftrightarrow$.*

*Proof.* (sketch) Transition systems $T_3$ and $T_4$ of Figure 2 are ready simulation equivalent. As $T_4$ has finitely many states, $T_3$ is regular w.r.t. all equivalences which lie under ready simulation equivalence in van Glabbeek's hierarchy. At the same time we may observe that $T_3$ can reach infinitely many states which are pairwise nonequivalent w.r.t. trace equivalence. Hence $T_3$ is not strongly regular w.r.t. any equivalence in van Glabbeek's hierarchy.

Similarly, $T_1$ and $T_2$ are 2-nested simulation equivalent, but $T_1$ can reach infinitely many states which are pairwise nonequivalent w.r.t. possible-futures equivalence. Hence $T_1$ is regular w.r.t. possible-futures equivalence and 2-nested simulation equivalence, but not strongly regular w.r.t. the mentioned equivalences.                    □

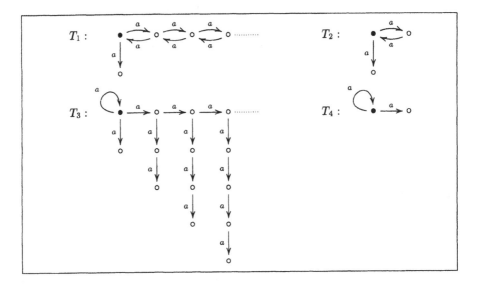

**Fig. 2.** Transition systems from the proof of Theorem 14

## 4   Future work

An open problem is whether the notions of regularity and strong regularity have different decidability features. However, this area seems to be quite unexplored. The notions of finite characterization and strong regularity surely deserve a deeper examination, and this is the subject we would like to work on in the future.

## Acknowledgement

Most of the presented work has originated in discussions with Mogens Nielsen during my stay at Aarhus University (BRICS). I am very grateful for his kind supervision and advice. Thanks are also due to anonymous referees for constructive comments.

# References

[BCS96] O. Burkart, D. Caucal, and B. Steffen. Bisimulation collapse and the process taxonomy. In *Proceedings of CONCUR'96*, volume 1119 of *LNCS*, pages 247–262. Springer-Verlag, 1996.

[BG96] D.J.B. Bosscher and W.O.D. Griffionen. Regularity for a large class of context-free processes is decidable. In *Proceedings of ICALP'96* [Ica96], pages 182–192.

[ČKK97] I. Černá, M. Křetínský, and A. Kučera. Bisimilarity is decidable in the union of normed BPA and normed BPP processes. *Electronic Notes in Theoretical Computer Science*, 6, 1997.

[Ica96] *Proceedings of ICALP'96*, volume 1099 of *LNCS*. Springer-Verlag, 1996.

[Jan97] P. Jančar. Bisimulation equivalence is decidable for one-counter processes. In *Proceedings of ICALP'97*, volume 1256 of *LNCS*, pages 549–559. Springer-Verlag, 1997.

[JE96] P. Jančar and J. Esparza. Deciding finiteness of Petri nets up to bisimilarity. In *Proceedings of ICALP'96* [Ica96], pages 478–489.

[Kuč96] A. Kučera. Regularity is decidable for normed PA processes in polynomial time. In *Proceedings of FST&TCS'96*, volume 1180 of *LNCS*, pages 111–122. Springer-Verlag, 1996.

[Kuč97] A. Kučera. How to parallelize sequential processes. In *Proceedings of CONCUR'97*, volume 1243 of *LNCS*, pages 302–316. Springer-Verlag, 1997.

[Mil89] R. Milner. *Communication and Concurrency*. Prentice-Hall, 1989.

[MM94] S. Mauw and H. Mulder. Regularity of BPA-systems is decidable. In *Proceedings of CONCUR'94*, volume 836 of *LNCS*, pages 34–47. Springer-Verlag, 1994.

[vG90] R.J. van Glabbeek. The linear time—branching time spectrum. In *Proceedings of CONCUR'90*, volume 458 of *LNCS*, pages 278–297. Springer-Verlag, 1990.

[vGW89] R.J. van Glabbeek and W.P. Weijland. Branching time and abstraction in bisimulation semantics. *Information Processing 89*, pages 613–618, 1989.

# Efficient Strong Sequentiality Using Replacement Restrictions*

Salvador Lucas

Departamento de Sistemas Informáticos y Computación
Universidad Politécnica de Valencia
Camino de Vera s/n, E-46071 Valencia, Spain.
e.mail: slucas@dsic.upv.es
URL: http://www.dsic.upv.es/users/elp/slucas.html

**Abstract.** Huet and Lévy defined the (orthogonal) *strongly sequential* term rewriting systems, for which *index reduction*, i.e., reduction of redexes placed at special positions called (strong) *indices*, is optimal and normalizing. Despite the fact that Huet and Lévy give an algorithm to compute indices for the general case, there are many proposals to define subclasses of strongly sequential rewrite systems for which this can be done more efficiently. In this paper, we show that sometimes it is possible to enlarge such classes by only introducing fixed replacement restrictions, without forcing any sensible modification of the corresponding index reduction strategy.

**Keywords:** functional programming, neededness, replacement restrictions, term rewriting.

## 1 Introduction

For orthogonal term rewriting systems (TRSs), Huet and Lévy gave a formal basis for the definition of efficient *sequential* strategies, i.e., reduction sequences in which only one needed redex is reduced in each step [2, 4]. The occurrences which is worthy to further explore to normalize a term are called (sequential) indices. Reduction of redexes placed at sequential indices (index reduction) is normalizing. In general, sequential indices are not computable, but Huet and Lévy define a class of orthogonal TRSs, the strongly sequential TRSs, for which this can be done. It is also decidable whether a TRS is strongly sequential.

The general algorithm to perform index reduction is rather complex, and several approximations have been developed in order to obtain more efficient, but less general methods [8, 9, 10]. A remarkable example are the left-normal TRSs ($LN$), a subclass of strongly sequential TRSs ($SS$), for which the leftmost-outermost strategy, an easily computable strategy, is indeed index reduction [8].

Context-sensitive rewriting (*csr*) [5] is rewriting with fixed restrictions on replacements which are completely determined by the symbols of the signature:

---

\* Work partially supported by Bancaixa (Bancaja-Europa grant) and CICYT (under grant TIC 95-0433-C03-03).

a mapping $\mu : \Sigma \rightarrow \mathcal{P}(\mathbb{N})$ called the *replacement map* indicates the argument positions which can be replaced for each symbol $f$. The arguments of $f$ indexed by $\mu(f)$ can be rewritten, and the others do not. The restrictions are lifted in the natural way from arguments of functions to arbitrary occurrences of terms.

**Example 1.1** *Let us consider the TRS $\mathcal{R}$:*

zero $\rightarrow$ 0                    first$(0, x) \rightarrow []$

from$(x) \rightarrow x ::$ from$(s(x))$    first$(s(x), y :: z) \rightarrow y ::$ first$(x, z)$

*We define* $\mu(s) = \mu(::) = \mu(\text{from}) = \emptyset$ *and* $\mu(\text{first}) = \{1, 2\}$. *We can perform the following context-sensitive derivation:*

first($\underline{\text{zero}}$, from$(0)) \rightarrow$ first$(0, \text{from}(0)) \rightarrow []$,

*but the infinite derivation*

first(zero, $\underline{\text{from}(0)}) \rightarrow$ first(zero, $0 :: \underline{\text{from}(s(0))}) \rightarrow \cdots$

*is avoided due to the restriction* $\mu(::) = \emptyset$.

In a computational system dealing with a certain subclass $CL$ of $SS$, we have an index reduction strategy $\mathbb{IF}$ which is normalizing for the TRSs in $CL$. In this paper we give a general method to incorporate replacement restrictions into the TRSs of any class $CL \subseteq SS$. This can lead to an effective extension of the considered class $CL$. We show how to compute the corresponding index reduction (restricted) strategy $\mathbb{IF}^{\mu}$ for each TRS by using the information in $\mathbb{IF}$.

In Section 2, we give some preliminary definitions. In Section 3, we introduce *csr*. Section 4 reviews the main concepts and results about strong sequentiality. Section 5 studies the general relations between replacement restrictions and strong sequentiality. Section 6 explains how to adapt the strategies.

## 2 Preliminaries

We denote as $V$ a countable set of variables and $\Sigma$ denotes a signature: a set of function symbols $\{f, g, \ldots\}$, each with a fixed arity given by a function $ar : \Sigma \rightarrow \mathbb{N}$. By $\mathcal{T}(\Sigma, V)$, we denote the set of terms. Denote as $\tilde{t}$ a $k$-tuple $t_1, \ldots, t_k$ of terms where $k$ will be clarified by the context. $Var(t)$ is the set of variables in $t$.

Terms are viewed as labelled trees in the usual way. Occurrences $u, v, \ldots$ are represented by chains of positive natural numbers used to address subterms of $t$. Occurrences are ordered by the standard prefix ordering: $u \leq v$ iff $\exists v'$ s.t. $v = u.v'$. Also, $u \parallel v$ means $u \not\leq v$ and $v \not\leq u$. We also use the *total* ordering $\leq_L$ given by: $\epsilon \leq_L u$ and $i.u \leq_L j.vv \Leftrightarrow i < j \lor (i = j \land u \leq_L v)$. Denote as $O(t)$ the set of occurrences of a term $t$. The occurrences of non-variable symbols in $t$ are $O_{\Sigma}(t)$, and $O_V(t)$ are the variable occurrences. The subterm at occurrence $u$ of $t$ is $t|_u$ and $t[s]_u$ is the term $t$ with the subterm at the occurrence $u$ replaced with $s$. The symbol labelling the root of $t$ is $root(t)$ and $\Sigma(t) = \{f \in \Sigma \mid \exists u \in O(t).root(t|_u) = f\}$ is the set of symbols from $\Sigma$ appearing in $t$. We refer to any term $C$, which is the same as $t$ everywhere except below $u$, i.e. there exists a term $s$ such that $C[s]_u = t$ (or just $C[s]$), as the *context* within the replacement occurs.

A rewrite rule is an ordered pair $(l, r)$, written $l \rightarrow r$, with $l, r \in \mathcal{T}(\Sigma, V)$, $l \notin V$ and $Var(r) \subseteq Var(l)$. The left-hand side (*lhs*) of the rule is $l$ and $r$ is the right-hand side (*rhs*). A TRS is a pair $\mathcal{R} = (\Sigma, R)$ where $R$ is a finite set of rewrite rules. Denote as $L(\mathcal{R})$ the *lhs*'s of $\mathcal{R}$. A TRS is left-linear if there are no multiple occurrences of a single variable in the *lhs*'s of the rules. Two rules $l \rightarrow r$ and $l' \rightarrow r'$ *overlap*, if there is a non-variable occurrence $u \in O(l)$, a substitution $\sigma$ such that $\sigma(l|_u) = \sigma(l')$. A TRS is non-ambiguous, if there are no overlapping *lhs*'s (trivial overlap in the same *lhs* is not considered). A left-linear, non-ambiguous TRS is called orthogonal. An instance $\sigma(l)$ of a *lhs* $l$ of a rule is a redex. The set of redex occurrences in $t$ is $O_{\mathcal{R}}(t) = \{u \in O(t) \mid \exists l \in L(\mathcal{R}) : t|_u = \sigma(l)\}$. We say that $t$ rewrites to $s$ (at the occurrence $u$), written $t \xrightarrow{u}_{\mathcal{R}} s$ (or just $t \rightarrow s$), if $t|_u = \sigma(l)$ and $s = t[\sigma(r)]_u$, for some rule $l \rightarrow r \in R$, $u \in O(t)$ and substitution $\sigma$. A term $t$ is in head-normal form (*hnf*) if there is no derivation $t = t_1 \rightarrow t_2 \rightarrow \cdots$ which reduces the root of a term $t_i$ $i \geq 1$.

$\mathbb{N}_k^+$ is an initial segment $\{1, 2, \ldots k\}$ of positive natural numbers, where $\mathbb{N}_0^+ = \emptyset$. $\mathcal{P}(\mathbb{N})$ is the powerset of natural numbers. We denote as $\mathcal{P}(\mathbb{N}_{>0}^*)$ the powerset of chains of positive numbers.

## 3 Context-sensitive rewriting

Given a signature $\Sigma$, a mapping $\mu : \Sigma \rightarrow \mathcal{P}(\mathbb{N})$ is a *replacement map* (or $\Sigma$-map) if for all $f \in \Sigma$. $\mu(f) \subseteq \mathbb{N}_{ar(f)}^+$. Hence, $\mu$ determines the *argument* positions which can be reduced for each symbol in the signature [5]. We define an ordering $\sqsubseteq$ on $M_\Sigma$, the set of all $\Sigma$-maps: $\mu \sqsubseteq \mu'$ if for all $f \in \Sigma$, $\mu(f) \subseteq \mu'(f)$. Then, $(M_\Sigma, \sqsubseteq, \mu_\perp, \mu_\top, \sqcup, \sqcap)$ is a lattice: for all $f \in \Sigma$, $\mu_\perp(f) = \emptyset$, $\mu_\top(f) = \mathbb{N}_{ar(f)}^+$, $(\mu \sqcup \mu')(f) = \mu(f) \cup \mu'(f)$ and $(\mu \sqcap \mu')(f) = \mu(f) \cap \mu'(f)$. Hence, $\mu \sqsubseteq \mu'$ means that $\mu$ considers less positions than $\mu'$ for reduction.

The set of $\mu$-*replacing* occurrences $O^\mu(t)$ is given by: $O^\mu(x) = \{\epsilon\}$, if $x \in V$ and $O^\mu(f(\bar{t})) = \{\epsilon\} \cup \bigcup_{i \in \mu(f)} i.O^\mu(t_i)$. The *non-replacing* occurrences are $\widetilde{O^\mu}(t) = O(t) \backslash O^\mu(t)$. In *csr*, we rewrite *replacing* occurrences: $t$ $\mu$-rewrites to $s$, written $t \hookrightarrow_{\mathcal{R}(\mu)} s$ (or simply $\hookrightarrow$), if $t \xrightarrow{u}_{\mathcal{R}} s$ and $u \in O^\mu(t)$. The set of *replacing redexes* is $O_{\mathcal{R}}^\mu(t) = O_{\mathcal{R}}(t) \cap O^\mu(t)$ and $t$ is a $\mu$-normal form, if $O_{\mathcal{R}}^\mu(t) = \emptyset$. The set of replacing variables is $Var^\mu(t) = \{x \in Var(t) \mid \exists u \in O(t). t|_u = x\}$.

A term $t \in \mathcal{T}(\Sigma, V)$ is $\mu$-*compatible* if every non variable occurrence of $t$ is a replacing occurrence: $O_\Sigma(t) \subseteq O^\mu(t)$. Given a term $t$, the *minimum* $\Sigma(t)$-map $\mu_t$ ensuring that $t$ is compatible is: if $t = x$, then $\mu_t(f) = \emptyset$ for all $f \in \Sigma$. If $t = f(t_1, \ldots, t_k)$, then $\mu_t = \mu_t^\epsilon \sqcup \mu_{t_1} \sqcup \cdots \sqcup \mu_{t_k}$, where $\mu_t^\epsilon$ is $\mu_t^\epsilon(f) = \{i \mid t_i \notin V\}$ and $\mu_t^\epsilon(g) = \emptyset$ for all $g \neq f$.

**Example 3.1** *If* $t = \text{first}(0, x)$, *then* $\mu_t(\text{first}) = \{1\}$. *Since* $\mu_{s(x)}(s) = \emptyset$ *and* $\mu_{y::z}(::) = \emptyset$, *for* $t' = \text{first}(s(x), y :: z)$, *we get* $\mu_{t'}(s) = \mu_{t'}(::) = \emptyset$ *and* $\mu_{t'}(\text{first}) = \{1, 2\}$.

The *canonical replacement map* for $\mathcal{R}$, $\mu_{\mathcal{R}}^{com} = \bigsqcup_{l \in L(\mathcal{R})} \mu_l$, is the minimum replacement map which makes the *lhs*'s of $\mathcal{R}$ compatible terms. For instance, $\mu$ in Example 1.1 is the canonical replacement map for the TRS $\mathcal{R}$ in the example.

# 4 Strong sequentiality

Indices point to unexplored parts of a term $t$ which is worthy to further develop to normalize $t$. To represent *unknown* parts of $t$, a new constant symbol $\Omega$ is introduced. Terms in $\mathcal{T}(\Sigma \cup \{\Omega\}, V)$ are said $\Omega$-terms. Denote as $O_\Omega(t)$ the set of occurrences of $\Omega$ in $t$: $O_\Omega(t) = \{u \in O(t) \mid t|_u = \Omega\}$. An ordering $\leq$ on $\Omega$-terms is given: $\Omega \leq t$ for all $t$, $x \leq x$ if $x \in V$, and $f(\tilde{t}) \leq f(\tilde{s})$ if $t_i \leq s_i$ for all $1 \leq i \leq ar(f)$. Thus $t \leq s$ means '$t$ is less defined than $s$'. An $\Omega$-normal form $t$ is s.t. $O_\mathcal{R}(t) = \emptyset$ and $O_\Omega(t) \neq \emptyset$. A redex scheme is a *lhs* in $L(\mathcal{R})$ where all variables are replaced by $\Omega$. Let $L_\Omega(\mathcal{R})$ be the set of *redex scheme* of $\mathcal{R}$.

To calculate an index we use a function $\omega$ defined by means of a reduction relation $\to_\Omega$ [4]: $C[t] \to_\Omega C[\Omega]$ if $t \neq \Omega$ and there is $l \in L_\Omega(\mathcal{R})$ such that $t \uparrow l$, i.e., there exists $s \in \mathcal{T}(\Sigma \cup \{\Omega\}, V)$, such that $t \leq s$ and $l \leq s$. Then, $\to_\Omega$ is confluent and terminating (see [2, 4]). Let $\omega(t)$ be the $\to_\Omega$-normal form of $t$.

**Definition 4.1 ([4])** *Let $t \in \mathcal{T}(\Sigma \cup \{\Omega\}, V)$ and $u \in O_\Omega(t)$. Let $\bullet$ be a fresh constant symbol, and $t' = t[\bullet]_u$. Then $u$ is an* index *of $t$, iff $\omega(t')|_u = \bullet$. Let us denote as $\mathcal{I}(t)$ the set of indices of $t$.*

A term $t$ is *rigid* if $\omega(t) = t$, *soft* if $\omega(t) = \Omega$ [3] and a *strong hnf* if $\omega(t) \neq \Omega$.

**Proposition 4.2** *If $t$ is a strong hnf, then $t$ is a hnf.*

**Definition 4.3 ([3])** *An orthogonal TRS is* strongly sequential *if every $\Omega$-normal form has an index.*

Any strategy which always reduces redexes pointed by indices is *index reduction*.

**Theorem 4.4 ([2])** *Index reduction is normalizing for orthogonal, strongly sequential TRSs.*

# 5 Strong sequentiality and replacement restrictions

In this section, we analyze the relation between replacement restrictions given by a replacement map $\mu$ and indices of terms w.r.t. a TRS $\mathcal{R}$. We consider *replacing* indices, $\mathcal{I}^\mu(t) = \mathcal{I}(t) \cap O^\mu(t)$, and *non-replacing* ones: $\widetilde{\mathcal{I}^\mu}(t) = \mathcal{I}(t) \cap \widetilde{O^\mu}(t)$.

**Proposition 5.1** *Let $\mathcal{R}$ be a TRS, and $\mu$ be such that $\mu_\mathcal{R}^{com} \sqsubseteq \mu$. Let $t \in \mathcal{T}(\Sigma \cup \{\Omega\}, V)$ be a soft term. Then $\widetilde{\mathcal{I}^\mu}(t) = \emptyset$.*

From Proposition 4.2, if $t$ is not a *hnf*, then $t$ is not a strong *hnf*. Hence $t$ is soft. Thus, Proposition 5.1 entails that, in reductions leading to a *hnf, only replacing redex occurrences must be considered*, since there are no non-replacing indices.

**Theorem 5.2** *Let $\mathcal{R}$ be an orthogonal, strongly sequential TRS, and $\mu$ be such that $\mu_\mathcal{R}^{com} \sqsubseteq \mu$. Let $t \in \mathcal{T}(\Sigma \cup \{\Omega\}, V)$ be an $\Omega$-normal form, such that $O_\Omega^\mu(t) \neq \emptyset$. Then $\mathcal{I}^\mu(t) \neq \emptyset$.*

This result means that an index reduction strategy can be refined by including replacement restrictions without damage its index reduction nature.

## 5.1 The contractive transformation

Now, we connect strong sequentiality of a TRS $\mathcal{R}$ and strong sequentiality of a TRS $\mathcal{R}^\mu$ obtained by means of a transformation called $\mu$-*contraction* which we have introduced in a previous work [7]: The $\mu$-*contracted signature* $\Sigma^\mu$ is defined as follows: $f_\mu \in \Sigma^\mu \wedge ar(f_\mu) = |\mu(f)| \Leftrightarrow f \in \Sigma$. Terms are related by means of a $\mu$-*contracting function* $\tau_\mu : \mathcal{T}(\Sigma, V) \rightarrow \mathcal{T}(\Sigma^\mu, V)$. This function *drops* the non-replacing immediate subterms of a term $t$ and constructs a '$\mu$-contracted' term by joining the (also transformed) replacing subterms below the corresponding operation of the $\mu$-contracted signature: $\tau_\mu(x) = x$ if $x \in V$, and $\tau_\mu(f(t_1, \ldots, t_k)) = f_\mu(\tau_\mu(t_{i_1}), \ldots, \tau_\mu(t_{i_p}))$ if $\mu(f) = \{i_1, \ldots, i_p\}$. The $i$-th immediate (replacing) subterm of $f$ in $t = f(\tilde{t})$ becomes the $|\mathbb{N}_i^+ \cap \mu(\mathsf{f})|$-th immediate subterm of $f_\mu$ in $\tau_\mu(t)$.

**Example 5.3** *Let $\mathcal{R}$ and $\mu = \mu_\mathcal{R}^{com}$ as given in Example 1.1.*
*Then $t = \mathtt{first}(\mathtt{s}(\mathtt{x}), \mathtt{y} :: \mathtt{z})$ contracts to $\tau_\mu(t) = \mathtt{first}_\mu(\mathtt{s}_\mu, ::_\mu)$ and $t' = \mathtt{y} :: \mathtt{first}(\mathtt{x}, \mathtt{z})$ contracts to $\tau_\mu(t') = ::_\mu$.*

Since $\tau_\mu$ is surjective, we are allowed to consider any $s \in \mathcal{T}(\Sigma^\mu, V)$ as the image by $\tau_\mu$ of some $t \in \mathcal{T}(\Sigma, V)$. We write $\tau_\mu(t)$ instead of $s$ when $s = \tau_\mu(t)$.

Replacing occurrences of $t$ and occurrences of $\tau_\mu(t)$ are related by $\nu_{\mu,t}$ : $O^\mu(t) \rightarrow O(\tau_\mu(t))$: $\nu_{\mu,t}(\epsilon) = \epsilon$, and $\nu_{\mu,f(\tilde{t})}(i.u) = |\mathbb{N}_i^+ \cap \mu(f)|.\nu_{\mu,t_i}(u)$ if $i \in \mu(f)$. We write $\nu_t$ instead of $\nu_{\mu,t}$ when $\mu$ is clear from the context. Note that $\nu_{\mu,t}$ is a bijection, and $\nu_{\mu,t}(O^\mu(t)) = O(\tau_\mu(t))$.

We use $\tau_\mu$ to connect TRSs. Let $\mathcal{R} = (\Sigma, R)$ be a TRS. The $\mu$-contraction of $\mathcal{R}$ is $\mathcal{R}^\mu = (\Sigma^\mu, R^\mu)$, where $R^\mu = \{\tau_\mu(l) \rightarrow \tau_\mu(r) \mid l \rightarrow r \in R\}$. We indicate as $\rho_\mu(l \rightarrow r)$ the rule of $\mathcal{R}^\mu$ which corresponds to a rule $l \rightarrow r$ of $\mathcal{R}$. $\rho_\mu$ is surjective.

**Example 5.4** *By considering $\mathcal{R}$ and $\mu$ as in Example 1.1, $\mathcal{R}^\mu$ is:*

$$\mathtt{zero}_\mu \rightarrow \mathtt{0}_\mu \qquad \mathtt{first}_\mu(\mathtt{0}_\mu, \mathtt{x}) \rightarrow []_\mu$$
$$\mathtt{from}_\mu \rightarrow ::_\mu \qquad \mathtt{first}_\mu(\mathtt{s}_\mu, ::_\mu) \rightarrow ::_\mu$$

The definition of $\mathcal{R}^\mu$ only has complete sense if $Var^\mu(r) \subseteq Var^\mu(l)$ for each $l \rightarrow r \in R$. In this way, $\mathcal{R}^\mu$ has no extra variables. For instance, in Example 5.4, we have obtained a valid TRS. However, if we define $\mu(::) = \{1\}$, the new contraction does not yield a TRS, because we obtain extra variables. Nevertheless, concerning the results in this paper, we are independent from this circumstance.

## 5.2 Inducing strong sequentiality

In the following, we use $\mathcal{J}(t)$ to denote indices obtained using $L_\Omega(\mathcal{R}^\mu)$.

**Proposition 5.5** *Let $\mathcal{R}$ be a TRS and $\mu$ be such that $\mu_\mathcal{R}^{com} \sqsubseteq \mu$. Let $t$ be an $\Omega$-term. Then $u \in \mathcal{I}^\mu(t)$ iff $\nu_t(u) \in \mathcal{J}(\tau_\mu(t))$, i.e., $\nu_t(\mathcal{I}^\mu(t)) = \mathcal{J}(\tau_\mu(t))$.*

Strong sequentiality is independent from $rhs$'s of rules. Since $\tau_\mu$ can introduce extra variables in the $rhs$'s, in order to formulate our results, we say that TRSs $\mathcal{R} = (\Sigma, R)$ and $\mathcal{R}' = (\Sigma', R')$ are *left-equivalent* ($\mathcal{R} \sim_l \mathcal{R}'$) if they have the same $lhs$'s, i.e., $R = \{l_i \rightarrow r_i \mid 1 \leq i \leq n\}$ and $R' = \{l_i \rightarrow r'_i \mid 1 \leq i \leq n\}$ [4].

**Theorem 5.6** *Let $\mathcal{R}$ be an orthogonal TRS. Let $\mu$ be such that $\mu_{\mathcal{R}}^{com} \sqsubseteq \mu$ and $\mathcal{R}^{\mu}$ is left equivalent to a strongly sequential TRS. Then $\mathcal{R}$ is strongly sequential.*

To illustrate Theorem 5.6, we recall the notion of left-normal TRS [8, 10]: $\mathcal{R}$ is left-normal if, for all $l \in L(\mathcal{R})$, the function symbols *precede* the variables according to $\leq_L$. Thus $\mathcal{R}$ in Example 1.1, is *not* left-normal. No easy syntactic characterization of strong sequentiality for $\mathcal{R}$ is available. However, $\mathcal{R}^{\mu}$, where $\mu = \mu_{\mathcal{R}}^{com}$, in Example 5.4 is left-normal, hence strongly sequential. Therefore, Theorem 5.6 entails strong sequentiality of $\mathcal{R}$. We give a further example.

**Example 5.7** *Let us consider the TRS $\mathcal{R}$ to manipulate prime numbers.*
$$\text{from}(x) \to x :: \text{from}(s(x))$$
$$\text{primes} \to \text{sieve}(\text{from}(s(s(0))))$$
$$\text{sieve}(x :: y) \to x :: \text{sieve}(\text{filter}(x, y))$$
$$\text{filter}(x, y :: z) \to \text{if}(x|y, \text{filter}(x, z), y :: \text{filter}(x, z))$$
$$\text{sel}(0, x :: y) \to x \qquad \text{if}(\text{true}, x, y) \to x$$
$$\text{sel}(s(x), y :: z) \to \text{sel}(x, z) \quad \text{if}(\text{false}, x, y) \to y$$
*Assume that $|$ is a built-in predicate and $x|y$ means $x$ divides $y$. If we let $\mu = \mu_{\mathcal{R}}^{com}$, then $\mu(s) = \mu(::) = \mu(\text{from}) = \emptyset$, $\mu(\text{sieve}) = \mu(\text{if}) = \{1\}$, $\mu(\text{filter}) = \{2\}$, $\mu(\text{sel}) = \{1, 2\}$. $\mathcal{R}$ is not left-normal. Then $\mathcal{R}^{\mu}$ is:*
$$\text{from}_{\mu} \to ::_{\mu}$$
$$\text{primes}_{\mu} \to \text{sieve}_{\mu}(\text{from}_{\mu})$$
$$\text{sieve}_{\mu}(::_{\mu}) \to ::_{\mu}$$
$$\text{filter}_{\mu}(::_{\mu}) \to \text{if}_{\mu}(|_{\mu})$$
$$\text{sel}_{\mu}(0_{\mu}, ::_{\mu}) \to x \qquad \text{if}_{\mu}(\text{true}_{\mu}) \to x$$
$$\text{sel}_{\mu}(s_{\mu}, ::_{\mu}) \to \text{sel}_{\mu}(x, z) \quad \text{if}_{\mu}(\text{false}_{\mu}) \to y$$
$\mathcal{R}^{\mu}$ *is not a TRS (it has extra variables), but it is immediate to have a left-equivalent TRS $\mathcal{S} \sim_l \mathcal{R}^{\mu}$. $\mathcal{S}$ must be a ground TRS (i.e., $l, r \in \mathcal{T}(\Sigma, \emptyset)$ for all rule $l \to r$ of $\mathcal{S}$), since the lhs's have no variable. Ground TRSs are left-normal, hence strongly sequential. Thus, $\mathcal{R}$ also is.*

**Theorem 5.8** *Let $\mathcal{R}$ be an orthogonal, strongly sequential TRS. Let $\mu$ be such that $\mu_{\mathcal{R}}^{com} \sqsubseteq \mu$. Then, $\mathcal{R}^{\mu}$ is left equivalent to a strongly sequential TRS.*

Hence, Theorems 5.6 and 5.8 amount to saying that proving strong sequentiality in a TRS $\mathcal{R}$ or its $\mu$-contraction $\mathcal{R}^{\mu}$ is completely equivalent, whenever $\mu_{\mathcal{R}}^{com} \sqsubseteq \mu$.

# 6 Induced strategies

We have shown that it is possible to conclude strong sequentiality of $\mathcal{R}$ from strong sequentiality of (a TRS left-equivalent to) $\mathcal{R}^{\mu}$. In this section, we show how to obtain an index reduction strategy for $\mathcal{R}$ by using an index reduction strategy for $\mathcal{R}^{\mu}$.

A rewriting strategy for a TRS $\mathcal{R} = (\Sigma, R)$ is a function $\mathbb{F}_{\mathcal{R}} : \mathcal{T}(\Sigma, V) \to \mathcal{T}(\Sigma, V)$ such that $\mathbb{F}_{\mathcal{R}}(t) = t$ if $O_{\mathcal{R}}(t) = \emptyset$, i.e, $t$ is a normal form, and $t \to_{\mathcal{R}}^{+} \mathbb{F}_{\mathcal{R}}(t)$ otherwise ($t \to_{\mathcal{R}} \mathbb{F}_{\mathcal{R}}(t)$ for sequential strategies) [3]. We drop the suffix

$\mathcal{R}$ when no confusion arises. Also, we split $\mathbb{F}$ into components $\mathbb{F} = (\mathbb{P}, \mathbb{R})$. The component $\mathbb{P} : \mathcal{T}(\Sigma, V) \to \mathcal{P}(\mathbb{N}_{>0}^*)$ selects the redex occurrences to be reduced (a singleton for sequential strategies), and $\mathbb{R}$ selects the rules which apply (see [1]). Since we only deal with orthogonal TRSs, the rule which applies to a redex is uniquely determined by the redex itself. Then, for a given $\mathbb{P}$, we get $\mathbb{R}$ fixed. Hence, we do not pay more attention to $\mathbb{R}$. Then, $\mathbb{P}$ and $\mathbb{F}$ are related by:

$\mathbb{P}(t) = \emptyset$ iff $O_{\mathcal{R}}(t) = \emptyset$.

$\mathbb{P}(t) = \{u_1, \ldots, u_n\} \subseteq O_{\mathcal{R}}(t)$ iff $u_i \parallel u_j$ for $1 \leq i < j \leq n$, $t|_{u_i} = \sigma(l_i)$ for a single $l_i \to r_i \in R$, $1 \leq i \leq n$, and $\mathbb{F}(t) = t[\sigma(r_1)]_{u_1} \cdots [\sigma(r_n)]_{u_n}$.

The notion of context-sensitive strategy or $\mu$-strategy $\mathbb{F}_{\mathcal{R}(\mu)}$ is analogous, but it uses $\hookrightarrow_{\mathcal{R}(\mu)}$ instead of $\to_{\mathcal{R}}$ and $O_{\mathcal{R}}^{\mu}$ instead of $O_{\mathcal{R}}$ (see [6]). However, note that, from $\mu$-strategies, we only can expect to $\mu$-normalize terms, rather than completely normalize them[2]. Now, we introduce the notion of *induced $\mu$-strategy*.

**Definition 6.1** *Let* $\mathcal{R} = (\Sigma, R)$ *be an orthogonal TRS and* $\mu$ *be such that* $\mu_{\mathcal{R}}^{com} \sqsubseteq \mu$. *Let* $\mathbb{F} = (\mathbb{P}, \mathbb{R})$ *be a rewriting strategy for* $S \sim_l \mathcal{R}^{\mu}$. *The $\mu$-rewriting strategy* $\mathbb{F}^{\mu} = (\mathbb{P}^{\mu}, \mathbb{R}^{\mu})$ *induced by* $\mathbb{F}$ *is given by* $\mathbb{P}^{\mu}(t) = \nu_t^{-1}(\mathbb{P}(\tau_{\mu}(t)))$ *for all* $t \in \mathcal{T}(\Sigma, V)$. *Since whenever* $\mu_{\mathcal{R}}^{com} \sqsubseteq \mu$ *the $\mu$-contraction preserves orthogonality,* $\mathbb{R}^{\mu}$ *is again fixed by* $\mathbb{P}^{\mu}$.

The definition of $\mathbb{P}^{\mu}$ should be obvious after Proposition 5.5. Now we prove that $\mathbb{F}^{\mu}$ in Definition 6.1 holds the expected property, i.e., it is an index reduction $\mu$-strategy for the TRS $\mathcal{R}$, whenever $\mathbb{F}$ is index reduction for (a TRS left-equivalent to) $\mathcal{R}^{\mu}$.

**Theorem 6.2** *Let* $S$ *be an orthogonal strongly sequential TRS. Let* $\mathbb{F}$ *be an index reduction strategy for* $S$. *Let* $\mathcal{R}$ *be a TRS and* $\mu$ *be such that* $\mu_{\mathcal{R}}^{com} \sqsubseteq \mu$ *and* $\mathcal{R}^{\mu} \sim_l S$. *Then,* $\mathbb{F}^{\mu}$ *is an index reduction $\mu$-strategy for* $\mathcal{R}$.

Sometimes, $\mathbb{P}$ can be further decomposed as follows: $\mathbb{P} = \widehat{\mathbb{P}} \circ O_{\mathcal{R}}$ where $\widehat{\mathbb{P}} : \mathcal{P}(\mathbb{N}_{>0}^*) \to \mathcal{P}(\mathbb{N}_{>0}^*)$ selects a set of occurrences from a non empty set of redex occurrences ($\widehat{\mathbb{P}}(\emptyset) = \emptyset$), and $O_{\mathcal{R}} : \mathcal{T}(\Sigma, V) \to \mathcal{P}(\mathbb{N}_{>0}^*)$ gives the redex occurrences of $t$. We also require $\widehat{\mathbb{P}}(U) \subseteq U$ for all $U \in \mathcal{P}(\mathbb{N}_{>0}^*)$. Some examples [3]: *Leftmost-outermost* ($\mathbb{F}_{lo}$): $\mathbb{P}_{lo}(t) = \{min_{\leq_L}(minimal_{\leq}(O_{\mathcal{R}}(t)))\} = \{min_{\leq_L}(O_{\mathcal{R}}(t))\}$. *Parallel-outermost* ($\mathbb{F}_{po}$): $\mathbb{P}_{po}(t) = minimal_{\leq}(O_{\mathcal{R}}(t))$. Note that in these strategies, the question of *which* rule should be applied to a selected redex is left open: In orthogonal TRSs the redex fixes the rule.

The calculus of the $\mathbb{P}^{\mu}$ component of an induced strategy (Definition 6.1), for a given term $t$, proceeds by using $\tau_{\mu}$ for moving to $\mathcal{R}^{\mu}$, next using the $\mathbb{P}$ component of the original strategy and coming back to $\mathcal{R}$ by means of $\nu_t^{-1}$. The decomposition $\mathbb{P} = \widehat{\mathbb{P}} \circ O_{\mathcal{R}}$ can lead to a simplification of this procedure.

---

[2] Nevertheless, it is easy to use $\mu$-normalizing strategies to define normalizing procedures [6].

**Proposition 6.3** *Let $\mathcal{R}$ be an orthogonal TRS, and $\mu$ be s.t. $\mu_{\mathcal{R}}^{com} \sqsubseteq \mu$. Let $\mathbb{F} = (\mathbb{P}, \mathbb{R})$ be a rewriting strategy for $S \sim_l \mathcal{R}^\mu$ and $\mathbb{F}^\mu = (\mathbb{P}^\mu, \mathbb{R}^\mu)$ be the induced $\mu$-strategy for $\mathcal{R}$. Let $\mathbb{P}$ be such that $\mathbb{P} = \widehat{\mathbb{P}} \circ O_{\mathcal{R}^\mu}$. If, for all $t \in \mathcal{T}(\Sigma, V)$ and $U \subseteq O^\mu(t)$, we have $\nu_t(\widehat{\mathbb{P}}(U)) = \widehat{\mathbb{P}}(\nu_t(U))$, then $\mathbb{P}^\mu(t) = \widehat{\mathbb{P}}(O_{\mathcal{R}}^\mu(t))$.*

The result on normalization by means of $\mathbb{F}_{lo}$ can be given by showing that an orthogonal, left-normal TRS is strongly sequential and the leftmost-outermost redex occurrence $min_{\leq_L}(O_{\mathcal{R}}(t))$ of a term $t$ is an index (O'Donnell, [8]).

We show that $min_{\leq_L}$ commutes with $\nu$. This allows us to give a simpler expression of the induced leftmost-outermost strategy $\mathbb{F}_{lo}^\mu$.

**Proposition 6.4** *Let $t \in \mathcal{T}(\Sigma, V)$. $\forall U \subseteq O^\mu(t)$, $\nu_t(min_{\leq_L}(U)) = min_{\leq_L}(\nu_t(U))$.*

Therefore, from Proposition 6.4 and Proposition 6.3, the leftmost-outermost strategy $\mathbb{F}_{lo}$ induces a $\mu$-strategy $\mathbb{F}_{lo}^\mu$ for a given $\mathcal{R}$ and $\mu$ such that $\mathcal{R}^\mu \sim_l S$ and $S$ is left-normal. The calculus is nicely simple: $\mathbb{P}_{lo}^\mu(t) = min_{\leq_L}(O_{\mathcal{R}}^\mu(t))$.

Concerning Example 5.7 we have even better possibilies. Since in a ground, orthogonal TRS, every redex occurrence is an index, *any strategy is index reduction*. We can use $\widehat{\mathbb{P}} = choose$. Clearly, for all $t$, and $U \subseteq O^\mu(t)$, $\nu_t(choose(U)) = choose(\nu_t(U))$ (because of non-determinism). Hence $\mathbb{P}^\mu = choose(O_{\mathcal{R}}^\mu(t))$.

# References

1. S. Antoy and A. Middeldorp. A Sequential Reduction Strategy. *Theoretical Computer Science* 165:75-95, 1996.
2. G. Huet and J.J. Lévy. Computations in orthogonal term rewriting systems. In *Computational logic: essays in honour of J. Alan Robinson*, MIT Press, 1991.
3. J.W. Klop. Term Rewriting Systems. In *Handbook of Logic in Computer Science*, volume 3, pages 1-116. Oxford University Press, 1992.
4. J.W. Klop and A. Middeldorp. Sequentiality in Orthogonal Term Rewriting Systems. *Journal of Symbolic Computation* 12:161-195, 1991.
5. S. Lucas. Context-sensitive computations in functional and functional logic programs. *Journal of Functional and Logic Programming*, 1997, *to appear*.
6. S. Lucas. Needed Reductions with Context-Sensitive Rewriting. In *Proc. of ALP'97*, LNCS *to appear*.
7. S. Lucas. Termination of Context-Sensitive Rewriting by Rewriting. In *Proc. of ICALP'96*, LNCS 1099:122-133, Springer-Verlag, 1996.
8. M.J. O'Donnell. Equational Logic as a Programming Language. The MIT Press, 1985.
9. R.C. Sekar, S. Pawagi and I.V. Ramakrishnan. Transforming Strongly Sequential Rewrite Systems with Constructors for Efficient Parallel Execution. In *Proc. of RTA'89*, LNCS 355:404-418, 1989.
10. Y. Toyama, S. Smetsers, M.C.J.D. van Eekelen and R. Plasmeijer. The Functional Strategy and Transitive Term Rewriting Systems. *Term Graph Rewriting - Theory and Practice*, pages 117-129, John Wiley & sons, 1993.

# Optimal Encodings

Massimo Marchiori

*CWI*

*P.O. Box 94079, NL-1090 GB Amsterdam, The Netherlands*

max@cwi.nl

**Abstract**

The basic notion of encoding is one of the most important present in computer science. So far, they have not been per se the subject of serious research, because of their apparent simplicity. In this paper we show how the realm of encodings, instead, deserves big attention. In particular, we address the fundamental question of optimality of encodings: data need a certain storage cost, so it seems natural to investigate whether some encodings are better than others, in the sense they waste less space. We give a precise formalization of this analysis, within the context of extendable families of encodings, and show that the structure is so rich that no optimal encoding can be found, viz. one can arbitrarily improve the data packing. Secondly, we raise the subtle point of the effect of encodings on the computational power of the device: although so far this problem has been passed over, it is not obvious at all whether or not an encoding affects the computational power of a machine. The subtlety of the point is formally shown, by proving that for almost all the machines encodings behave nicely. However, things become deeply involved just in the most basic case of the 2-register machine, where only particular encodings are safe. The analysis then reveals than in this context not only there are optimal elements, but even a best one, which rather intriguingly is shown to be the first encoding system ever developed.

## 1 Introduction

Encodings play a major role in nowadays computer science. They constitute the basic concept on which basic areas like recursion theory, machine theory, protocol theory and logical theories rely. For instance, every machine, in order to interact with the external world, and to be compared with other machines and/or with the recursive functions, has necessarily to employ encoding (and decoding) functions that act as knowledge translators. Such a simple yet fundamental concept, however, has so far been quite neglected as a topic of study in its own, On the one hand, this is seemingly justified by the fact that its definition is so simple not to deserve much attention. On the other hand, the very few studies on the subject have shown no interesting results (for instance, [14] shows that all the encodings are in a sense equivalent). In this paper, we start devoting to encodings the attention that such fundamental objects deserve, by studying their optimality in connection with machines and protocols. Given a machine, are there "optimal" encodings for it, in the sense that they waste a minimal amount of resources? In order to answer this fundamental question, we start by considering general encoding families. This concept is refined by the practical need to have encoding families that are in a sense extendable (and, dually, backward compatible): these notions are formalized within the natural concept of a so-called encoding system. In this setting, we consider the cost of such a system as, roughly speaking, the cost of storing the translated data: the less this cost, the better the encoding system. We then turn to the analysis of optimality, not sticking to a fixed measure of data cost, but keeping the analysis quite general, imposing only the condition that in order to store bigger numbers one needs more resources than to store smaller numbers. The formal analysis reveals than, quite surprisingly, there is no optimal encoding system, that is to say one can arbitrarily improve the cost of the data translation. However, a point that does not have to be passed over is that when using encodings, we are actually changing the considered machine; this change is seemingly negligible, but are we sure that by employing an encoding we do not affect the computational power of the given machine? This notion is formalized by explicitly requiring an encoding system to be *safe*, i.e., not to diminish the computational power of a machine. This rather subtle point leads then to the big surprise. In a sense, the fact that so far no one considered this issue is shown to be justified, since for almost any machine, every encoding system is automatically safe. However, in the most basic case of the 2-register machine the situation reveals to be dramatically more complex: only encoding systems of a particular form do not affect the computational power of the machine. Even more surprisingly, the analysis of optimality reveals that in this context not only there are optimal encoding systems, but even a best one, which rather intriguingly is shown to be just the original encoding system $G$ developed by Gödel in order to prove his famous theorem on the incompleteness of formal theories.

## 2 Preliminaries

Given a function $f : A \to B$ and an element $b \in B$, $f^{-1}(b)$ denotes the set $\{a \in A : f(a) = b\}$. We will write $f(C)$ to denote the set $\{f(c) : c \in C\}$. As far as machines are concerned, we consider three (fixed) symbol sets: $\mathcal{F}$ (the functions), $\mathcal{P}$ (the predicates) and $\mathcal{L}$ (the labels). $\mathcal{L}$ is also required to be infinite, and to have a distinguished element named START. A machine assigns to every function symbol $f \in \mathcal{F}$ (resp. to every predicate symbol $p \in \mathcal{P}$) a partial function $M_f$ (resp. a partial predicate $M_p$) over the set $|M|$, which is said the *memory set* of the machine $M$. The strings of symbols of the form

i)    START: GOTO $L$;    iii) $L$: IF $p$ THEN GOTO $L'$ ELSE GOTO $L''$;
ii)   $L$: DO $f$ GOTO $L'$;    iv) $L$: HALT

with $f \in \mathcal{F}$, $p \in \mathcal{P}$ and $L, L' \in \mathcal{L}$, are said the *instructions*. A program $P$ is then a finite set of instructions that has exactly one instruction of type i), that is only one start instruction, and for every label $L$ at most one instruction beginning with that label.

Given a machine $M$, every program $P$ defines a partial function $M_P$ over $|M|$ defined in the obvious way, simply 'computing' the program $P$ starting from the (unique) instruction of type i) present in $P$, and using the function $M_f$ in case an instruction of type ii) is found, and the predicate $M_p$ in case a test instruction of type iii) is found. The computation ends when an halt instruction of type iv) is encountered. The formal definition is trivial but lengthy, so we omit it (see e.g. [5]). This way of defining machines is absolutely general, since it can represent all of the usual machines, like Turing ones, pushdown automata and so on (see [16, 5]). The most basic family of machines is that of the register machines: the *n-register machine* $SR_n$ has as memory set $n$ registers $R_1, \ldots, R_n$, each capable of holding a natural number (viz. $|SR_n| = \mathbb{N}^n$), instructions $R_i \leftarrow R_i - 1$ and $R_i \leftarrow R_i + 1$ and predicates $R_i = 0$? ($1 \le i \le n$), with the obvious meaning of decrementing/incrementing a register by one and testing if a register has a zero value.

For every program $P$, the function $M_P$ is defined over $|M|$: if we want to make the machine interact with the outer world (and/or to compare the functions computable with $M$ with other classes of functions having different domain and/or codomain), we need so called *encoding* and *decoding* functions: an encoding function is a computable function with codomain $|M|$, and a decoding function is a computable function with domain $|M|$. This way, we can compare some functions from $A$ to $B$ with the corresponding functions of a machine $M$ by means of an encoding function $e : A \to |M|$ and of a decoding function $d : |M| \to B$, taking for every program $P$ the function $d \circ M_P \circ e$.

For every machine $M$, if $e$ and $d$ are an encoding and decoding function for $M$, let us indicate with $\mathcal{F}(M, e, d)$ the set of functions $\{d \circ M_P \circ e \mid P \text{ is a program}\}$. Hence, saying that a machine $M$ has *full computational power* equals to say that $\forall k \in \mathbb{N}$ there are functions $e_k : \mathbb{N}^k \to |M|$, $d_k : |M| \to \mathbb{N}$ such that $\mathcal{F}(M, e_k, d_k) = REC_k$ (the symbol $REC_k$ stands for all the (partial) recursive functions from $\mathbb{N}^k$ to $\mathbb{N}$). The register machines $SR_0$ and $SR_1$ are readily too poor, but if $n \ge 2$ then $SR_n$ has full computational power. Hence the most 'economical' register machine having full computational power is $SR_2$. Incidentally, we note how the importance of the 2-register machine is not only purely theoretical, but even practical since, just to mention a few recent applications, it has been used to provide the link between many seemingly unrelated topics in computer science [10], and to give the simplest known proof of undecidability of termination for term rewriting systems [4],

## 3 Encoding Systems

While a single encoding formalizes the concept of a static and fixed situation, there is often the need to reason about encodings in a *dynamic* way, in situations where the input domain can change with time, being extended. The proper formalization of this concept is the following:

**Definition 3.1** An *encoding system* $f_{[]}$ for the machine $M$ is an assignment to each $k \in \mathbb{N}$ of an encoding $f_{[k]} : \mathbb{N}^k \to |M|$, such that $\forall k \in \mathbb{N}, \forall (x_1, \ldots, x_k) \in \mathbb{N}^k . f_{[k+1]}(x_1, \ldots, x_k, 0) = f_{[k]}(x_1, \ldots, x_k)$. □

Encoding systems formalize the concept of a family of encodings like recursive functions has been formalized using programming systems [13, 11, 15], and natural numbers has been formalized into the $\lambda$-calculus using numeral systems [2]. However, they also differ from such notions, that in general do not say anything about the relation among the various objects of the family, that can be in principle completely unrelated. Here the idea instead, as said, is that the initial input domain (of an $f_{[m]}$) can be *extended* into a bigger input domain (those of an $f_{[n]}$, with $n > m$): so, the above definition formalizes the concept that every encoding $f_{[k]} : \mathbb{N}^k \to \mathbb{N}$ encompasses all the previous encodings $f_{[i]}$ ($0 \le i < k$). This means that we can arbitrarily enlarge the input domain in a smooth way, i.e. without changing everything regarding the domain; more technically, we are requiring *backward compatibility*: if the input domain is extended with some optional parameters, the old algorithm for the given machine has only to be integrated to process the optional parameters when present, and not to be rewritten as a whole algorithm from the beginning, since the meaning of the input data *does not change*. These concepts have tight connections with the setting of data protocols, as we will see in the next section.

However, there is yet another point that should be kept into account: the *computational power*. Indeed, the minimum requirement for a machine is to have full computational power; however, when using an encoding function we are modifying the original machine: is this modification safe? That is to say, do we still have full computational power? The following notion formalizes the requirement that an encoding system should not diminish the full computational power of a machine:

**Definition 3.2** An encoding system $e_{[]}$ is *safe* for a machine $M$ if $\forall k \in \mathbb{N}, \exists d_k . \mathcal{F}(M, e_{[k]}, d_k) = REC_k$ □

## 4 Data Protocols

Abstractly, we consider a *(data) protocol* $P$ to be a computable map from the set $\mathcal{M}_P$ of *P-messages* to the set $\mathcal{M}$ of *messages* (usually some structure is required on a protocol, but this is unrelevant for our present purposes). Each generic agent $\mathcal{A}$ receives a message and produces an output message: we write $\mathcal{A}(m)$ to indicate the resulting output of the agent $\mathcal{A}$ once it has been provided with a message $m$.

**Definition 4.1** $\{P_i\}_{i<k}$ ($k$ an ordinal) is a *hierarchical family of protocols (with gap-power $\delta$)* if

- $\forall i < k$, $P_i$ is a protocol

- there are *extension operators* $\mathcal{E}_i$ ($i < k$) such that each $\mathcal{E}_i$ is an injective computable map from $\mathcal{M}_{P_i}$ to $\mathcal{M}_{P_{i+1}}$

- $\forall i < k.\ card(P_{i+1} \setminus \mathcal{E}_i(P_i)) \ge \delta$

- for every agent $\mathcal{A}$, and for every $i < k$ and $m \in \mathcal{M}_{P_i}$, $\mathcal{A}(P_i(m)) = \mathcal{A}(P_{i+1}(\mathcal{E}_i(m)))$ □

Here, the cardinality of a set $A$ is indicated by $card(A)$.

The concept of hierarchical family of protocols has tight relationships with hierarchical development in software engineering (cf. [17, 9]), and with layered systems as used in protocol theory (see [12]).

More important, it encompasses the concept of *data reusability* which is a fundamental topic in nowadays hardware/software development (see e.g. [3, 18, 1]).

The interesting fact is that, as it is easy to check, every encoding system $f_{[]}$ can naturally be seen as a hierarchical family of protocols (with gap-power $\aleph_0$) for a machine $M$, by letting $\mathcal{M} = \mathbb{N}$, $\mathcal{M}_{P_i} = \mathbb{N}^i$, $P_i = f_{[i]}$ and $\mathcal{E}_i = \epsilon_i$ (where $\epsilon_i : \mathbb{N}^i \to \mathbb{N}^{i+1}$ is the canonical embedding $(x_1, \ldots, x_i) \mapsto (x_1, \ldots, x_i, 0)$). Note that the last condition of Definition 4.1 is verified by definition of encoding system (thus, encoding systems incorporate in a sense the corresponding notions of *extendability* and *backward-compatibility* for data protocols).

## 5  Canonical Structure

An *encoding* for a machine $M$ is a computable injective mapping from $\mathbb{N}$ to $|M|$. That is to say, encodings are needed in order to translate data from the outer world ($\mathbb{N}$) into the machine data representation ($|M|$). Just like the outer world data domain can be set w.l.o.g. to a fixed general structure ($\mathbb{N}$), the codomain $|M|$, which is machine-dependent and therefore not suitable as such of mathematical analysis, can be abstracted into $\mathbb{N}$, by fixing an embedding $\epsilon_M$ (i.e. an injective function) from $\mathbb{N}$ to $|M|$. This way we can boil down the problem of studying an encoding from $\mathbb{N}$ to $|M|$ by studying only the encodings from $\mathbb{N}$ to $\mathbb{N}$ (which are also called numberings, see e.g. [14]), since each such encoding can be turned into an encoding for $M$ via composition with $\epsilon_M$. In the case of register machines, since $|SR_n| = \mathbb{N}^n$, we can simply use as $\epsilon_{SR_n}$ the natural embedding $\imath_n : \mathbb{N} \hookrightarrow \mathbb{N}^n$ given by $m \mapsto (m, 0, \ldots, 0)$. The nice fact is that this choice is not restrictive, since this natural embedding enjoys the following universal property:

**Theorem 5.1**  *Given an encoding system $e_{[]}$, for every $n$ and for every embedding $\epsilon_{SR_n}$, if the encoding system $e_{[]} \circ \epsilon_{SR_n}$ is safe then the encoding system $e_{[]} \circ \imath_n$ is safe as well.*

Henceforth, when talking about encodings we will thus mean encodings with codomain $\mathbb{N}$, that can be considered as encodings for the given machine by composing them with a fixed suitable embedding (as seen, $\imath_n$ in the case of the register machine $SR_n$).

## 6  Cost

Now that we have fixed a common codomain for encodings ($\mathbb{N}$), we can accordingly turn our attention to their comparison. In order to define whether or not an encoding is better than another encoding, we have first of all to define what is the *cost* of a certain (finite) set of natural numbers $D$, that is to say we need a *cost function* from the set of the finite sets of natural numbers (denoted by $\mathcal{P}_{fin}(\mathbb{N})$) to $\mathbb{N}$ (w.l.o.g. we use natural numbers to denote costs). For instance, we could use as cost function $\nu$ the function $\nu(D) = \sum_{x \in D} x$, or some other more involved measure.

In this paper, we do not stick to a particular cost function, but instead develop the analysis in full generality, requiring only that the chosen cost function $\nu$ is *increasing*: $\nu(A) \leq \nu(B), r < s, r \notin A, s \notin B \Rightarrow \nu(A \cup \{r\}) < \nu(B \cup \{s\})$ (that is to say, we only require that greater numbers have greater cost).

It is well known (see e.g. [6]) that from every (partial) order relation $\trianglelefteq$ one can recover the corresponding strict order $\lhd$, and vice versa. Hence, in the following we will arbitrarily use orders or strict orders.

Once given a cost function, if we have an encoding $f : \mathbb{N}^k \to \mathbb{N}$, the cost of encoding a certain set $A \subseteq \mathcal{P}_{fin}(\mathbb{N}^k)$ can be defined as $\nu(\{f(\bar{x}) : \bar{x} \in A\})$ or, briefly, $\nu(f(A))$. So, we can now compare two encodings with respect to their cost: we say that an encoding $f : \mathbb{N}^k \to \mathbb{N}$ is *no more expensive* (w.r.t. a cost function $\nu$) than an encoding $g : \mathbb{N}^k \to \mathbb{N}$ ($f \preceq g$) if $\forall A \in \mathcal{P}_{fin}(\mathbb{N}^k). \nu(f(A)) \leq \nu(g(A))$. Accordingly, we have to define the corresponding cost ordering for encoding systems. What is the minimal requirement we can impose on such a relation? If an encoding system $f_{[]}$ is 'less expensive' than $g_{[]}$, then there should be at least a $k$ such that $f_{[k]} \prec g_{[k]}$ (i.e. at least an encoding of $f_{[]}$ should be less expensive than the corresponding encoding of $g_{[]}$). On the other hand, the reverse should not be true, that is to say there should be no $k$ such that $g_{[k]} \prec f_{[k]}$. This is formalized into the following definition:

**Definition 6.1** An encoding system $f_{[]}$ is *less expensive* than an encoding system $g_{[]}$ (notation $f_{[]} \lhd g_{[]}$) if $(\exists k. f_{[k]} \prec g_{[k]}) \wedge (\neg \exists k. g_{[k]} \prec f_{[k]})$. $\square$

Now we turn our attention to the properties that these expensiveness relations satisfy: for instance, do they define an ordering? It is easy to see that the relation $\preceq$ defines an order on encodings. As far as encoding systems are concerned, again it can be proved that the relation $\lhd$ is a strict order. Thus, the following definition makes sense:

**Definition 6.2** We call *optimal* an encoding system which is minimal w.r.t. $\lhd$. If there is only one optimal encoding system (i.e., it is the minimum w.r.t. $\lhd$), we say it is the *best* encoding system. $\square$

## 7 Optimality and Computational Power

The big question of what form can have the optimal encodings is answered in the general case by the following main result:

**Theorem 7.1** *There is no optimal encoding system.*

So, the structure of encoding systems is so rich that we can continuously improve, via suitable "packings", every encoding system. However, as seen we should also take into account the safeness of the encoding system, caring not to diminish the computational power of the machine. We begin this analysis from the most basic case, the 2-register machine. Call a function $f : \mathbb{N}^k \to \mathbb{N}$

$I$-injective ($I = \{i_1, \ldots, i_m\}$ a subset of $\{1, \ldots, k\}$) if $(x_{i_1}, \ldots, x_{i_m})$ different from $(x'_{i_1}, \ldots, x'_{i_m})$ implies that $f(x_1, \ldots, x_{i_1}, \ldots, x_{i_m}, \ldots, x_k)$ is different from $f(x_1, \ldots, x'_{i_1}, \ldots, x'_{i_m}, \ldots, x_k)$. Via a not easy proof, we can provide the following complete characterization of the encoding systems safe for the 2-register machine.

**Theorem 7.2** *An encoding system $f_{[]}$ is safe for the 2-register machine if and only if for every $k \in \mathbb{N}$ $f_{[k]}(x_1, \ldots, x_k) = f_{[k]}(\bar{x})$ is of the form $q_1^{h_1(\bar{x})} \cdot q_2^{h_2(\bar{x})} \cdots q_{m_k}^{h_{m_k}(\bar{x})} \cdot a_k + b_k$, with $m_k \geq 1 \leq a_k$, the $q_i$ are coprimes, $b_k$ is an integer, the $h_i$ are computable functions and there is a mapping $\Im_k : [1, k] \to [1, m_k]$ such that $\forall j \in [1, m_k]$ $h_j$ is $\Im_k^{-1}(j)$-injective.*

Once established by the above important result the form of the safe encoding system, we can go on with analyzing optimality. Let us define the following encoding system:

**Definition 7.3** *The encoding system $G_{[]}$ is defined as $G_{[k]}(x_1, \ldots, x_k) = \pi_1^{x_1} \cdot \pi_2^{x_2} \cdots \pi_k^{x_k} - 1$ ($\pi_i$ indicates the $i$-th prime, i.e. $\pi_1 = 2$, $\pi_2 = 3$, $\pi_3 = 5$, etc.).* □

This encoding system is named $G_{[]}$ because it has been initially used by Gödel to prove his famous theorem on the incompleteness of formal theories, cf. [7] (actually, his original encoding was $G_{[]} + 1$, since he counted from 1 onwards, while we count from 0 onwards). It is easy to see that $G_{[]}$ is indeed an encoding system; but far more can be proved:

**Theorem 7.4** *The encoding system $G_{[]}$ is the best encoding system safe for the 2-register machine.*

Hence, $G_{[]}$ is less expensive than *every other* encoding system safe for $SR_2$.

This result is in a sense a big exception, revealing the fundamental character of $G_{[]}$, since when we consider the other cases we obtain:

**Theorem 7.5** *Every encoding system is safe for the $n$-register machine, when $n \geq 3$.*

Therefore, combining the above result with Theorem 7.1, we get right away:

**Corollary 7.6** *There are no optimal encoding systems for the $n$-register machine, when $n \geq 3$.*

So, the minimal requirement of safeness, but for the very basic case of the 2-register machine, is always satisfied by encoding systems, which may *a posteriori* explain why this notion was not so far explicitly considered.

A similar analysis can be performed, following the same headlines, for all the other usual machines like Turing machines, Post machines and so on (see e.g. [5] and [8]): analogously, it can be proved that for all these machines there are no optimal encoding systems.

Because of the aforementioned correspondence between encoding systems and hierarchical families of protocols, these results can be also be re-expressed in terms of abstract data protocols. When the hierarchical families of protocols are of bounded size (i.e. $\{P_i\}_{i \leq k}$ with $k$ finite), there are obviously ad-hoc optimal families. However, when it is not the case (i.e. the data protocol is required to be *infinitely extendable*), then Theorem 7.4 roughly says that the Gödel encoding system $G_{[]}$ can be considered as the best hierarchical family of protocols for the 2-register machine, and, accordingly, the other nonexistence results say that when other agents are considered, there is not such best/optimal family.

## References

[1] R. Balzer. Evolution as a new basis for reusability. In *Reusability in Programming*, pages 80–82. Newport, RI, 1983.

[2] H.P. Barendregt. *The Lambda Calculus: its Syntax and Semantics*. North–Holland, 1981.

[3] V.R. Basili and A.J. Turner. Iterative enhancement: A practical technique for software development. *IEEE Transactions on Software Engineering*, 4:390–396, December 1975.

[4] M. Bezem, J.W. Klop, and R.C de Vrijer. *Term Rewriting Systems*, volume 25 of *Cambridge Tracts in Theoretical Computer Science*. Cambridge University Press, 1997. To appear.

[5] K. Clark and D. Cowell. *Programs, machines, and computation*. McGraw-Hill, 1976.

[6] B.A. Davey and H.A. Priestley. *Introduction to Lattices and Order*. Cambridge University Press, 1990.

[7] K. Gödel. *On Formally Undecidable Propositions of Principia Mathematica and Related Systems*. Oliver&Boyd, Edinburgh and London, 1962.

[8] J.E. Hopcroft and I. Ullman. *Introduction to Automata Theory, Languages and Computation*. Addison-Wesley, 1979.

[9] I.V. Horebeek and J. Lewi. *Algebraic Specifications in Software Engineering*. Springer–Verlag, 1989.

[10] M. Kanovich. Petri nets, horn programs, linear logic, and vector games. In M. Hagiya and J.C. Mitchell, editors, *International Symposium on Theoretical Aspects of Computer Software*, volume 789 of *LNCS*, pages 642–666. Springer–Verlag, 1994.

[11] A.J. Kfoury, R.N. Moll, and M.A. Arbib. *A programming approach to computability*. Springer–Verlag, New York, 1982.

[12] S.S. Lam and A.U. Shankar. A composition theorem for layered systems. In B. Jonsson, J. Parrow, and B. Perhson, editors, *11th International Symposium on Protocol Specification*, pages 93–108. North-Holland, 1991.

[13] M. Machtey and P. Young. *An Introduction to the General Theory of Algorithms*. North–Holland, 1978.

[14] Yu.I. Manin. *A Course in Mathematical Logic*. Springer–Verlag, 1977.

[15] I.C.C. Phillips. Recursion theory. In S. Abramsky, Dov M. Gabbay, and T.S.E. Maibaum, editors, *Handbook of Logic in Computer Science*, volume 1, chapter 1, pages 79–188. Clarendon Press, Oxford, 1992.

[16] D. Scott. Some definitional suggestions for automata theory. *Journal of Computer and System Sciences*, 1:187–212, 1967.

[17] J.D Warnier. *Logical Constructions of Programs*. H.E. Stenfert Kroese, 1974.

[18] P. Wegner. Varietes of reusability. In *Reusability in Programming*, pages 30–44. Newport, RI, 1983.

# Monotonic Rewriting Automata with a Restart Operation*

František Mráz[1] and Martin Plátek[1] and Petr Jančar[2] and Jörg Vogel[3]

[1] Charles University, Department of Computer Science, Malostranské nám. 25, 118 00 PRAHA 1, Czech Republic, e-mail: mraz@ksvi.mff.cuni.cz, platek@ksi.mff.cuni.cz
[2] Univ. of Ostrava, Techn. Univ. of Ostrava, e-mail: jancar@osu.cz
[3] Friedrich Schiller University, Computer Science Institute, 07740 JENA, Germany, e-mail: vogel@informatik.uni-jena.de

**Abstract.** We introduce a hierarchy of monotonic rewriting automata with a restart operation and show that its deterministic version collapses in a sequence of characterizations of deterministic context-free languages. The nondeterministic version of it gives a proper hierarchy of classes of languages with the class of context-free languages on the top.

## 1  Introduction

We introduce rewriting automata with a restart operation ($RRWW$-automata) which generalize in two ways the restarting automata with rewriting ($RW$-automata) introduced in [2].

A $RW$-automaton can be roughly described as follows. It has a finite control unit, a head with a lookahead window attached to a list, and it works in certain cycles. In a cycle, it moves the head from left to right along the word on the list; according to its instructions, it can at some point rewrite the contents of its lookahead by a shorter string of input symbols and "restart" – i.e. reset the control unit to the initial state and place the head on the left end of the shortened word. The computation halts in an accepting or a rejecting state.

A $RRWW$-automaton can after the rewriting check the remaining part of the list, and in the rewriting instruction it can use also some noninput symbols.

As usual, we define some subclasses of the automata. Similarly as in [1] we study a natural property of monotonicity (during a monotonic computation, "the places of rewriting do not increase their distances from the right end"). We show that all introduced types of deterministic monotonic $RRWW$-automata recognize exactly deterministic context-free languages ($DCFL$). On the other side the nondeterministic version of the introduced hierarchy of $RRWW$-automata gives a proper hierarchy of classes of languages with the class of context-free languages ($CFL$) on the top.

Our motivation for introducing the restarting automata with rewriting in [3] was to model (elementary) syntactic analysis of natural languages in a similar way as in [5]. Such syntactic analysis consists in stepwise simplification of an

* Supported by the Grant Agency of the Czech Republic, Grant-No. 201/96/0195

extended sentence so that the (in)correctness of the sentence is not affected. Thus after some number of steps a simple sentence is got or an error is found. Such computations can be done by so called *RRW*-automata, which are the *RRWW*-automata using only the input symbols in rewriting. Formally is this property expressed by the so called "error preserving property". The *RRWW*-automata using non input symbols do not ensure this error preserving property automatically, but on the other side they are able to recognize all *CFL*.

## 2 Definitions and basic properties

We present the definitions informally; the formal technical details could be added in a standard way of the automata theory.

A *RRWW-automaton M* (with bounded lookahead) is a device with a finite state control unit and one head moving on a finite linear (doubly linked) list of items (cells). The first item always contains a special symbol $\mathbb{c}$, the last one another special symbol $, and each other item contains a symbol from a finite alphabet (not containing $\mathbb{c}$, $). The head has a lookahead "window" of length $k$ (for some $k > 0$) – besides the current item, $M$ also scans the next $k$ right neighbour items (or simply the end of the word when the distance to $ is less than $k$). In the *initial configuration*, the control unit is in a fixed, initial, state and the head is attached to the item with the left sentinel $\mathbb{c}$ (scanning also the first $k$ symbols of the input word).

The *computation* of $M$ is controlled by a finite set of *instructions* of the following three types:

(1) $(q, au) \to (q', MVR)$
(2) $(q, au) \to (q', REWRITE(v))$
(3) $(q, au) \to RESTART$

The left-hand side of an instruction determines when it is applicable – $q$ means the current state (of the control unit), $a$ the symbol being scanned by the head, and $u$ means the contents of the lookahead window ($u$ being a string of length $k$ or less if it ends with $). The right-hand side describes the activity to be performed.

In case (1), $M$ changes the current state to $q'$ and moves the head to the right neighbour item of the item containing $a$.

In case (2), the activity consists of deleting (removing) some items of the just scanned part of the list (containing $au$), and of rewriting some of the nondeleted scanned items (in other words $au$ is replaced with $v$, where $v$ must be shorter than $au$). After that, the head of $M$ is moved right to the item containing the first symbol after the lookahead and the current state of $M$ is changed to $q'$. There is one exception: if $au$ ends by $ then $v$ also ends by $ (the right sentinel cannot be deleted or rewritten) and after the rewriting the head is moved to the item containing $.

In case (3), *RESTART* means entering the initial state and placing the head on the first item of the list (containing $\mathbb{c}$).

We will suppose that the control unit states are divided into two groups – the *nonhalting states* (an instruction is always applicable when the unit is in such a state) and the *halting states* (any computation finishes by entering such a state); the halting states are further divided into the *accepting states* and the *rejecting states*.

Any computation of an *RRWW*-automaton is naturally divided into certain phases or *cycles* by performed *RESTART*-instructions: in one cycle, the head moves right along the input list (with a bounded lookahead) until a halting state is entered or the computation is resumed in a initial configuration (thus a new cycle starts). We demand that the automaton makes exactly one *REWRITE*-instruction in each cycle ending by *RESTART* – i.e. new cycle starts on a shortened word.

It immediately implies that any computation of any *RRWW*-automaton is finite (finishing in a halting state).

In general, a *RRWW*-automaton is *nondeterministic*, i.e. there can be two or more instructions with the same left-hand side $(q, au)$. If it is not the case, the automaton is *deterministic*.

An input *word $w$ is accepted by $M$* if there is a computation which starts in the initial configuration with $w$ (bounded by sentinels ¢,\$) on the list and finishes in an *accepting configuration* where the control unit is in one of the accepting states. $L(M)$ denotes the language consisting of all words accepted by $M$; we say that *$M$ recognizes the language $L(M)$*.

By *RRW*-automata we mean *RRWW*-automata which use only input symbols in rewriting instructions (i.e. in instructions of the form (2) above, the string $v$ contains symbols from the input alphabet only).

By *RR*-automata we mean *RRW*-automata for which rewriting can be replaced by deleting (i.e. in instructions of the form (2) above, the string $v$ can be obtained by deleting some symbols from $au$).

By *R*-automata we mean *RR*-automata which do restart immediately after any *REWRITE*-instruction.

The next obvious claim express the mentioned lucidness of computations of *RRW*-automata. The notation $u \Rightarrow_M v$ means that there exists a cycle of $M$ starting in the initial configuration with the word $u$ and finishing in the initial configuration with the word $v$; the relation $\Rightarrow_M^*$ is the reflexive and transitive closure of $\Rightarrow_M$.

**Claim 1 The error preserving property.** *Let $M$ be an RRW-automaton, $u$, $v$ some words in the alphabet of $M$. If $u \Rightarrow_M^* v$ and $v \in L(M)$, then $u \in L(M)$.*

In general *RRWW*-automata do not fulfill the error preserving property, because of using symbols not belonging to the input alphabet during their computations.

Next we introduce the monotonicity property of computations of *RRWW*-automata. Let for any cycle $Cyc$, in which a *REWRITE*-instruction is performed, $Dist(Cyc)$ denote the distance of the last item in the lookahead window at the place of rewriting from the right sentinel in the current list. We say that a computation $C$ of an $RRWW$-automaton $M$ is monotonic if for the sequence of its cycles $Cyc_1, Cyc_2, ..., Cyc_n$ the sequence $Dist(Cyc_1), Dist(Cyc_2), ..., Dist(Cyc_n)$ is monotonic (not increasing).

By a *monotonic RRWW-automaton* we mean a $RRWW$-automaton for which all its computations are monotonic.

For brevity, prefix *det-* denotes the deterministic versions of $RRWW$-automata similarly *mon-* the monotonic versions. $\mathcal{L}(A)$, where $A$ is some class of automata, denotes the class of languages recognizable by automata from $A$. E.g. the class of languages recognizable by deterministic monotonic $R$-automata is denoted by $\mathcal{L}(det\text{-}mon\text{-}R)$.

**Theorem 2.** *There is an algorithm which for any $RRWW$-automaton $M$ decides whether $M$ is monotonic or not.*

**Proof:** The proof can be made in a quite similar way as for the $RW$-automata in [3].

$\square$

## 3   Monotonic $RRWW$-automata

For technical reasons we introduce a special form of $RRWW$-automata. We say that a $RRWW$-automaton $M$ is in a *det-MVR-form* if for any couple $(q, v)$, where $q$ is a state of $M$ and $v$ is a string of its input symbols, there is at most one instruction of the form $(q, v) \rightarrow (q', MVR)$.

The following special form statement holds:

**Lemma 3.** *For any mon-RRWW-automaton $M$ an equivalent mon-RRWW-automaton $M'$ in the det-MVR-form can be constructed.*

The proof can be done in a similar way as the construction of equivalent deterministic finite automaton for a given nondeterministic one.

The following theorem states the main result of this paper. Any (deterministic) monotonic $RRWW$-automaton recognizes a (deterministic) context-free language.

**Theorem 4.** $\mathcal{L}(mon\text{-}RRWW) \subseteq CFL$ *and* $\mathcal{L}(det\text{-}mon\text{-}RRWW) \subseteq DCFL$.

**Proof:** Let $L$ be a language recognized by a *mon-RRWW*-automaton $M$, with a lookahead of length $k$. W.l.o.g. we suppose that $M$ in each cycle scans the whole input list until the right sentinel. Scanning the right sentinel \$, $M$ either restarts or accepts or rejects. Moreover we will suppose $M$ being in the *det-MVR*-form, i.e. in each configuration at most one *MVR*-instruction is applicable (and any number of *REWRITE*-instructions).

Each cycle by $M$ containing *REWRITE*-instruction can be naturally divided into three parts: the left part – steps until the *REWRITE*-instruction, the middle part – the *REWRITE*-instruction itself and the right part – steps after the

*REWRITE*-instruction. Cycles without any *REWRITE*-instruction are considered as having the left part only (such cycles end by halting). The particular parts of a cycle determine the corresponding left, middle, and the right part of the corresponding list. The central part begins with the item scanned by the *REWRITE*-instruction and ends by the item scanned after the performance of the *REWRITE*-instruction.

We show how to construct a (nondeterministic) pushdown automaton $P$ which simulates $M$ and accepts $L(M).\{\$\}$. Thus $L(M) = L(P)/\{\$\}$, where $/$ denotes the right quotient. Because the class *CFL* is closed on the right quotient with any regular language, thus $L(M)$ is a context-free language. The construction of $P$ is based on the construction of a pushdown automaton equivalent to a monotonic restarting automaton with rewriting in [3]. Modification must be made for simulating also the right parts of cycles. Actually, if $M$ rewrites in a cycle $C_1$ in which it enters a halting state (at the right sentinel) and $w_1$ is the contents of its working list at the end of the cycle $C_1$ then $P$ will simulate also computation of $M$ which starts with $w_1$ on its list. $P$ will simulate further cycles of $M$ – let $C_2$ denote the next cycle in which $M$ enters a halting state. If $M$ rewrites in a cycle $C_2$ and $w_2$ is the contents of the list at the end of $C_2$ then $P$ will simulate also computation of $M$ which starts with $w_2$ on the list and so on until $M$ enters a halting state in a cycle with the left part only (a cycle without rewriting). At the end of simulation $P$ must decide which was the first halting state entered by $M$.

At any time, $P$ simulates left or middle part of a cycle denoted by *Cycle* and right parts of all cycles preceding *Cycle*. This is enabled by the monotonicity property of $M$ – the places of *REWRITE* do not increase their distances from the right sentinel. Simultaneously, in the pushdown store of $P$ an auxiliary information will be kept for simulation of (left parts of) cycles following after *Cycle*.

The control unit of $P$ has several components for storing finite information. The component $CSt$ contains the current state of $M$ in the corresponding step of *Cycle*. The component $B$ contains a string of input symbols of length at most $(2k + 1)$ – it will contain the scanned item and the lookahead of $M$ in *Cycle* and will be used in *REWRITE*-instruction simulation. The next component $R$ contains a string of at most $k + 1$ input symbols – it will contain the scanned symbol and lookahead for simulation of the right parts of cycles preceding *Cycle*. Let $n$ denote the number of states of $M$. The last component is a vector $RS$ of length $n$. The element $RS_i$, for $1 \leq i \leq n$, can be empty or can contain a state of $M$ – this vector will be used in simulation of the right parts of cycles preceding *Cycle*.

The pushdown store of $M$ will contain symbols already scanned by $M$ in *Cycle* with some auxiliary information. Actually the symbols stored in the pushdown are composed of the input symbol from the list and a state in which the symbol was entered (from the left) in *Cycle*. Because of the *det-MVR*-form of $M$, this information can be used in simulation of left parts of the cycles following after *Cycle*.

Any reading of input symbol by $P$ is done using the following procedure:

*Get_input_symbol:* If $R=\$$ then do nothing (we are at the end of the list). Otherwise for each $RS_i$ different from $\lambda$ $(i = 1, 2, \ldots, n)$ simulate one step in the right part of some cycle preceding *Cycle* – when the head of $M$ scans the first symbol of $R$ and its lookahead window contains the rest of $R$ and the control unit is in the state $RS_i$. Because of the *det-MVR*-form there is exactly one instruction of the form $(RS_i, R) \to (q_i, MVR)$ (for $i = 1, 2, \ldots, n$). $P$ replaces the contents of each $RS_i$ by $q_i$. If $RS_i = RS_j \neq \lambda$, for some $1 \leq i < j \leq n$, it means that in right parts of some two cycles (preceding *Cycle*) $M$ enters the same state $q_i = q_j$ at the same symbol (corresponding to the second symbol in $R$). But then due to *det-MVR*-form of $M$, these two cycles would continue by the same steps and both end in the same state. Accepting/rejecting depends on the first halting state entered by $M$ at the right sentinel, thus we need not simulate both cycles, it is enough to continue in simulation of the former cycle – we discard $RS_j$ by shifting the contents of $RS_{j+1}, \ldots, RS_n$ to $RS_j, \ldots, RS_{n-1}$ and entering $\lambda$ into $RS_n$.
Next $P$ removes the first symbol from $R$ and appends it to $B$, shifts the contents of $R$ to the left (the contents of $R$ is shortened) if the last symbol of $R$ is not $\$$ then $P$ reads next input symbol and appends it to $R$.

This procedure simulates stepwise the right parts of cycles preceding *Cycle*. The following *RS-invariant* is kept:

If $RS_i$ contains a state $q$ of $M$ then all $RS_1, \ldots, RS_{i-1}$ contain states of $M$. Then $q$ is the state in which $M$ scans the first symbol $x$ stored in $R$ in the right part of some cycle $C$ preceding the current cycle *Cycle* such, that there is no cycle preceding $C$ in which this item was scanned in the same state $q$ and for any $j$, $1 \leq j < i$, there is a cycle preceding $C$ in which $x$ is scanned in the state $RS_j$.

**The simulation algorithm of $P$:**

*Initialization:* $P$ starts by storing the initial state of $M$ into $CSt$, pushing $\phi$ (the left endmarker of $M$) into the first cell of the buffer $B$ and the first $k$ symbols of the input word of $M$ into the next $k$ cells of the buffer $R$ (cells $2, 3, \ldots, k + 1$). All $RS_i$, for $i = 1, 2, \ldots, n$, are initialized to $\lambda$. $P$ initializes $B$ by performing the procedure *Get_input_symbol* $(k + 1)$-times.

*The main cycle:* During the simulation, the following conditions will hold invariantly:
  - $CSt$ contains the state of $M$ in which $M$ can be visiting the simulated (currently scanned) item in the simulated cycle *Cycle*,
  - the first cell of $B$ contains the current symbol of $M$ (scanned by the head in *Cycle*) and the rest of $B$ concatenated with $R$ contains $m$ right neighbour symbols of the current one (lookahead of length $m \leq 3k + 1$),
  - the pushdown contains the left-hand side (w.r.t. the head) of the list in *Cycle*, the leftmost symbol ($\phi$) being at the bottom. In fact, any pushdown

symbol will be composed – it will contain the relevant symbol of the working list and the state of $M$ in which this symbol could be entered in *Cycle*,
- for $RS$ the $RS$-invariant holds.

The mentioned invariants will be preserved by the following simulation of instructions of $M$. The left-hand side $(q, au)$ of the instruction to be simulated is determined by the information stored in the control unit. The activity to be performed depends on the right-hand sides of applicable instructions of $M$:

1. At most one possible instruction of the form $(q, au) \rightarrow (q', MVR)$:
   $P$ puts the contents of the first cell of $B$ and $CSt$ as a composed symbol on the top of the pushdown, stores $q'$ into $CSt$, and shifts the contents of $B$ to the left. If the length of $B$ is less than $k + 1$ then $P$ executes *Get_input_symbol*.

2. An instruction of the form $(q, au) \rightarrow (q', REWRITE(v))$:
   The first $|au|$ symbols from $B$ are replaced by the shorter sequence $v$. The topmost $k + 1$ symbols are successively popped from the pushdown and the relevant symbols are added from the left to $B$ (shifting the rest to the right). The state parts of (composed) symbols are forgotten, the state part of the $(k + 1)$-th symbol (the leftmost symbol in $B$) is stored in $CSt$. Thus not only the $REWRITE(v)$-instruction is simulated but also the beginning of the left part of the next cycle, which is the new *Cycle*.
   It should be clear that because of monotonicity of $M$, at the time of simulating rewriting the first symbol after the lookahead $u$ (or \$) is the first symbol of $R$ and the simulation of the corresponding right part will start on it in the state $q'$. Let $l = \max\{i \mid 1 \le i \le s : RS_i \neq \lambda\}$. If there is no $RS_i$ ($1 \le i \le l$) equal to $q'$, then $P$ stores $q'$ in $RS_{l+1}$.

*End of the simulation:* At this point $B$ is empty, i.e. the head of $M$ is scanning the right sentinel \$ in the current cycle *Cycle* and the vector $RS$ contains states in which ended right parts in all the cycles preceding *Cycle*. At this point $P$ using $RS$ decides which was the first halting state state entered by $M$. Let $i$ is minimal such that $RS_i$ contains a halting state. If $RS_i$ is accepting, then $P$ accepts, otherwise rejects.

It should be clear that due to monotonicity of $M$ the second half of $B$ (cells $k + 2, k + 3, \ldots, 2k + 1$) is empty at the time of simulating a $RESTART(v)$-operation. Hence the described construction is correct which proves the inclusion $\mathcal{L}(mon\text{-}RRWW) \subseteq CFL$.

Obviously, deterministic *mon-RRWW*-automaton is in *det-MVR*-form and the above construction applied to a *det-mon-RRWW*-automaton yields a deterministic pushdown automaton recognizing $L(P) = L(M).\{\$\}$. Because $DCFL$ is closed under quotient with any regular language, the language $L(M) = L(P)/\{\$\}$ is a deterministic context-free language – this proves the second part of the statement. □

The following lemma was proved in [1].

**Lemma 5.** $DCFL \subseteq \mathcal{L}(det\text{-}mon\text{-}R)$

From Lemma 5 and Theorem 4 follows that all deterministic monotonic subclasses of $\mathcal{L}(RRWW)$ collapse to $DCFL$.

**Theorem 6.** $\mathcal{L}(det\text{-}mon\text{-}R) = \mathcal{L}(det\text{-}mon\text{-}RR) = \mathcal{L}(det\text{-}mon\text{-}RRW) = \mathcal{L}(det\text{-}mon\text{-}RRWW) = DCFL.$

Despite the previous theorem we are able to prove that the corresponding nondeterministic classes of languages create a proper hierarchy inside $CFL$. Due to the lack of space we omit the proofs of the next theorems in this draft. The full proofs can be found in [4].

**Theorem 7.** $\mathcal{L}(mon\text{-}RRWW) = CFL$

**Theorem 8.** $\mathcal{L}(mon\text{-}R) \subset \mathcal{L}(mon\text{-}RR) \subset \mathcal{L}(mon\text{-}RRW) \subset \mathcal{L}(mon\text{-}RRWW).$
($\subset$ *denotes the proper subset relation*).

**Conclusion remark:** In future work we will consider also nonmonotonic classes of $RRWW$-automata. We will also consider the $RRWW$-automata in an explicit way as reduction systems (see [3]).

# References

1. P. Jančar, F. Mráz, M. Plátek, J. Vogel: *Restarting Automata;*, in Proc. FCT'95, Dresden, Germany, August 1995, LNCS 965, Springer Verlag, 1995, pp. 283–292
2. P. Jančar, F. Mráz, M. Plátek, J. Vogel: *Restarting Automata with Rewriting;*, in Proc. SOFSEM'96: Theory and practice of informatics, Milovy, Czech Republic, November 1996, LNCS 1175, Springer Verlag, 1996, pp. 401–408
3. P. Jančar, F. Mráz, M. Plátek, J. Vogel: *On Restarting Automata with Rewriting*, in New Trends in Formal Languages (Control, Cooperation and Combinatorics), Eds. G. Paun, A. Salomaa, LNCS 1218, Springer Verlag, 1997, pp. 119–136
4. F. Mráz, M. Plátek, P. Jančar, J. Vogel: *On Monotonic Rewriting Automata with a Restart Operation*, TR 97/6, Department of Computer Science, Charles University, 1997
5. M.Novotný: *S algebrou od jazyka ke gramatice a zpět*, Academia, Praha, 1988, (*in Czech*)

# Kahn's Fixed-Point Characterization for Linear Dynamic Networks

Shan-Hwei Nienhuys-Cheng    Arie de Bruin

cheng@cs.few.eur.nl    arie@cs.few.eur.nl

Dept of Comp. Science, Erasmus University Rotterdam
P.O. Box 1738, 3000DR, Rotterdam, the Netherlands

**Abstract.** We consider dynamic Kahn-like dataflow networks defined by a simple language **L** containing the fork-statement. The first part of the Kahn principle states that such networks are deterministic on the I/O level : for each network, different executions provided with the same input deliver the same output. The second part of the principle states that the function from input streams to output streams (which is now defined because of the first part) can be obtained as a fixed point of a suitable operator derived from the network specification. The first part has been proven by us in [BN96, BN97]. To prove the second part, we will use the metric framework. We introduce a nondeterministic transition system **NT** from which we derive an operational semantics $O_n$. We also define a deterministic transition system **DT** and prove that the operational semantics $O_d$ derived from **DT** is the same as $O_n$. Finally, we define a denotational semantics $D$ and prove $D = O_d$. This implies $O_n = D$. Thus the second part of the Kahn principle is established.

## 1 Introduction

A dataflow network consists of a number of parallel processes which are interconnected by directed channels. Processes communicate with each other only through these channels, they share no variables. The channels act as possibly infinite FIFO queues. The Kahn principle concerns networks with deterministic nodes: each process selects the channel it uses next in a deterministic way. However, which process gets the chance to execute a step at a certain moment depends on the underlying transition system. Such a system can be deterministic or nondeterministic. In a nondeterministic system, which process will execute next is not pre-determined. Thus different executions (computations) are possible.

The Kahn principle [K74] contains two parts. Part 1: The nondeterminism caused by the asynchronicity of the computing processes does not lead to global nondeterminism in the history level I/O-behaviour of the network. This means that given an initial state of the variables in the processes and a set of input streams, the set of output streams is uniquely determined, independent of the computations. This function can be used to define the operational semantics for the program which has induced these processes. Part 2: A meaning can be given to the language (denotational semantics) by defining it as a special solution of a set of equations and furthermore, the function defined in part 1 is in fact this solution. In other words, the operational semantics is equal to the denotational semantics.

Here we investigate a simple dynamic language **L** in which it is possible to dynamically create linear arrays of processes, not unlike a unix pipeline which can be built up using the unix primitive 'pipe' and 'fork'. In [BN96, BN97], we have defined a nondeterministic transition system **NT** for **L** describing all possible interleavings and proved that all computations either give the same maximal output (when the computation is fair), or a prefix of it. Thus we have established the first part of the Kahn principle for linear dynamic networks.

In this article, we establish the second part of the Kahn principle. Due to lack of space, proofs are omitted here (cf. [BN96, BN97, NB97, BV96]). For related work, we refer to [BBB93, BV96, C72, A81, LS89].

## 2 Operational Semantics in NT

We first introduce some statements in **L** informally. For a given input stream and an initial state, the execution of a program in **L** produces an output stream. Initially the program is executed by one process, using precisely one input and one output channel. Execution of the statement $read(x)$ will fetch the next value from the input channel and assign it to the variable $x$. Execution of the statement $write(e)$ will evaluate the expression $e$ and write the resulting value to the output channel. A process can split up into two new, nearly identical subprocesses, a mother process and a daughter process. This effect is achieved by a statement of the form $fork(w)$. The original input channel becomes the mother's input channel and likewise the original output channel becomes the daughter's output channel. Between mother and daughter a new channel is created, which is the mother's output and the daughter's input channel. This channel will originally be empty. Both processes proceed by executing the statement following $fork(w)$. We distinguish between mother and daughter by giving the variable $w$ in the mother process the value 0, in the daughter the value 1.

### 2.1 Preliminaries

- Let $(v \in)$ **Var**, $(\alpha, \beta, \gamma \in)$ **Val**, $(e \in)$ **Exp** and $(b \in)$ **Bexp** be given sets. They are called the set of *variables*, the set of *values*, the set of *expressions* and the set of *boolean expressions*, respectively. A *state* is a function $\sigma :$ **Var** $\to$ **Val**. Let **State** be the set of all states. The notation $\sigma\{\beta/v\}$ is used to denote a state equal to $\sigma$, except that now $\sigma(v) = \beta$. Two functions $V :$ **Exp** $\to$ (**State** $\to$ **Val**) and $B :$ **Bexp** $\to$ (**State** $\to \{true, false\}$) are assumed to be available.
- Let **L** be the language which contains the following statements:
  $s ::= v := e \mid skip \mid write(e) \mid read(v) \mid fork(v) \mid s; s \mid$
  **if** $b$ **then** $s$ **else** $s$ **fi** $\mid$ **while** $b$ **do** $s$ **od**
- Consider a value $\tau \notin$ **Val**. Let **V** $=$ **Val** $\cup \{\tau\}$. Let $(\eta, \zeta, \xi \in)$ **V**$^\infty$ be the set of finite or infinite sequences of elements from **V**. Such a sequence is called

a *stream*. Let $\epsilon$ be the *empty stream*. For two streams $\xi$ and $\zeta$, let $\xi \cdot \zeta (= \xi\zeta)$ be their concatenation. If $\zeta = \alpha\eta, \alpha \in \mathbf{V}$, then $first(\zeta) = \alpha$ and $rest(\zeta) = \eta$.

- A process is a function $P : \mathbf{State} \to (\mathbf{V}^\infty \to \mathbf{V}^\infty)$. The set of all processes is **Proc**
- Let $E$ denote termination. A resumption is recursively defined by $r ::= E \,|\, s : r$, where $s$ in **L**. Let **Res** be the set of all resumptions.

## 2.2 Configurations

We will use 'configurations' to describe snapshots of a network at work.

$a :\xleftarrow{\epsilon} \boxed{read(y) : E, \{v = 1\}} \xleftarrow{\epsilon} \boxed{write(x) : E, \{x = 3, v = 0\}} \xleftarrow{4 \cdot 5}$

$b :\xleftarrow{\epsilon} \boxed{read(y) : E, \{v = 1\}} \xleftarrow{3} \boxed{E, \{x = 3, v = 0\}} \xleftarrow{4 \cdot 5}$

$b' :\xleftarrow{\epsilon} \boxed{read(y) : E, \{v = 1\}} \xleftarrow{\epsilon} \boxed{write(x) : E, \{x = 3, v = 0\}} \xleftarrow{4} \boxed{0^{th}\text{process}} \xleftarrow{5}$

For example, consider the situation shown in item $a$ above. A box represents a subprocess and it contains information about the state and the program still to be executed. They are connected by arrows representing I/O channels. We allow 3 to be written without bothering whether it is needed by the next subprocess (item $b$ above). To describe such situations by configurations, we need a buffer stream containing what is written and not yet used in the input channel of a subprocess. We extend this model slightly in the following way. For consistency we also introduce a buffer stream for the first process from the right. Initially this stream equals $\epsilon$. We can view this situation as if there is still another subprocess (the 0-th subprocess) which contains hidden *read* + *write* statements. All it does, is to fetch the next element from the global input stream and to write it into the buffer of the first subprocess (see item $b'$ above). The configuration we use to describe item $a$ is given below in $\underline{a}$. (Notice that the buffer stream of the first subprocess is given as $\epsilon$.) We can consider a transition from $\underline{a}$ to $\underline{b}$ caused by the 1st subprocess and a transition to $\underline{b'}$ caused by the 0-th subprocess.

$\underline{a} : \langle read(y) : E, \{v = 1\}, \epsilon, \langle write(x) : E, \{x = 3, v = 0\}, \epsilon, 4 \cdot 5\rangle\rangle$

$\underline{b} : \langle read(y) : E, \{v = 1\}, 3, \langle E, \{x = 3, v = 0\}, \epsilon, 4 \cdot 5\rangle\rangle$

$\underline{b'} : \langle read(y) : E, \{v = 1\}, \epsilon, \langle write(x) : E, \{x = 3, v = 0\}, 4, 5\rangle\rangle$

**Definition 1.** A configuration $\rho$ is recursively defined as $\rho ::= \zeta | \langle r, \sigma, \eta, \rho'\rangle$, where $r$ is a resumption, $\zeta \in \mathbf{Val}^\infty, \eta \in \mathbf{V}^\infty, \sigma \in \mathbf{State}$ and $\rho'$ is a configuration. Let **Config** be the set of all configurations. A configuration of nesting number $n > 0$ has the form $\langle r_n, \sigma_n, \eta_n, \langle r_{n-1}, \sigma_{n-1}, \eta_{n-1}, \langle \ldots \langle r_1, \sigma_1, \eta_1, \zeta\rangle \ldots\rangle$. A stream $\zeta$ has nesting number 0.

## 2.3 A nondeterministic transition system NT for L

A transition system is a relation $\mapsto \subseteq$ **Config** $\times$ **Config**. If $(\rho, \rho') \in \mapsto$, then it is called a *transition* from $\rho$ to $\rho'$, denoted by $\rho \longrightarrow \rho'$. A configuration $\rho$ is called a *terminal* if there is no $\rho'$ such that $(\rho, \rho') \in \mapsto$. This is encoded by $\rho \longrightarrow \otimes$. For convenience we call it an *empty transition*. A transition can be accompanied by a value $\alpha \in \mathbf{V}$, in which case we write $\rho \xrightarrow{\alpha} \rho'$. We say $\alpha$ is the output of the transition. We define this relation by induction on the nesting number of configurations. The following transition system is based on the principle that every subprocess is a candidate to take the next step. Thus different executions are possible.

1. $\epsilon \longrightarrow \otimes$
   If $\rho \longrightarrow \otimes$, then $\langle E, \sigma, \eta, \rho \rangle \longrightarrow \otimes$
   If $\rho \longrightarrow \otimes$, then $\langle read(v) : r, \sigma, \epsilon, \rho \rangle \longrightarrow \otimes$

2. $\beta\zeta \xrightarrow{\beta} \zeta$

3. If $\rho \xrightarrow{\alpha} \rho'$, then $\langle r, \sigma, \eta, \rho \rangle \longrightarrow \langle r, \sigma, \eta\alpha, \rho' \rangle$
   If $\rho \longrightarrow \rho'$, then $\langle r, \sigma, \eta, \rho \rangle \longrightarrow \langle r, \sigma, \eta, \rho' \rangle$

4. $\langle (v := e) : r, \sigma, \eta, \rho \rangle \xrightarrow{\tau} \langle r, \sigma\{\beta/v\}, \eta, \rho \rangle$, where $\beta = V(e)(\sigma)$

5. $\langle skip : r, \sigma, \eta, \rho \rangle \xrightarrow{\tau} \langle r, \sigma, \eta, \rho \rangle$

6. $\langle write(e) : r, \sigma, \eta, \rho \rangle \xrightarrow{\beta} \langle r, \sigma, \eta, \rho \rangle$, where $\beta = V(e)(\sigma)$

7. For $\eta = \beta\eta', \beta \neq \tau$, $\langle read(v) : r, \sigma, \eta, \rho \rangle \xrightarrow{\tau} \langle r, \sigma\{\beta/v\}, \eta', \rho \rangle$
   For $\eta = \tau\eta'$, $\langle read(v) : r, \sigma, \eta, \rho \rangle \xrightarrow{\tau} \langle read(v) : r, \sigma, \eta', \rho \rangle$

8. $\langle fork(v) : r, \sigma, \eta, \rho \rangle \xrightarrow{\tau} \langle r, \sigma\{1/v\}, \epsilon, \langle r, \sigma\{0/v\}, \eta, \rho \rangle \rangle$

9. $\langle (s_1; s_2) : r, \sigma, \eta, \rho \rangle \xrightarrow{\tau} \langle s_1 : (s_2 : r), \sigma, \eta, \rho \rangle$

10. If $B(b)(\sigma)$, then $\langle \textbf{if } b \textbf{ then } s_1 \textbf{ else } s_2 \textbf{ fi} : r, \sigma, \eta, \rho \rangle \xrightarrow{\tau} \langle s_1 : r, \sigma, \eta, \rho \rangle$
    If $\neg B(b)(\sigma)$, then $\langle \textbf{if } b \textbf{ then } s_1 \textbf{ else } s_2 \textbf{ fi} : r, \sigma, \eta, \rho \rangle \xrightarrow{\tau} \langle s_2 : r, \sigma, \eta, \rho \rangle$

11. $\langle \textbf{while } b \textbf{ do } s \textbf{ od} : r, \sigma, \eta, \rho \rangle \xrightarrow{\tau}$
    $\langle \textbf{if } b \textbf{ then } s; \textbf{ while } b \textbf{ do } s \textbf{ od else } skip \textbf{ fi} : r, \sigma, \eta, \rho \rangle$

*Example 1.* The following is a possible transition sequence in **NT**.
$\rho = \rho_0 = \langle read(y) : (fork(v)) : E), \{\}, \alpha\beta, \gamma \rangle \longrightarrow$
$\rho_1 = \langle read(y) : (fork(v) : E), \{\}, \alpha\beta\gamma, \epsilon \rangle \xrightarrow{\tau}$
$\rho_2 = \langle fork(v) : E, \{y = \alpha\}, \beta\gamma, \epsilon \rangle \xrightarrow{\tau}$
$\rho_3 = \langle E, \{y = \alpha, v = 1\}, \epsilon, \langle E, \{y = \alpha, v = 0\}, \beta\gamma, \epsilon \rangle \rangle \; (\longrightarrow \otimes)$

## 2.4 Enabledness and computations

Let $\rho = \langle r_m, \sigma_m, \eta_m, \ldots, \langle r_i, \sigma_i, \eta_i, \langle \ldots \langle r_1, \sigma_1, \eta_1, \zeta \rangle \ldots \rangle \ldots \rangle$. The *input stream* $\zeta$ is defined as the *0-th subconfiguration*. For $1 \leq i \leq m$, the *i-th subconfiguration* is $\langle r_i, \sigma_i, \eta_i, \langle \ldots \langle r_1, \sigma_1, \eta_1, \zeta \rangle \ldots \rangle \rangle$. Informally we say that the $(r_i, \sigma_i)$ pair is the *i-th subprocess* of $\rho$. Here $r_i, \sigma_i$ and $\eta_i$ are called the *resumption, state* and *buffer* of the *i*-th subprocess, respectively. The transition $\langle \ldots \eta_{i+1}, \langle r_i, \sigma_i, \eta_i, \langle \ldots \rangle \ldots \rangle \longrightarrow \langle \ldots \eta'_{i+1}, \langle r'_i, \sigma'_i, \eta'_i, \langle \ldots \rangle \ldots \rangle \ldots \rangle$ is said to be *caused by* the *i*-th subprocess. The *i*-th subprocess can cause a transition only if it is enabled. We define that the 0-th subprocess is *enabled* if the input stream is not empty. The *i*-th subprocess is *enabled* if one of the rules from 4 to 11 can be applied to $\langle r_i, \sigma_i, \eta_i, \epsilon \rangle$. If a subprocess is not enabled, then it is called *disabled*. For an initial $\rho$, a finite or infinite transition sequence is called a *computation* of $\rho$:
$$c(\rho) : \rho = \rho_0 \longrightarrow \rho_1 \longrightarrow \ldots \rho_{n-1} \longrightarrow \rho_n \longrightarrow \ldots$$

The transition from $\rho_{n-1}$ to $\rho_n$ is called the $n$-th *step* and $\rho_n$ is called the $n$-th *stage* of the computation. The *output stream* of $c(\rho)$ is the sequence of values produced by $c(\rho)$.

## 2.5 Operational semantics

To define our operational semantics we first introduce the notion *canonical computation* $C_0(\rho)$. The computation $C_0(\rho)$ begins with the transition caused by the 0-th subprocess if it is enabled. In the next steps, the transitions are successively caused by the subprocesses from right to left (if they are enabled). After the transition caused by the leftmost subprocess we start anew with the 0-th subprocess. This procedure is repeated as often as possible. Example 1 described such a computation. The canonical computation $C_0(\rho)$ corresponds with the computation defined by a very fair scheduler, which allows each subprocess, in a right-to-left fashion, to take one step.

For a given input stream $\zeta$, a statement $s$ and a state $\sigma$, we can consider different computations starting from $\rho = \langle s : E, \sigma, \epsilon, \zeta \rangle$. In [BN96, BN97] while proving the first part of the Kahn principle, we showed that $C_0(\rho)$ delivers the maximal output stream, i.e., the output stream of any other computation of $\rho$ is a prefix of the output stream of $C_0(\rho)$. This maximal output stream can be defined as the operational semantics of **NT**.

**Definition 2.** The operational semantics based on **NT** is defined as the function $O_n:\mathbf{L} \to \mathbf{Proc}$ such that for any $s \in \mathbf{L}, \sigma \in \mathbf{State}$ and $\zeta \in \mathbf{Val}^\infty, O_n(s)(\sigma)(\zeta) =$ the output stream of $C_0(\langle s : E, \sigma, \epsilon, \zeta \rangle)$.

# 3 A deterministic Synchronous Transition System DT and its Relation to NT

In this section, we will introduce a computation $C(\rho)$ in **NT** which also delivers the maximal output stream. We will first define a deterministic transition system **DT**. In general, a transition in **DT** lets all subprocesses take a step simultaneously. The operational semantics $O_d$ is determined by the output streams of the maximal transition sequences in **DT**. By comparing the maximal transition sequence starting from $\rho$ in **DT** with $C(\rho)$ in **NT** we can establish the equivalence between $O_d$ and $O_n$.

## 3.1 A deterministic synchronous transition systems DT on L

In **DT** we use $\otimes$ and $\Longrightarrow$ the same way as we use $\otimes$ and $\longrightarrow$ in **NT**. Every transition in **DT** from a nonterminal delivers a value to the leftmost output channel. The transition from $\langle r, \sigma, \eta, \rho \rangle$ is defined by the transition from $\rho$. In most of the cases below, the right hand side of a transition will feature a buffer stream $\eta$ and a configuration $\rho$. These $\eta, \rho$ depend on the transition from $\rho$. We define $\underline{\rho} = \rho', \underline{\eta} = \eta\alpha$ if $\rho \overset{\alpha}{\Longrightarrow} \rho'$ and $\underline{\rho} = \rho, \underline{\eta} = \eta$ if $\rho \Longrightarrow \otimes$.

1. $\epsilon \Longrightarrow \otimes$
   For all $\rho, \langle E, \sigma, \eta, \rho \rangle \Longrightarrow \otimes$
   $\langle read(v) : r, \sigma, \epsilon, \rho \rangle \Longrightarrow \otimes$ if $\rho \Longrightarrow \otimes$

2. $\beta\zeta \overset{\beta}{\Longrightarrow} \zeta$

3. $\langle (v := e) : r, \sigma, \eta, \rho \rangle \overset{\tau}{\Longrightarrow} \langle r, \sigma\{\beta/v\}, \underline{\eta}, \underline{\rho} \rangle$, where $\beta = V(e)(\sigma)$.

4. $\langle skip : r, \sigma, \eta, \rho \rangle \overset{\tau}{\Longrightarrow} \langle r, \sigma, \underline{\eta}, \underline{\rho} \rangle$

5. $\langle write(e) : r, \sigma, \eta, \rho \rangle \overset{\beta}{\Longrightarrow} \langle r, \sigma, \underline{\eta}, \underline{\rho} \rangle$, where $\beta = V(e)(\sigma)$

6. $\langle read(v) : r, \sigma, \eta, \rho \rangle \overset{\tau}{\Longrightarrow} \langle r, \sigma\{\beta/v\}, rest(\underline{\eta}), \underline{\rho} \rangle$, for $\eta \neq \epsilon$ and $\beta = first(\eta) \neq \tau$
   $\langle read(v) : r, \sigma, \eta, \rho \rangle \overset{\tau}{\Longrightarrow} \langle read(v) : r, \sigma, rest(\underline{\eta}), \underline{\rho} \rangle$, for $\eta \neq \epsilon$ and $\tau = first(\eta)$

7. $\langle fork(v) : r, \sigma, \eta, \rho \rangle \overset{\tau}{\Longrightarrow} \langle r, \sigma\{1/v\}, \epsilon, \langle r, \sigma\{0/v\}, \underline{\eta}, \underline{\rho} \rangle \rangle$

8. $\langle (s_1; s_2) : r, \sigma, \eta, \rho \rangle \overset{\tau}{\Longrightarrow} \langle s_1 : (s_2 : r), \sigma, \underline{\eta}, \underline{\rho} \rangle$

9. If $B(b)(\sigma)$, then $\langle$ **if** $b$ **then** $s_1$ **else** $s_2$ **fi** $: r, \sigma, \eta, \rho \rangle \overset{\tau}{\Longrightarrow} \langle s_1 : r, \sigma, \underline{\eta}, \underline{\rho} \rangle$
   If $\neg B(b)(\sigma)$, then $\langle$ **if** $b$ **then** $s_1$ **else** $s_2$ **fi** $: r, \sigma, \eta, \rho \rangle \overset{\tau}{\Longrightarrow} \langle s_2 : r, \sigma, \underline{\eta}, \underline{\rho} \rangle$

10. $\langle$ **while** $b$ **do** $s$ **od** $: r, \sigma, \eta, \rho \rangle \overset{\tau}{\Longrightarrow}$
    $\langle$ **if** $b$ **then** $s$; **while** $b$ **do** $s$ **od else** $skip$ **fi** $: r, \sigma, \underline{\eta}, \underline{\rho} \rangle$

*Example 2.* A transition sequence in **DT** from the same $\rho$ as in Example 1:
$\rho = \langle read(y) : (fork(v) : E), \{\}, \alpha\beta, \gamma \rangle \overset{\tau}{\Longrightarrow} \langle fork(v) : E, \{y = \alpha\}, \beta\gamma, \epsilon \rangle \overset{\tau}{\Longrightarrow}$
$\langle E, \{y = \alpha, v = 1\}, \epsilon, \langle E, \{y = \alpha, v = 0\}, \beta\gamma, \epsilon \rangle \rangle \Longrightarrow \otimes$

*Example 3.* The following are two transition sequences in **DT**.
  (a) $\langle read(x) : E, \sigma, \epsilon, \alpha\beta \rangle \overset{\tau}{\Longrightarrow} \langle E, \sigma\{\alpha/x\}, \epsilon, \beta \rangle \Longrightarrow \otimes$.
  (b) $\langle read(w) : E, \sigma_1, \epsilon, \langle read(w) : E, \sigma_0, \alpha, \epsilon \rangle \rangle$.
      $\overset{\tau}{\Longrightarrow} \langle read(w) : E, \sigma_1, \epsilon, \langle E, \sigma_0\{\alpha/w\}, \epsilon, \epsilon \rangle \rangle \Longrightarrow \otimes$

*Example 4.* Notice that a terminal in **NT** is also a terminal in **DT**. The terminals in **DT** are not always terminals in **NT**. E.g. $\langle E, \sigma, \eta, \langle (v := e) : E, \sigma', \epsilon, \alpha \rangle \rangle$ and $\langle read(x) : E, \sigma, \epsilon, \langle read(y), \sigma, \epsilon, \langle E, \sigma', \epsilon, \alpha \rangle \rangle \rangle$ are terminals in **DT** but not in **NT**.

### 3.2   The operational semantics derived from DT

We define the operational semantics $O_d$, based on **DT**, as the unique fixed point $o$ of a contraction $\Phi$ on the complete ultrametric space $\mathbf{SemO} = \mathbf{Config} \to \mathbf{V}^\infty$. This is rather standard, for more background, definitions and proofs, cf. [BV96], in particular Section 9.2.

**Definition 3.** Let $\mathbf{SemO} = \mathbf{Config} \to \mathbf{V}^\infty$. Let $\Phi : \mathbf{SemO} \to \mathbf{SemO}$ be defined by $\Phi(f)(\rho) = \epsilon$ if $\rho \Longrightarrow \otimes$ and $\Phi(f)(\rho) = \beta f(\rho')$ if $\rho \overset{\beta}{\Longrightarrow} \rho'$ in **DT**.

**Theorem 4.** $\Phi : \mathbf{SemO} \to \mathbf{SemO}$ is a $1/2$-contraction and hence $\Phi$ has a unique fixed point $o$.

By Definition 3, $o(\rho) = \epsilon$ if $\rho \Longrightarrow \otimes$ and $o(\rho) = \beta o(\rho')$ if $\rho \overset{\beta}{\Longrightarrow} \rho'$. Thus $o(\rho)$ is the sequence of values produced by the unique maximal transition sequence starting from $\rho$. For instance, in Example 2, we have $o(\rho) = \tau^2$. The operational semantics $O_d : \mathbf{L} \to \mathbf{Proc}$ is defined by $O_d(s) = \lambda\sigma.\lambda\zeta.o(\langle s : E, \sigma, \epsilon, \zeta \rangle)$.

### 3.3 The relation between transitions in NT and DT

The operational semantics in **NT** is defined by the canonical computation $C_0(\rho)$ or other computations which delivers the same maximal output stream (cf. [BN96, BN97]). On the other hand, the operational semantics in **DT** is based on $o$ which delivers the output stream of the maximal transition sequence of synchronous steps from a configuration. In this section we want to show that a step in **DT** corresponds with a finite sequence of steps in **NT**. Such a sequence is called a *big step*. A combination of such big steps will be a computation $C(\rho)$ in **NT**. $C(\rho)$ delivers also the maximal output stream.

**An example of $C(\rho)$ as a combination of big steps.** Given a configuration $\rho$, we can define a computation $C_j(\rho)$ which leaves all $k$-th ($k < j$) subconfigurations of $\rho$ unchanged. $C_j(\rho)$ is a computation defined by the following procedure. It begins with the transition caused by the $j$-th subprocess if it is enabled. In the next steps, the subprocesses from right to left cause successively transitions and begin again with the $j$-th subprocess. This procedure is repeated as often as possible. The canonical computation $C_0(\rho)$ is a special case for $j = 0$. A *big step* of $C_j(\rho)$ is a sequence of transition steps in $C_j(\rho)$ where the transitions start from the $j$-th subprocess and end at the $j$-th subprocess again.

*Example 5.* Suppose $\rho = \langle write(1) : (write(2) : E), \sigma, \epsilon, \langle write(3) : E, \sigma', \epsilon, 4.5\rangle\rangle$. We first use the following big step induced by $C_0(\rho)$ to define the first steps in $C(\rho)$ in **NT**: $\rho \longrightarrow \langle write(1) : (write(2) : E), \sigma, \epsilon, \langle write(3) : E, \sigma', 4, 5\rangle\rangle \longrightarrow \langle write(1) : (write(2) : E), \sigma, 3, \langle E, \sigma', 4, 5\rangle\rangle \overset{1}{\longrightarrow} \langle write(2) : E, \sigma, 3, \langle E, \sigma', 4, 5\rangle\rangle = \rho_1$. Notice that $\langle E, \sigma', 4, 5\rangle$ is a terminal in **DT** but not in **NT**. This part will stay the same in a transition sequence from $\rho_1$ in **DT**. Hence we use a big step in $C_2(\rho_1)$ for defining $C(\rho)$ in **NT** further: $\rho_1 \overset{2}{\longrightarrow} \langle E, \sigma, 3, \langle E, \sigma', 4, 5\rangle\rangle = \rho_2 \longrightarrow \otimes$ The combination of the transitions in **NT** from $\rho$ to $\rho_1$ to $\rho_2$ to $\otimes$ is defined as $C(\rho)$. We denote the big steps by $\rho \overset{*1}{\longrightarrow} \rho_1 \overset{*2}{\longrightarrow} \rho_2 \longrightarrow \otimes$. Furthermore, $\rho \overset{1}{\Longrightarrow} \rho_1 \overset{2}{\Longrightarrow} \rho_2 \Longrightarrow \otimes$ is the unique maximal transition sequence from $\rho$ in **DT**.

For a given $\rho$, usually a big step of $C_0(\rho)$ in **NT** corresponds with a synchronous transition step in **DT**. However, if a subconfiguration of $\rho$ is $\langle E, \sigma, \eta, \rho'\rangle$, then this subconfiguration is a terminal in **DT** but may not be a terminal in **NT**. Fortunately, the transitions of $\rho'$ in **NT** will not influence the global output stream because of the obstruction by $E$. We will define $C(\rho)$ as a sequence of different types of big steps. The lemma given afterwards will imply that $C(\rho)$ delivers the same maximal stream as $C_0(\rho)$.

**Terminals in the transition system DT.** The transition system **DT** is complete in the sense that $\rho \overset{\alpha}{\Longrightarrow} \rho'$ or $\rho \Longrightarrow \otimes$ for all $\rho \in$ **Config**. Let $T_0 = \{\epsilon\} \cup \{\rho| \rho = \langle E, \sigma, \eta, \rho'\rangle\}$ and $T_i = \{\rho| \rho = \langle read(v) : r, \sigma, \epsilon, \rho'\rangle, \rho' \in T_{i-1}\}$. Then $T = \bigcup T_i$ is precisely the set of all terminals in this transition system **DT**. The terminals in **DT** which are not terminals in **NT**, are either of the form

    *type 1* : $\langle E, \sigma_m, \eta_m, \ldots, \langle r_i, \sigma_i, \eta_i, \langle \ldots, \langle r_1, \sigma_1, \eta_1, \zeta\rangle \ldots\rangle\rangle$, where some $i$-th subprocess is enabled in **NT**, or of the form

$$type\ 2 : \langle r_m, \sigma_m, \epsilon, \langle r_{m-1}, \sigma_{m-1}, \epsilon, \langle \ldots, \langle r_1, \sigma_1, \epsilon, \rho' \rangle \ldots \rangle,$$

where all $r_m, \ldots, r_1$ begin with a read-statement and $\rho'$ is of type 1.

**Lemma 5.** Let $\rho = \langle r_m, \sigma_m, \eta_m, \ldots, \langle r_1, \sigma_1, \eta_1, \rho \rangle \rangle$ be a non-terminal in **NT**. Suppose $\rho$ has nesting numer $k$ and is the leftmost **DT**-terminal of type 1 or 2 in $\rho$. If $k < m$, then a big step $\rho \xrightarrow{*\alpha} \underline{\rho}_1$ of $C_{k+1}(\rho)$ in **NT** has the same effect as a single step $\underline{\rho} \xRightarrow{\alpha} \underline{\rho}_1$ in **DT**. If $k = m$, then $\underline{\rho} \xrightarrow{*} \underline{\rho}_1$ has empty global output in **NT** and thus it corresponds with the empty transition $\underline{\rho} \Longrightarrow \otimes$ in **DT**.

**Remark.** If $\rho$ does not contain such a **DT**-terminal, then let $k = -1$. In this situation, $\underline{\rho} \xrightarrow{*\alpha} \underline{\rho}_1$ means a big step of $C_0(\rho)$.

**Definition 6.** Given $\rho$, we define the computation $C(\underline{\rho})$ in **NT** as follows:

$$C(\underline{\rho}) : \underline{\rho} = \underline{\rho}_0 \xrightarrow{*\alpha_0} \underline{\rho}_1 \xrightarrow{*\alpha_1} \ldots \xrightarrow{*\alpha_{i-1}} \underline{\rho}_i \xrightarrow{*\alpha_i} \underline{\rho}_{i+1} \ldots,$$

If $\underline{\rho}_i$ is not a terminal in **NT** and the $k$-th subconfiguration is the leftmost **DT**-terminal of type 1 or 2 in $\underline{\rho}_i$, then $\underline{\rho}_i \xrightarrow{*\alpha_i} \underline{\rho}_{i+1}$ is the sequence in a big step in **NT** induced by $C_{k+1}(\underline{\rho}_i)$.

**Proposition 7.** Given $\underline{\rho}$, the unique maximal transition sequence in **DT** from $\underline{\rho}$ delivers the same global output as $C(\underline{\rho})$ in **NT**.

Now we want to compare $C(\underline{\rho})$ and $C_0(\underline{\rho})$ by using the following lemma and Proposition 7.

**Lemma 8.** Let $\rho = \langle r_m, \sigma_m, \eta_m, \ldots, \langle r_1, \sigma_1, \eta_1, \rho \rangle \rangle$, where $\rho$ is a terminal in **DT** with nesting number $k \geq -1$ and $\rho' = \langle r_m, \sigma_m, \eta_m, \ldots, \langle r_1, \sigma_1, \eta_1, \rho' \rangle \rangle$, where $\rho'$ is a terminal in **DT** with nesting number $i \geq -1$. If for some $j \leq k + 1$, the stage after a big step of $C_j(\underline{\rho})$ from $\rho$ is $\underline{\rho}_1 = \langle r'_n, \sigma'_n, \eta'_n, \ldots, \langle r'_1, \sigma'_1, \eta'_1, \rho_1 \rangle \rangle$, then for every $s \leq i + 1$, the stage after a big step of $C_s(\rho')$ from $\rho'$ has the form $\underline{\rho}'_1 = \langle r'_n, \sigma'_n, \eta'_n, \ldots, \langle r'_1, \sigma'_1, \eta'_1, \rho'_1 \rangle \rangle$. Moreover, both big steps deliver the same global output (if any) and both $\rho_1$ and $\rho'_1$ are terminals in **DT**.

Lemma 8 guarantees that we can compare $\underline{\rho}_1$ and $\underline{\rho}'_1$ in the same way as $\underline{\rho}$ and $\underline{\rho}'$. Therefore the global output of $C_0(\underline{\rho})$ and $C(\underline{\rho})$ are the same.

**Theorem 9.** $O_d = O_n$.

## 4 Denotational Semantics and Equivalence of Semantics

In this section we will define a denotational semantics by using the unique fixed point $m$ of some contraction $\Psi$. Let $\mathbf{V}^\infty \to_n \mathbf{V}^\infty$ be the set of all nonexpansive functions from $\mathbf{V}^\infty$ to $\mathbf{V}^\infty$. We further define the set of all *continuations*, denoted by **Cont**, as the set of all functions $\theta : \mathbf{State} \to (\mathbf{V}^\infty \to_n \mathbf{V}^\infty)$. The set of all nonexpansive functions from **Cont** to **Cont**, is denoted by $\mathbf{Cont} \to_n \mathbf{Cont}$. The set **SemD** denotes the set of all functions $\psi : \mathbf{L} \to (\mathbf{Cont} \to_n \mathbf{Cont})$. Also in this section the approach is standard, we refer again to [BV96].

## 4.1 Using a contraction $\Psi$ to define the denotational semantics

**Definition 10.** Let $\Psi$ be defined as follows.

1. $\Psi(\psi)(v := e)(\theta)(\sigma)(\zeta) = \tau\theta(\sigma\{\beta/v\})(\zeta)$, where $\beta = V(e)(\sigma)$
2. $\Psi(\psi)(\text{skip})(\theta)(\sigma)(\zeta) = \tau\theta(\sigma)(\zeta)$
3. $\Psi(\psi)(\text{write}(e))(\theta)(\sigma)(\zeta) = \beta\theta(\sigma)(\zeta)$, where $\beta = V(e)(\sigma)$
4. $\Psi(\psi)(\text{read}(v))(\theta)(\sigma)(\epsilon) = \epsilon$
   If $\zeta \neq \epsilon$, first$(\zeta) = \tau$, then $\Psi(\psi)(\text{read}(v))(\theta)(\sigma)(\zeta) = \tau\psi(\text{read}(v))(\theta)(\sigma)(\text{rest}(\zeta))$
   If $\zeta \neq \epsilon$, first$(\zeta) = \alpha \neq \tau$, then $\Psi(\psi)(\text{read}(v))(\theta)(\sigma)(\zeta) = \tau\theta(\sigma\{\alpha/v\})(\text{rest}(\zeta))$
5. $\Psi(\psi)(\text{fork}(v))(\theta)(\sigma)(\zeta) = \tau\theta(\sigma\{1/v\})(\theta(\sigma\{0/v\})(\zeta))$
6. $\Psi(\psi)(s_1; \ s_2)(\theta)(\sigma)(\zeta) = \tau\Psi(\psi)(s_1)(\Psi(\psi)(s_2)(\theta))(\sigma)(\zeta)$
7. If $B(b)(\sigma)$, then $\Psi(\psi)(\textbf{if } b \textbf{ then } s_1 \textbf{ else } s_2 \textbf{ fi})(\theta)(\sigma)(\zeta) = \tau\Psi(\psi)(s_1)(\theta)(\sigma)(\zeta)$
   If $\neg B(b)(\sigma)$, then $\Psi(\psi)(\textbf{if } b \textbf{ then } s_1 \textbf{ else } s_2 \textbf{ fi})(\theta)(\sigma)(\zeta) = \tau\Psi(\psi)(s_2)(\theta)(\sigma)(\zeta)$
8. $\Psi(\psi)(\textbf{while } b \textbf{ do } s \textbf{ od})(\theta)(\sigma)(\zeta)$
   $= \tau\psi(\textbf{if } b \textbf{ then } s; \ \textbf{while } b \textbf{ do } s \textbf{ od else } skip \textbf{ fi})(\theta)(\sigma)(\zeta)$

**Theorem 11.** $\Psi : \textbf{SemD} \rightarrow \textbf{SemD}$ is well defined and it is a $1/2$-contraction.

**Definition 12.** Let $\theta_0 \in \textbf{State} \rightarrow (\mathbf{V}^\infty \rightarrow \mathbf{V}^\infty)$ be the trivial continuation defined by $\theta_0(\sigma)(\zeta) = \epsilon$. Let $m$ be the unique fixed point of $\Psi$, i.e. $\Psi(m) = m$. The *denotational semantics* $\mathbf{D}$ of $\mathbf{L}$ is defined as $D(s)(\sigma)(\zeta) = m(s)(\theta_0)(\sigma)(\zeta)$ for all $\sigma \in \textbf{State}$ and $\zeta \in \mathbf{V}^\infty$.

## 4.2 Equivalence of $O_n, O_d$ and $D$

We will define a mapping $I$ using the fixed point $m$ of $\Psi$. We can prove that $I$ is also a fixed point of $\Phi$. From the uniqueness of the fixed point of $\Phi$, we can conclude $I = o$ and then $D = O_d$. The proofs of Lemma 14 and 15 are by induction on the nesting number of $\rho$. Lemma 16 follows immediately from the previous ones.

**Definition 13.** The mapping $H : \textbf{Res} \rightarrow \textbf{Cont}$ is defined by $H(E) = \theta_0$ and $H(s : r) = m(s)H(r)$. The mapping $I : \textbf{Config} \rightarrow \mathbf{V}^\infty$ is defined by $I(\zeta) = \zeta$ and $I(\langle r, \sigma, \eta, \rho \rangle) = H(r)(\sigma)(\eta I(\rho))$.

*Example 6.* Consider Example 3($a$). We have:
$o(\langle \text{read}(x) : E, \sigma, \epsilon, \alpha\beta \rangle) = \tau$,
$I(\langle \text{read}(x) : E, \sigma, \epsilon, \alpha\beta \rangle) = H(\text{read}(x) : E)(\sigma)(\epsilon.\alpha\beta) = m(\text{read}(x))(H(E))(\sigma)(\alpha\beta)$
$= \tau H(E)(\sigma\{\alpha/x\})(\beta) = \tau\theta_0(\sigma\{\alpha/x\})(\beta) = \tau$. Thus $I(\rho) = o(\rho)$.

**Lemma 14.** If $\rho \Longrightarrow \otimes$ in **DT**, then $I(\rho) = \epsilon$ .

**Lemma 15.** For all $\rho \overset{\beta}{\Longrightarrow} \rho'$ in **DT**, we have $I(\rho) = \beta I(\rho')$.

**Lemma 16.** $\Phi(I) = I$.

Since $\Phi$ has only one fixed point, we have $I = o$. We can then prove the equivalence of all semantics defined in this article. This establishes the second part of the Kahn principle:

**Theorem 17.** $D(s) = O_d(s) = O_n(s)$ for all $s \in \mathbf{L}$.

# 5 Summary

Let a simple language **L** be given. In [BN96, BN97] we have defined a nondeterministic interleaving transition system without $\tau$'s. In these papers we also proved the first part of the Kahn principle, namely that every computation delivers the maximal output stream or a prefix of it. In this paper, we first defined a nondeterministic interleaving transition system **NT** with $\tau$'s. The proof in [BN96, BN97] can easily be adapted to prove the first part of Kahn's principle with respect to **NT**. Thus we can assume that all computations in **NT** either give a maximal stream or a prefix of it. This can be used to define the operational semantics $O_n$. We have concentrated on the second part of Kahn's principle in this paper. We do not prove $O_n = D$ directly. Instead we define **DT** and use $O_d$ as a bridge to prove $O_n = D$.

# References

[A81] A. Arnold, Sémantique des processus communicants, RAIRO Theor.Inf. 15,2, pp.103-139, 1981.

[BBB93] J. W. de Bakker, F. van Breugel and A. de Bruin, Comparative semantics for linear arrays of communicating processes, a study of the UNIX fork and pipe commands, in: A.M. Borzyszkowski and S. Sokolowski (eds.) Proc. Math. Foundations of Computer Science, LNCS 711, pp.252-261, Springer, 1993.

[BV96] J. W. de Bakker, E. de Vink, *Control Flow Semantics*, MIT press, 1996.

[BN96] A. de Bruin, S. H. Nienhuys-Cheng, Linear dynamic Kahn networks are deterministic, in: W. Penczek, A. Szalas (Eds.), Proc. Math. Foundations of Computer Science, LNCS 1113, pp.242-254, Springer, 1996.

[BN97] A. de Bruin, S. H. Nienhuys-Cheng, Linear dynamic Kahn networks are deterministic, complete version of [BN96]. Accepted by *Theoretical Computer Science*, 1997, special issue for MFCS96. It is based on a technical report, Erasmus University Rotterdam.
(http://kaa.cs.few.eur.nl/few/inf/publicaties/rapporten/eur-few-cs-94-06.html).

[C72] J.M. Cadiou, Recursive definitions of partial functions and their computations, Ph.D. thesis, Stanford Univ., 1972.

[K74] G. Kahn, The semantics of a simple language for parallel programming, in: Proc. IFIP74, J.L. Rosenfeld (ed.), North-Holland, pp.471-475, 1974.

[LS89] N. Lynch and E. Stark, A proof of the Kahn principle for input/output automata, Information and Computation, 82, 1, pp.81-92, 1989.

[NB97] S. H. Nienhuys-Cheng and A. de Bruin. Kahn's fixed-point characterization for linear dynamic networks. Technical report, Erasmus University Rotterdam.
(http://kaa.cs.few.eur.nl/few/inf/publicaties/rapporten/eur-few-cs-97-06.html).

# DESAM – Annotated Corpus for Czech

Karel Pala, Pavel Rychlý and Pavel Smrž

Faculty of Informatics, Masaryk University Brno
Botanická 68a, 602 00 Brno, Czech Republic
E-mail: {pala,pary,smrz}@fi.muni.cz

**Abstract.** This paper deals with Czech disambiguated corpus DESAM. It is a tagged corpus which has been manually disambiguated and can be used in various applications. We discuss the structure of the corpus, tools used for its managing, linguistic applications, and also possible use of machine learning techniques relying on the disambiguated data. Possible ways of developing the procedures for complete automatic disambiguation are considered.

## 1 Introduction

In computational linguistics, "corpus" is a collection of written (or sometimes spoken) texts. Corpora could be used in several application areas: building dictionaries, general linguistic research, natural language processing, information retrieval, machine translation etc.

In corpus exploration, a user must be able to express the query as precisely as possible in order to minimize the number of concordance items searched for. It should be possible to refer to linguistic or structural information in corpus. We use the term "tagged (annotated) corpus" for a corpus which contains not only of a sequence of words but also comprises an additional information. Typically, this includes linguistic information which is associated with the particular word forms in corpus: the most common linguistic tags are *lemma* (the basic word form), *part of speech (POS)* and the respective *grammatical categories*. Another level of annotations concerns *structural information* which identifies a metatext structure of the text in corpus. For example, we can mark (annotate) that the sequence of word forms is a part of the headline or a regular sentence in a paragraph [1].

The most reasonable way how to build large annotated corpora is an automatic tagging of the texts by computer programmes. However, natural languages display rather complex structure and therefore it is no surprise that the attempts to process them by the simple deterministic algorithms do not always yield satisfactory results. The result is that the present tagging programmes are not able to give fully reliable results and there are many ambiguities in their output.

Various strategies trying to resolve the ambiguities in the tagged corpora have been developed and applied within the field of corpus linguistics. The most frequently used are the following:

1. *probabilistic techniques* like the ones used in CLAWS tagger [2] or Cutting's tagger [3];
2. *deterministic rule-based techniques* using CF-like formalisms which may be enhanced with some heuristic rules;
3. *various combinations* of the former two approaches, e. g. constraint grammars approach [4];
4. attempts to apply *learning* techniques that would make use of previous experience in the process disambiguating.

## 2   The annotated Czech corpus – DESAM

The DESAM corpus has been built at the Faculty of Informatics, Masaryk University, as a part of the complex project whose main purpose is to build Czech National Corpus (200 mil. word forms by the end of 1998). The DESAM corpus is:

- a **Czech** corpus: included texts are written in Czech language.
- a **general** corpus: subject field is not specifically restricted. Texts are taken from newspapers and scientific magazines.
- a **tagged** corpus: a lemma and a corresponding tag is stored for each word form in the corpus.

DESAM corpus is the first annotated and fully **disambiguated** corpus for Czech language that can be run under the corpus manager CQP (see below) and its presented version will be later included in Czech National Corpus as its annotated part.

The LEMMA programme ([5], [6]) has been used for tagging texts included in DESAM. This programme is able to perform full morphological analysis of the arbitrary raw Czech text and for each Czech word form it yields the following output:

- its lemma or lemmata, i.e. as it is usual in Czech grammar, nominative singular for nouns, adjectives, pronouns and numerals, and infinitive for verbs
- its corresponding POS symbol: presently we work with 10 basic parts of speech that are normally distinguished in Czech grammars, however, the complete list of POS tags contains about 30 items (including subclassifications). If a word form can be associated with two or more POS symbols LEMMA offers all of them
- all the grammatical categories associated with a word form, i.e. in Czech this includes for nouns, adjectives, pronouns and numerals: case, number and gender; for verbs: person, number, tense, mode, voice, aspect and also gender
- optionally, also all word forms which can be generated from a given lemma.

Moreover, in the course of corpus tagging LEMMA gives all the possible combinations of lemmata and the respective tags for each input word form. A tag is conceived as a character string carrying information about POS and the respective grammatical categories using attribute–values coding convention. The total number of all possible tags for Czech language is about 1800. In the following example the tag **k1gMnSc1** means: part of speech (**k**) = noun (**1**), gender (**g**) = male animate (**M**), number (**n**) = singular (**S**) and case (**c**) = first (nominative) (**1**).

Example of the LEMMA output:

```
Václav <l>Václav <c>k1gMnSc1
Havel <l>Havel <c>k1gMnSc1
přišel <l>přijít <c>k5eApMnStMmPaP,k5eApInStMmPaP
naopak <l>naopak <c>k6xMeA
s <l>s <c>k7c7
vlastním <l>vlastní <c>k2eAgMnSc67d1,k2eAgXnPc3d1,k2eAgUnSc67d1
         <l>vlastnit <c>k5eAp1nStPmIaI
volebním <l>volební <c>k2eAgMnSc67d1,k2eAgXnPc3d1,k2eAgUnSc67d1
programem <l>program <c>k1gInSc7
,
který <l>který <c>k3xQgMnSc15,k3xQgInSc145
nikomu <l>nikdo <c>k3xNnSc3
neubližuje <l>ubližovat <c>k5eNpMnStPmTaI,k5eNp3nStPmIaI
.
```

(English translation: *Václav Havel, on the contrary, came with his own "election programme" which does not hurt anybody.*)

The original version of LEMMA processed only individual word forms without considering any context. Its present version, however, is already able to process the set of basic Czech collocations (about 500 items). If we have look at the output of LEMMA, we can see that about 50% of the processed word forms have more than one possible lemma and respective tag associated with them which means that we have to face the problem of disambiguation and look for a way how to solve it. The reasons for disambiguation are evident:

- if we want to perform syntactic analysis of the tagged Czech text we need to remove as many ambiguities as possible
- if we want to have a reliably tagged corpus we also need a successful disambiguation procedure.

Therefore we had to make a decision: either to start with the poorly disambiguated corpus using the output from LEMMA as it stood or to begin with the manual disambiguation. We took the latter direction and all texts in the DESAM corpus were disambiguated manually. A programme DES [7] which navigates and helps users in manual disambiguating has been developed for this purpose. In [7], one can find a first basic analysis of ambiguity measures for Czech. The results are summarized in Table 1.

|                              | before analysis | after analysis |
|------------------------------|----------------:|---------------:|
| Total word forms             | 10,300          | 10,300         |
| Ambiguous word forms         | 5,200           | 2,950          |
| Total tags                   | 33,360          | 16,680         |
| Tags per ambiguous word form | 5.93            | 2.73           |

**Table 1.** Measures of disambiguation

## 3 Corpus tools

It is obvious that large corpora should be easily accessible and the users should be equipped with a friendly environment enabling them to ask as many various queries as possible. For this purpose the IMS Corpus Workbench [8] has been chosen and installed at one of our SUN workstations. It is a set of tools for the administration and representation of large text corpora and retrieval of the information from the corpora. One of the tools is a query processor CQP [8] which evaluates given queries and returns the result on the screen or to another output that can be used in further processing. For more comfortable interaction or presentation there is XKWIC [9], a graphical user interface running in X–Window system.

Within the IMS Corpus Workbench, a corpus is represented as a sequence of *positions*. Each position is a set of *attributes* and each attribute contains a character information. Presently, we work with three attributes in the DESAM corpus: **word** which represents the particular word form at this position, **lemma** representing the corresponding basic word form and **tag** associated with this position.

As we said above, we can also store some *structural tags* in the corpus. At the present moment we use two structural tags in DESAM corpus: **doc** for documents and **p** for paragraphs. Apart from this the whole corpus is also divided into many documents using the **doc** tags. In most cases, a document represents a newspaper article. Each document is divided into paragraphs using the **p** tags. Some small articles may consist of one paragraph only.

## 4 Linguistic results

Information about the current size of the DESAM corpus is displayed in Table 2. We would like to stress that the presented tables yield new and quite interesting data about Czech language.

First of all, total frequencies of words, lemmata and tags are presented in the table. That shows, for example, that 60% of different words, 44% of different lemmata, 25% of different tags are hapax legomena, i.e. they occur only once in our corpus texts. This is in good agreement with some previous statistical find-

| Documents | 520 |
|---|---|
| Paragraphs | 4,802 |
| Positions | 263,042 |
| Word forms | 209,921 |
| Different word forms | 49,211 |
| Different lemmata | 19,621 |
| Different tags | 1,097 |

**Table 2.** Counts of the DESAM corpus

ings, particularly with Czech Frequency Dictionary [10]. The most frequent word forms and lemmata and their respective frequencies are displayed in Table 3.

| Word forms | | Lemmata | | Tags | |
|---|---|---|---|---|---|
| a | 5642 | a | 5874 | k7c6 | 9105 |
| v | 4237 | být | 5362 | k8xC | 8410 |
| se | 4046 | v | 4844 | k5eAp3nStPmIaI | 6029 |
| na | 3547 | sebe | 4589 | k7c2 | 5645 |
| je | 2508 | na | 3858 | k7c4 | 5484 |
| že | 1926 | ten | 2134 | k1gFnSc2 | 5369 |
| z | 1567 | on | 1982 | k1gFnSc1 | 5126 |
| s | 1552 | že | 1930 | k9 | 4905 |
| o | 1388 | z | 1718 | k1gInSc2 | 4168 |
| i | 1304 | s | 1649 | k8xS | 4087 |
| do | 1300 | který | 1506 | k1gInSc1 | 3927 |
| to | 1000 | o | 1430 | k1gFnSc4 | 3835 |
| pro | 945 | do | 1363 | k3xXnSc4 | 3779 |
| ve | 891 | i | 1360 | k1gMnSc1 | 3576 |
| k | 863 | mít | 1319 | k1gInSc4 | 3124 |

**Table 3.** Most frequent word forms, lemmata and tags

The frequencies presented in the table again fit well into the picture one can find in the existing Czech frequency dictionaries, however, the occurrence of the two verbs *být* a *mít* among 15 most frequented items is quite interesting and could be perhaps most plausibly explained by the fact that about one third of DESAM consists of the scientific text.

Using XKWIC, it is possible to make more sophisticated queries. For example, we can select all contexts where word forms **s** and **se** (two forms of "with" in English) are used as prepositions be query:

```
[word="se?"%c & pos="k7.*"]
```

Then we can compute the frequency distribution (Table 4).

| s  | 1323 |
|----|------|
| se | 234  |

**Table 4.** Frequencies of **s/se** word forms

In the following example we have selected all the word forms **kolem** (meaning either *a wheel* or *around*) and computed the frequency distribution of the tags for each value of lemma (Table 5).

$$[\texttt{word="kolem"\%c}]$$

Czech word form "kolem" could be:

- noun – lemma = "kolo" ("wheel" in English)
- preposition – lemma = "kolem" ("around" in English)
- adverb – lemma = "kolem" ("round" in English)

| kolem | k7c2     | 64 |
|-------|----------|----|
|       | k6xL     | 1  |
|       | k6xM     | 1  |
| kolo  | k1gNnSc7 | 2  |

**Table 5.** Frequencies of tags for **kolem** word forms

# 5  Machine Learning Techniques for Automatic Disambiguation

We would like to mention some experiments we have tried with machine learning methods with regard to corpus disambiguation.

## 5.1  Statistical methods

We have employed a simple statistical method for disambiguation. Similarly to [11], we have used the basic source channel model. The tagging procedure selects a sequence of tags $T$ for the sentence $W$:

$$\Phi : W \to T. \tag{1}$$

The optimal tagging procedure maximises the product $P(W|T)P(W)$:

$$\Phi(W) = \text{argmax}_T\, P(T|W) = \text{argmax}_T\, P(W|T)P(T). \qquad (2)$$

The basic methods of trigrams and maximum likelihood are employed to estimate the probabilities.

The results of our experiments are summarised in Table 6. The results are comparable to those published in [11] taking into consideration slightly different conditions. As stated in [11] the trigram tag prediction model needs much more data in order to get better results.

| Training data | 62,451 word forms |
|---|---|
| Test data | 10,328 word forms |
| Tagging accuracy | 67,42% |

**Table 6.** Results of probabilistic disambiguation

## 5.2 Connectionist methods

Recently, to avoid problems with the simple probabilistic approach described above we have started to experiment with neural networks for disambiguation. The architecture of the networks used in the experiments is similar to that of NETTALK [12]. The input of the network is a series of tag-sets of seven consecutive words from one of the training sentences. The central tag-set in the sequence is the current one for which the output is to be produced. Three sets of possible tags on either side of this central position provide context that helps to choose the correct tag for the central word. Sentences are moved through the window so that each word with possible tags in the sentence can be seen. Blanks are added before and after the sentence as needed.

One type of trained networks uses unary encoding. For each of the seven word positions in the input, the network has a set of $1097 + 1$ input units: one for each of 1097 different tags and one for blank. Thus, there are 1098 x 7 = 7686 input units. Other tested type uses compressed version of input according to the inner representation of tags in the programme LEMMA. Output is coded in the same way as input, the networks have one set of neurons for the actual position only.

Preliminary experiments suggest that the results obtained using neural networks could be better than those obtained by the simple probabilistic approach. It is true especially when the unary coding of inputs and outputs is used. The biggest problem despite the lack of training data is the extremely long time needed for training the neural network even if the supercomputer Silicon Graphics POWER Challenge L is used. The time necessary for training takes several days.

# 6 Conclusions

The most important result consists in building the fully annotated and disambiguated Czech corpus DESAM at FI MU. The whole process of its building took approximately 10 man-months. At the present moment DESAM runs under IMS Workbench and is accessible for all people who are interested in corpus applications within NLP field. DESAM is already serving as a training corpus in two different ways:

- as indicated above – when using statistical approaches to disambiguation,
- for building rule-based parsing algorithms for Czech. The first results in this respect can be found in [7] and they have already been used in designing a disambiguating programme processing Czech noun groups in raw text.

# References

1. K. Pala. Desambiguating syntactic constructions from tagged corpus. In *Workshop on AI Methods in Machine Learning*, 1996.
2. R. Garside. *The CLAWS word-tagging system, The computational analysis of English*. Longman, London, 1987.
3. D. Cutting. A practical part-of-speech tagger. In *Proceedings of the 3rd Conference on Natural Language Processing*, Trento, Italy, March–April 1992.
4. F. Karlsson, A. Voutilainen, J. Heikkilä, and A. Anttila. *Constraint Grammars*. Mouton de Gruyter, Berlin, 1995.
5. P. Ševeček. *LEMMA – a lemmatizer for Czech*. Brno, 1996. (manuscript).
6. K. Osolsobě. *Algorithmic description of Czech morphology*. PhD thesis, Masaryk University, Brno, 1996.
7. V. Puža. Syntactic analysis of natural language with a view to a corpora tagging. Master's thesis, Faculty of Informatics, Masaryk University, Brno, 1997.
8. B. M. Schulze and O. Christ. *The CQP User's Manual*.
9. O. Christ. *The XKWIC User Manual*.
10. J. Jelínek, J. V. Bečka, and M. Těšitelová. *Frequency Dictionary of Czech*. Academia, Praha, 1961.
11. J. Hajič and B. Hladká. Probabilistic and rule-based tagging of an inflective language — a comparison. Technical Report 1, Institute of Formal and Applied Linguistics, Faculty of Mathematics and Physics, Charles University, November 1996.
12. T. J. Sejnowski and C. R. Rosenberg. Parallel Networks that Learn to Pronounce English Text. *Complex Systems*, 1:145–168, 1987.

# Mobility Management in CORBA: A Generic Implementation of the LifeCycle Service

Yvan Peter

CNET de Caen
42, rue des Coutures
BP 6243, F-14066 Caen cedex
Yvan.Peter@cnet.francetelecom.fr

**Abstract.** CORBA enables objects interaction in a transparent manner according to location and heterogeneity. However, this kind of platforms lack mobility functions. That is why we propose a solution based on a generic implementation of CORBA's *LifeCycle Service*. Our mechanism relies on intermediary objects that are used to encapsulate mobility management.
**Keywords:** CORBA, LifeCycle Service, mobile object, proxy.

## 1 Introduction

CORBA (*Common Object Request Broker Architecture*) is a new standard for distributed environments which enables object components to interact in a transparent manner according to location and heterogeneity. Mobility functions are needed on these platforms for classical concerns such as administration, fault tolerance or load balancing or for newer research fields such as mobile computing. We propose a generic mechanism to introduce these functions following CORBA's mobility model based on the *LifeCycle Service*. Much research work has been done about process migration and more recently object migration. Most projects consider an homogeneous environment (e.g., [4], [6]). This cannot be the case with CORBA which is middleware and thus works on a variety of operating systems, hence heterogeneity is a primary concern. Migration transparency is also important and difficult to achieve in an open environment (i.e., in which objects do not know their clients *a priori*). It poses mainly two problems: managing *object references* so that clients are unaware of location changes of servers and actually migrating the server while it is invoked by clients. After a short overview of CORBA, we review in part 2 the problems related to mobility on this kind of platforms. Part 3 introduces CORBA's mobility model and presents our generic solution which can handle these problems in a CORBA compliant way. We will then relate our solution to existing work and consider future extensions.

## 2 The CORBA standard

The CORBA standard has been defined by the OMG *(Object Management Group)* to enable interaction of distributed components. It is part of OMG's object reference model *OMA (Object Management Architecture)*. The ORB (*Object Request Broker*) defined by CORBA is the central part of the OMA. In the following, we will present an overview of the main parts of the ORB's architecture of which a detailed description can be found in [2].

The ORB is a "software bus" which enables a client to access a service in a transparent manner according to location and heterogeneity. This transparency is achieved by using *interfaces* which describe the service supplied by a server regardless of its implementation. An interface is written in *IDL (Interface Definition Language)*, a descriptive (C++ like) language. The ORB architecture lays on the *ORB Core* which is the communication layer and has elements on the client side and on the server side (Fig. 1).

On the client side one can find *stubs* which are generated from the IDL interface of the server and which encapsulate *marshalling* and *unmarshalling* of requests and results. A *stub* acts as a *proxy object* hiding to the client the fact that the server it invokes is remote.

On the server side one can find :

- the *skeleton* which is generated from the IDL interface and has the same role as the client stub.
- the *Object Adapter* which serves as a run time environment for the servers (activation, deactivation, ...). It assigns to each server an *object reference* which is unique across an ORB. These references are interpreted by the ORB to locate the server which is invoked. They are opaque objects in the sense that their implementation is not defined and can be different in each product.

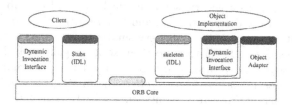

**Fig. 1.** Common Object Request Broker Architecture

There also exists a dynamic invocation mechanism that enables a client to build dynamically a request to invoke a previously unknown server. However, in the rest of this paper we will focus on the static invocation mechanism.

## 3 The problems of object mobility on CORBA

The main feature of CORBA is its heterogeneity. It works on a variety of systems and many products exist for which, although they bear the same standard interfaces, we have no knowledge of how they work internally. Therefore we are bound to build our mechanisms on top of CORBA using only the standard parts of it. The CORBA environment induces the following problems:

**Communication problem** static invocation mechanism is synchronous (asynchronous *oneway* methods exist, but they cannot return any result). Moving an object during an invocation means breaking the communication link and reconnecting transparently after migration. RPC (*Remote Procedure Call*) protocols have been designed to handle this in the field of mobile computing (e.g., [3]) but it is not useful in our case since we cannot have access to the communication layer.

**Reference problem** object references are managed by the ORB and the object adapter. Among Object Services, the *LifeCycle Service* is used to manage objects (removal, copy, migration). The specification requires that an *object reference* remains valid through the use of the *move* operation [1]. However, Object Services are not mandatory and each product can handle this requirement in its own way.

**State management problem** the externalization service is a convenient way to save and restore an object's state to and from a stream object. Although this service is not widespread, it is easy to develop a basic implementation using IDL types to avoid heterogeneity problem.

## 4  A generic mechanism for object mobility

We have seen that a mechanism to manage object mobility on CORBA could neither rely on a particular product's features nor on the underlying operating system. So we propose a generic solution which is based on two intermediary objects called *proxy* and *representative* and a *controller* object to handle migration. These intermediary objects are very similar to *stubs* and *skeletons* and in fact they are used as a mean to wrap them so as to gain control over invocation mechanisms and to provide transparent object reference management. As *stubs* and *skeletons*, *proxy* and *representative* objects are generated from the IDL interface of the server. We will first introduce CORBA's mobility model based on the *LifeCycle Service* and then we will explain how our solution fits in this model.

### 4.1  CORBA mobility model

CORBA mobility model is based on the *LifeCycle Service* [1] (cf. IDL 1). There is no standard implementation because objects may be of very different kinds. Therefore, an object that provides this service must support it by implementing the methods for copying, moving and removing itself. The mechanism used to copy and move an object is illustrated figure 2. Objects are created by *factories*. A factory is an object that knows how to create certain kinds of objects. The *client* (i.e, the user of the LifeCycleObject interface) must provide to the *target* object (i.e, the object to be moved, copied or removed) a reference to a *factory finder* (1). It is up to the target to obtain a list of proper factories using this factory finder (2) and to create the new instance of itself by invoking the factory it has chosen (3). The choice of a factory is based on *criteria* passed to the target by the client. A factory finder is bound to a search area, and, as a consequence, defines an abstract location for the object's destination.

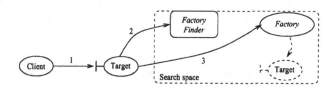

**Fig. 2.** object mobility model in CORBA

```
interface LifeCycleObject {
  LifeCycleObject copy(in FactoryFinder there,
                       in Criteria the_criteria)
        raises(NoFactory, NotCopyable, InvalidCriteria,
                       CannotMeetCriteria);
  void move(in FactoryFinder there,
            in Criteria the_criteria)
        raises(NoFactory, NotMovable, InvalidCriteria,
                  CannotMeetCriteria);
  void remove()
        raises(NotRemovable);
};
```

**IDL 1:** LifeCycle Service's interface

## 4.2  Definition of a generic mechanism

As stated before, we propose a generic mechanism to achieve mobility following CORBA's model by using intermediary objects which encapsulate migration mechanisms.

**The proxy object.** This object is on the client side and, as the stub, is generated from the IDL interface of the server. Its role is to encapsulate the stub to handle stale object references and broken invocations. This object is used in the same way as the stub in the client code and, thus, there is no change in the programming model of a client.

**The representative object.** A server which must be mobile must have an interface that inherits from *LifeCycleObject*. The *representative* object is on the server side and is also generated from its interface. It encapsulates the server object via inheritance and is used as a mean to intercept methods and to perform some extra operations to handle invocation management before calling the server's actual methods.

**The controller object.** This object actually implements the methods of the *LifeCycleObject* interface providing different migration behaviours. This way, behaviours can be changed without modifying the application object. The following behaviours have already been implemented :

**Passive** Migration can occur only when the object is quiescent. The *controller* waits for running activities (i.e, invocations) to terminate before migrating.

**Copy and differed deletion** Upon migration the server is replicated on the destination site and the instance of the server on the original site is deleted only when it becomes quiescent.

**Active** In this case, the server must use the *checkpoint()* method provided by the *controller* to enable the server to be stopped during invocations. When every invocation in the server has been stopped, the server is migrated and an exception is raised for each interrupted invocation. It is up to the *proxies* on the clients' side to restart these invocations. This level of mobility is interesting for instance when the machine on which a server is executed is about to be turned off.

Handling of invocations that occur during the migration process is done the following way : they are trapped by the *representative* until migration is done. These invocations are then released with a MOVED exception raised by the *representative*. This exception comprises the object reference of the new instance of the server that the *proxy* can use.

Figure 3a, illustrates how the mechanism works in case of active migration. An invocation (1) goes through the representative and is registered by the controller before the actual server's method is called (2). The server checks if it should migrate using the controller (3) (checkpoint method). If a migration is to occur, a new instance of the object is created at the destination site (4). This new instance then invokes the former server (5) to get its state. The former server's representative then raises a *MOVED* exception for each invocation (6) before the server is destroyed. When a client's proxy receives the exception, it restarts the invocation on the new server (7). Figure 3b illustrates how the mechanism fits in the LifeCycle architecture. When a move invocation is received by the representative (1), it is passed to the controller which invokes the factory finder

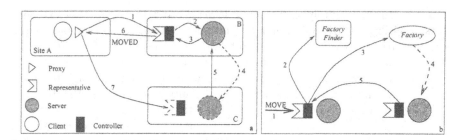

**Fig. 3.** Architecture and working of the migration mechanism

to obtain a list of available factories (2). It then invokes the chosen factory (3) to create a new instance (4). Upon creation, the new representative invokes the former one to obtain the server's state and restore it in the new server(5).

### 4.3 Implementation

The two intermediary objects and the controller are generated from the server's IDL interface. When a server has to migrate, we must stop current activities (i.e., invocations) before saving the state. It is the programmer's duty to place explicit breakpoints where migration can occur. State management is done through save_state() and restore_state() methods. These methods are generated by a preprocessing tool from the object's definition in native language. The tool is yet basic and can handle only simple and constructed types and sinple pointers such as char*. The state is saved in a stream object which is accessed with methods that complies with the externalization service's interface.

Binary executables differ according to machines and systems. They may also differ according to ORBs because of differences in the way objects and the ORB interact in different products although the OMG is making an effort to enhance server portability. For this reason, the network is divided in domains according to machines, systems and ORBs and a binary executable is generated for each domain. At the moment, the code is accessible via distributed file systems such as NFS (Networked File System) but it could be copied when an object migrates to a node which has no access to a binary executable.

## 5 Related work

VisiBroker [9] from Visigenic is an ORB with support for fault tolerance. It implements what we have called copy and differed deletion behaviour and the ORB handles rebinding of object references transparently. For servers which have a state, the clients must register to an event manager and implement a handler to update their view of the server. Since rebinding is an internal mechanism of the ORB, it does not need any other mechanism at application level. However, this solution does not rely on the *Lifecycle Service* and, thus, cannot be extended outside Visibroker; Orbix[7] from IONA proposes *Smart Proxies* that allow a programmer to inherit from generated stubs so as to change the default behaviour. This can be very convenient to implement transparent reference lookup in the naming service when a reference is stale. However this is not enough to implement all the functions of our model. In [8], the author studies object mobility on CORBA for graphics applications. His work is targeted at fine grain objects and the solution does not consider active objects. It provides an implementation of the LifeCycle Service but management of object references relies on the ability of ORBeline (now Visibroker) to rebind object references transparently.

The use of preprocessing tools to save objects state at application level has been done in *DOME* [5] to implement fault tolerance for scientific programs on PVM[1]. This implementation relies on the use of predefined classes of objects (scalar, vector, ...) which can save their value. But for standard data types, saving is not automatic and must be added by the programmer. Moreover, invocation management is not an issue since communications are only data exchanges and state saving is not done during communication steps.

## 6 Future work

We have already implemented our mechanism and the *LifeCycle Service* (factory and factory finder) on top of COOL-ORB from Chorus Systems. Although we have no performance measures yet we do not expect these additional mechanisms to be too costly when there is no migration since we only add an extra method call for each incoming invocation and the checkpoint method amounts to a single test. Yet, an object which is asked to move while it is invoking (nested invocations) returns a *NotMoveable* exception. To be able to interrupt a nested invocation, would imply to provide a kind of transaction mechanism so as to be able to undo any effect of the outgoing invocation. We plan to implement the *Compound LifeCycle* so as to be able to migrate graphs of objects. With this facility we will be able to study dynamic clustering of objects so that objects that interact intensively can be automatically migrated together. Another point we are interested in is the use of *Criteria* defined in the LifeCycle Service to choose the right factory. Nothing has been specified about these criteria and we plan to explore their use to express requirements such as:

| Requirement | Possible criteria |
|---|---|
| Resource availability | type of OS, ORB, binary executable; special device |
| Performance | machine load, destination on LAN or WAN |
| Security | trusted hosts, domains, ... |
| Cost | destination on LAN or WAN, ... |

---

[1] Parallel Virtual Machine is an environment for process communication in heterogeneous network of workstations.

# 7 Conclusion

We propose a generic mechanism to introduce object mobility in a CORBA compliant way which relies on intermediary objects generated from the server's IDL interface. An interesting point in this approach is that it can be applied to a variety of languages. The aim, which is to provide object migration in an environment and across different ORB products, fits well in the effort from OMG to achieve interoperability. A first implementation has been realized on top of COOL-ORB and is being extended to some other ORBs (Visibroker, Orbix) to validate our model and to study performances with different products. The next step will be to implement the *Compound Lifecycle Service* to allow migration of objects graphs so as to explore dynamic clustering of objects to enhance migration performance and to study the use of *criteria* mostly according to security and object positioning

# References

1. Common Object Services Specification, Volume I. Technical Report 94-1-1 Revision 1.0, Object Management Group, Mars 1994.
2. Common Object Request Broker Architecture and Specification. Technical Report 96-03-04 Revision 2.0, Object Management Group, Juillet 1995.
3. Ajay Bakre and B. R. Badrinath. M-rpc: A remote procedure call service for mobile clients. Technical report, Rutgers, State University of New Jersey, 1995.
4. Amnon Barak, Oren Laden, and Yuval Yarom. The NOW MOSIX and its Preemptive Process Migration Scheme. *Bulletin of the IEEE Technical Committee on Operating Systems and Application Environments*, 7(2):5–11, 1995.
5. Adam Beguelin, Erik Seligman, and Michael Starkey. Dome: Distributed object migration environment. Technical Report CMU-CS-94-153, Carnegie Mellon University, May 1994.
6. Eric Jul, Henry Levy, Norman Hutchinson, and Andrew Black. Fine-Grained Mobility in the Emerald System. *ACM Transactions on Computer Systems*, 6(1):109–133, fvrier 1988.
7. IONA Technologies Ltd. *Orbix 2 - Programming Guide*, 1996.
8. Vijay Machiraju. Object Migration for Distributed Graphics Applications. Master's thesis, University of Utah, Novembre 1996.
9. Visigenic. *Visibroker for C++ - Programmers Guide*, 1996.

# A Theory of Game Trees, Based on Solution Trees

Wim Pijls, Arie de Bruin

Department of Computer Science, Erasmus University,
P.O.Box 1738, 3000 DR Rotterdam, The Netherlands.
e-mail *{pijls,adebruin}@few.eur.nl*

**Abstract.** In this paper, a theory of game tree algorithms is presented, entirely based upon the concept of solution tree. During execution of a game tree algorithm, one may distinguish between so-called alive and dead nodes. It will turn out, that only alive nodes have to be considered, whereas dead nodes should be neglected. The algorithm may stop, when every node is dead. Further, it is proved that every algorithm needs to build a critical tree. Finally, we show, that some common game tree algorithms agree with this theory.
**Keywords:** Game tree search, Minimax search, Solution trees, Alpha-beta, SSS*, MTD, (Nega)Scout, Proof Number Search.

## 1 Introduction

A game tree models the behavior of a two-player game. Each node $n$ in such a tree represents a position in a game. The players are called MAX and MIN. Accordingly, *max* and *min* nodes are distinguished. In every terminal (end position) $p$ of a game tree, a function value $f(p)$ is assumed to be defined, representing the profit or the payoff for MAX, or equivalently, the loss for MIN. MAX aims at maximizing this payoff, whereas MIN aims at minimizing it. The guaranteed payoff for MAX in a non-terminal $n$ is determined by the minimax function $f$, defined as:

$$f(n) = \max\{f(c) \mid c \ \text{a child of } n\}, \text{if } n \text{ is a max node,}$$
$$= \min\{f(c) \mid c \ \text{a child of } n\}, \text{if } n \text{ is a min node.}$$

The minimax value in an initial position indicates, whether one player has a forced win in that position or not. For most games (Hex, Othello, Checkers, Chess, Go, Gomuko), this problem is known to be PSPACE-complete. Over the years, many algorithms have been designed, which compute the minimax value (also called *game value*) in the root, given a payoff value in the terminals. In this paper, we will capture the best-known algorithms in a unifying view. General properties of them will be derived. An earlier attempt to create such a unifying view was made by Ibaraki[3, 4].

The root of any given game tree is denoted by $r$ in this paper. The minimax value $f(T)$ of a tree $T$ is defined is as the minimax value $f(r)$ of the root $r$ of $T$. Given a statement related to a game tree, replacing the terms max/min by min/max yields the so-called *dual* version of that statement.

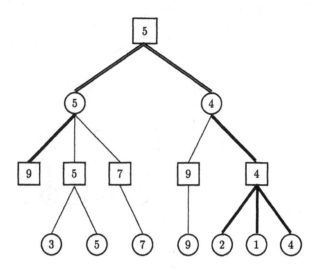

**Fig. 1.** A max solution tree in a game tree with $f$-values.

As already indicated in [2], *solution trees* play a key role in the theory of game trees. By definition, a solution tree is either a max tree $T^+$ or a min tree $T^-$. The shape of a max tree $T^+$ is defined by the rule, that all children should be included for every non-leaf max node and exactly one child should be included for every non-leaf min node. A max solution tree can be viewed as a strategy of MIN, including in each min position exactly one continuation and in each max node all continuations (all countermoves to MIN). Dually to a max tree, we have a min tree. In Figure 1, an example of a game tree is shown, labeled with its $f$-values. The bold edges generate a max tree.

The minimax value restricted to a solution tree $T^+$ or $T^-$ is denoted by $g$. The importance of solution trees is a result of Stockman's theorem[10]:

$$f(n) = \min\{g(T^+) \mid T^+ \text{ is a max tree with root } n\} \tag{1.1}$$
$$= \max\{g(T^-) \mid T^- \text{ is a min tree with root } n\} \tag{1.2}$$

Given a game tree $G$, a max or min tree $T$ in $G$ is called *optimal*, if $g(T) = f(G)$. The combination of an optimal max tree and an optimal min tree is called a *critical tree*.

This paper is organized as follows. Section 2 presents some preliminary definitions. The theory itself is discussed in Section 3. Section 4 gives an overview of some common game tree algorithms and their relationship to the theory. The reader is referred to [7] for proofs and other details.

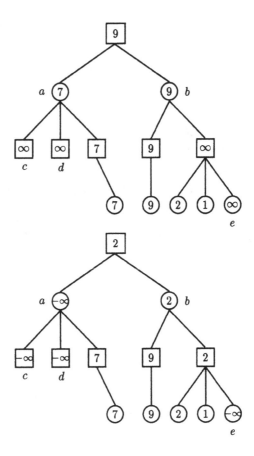

**Fig. 2.** A search tree with $f^+$-values (top) and $f^-$-values (bottom).

## 2 The Search Tree

In all game tree algorithms, the game tree is built up step by step. At any time during execution, a subtree of the game tree has been built up. Such a subtree is called a *search tree*. We assume, that, as soon as at least one child of a node $n$ is generated, all other children of $n$ are also added to the search tree. If the children of a node $n$ have been generated, $n$ is called *expanded* or *closed*. If a non-terminal $n$ has no children in a search tree (and hence, $n$ is a leaf in this search tree), then $n$ is called *open*. A terminal $n$ is called *closed* or *open* respectively, according to whether its payoff value has been computed or not. Notice, that a closed leaf in a search tree is always a terminal in the game tree.

In the open nodes of a search tree $S$ we can define two provisional payoff values, $+\infty$ and $-\infty$. By introducing these payoff values to $S$, we introduce two other game trees $S^+$ and $S^-$ with minimax values $f^+$ and $f^-$ respectively. The following inequality holds for every $x \in S$:

$$f^-(x) \le f(x) \le f^+(x)$$

A search tree, derived from Figure 1, can be found in Figure 2, where $c$, $d$ and $e$ are open nodes. In the trees, the $f^+$- and $f^-$-values respectively are shown.

We try to apply Stockman's theorem to a search tree. A solution tree $T$ is called *closed*, if every leaf of $T$ is a closed terminal. Hence, a closed solution tree in a search tree is also a solution tree in the game tree. For closed solution trees, the minimax value $g$ reduces to:

$$g(T^+) = \max\{f(p) \mid p \text{ a terminal in } T^+\} \tag{2.1}$$

$$g(T^-) = \min\{f(p) \mid p \text{ a terminal in } T^-\} \tag{2.2}$$

If a solution tree in a search tree has at least one open leaf, it is called an *open* solution tree. As a result of (2.1) and (2.2), any open solution tree has an infinite $g$-value. In order to assign a finite value to any (open or closed) solution tree, we introduce the $c$-function. For a max and a min tree $T^+$ and $T^-$ respectively, we define:

$$c(T^+) = \max\{f(p) \mid p \text{ a closed terminal in } T^+\}$$

$$c(T^-) = \min\{f(p) \mid p \text{ a closed terminal in } T^-\}$$

Notice, that, if $T$ is a closed solution tree, then $c(T) = g(T)$. Applying Stockman's theorem to $S^+$ and $S^-$ yields the following relations for $f^+(x)$ and $f^-(x)$ respectively, with $x$ any node in a search tree $S$.

$$f^+(x) = \min\{g(T^+) \mid T^+ \text{ a closed max tree with root } x\} \tag{2.3}$$

$$= \max\{c(T^-) \mid T^- \text{ a min tree with root } x\} \tag{2.4}$$

$$f^-(x) = \max\{g(T^-) \mid T^- \text{ a closed min tree with root } x\} \tag{2.5}$$

$$= \min\{c(T^+) \mid T^+ \text{ a max tree with root } x\} \tag{2.6}$$

It follows from (2.3) and (2.5), that, when the equality $f^+(r) = f^-(r)$ is obtained, also a critical tree is obtained.

## 3 A General Theory

In this section, we present a general paradigm underlying almost every game tree algorithm. The following definition is crucial to this theory.

**Definition** *A node $n$ in a search tree $S$ is called alive, if $h^-(n) < h^+(n)$, where $h^-(n)$ and $h^+(n)$ are defined as:*

$$h^-(n) = \min\{c(T^+) \mid T^+ \text{ a max tree in } S \text{ through } r \text{ and } n\} \tag{3.1}$$

$$h^+(n) = \max\{c(T^-) \mid T^- \text{ a min tree in } S \text{ through } r \text{ and } n\} \tag{3.2}$$

*A node $n$ that is not alive, is called dead.*

The following formulas for a node node in a search tree $S$ are of highly practical significance. See [7] for a full proof.

$$h^-(n) = \max\{f^-(x) \mid x = n \text{ or } x \text{ an ancestor of } n\} \tag{3.3}$$
$$= \max\{f^-(x) \mid x = n \text{ or } x \text{ a max type ancestor of } n\}$$
$$h^+(n) = \min\{f^+(x) \mid x = n \text{ or } x \text{ an ancestor of } n\} \tag{3.4}$$
$$= \min\{f^+(x) \mid x = n \text{ or } x \text{ a min type ancestor of } n\}$$

These equalities are utilized implicitly in several algorithms. As a consequence of the equalities, the following properties are trivial: every alive (non-terminal) node has at least one alive child, and every dead node has solely dead children.

**Alive nodes.** Given an alive node $n$ in a search tree $S$, we can define a game tree $G_n \supseteq S$, whose game value can be obtained, only if an open descendant[1] of $n$ is expanded. The construction of $G_n$ is as follows. Let $T^+$ and $T^-$ be solution trees in $S$ with $c(T^+) = h^-(n) < h^+(n) = c(T^-)$. The leaf $p_0$ at the end of the common path must be open, since, if it was not, we would have $f(p_0) \leq c(T^+) < c(T^-) \leq f(p_0)$. Since $c(T^+) < c(T^-)$ and $p_0$ belongs to both $T^+$ and $T^-$, the above definition implies, that $p_0$ is alive. Notice, that the path from $r$ to $p_0$ crosses $n$. Choose a value $f_0$ with $c(T^+) \leq f_0 \leq c(T^-)$. Define $f(p_0) = f_0$ and $f(p) \leq h^-(n)$ for any open node $p \neq p_0$ in $T^+$ and $f(q) \geq h^+(n)$ for any open node $q \neq p_0$ in $T^-$. The extended solution trees are denoted by $\bar{T}^+$ and $\bar{T}^-$ respectively. To complete $G_n$, the other open nodes in $S$ (if any) are closed arbitrarily. We see, that $\bar{T}^+ \cup \bar{T}^-$ is a critical tree in $G_n$ and $f(G_n) = f_0$. In any search tree $S'$ of $G_n$ with $p_0$ unexpanded, the inequality $f^-(r) \leq c(\bar{T}^+ \cap S') < c(\bar{T}^- \cap S') \leq f^+(r)$ holds. This shows, that, in order to be able to determine $f(G_n)$, one must expand $p_0$.

The above construction of $G_n$ is illustrated using Figures 2 and 3. The relations (3.3) and (3.4) tell us, that $h^-(a) = 2$ and $h^+(a) = 7$ in Figure 2. The game tree $G_a$ is constructed, by defining $f(d) = f_0$ with $f_0 \in [2, 7]$, $f(e) \leq 2$ and $f(c) \geq 7$.

**Main theory.** We now come to the main points of our theory.
a) The above construction shows, that, as long as a search tree contains at least one alive node $n$, the minimax value of the game tree under consideration is still unknown. We conclude, that the algorithm must be continued, as long as any node in the search tree is alive. The algorithm may only stop, when the root (and hence any other node) is dead.
b) The root $r$ is dead, iff $f^-(r) = f^+(r)$. In turn, this condition is equivalent to the availability of a critical tree in the search tree. We conclude, that every game tree algorithm must build a critical tree.
c) In order to achieve a critical tree, the dead open nodes in a search tree can be discarded. This will be proved and illustrated in the rest of this section.

---

[1] Here, $n$ is assumed to be its own descendant.

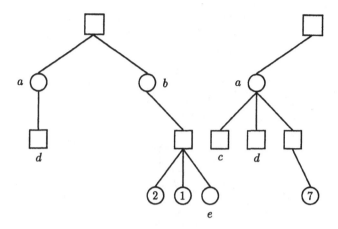

**Fig. 3.** A max and a min tree, derived from Figure 2.

**Dead nodes.** Suppose a search tree $S$ is extended to a game tree $G$ with $f^+(r) = f^-(r)$. We will prove, that $G$ contains at least one critical tree $C$, that avoids any node, that is open and dead in $S$.

Suppose an optimal max tree $T^+$ of $G$ includes a node $n$, that is open and dead in $S$. (The case, that an optimal min tree includes an open dead node of $S$, is dual). By (3.4), there is a node $u$, $u = n$ or $u$ an ancestor of $n$, with the property $f^+(u) = h^+(n)$ in $S$. By (2.3), $f^+(u)$ is the $g$-value of a closed max tree $T'$, rooted in $u$. Now, we have in $S$ the following series of (in)equalities: $g(T') = f^+(u) = h^+(n) \leq h^-(n) \leq c(T^+ \cap S) \leq g(T^+)$. Since $T'$ is a closed, $n \notin T'$. When we replace the subtree of $T^+$ below $u$ by $T'$, another optimal max tree $T_1^+$, which avoids $n$, is constructed. If $T_1^+$ still includes open dead nodes, the above procedure is repeated, until a closed max tree is obtained.

To illustrate the foregoing, see Figure 4. The nodes to the right of the path from $r$ to $y$ are open, as is node $y$. Since $h^+(v) = 3$ and $h^-(v) = 4$, $v$ is dead. Due to the given values $f^+(q)$ and $f^-(x)$ respectively, there is a closed max tree $T'$, rooted in $q$, with $g(T') = 3$ and a closed min tree $T''$, rooted in $x$, with $g(T'') = 4$. Suppose, that an optimal max tree $T^+$ crosses $v$. Then $T^+$ can be replaced by a related max tree through $q$. Dually, an optimal min tree through $v$ can be chosen, such that $x$ is included, rather than $y$.

## 4  Game Tree Algorithms

In this section, we show, how some well-known game tree algorithm fits our theory. See [7] for details.

**The alphabeta algorithm.** Alphabeta is the most classical game tree algorithm. The first formal description was given in [6]. The algorithm searches the

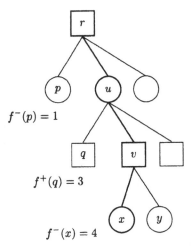

**Fig. 4.** An example.

game tree from left to right. The code consists of a recursive function with a node parameter $n$ and two real valued parameters $\alpha$ and $\beta$. When a nested call $alphabeta(n, \alpha, \beta)$ is performed, the relations $h^-(n) = \alpha$ and $h^+(n) = \beta$ hold, and $n$ is the leftmost open alive node in the search tree. Hence, alphabeta may be characterized as *the algorithm expanding the leftmost open alive node* in each step.

**SSS\* and MT-SSS.** The SSS\*-algorithm was published in 1979 by Stockman [10]. The original code is rather opaque. Recently[8, 9], it was shown, that SSS\* is equivalent to a series of alphabeta calls with $\beta = \alpha + 1$. This new formulation is called MT-SSS and has been proven a very feasible and convenient algorithm. We show in [7], that, whenever a node is expanded in MT-SSS, this node is *the leftmost open alive node with maximal $h^+$-value.* So, a characterization of MT-SSS is obtained.

There exist also dual versions, named DUAL and MT-DUAL respectively, which select the leftmost open alive node with minimal $h^-$-value in each step.

**Negascout.** We have explained in [2], how Negascout builds a critical tree. It is proved in [7], that Negascout expands only alive nodes.

**Proof number search.** This algorithm along with applications has been published in [1]. Due to this algorithm, the games Qubic and Connect-Four have been solved. The working can be understood best in terms of solution trees, although the original description does not mention this notion. The algorithm is designed for game trees with payoff values of just -1 and 1.

In a search tree, a min tree $T^-$ with $c(T^-) = 1$ is called a proof tree, and a max tree $T^+$ with $c(T^+) = -1$ is called a disproof tree. A (dis)proof tree with a minimal number of open nodes is called minimal. In each iteration, a minimal proof and a minimal disproof tree are considered and the node $p_0$ at the end of the common path (which is open and alive according to Section 3) is expanded. In order to trace easily this node $p_0$, so-called proof numbers and disproof numbers are introduced.

# References

1. L. Victor Allis, Maarten van der Meulen, and H. Jaap van den Herik, *Proof-number search*, Artificial Intelligence 66 (1994), pp. 91-124.
2. A. de Bruin, W. Pijls, *Trends in Game Tree Search*, Proceedings Sofsem '96, Theory and Practice of Informatics (Keith G. Jeffery et al. eds.), LNCS vol. 1175, 1996, pp. 255-274.
3. Toshihide Ibaraki, *Generalization of alpha-beta and SSS\* search procedures*, Artificial Intelligence 29 (1986), pp. 73-117.
4. Toshihide Ibaraki, *Search Algorithms for Minimax Game Trees*, Conference paper at *Twenty years NP-completeness*, Sicily, June 1991.
5. V. Kumar and L.N. Kanal, *A General Branch and Bound Formulation for Understanding and Synthesizing And/Or Tree Search Procedures*, Artificial Intelligence 21 (1983), pp. 179-198.
6. Donald E. Knuth and Ronald W. Moore, *An analysis of alpha-beta pruning*, Artificial Intelligence 6 (1975), no. 4, pp. 293-326.
7. Wim Pijls, Arie de Bruin, Aske Plaat, *A theory of game trees, based on solution trees*, Tech. Report EUR-CS-96-06, Erasmus University Rotterdam, December 1996, also available as:
   *http://www.cs.few.eur.nl/few/inf/publicaties/rapporten.eur-few-cs-96-06.ps*
8. Aske Plaat, Jonathan Schaeffer, Wim Pijls and Arie de Bruin, *Best-first fixed depth game tree search in practice*, Proceedings of the 14th International Joint Conference on Artificial Intelligence (IJCAI-95), vol. 1, pp. 273-279.
9. Aske Plaat, Jonathan Schaeffer, Wim Pijls and Arie de Bruin, *A Minimax Algorithm Better than SSS\**, Artificial Intelligence 84 (1996) pp. 299-337.
10. G. Stockman, *A minimax algorithm better than alpha-beta?*, Artificial Intelligence 12 (1979), no. 2, pp. 179-196.

# Approximation Algorithms for the Vertex Bipartization Problem

H.Schröder[1] A.E.May[1] I.Vrťo[2] O.Sýkora[2]

[1] PARC, Loughborough University, Loughborough, UK
[2] Institute for Informatics, Slovak Academy of Sciences, Bratislava

**Abstract.** To guarantee the optimal bipartite vertex coloring (bipartization) of a connected graph requires a coloring algorithm that is NP-complete, effectively preventing bipartization of even modest sized graphs. We present some approximation algorithms that run in *polynomial* time and lead to very good (but not necessarily optimal) colorings.

## 1 Introduction

Given an undirected graph $G = (V, E)$ we may partition the nodes into two sets $V_1$ and $V_2$ such that the number of edges in $(i, j) \mid (i, j) \in E \wedge i \in V_1 \wedge j \in V_2$ is maximized. This is called the *max-cut* problem, and corresponds to minimizing the number of edges in $(V_1 \times V_1 \cup V_2 \times V_2) \cap E$. We shall call the latter set $VS$, the violation set. The problem is equivalent to finding a coloring of the nodes such that the number of edges that connect nodes of the same color is minimized (and, of course, the number of edges connecting nodes of different colors is maximized).

The process of making a graph bipartite (and hence $|VS| = 0$) cannot be effected by node coloring alone unless the graph consists only of even cycles. If odd cycles are present then the structure of the graph must be altered before $|VS| = 0$ can be accomplished, although the overall topology of the graph can be retained.

One approach to making a graph containing one or more odd cycles bipartite is by the removal of the minimal number of edges required to effect $|VS| = 0$. Another approach would be to replace as few edges as possible by length 2 paths through new, unused nodes (the new nodes having degree 2). The latter node would have an inverse coloring to the nodes that supported the violating edge. We shall call this dual problem to the max-cut problem the *min-violation problem*.

The min-violation problem occurs in many practical situations. One application is in the design of fast asynchronous logic: here the problem is that only bipartite graphs (in this case state transition diagrams) can be implemented and thus general graphs have to be made bipartite by removing or expanding odd cycles by introducing additional nodes on edges within odd cycles. And as extra nodes introduce delays into the hardware, and increase hardware costs, adding a minimum of such nodes is highly desirable. We see here the dual to the max-cut problem.

We will treat the min-violation problem using the terminology from node colorings of graphs. We are only interested in 2-colorings (graphs with a 2-coloring are bipartite graphs). As colors we will use black and white throughout this paper. An optimal solution to the max-cut problem is an optimal solution to the min-violation problem. But what might be regarded as a good solution (relative to the size of the cut) to the max-cut problem is not necessarily regarded a good solution (relatively) to the min-violation problem: in fact there are algorithms that guarantee that the size of the cut they produce is on average at most 15% smaller than the maximum cut. Such algorithms would not guarantee that the size of the violation set they produce is only a constant factor away from the minimum size of the violation set. We will in fact show corresponding examples, where the size of the cut set is $O(n)$ while the size of the min-violation set is constant so that a relatively small change in the size of the cut set results in a relatively large change in the size of the set of violations.

Our simulation results demonstrate that the heuristics we developed are actually much closer to the optimum than the proven bounds.

## 2 Algorithms

### 2.1 Breadth first search (BFS)

Any tree can be made a bipartite graph by coloring alone. Thus Let $G = (V, E)$ be an arbitrary graph with $|V| = n$ and let $T = (V, E')$ with $E' \subseteq E$ be a spanning tree of $G$. Then the nodes of $G$ can be colored alternately black and white starting from the root of $T$ so that all edges from $E'$ connect a black node with a white node. The edges of (or two black) nodes are then the edges outside the cut and are replaced by a path of length 2 with a new black (white) node.

Thus we have a simple algorithm to produce a bipartite graph. This solution is not necessarily optimal even though there is always a spanning tree that produces an optimal coloring. Thus it makes sense to pick the best solution resulting from a given set of spanning trees. We decided to pick as a set the n possible breadth-first-search spanning trees (choosing n different roots). Note: the breadth-first spanning tree itself is not uniquely defined specifying its root, but its coloring is.

It is easy to produce examples of graphs where this algorithm is highly suboptimal. Instead of presenting such an example here we shall reveal later a graph that is colored highly non-optimally by all algorithms presented in this paper.

The BFS algorithm is as follows:

1. Produce n breadth-first-searches spanning trees of $G = (V, E)$ with n different roots.
2. For each of these spanning trees produce a 2-chromatic coloring of $V$: nodes that have an even length path back to the root node (during the traversal) are given one color, and vice-versa.
3. For each spanning tree determine $|VS|$, i.e. the number of edges in $E$ that connect either 2 black or 2 white nodes.
4. Pick a coloring with minimal $|VS|$.

## 2.2 Depth first search (DFS)

As for the BFS traversal, except a depth-first-search spanning tree is used.

## 2.3 The odd cycle number algorithm

An edge shared by many odd cycles is an obvious candidate for replacement (or removal). This is the basic idea for the following algorithm:

1. Produce the shortest pair-wise distances of all nodes of $G$ (e.g. using Warshall's algorithm).
2. For every edge $(i, j)$ in $G$ determine the odd-cycle-number, i.e. the number of nodes that have equal distance to i and j.
3. Remove an edge with maximal odd-cycle-number.
4. If there are still odd cycles in the remaining graph, i.e. edges with odd-cycle number unequal to zero then repeat the algorithm on the remaining graph.

## 2.4 The re-coloring algorithm (a greedy approach)

In this approach we start with any 1 or 2-chromatic coloring of the graph G and then look for white (black) nodes that are connected to more white (black) nodes than black (white) nodes. We change the color of such nodes as this will reduce the number of violations.

1. Label each node with $B$, where $B$ is the number of neighbors of the same color minus the number of neighbors of a different color.
2. Invert the color of a single node with highest $B$
3. Repeat from step 1 until all nodes have $B \leq 0$ (and thus $|VS|$ cannot be decreased further).

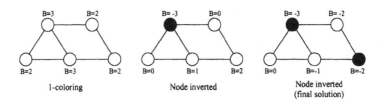

**Fig. 1.** Example showing the steps in the recoloring algorithm

## 3 Simulation Results

In order to establish the quality of heuristics we preferably would like to prove that the results delivered by the heuristics are within a certain small percentage range that is close to the optimal. In the case of the max-cut problem it is known that there are no polynomial time approximation algorithms that generate solutions close to the optimal (see [6]). Thus we used simulation to get a good indication of the average performance of the various heuristics presented in this paper.

Although during development it is satisfactory to *test* algorithms by using graphs generated by hand, this method does not lend itself to quantitative analysis: one may deliberately choose graphs that an algorithm will perform well on, or very sub-optimally. To this end the algorithms described herein were tested over a large set of automatically-generated random graphs. We produce random connected graphs keeping $|V|$ constant and varying the average degree of nodes over the entire graph. We test from a tree up to $K_{|V|}$ (using hundreds of graphs at each degree step to average out the randomness) plotting average degree versus $|VS|$.

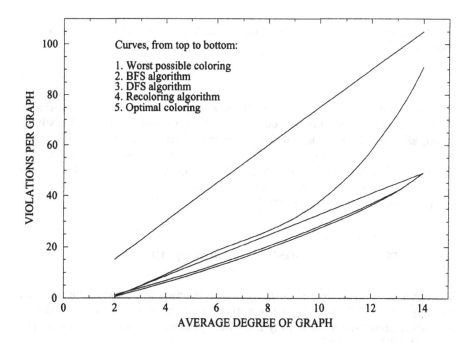

**Fig. 2.** Algorithm performance on 15-node graphs, averaged over 10 000 graphs per degree step, (except the optimal coloring which is averaged over 1000 graphs only)

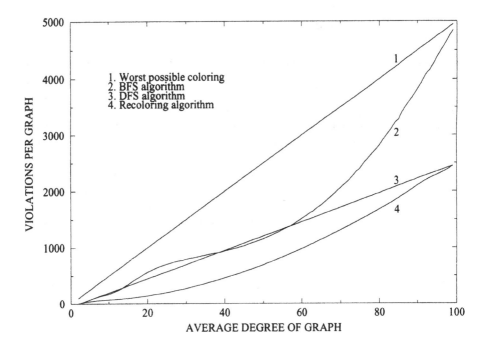

**Fig. 3.** Algorithm performance on 100-node graphs, averaged over 1000 graphs per degree step

To put our results in perspective, the performance of an optimum coloring is shown as a lower bounding line on figure 2. This was obtained by exhaustively trying all possible colorings. A higher bound is also plotted, assuming every edge is in violation (i.e. $(V_1 = V) \wedge (V_2 = \emptyset)$). The plot shown incorporating the optimum coloring is for 15-node graphs only – trying all the possible colorings in order to find the optimum coloring for significantly larger graphs becomes too computationally expensive to permit performance analysis.

## 4   An example demonstrating non-optimality

The example graph presented in this section is in terms of the size of the maximal cut only about 25% away from the optimal while in terms of min-violation it is a factor proportional to $n$ away from the optimum – no matter which of the above algorithms is used.

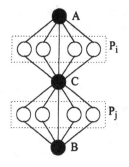

Let $G = (V, E)$ with $V = \{A, B, C\} \cup \{P_i \mid 1 \leq i \leq m\}$ where $m$ is even and $E = \{(A, C), (B, C)\} \cup \{(A, P_i), (C, P_i), (B, P_j), (C, P_j) \mid 1 \leq i \leq m/2, m/2 < j \leq m\}$ The optimal coloring is where nodes A, B and C have the same color and all other nodes have the different color, thus there are only 2 edges in the violating set.

Algorithm BFS will, whichever node we start at, either give different colors to A and C or to B and C. This results in $|VS| \geq m$. The odd-cycle number algorithm will for all edges $(A, P_i)$ and all edges $(B, P_j)$ (range of i and j as above) determine an odd-cycle number of $m+2$, while the edges $(A, C)$ and $(B, C)$ have an odd-cycle number of $m$. Thus the algorithm will result in a violation set of size $2m$. The re-coloring algorithm if started with the following coloring will not re-color a single node and thus will produce a violation set of size $m + 1$. The initial coloring is: $B, C$ and all $P_i$ white and A and all $P_j$ black (ranges as above).

Thus in this example all three algorithms deliver a violation set that is too large by a factor proportional to m. Note that the corresponding cut-set is in each case at most by a factor 2 away from the optimal.

## 5 Conclusions

It might be that for the min-violation problem even good (i.e. constant factor) approximation algorithms are NP-hard. But the simulation results show clearly that our algorithms are, on average, very close to optimal.

We have tried several improvements of the given algorithms: It certainly makes sense (and it does help considerably for graphs of small average node degree) to start the re-coloring algorithm with a coloring produced by a spanning tree algorithm. This is because there are colorings of trivial graphs that the re-coloring algorithm cannot improve: one example is a simple black-white string: $b, b, w, b, b, w$, etc. Changing the color of any single node in this context would not decrease $|VS|$. Pre-coloring the graph by using a spanning-tree technique would avoid this shortcoming, as a string is simple to color in an alternating fashion.

We have also implemented an algorithm that changes the colors of groups of nodes instead of single nodes. If we look at the aforementioned $b, b, w, b, b, w$ example, changing the color of a single black node does not decrease $|VS|$. We call this a *zero-gain* move. Now if we execute several zero-gain moves on our example string, we can eventually color it optimally to $b, w, b, w, b, w$, seeing a reduction in $|VS|$. The algorithm is more complex and yields slightly better results than the re-coloring version, but lacks a strict termination condition: if we can execute zero-gain moves (such that $|VS|$ is not always decreasing between iterations) we cannot be sure we will encounter a lower $|VS|$ later on. An 'artificially' applied termination condition must be applied, for example by limiting the number of color inversions each node can undergo.

This is a strong indication that techniques like simulated annealing could be applied successfully in particular in order to improve solutions that have been produced by one of the polynomial deterministic algorithms.

Our algorithms yield excellent results on average, and in polynomial time. If the user can tolerate the occasional bad solution, then our algorithms will suffice: indeed, in many situations there may be no choice, as the optimal solution is impossibly slow for even the smallest graphs. For exceptionally large graphs (millions of nodes) a faster algorithm such as the DFS may be desirable to attain tolerable run-time ($O(n^2)$). But certainly for graphs of thousands of nodes the re-coloring algorithm yields admirable results in good time. Conversely, the odd-cycle membership algorithm provides such slow speed that even 1000-node graphs may prove intractable.

Further investigations into the trade-off between quality of the solution and complexity and run-time of the algorithm are desirable.

# 6   References

1. Goemans, M. X., Williamson, D. P.: *0.878 approximation algorithm for max-cut and max-2-sat.* Proc. 26th Annual ACM Symposium on the Theory of Computing (May 1994) 422-431.

2. Hadlock, F. O.: *Finding a maximum cut of a planar graph in polynomial time.* SIAM J. Comput. 4 (1975) 221-225.

3. Haglin, D. J., Venkatesan, S. M.: *Approximation and intractability results for the maximum cut problem and its variants.* IEEE Transactions on Computers 40 (1991) 110-113.

4. Poljak, S., Tuza, Z.: *The max-cut problem - a survey. (Technical Report)*, Academia Sinica, Taipei, Taiwan, January 1994 (a shorter version under title: *Maximum cuts and largest bipartite subgraphs* was published in Combinatorial Optimization, Eds.: Cook, W.,Lovasz, L. and Seymour, P., Providence, RI (1995) 181-244).

5. Sahni, S., Gonzales, T.: *P-complete approximation problems.* Journal of the ACM 23 (1976) 555-565.

6. Sanjeev Arora, Carsten Lund, Rajeev Motwani, Madhu Sudan and Mario Szegedy: *Proof verification and intractability of approximation problems.* Proceedings of the 33rd Annual IEEE Symposium on the Foundations of Computer Science (1992) 14-23.

In 1976, Sahni and Gonzales presented a 1/2-approximation algorithm. Since then there were only small improvements of the result, e.g. Haglin and Venkatesan show a $1/2 + 1/(2n)$ - approximation algorithm.

Substantial progress was made by Goemans and Williamson who designed a 0.87856-approximation algorithm. The result is very interesting and it looks as if an implementation of it should work very well.

A very good survey of results on maximum cut problem was made by Poljak and Tuza.

# Optical All-to-All Communication for Some Product Graphs

## (Extended Abstract)

Heiko Schröder[1], Ondrej Sýkora[2], Imrich Vrťo[2,*]

[1] Department of Computer Studies
Loughborough University of Technology
Loughborough, Leicestershire, LE11 3TU, United Kingdom
[2] Institute for Informatics, Slovak Academy of Sciences,
P.O.Box 56, 840 00 Bratislava, Slovak Republic

**Abstract.** The problem of all-to-all communication in a network consists of designing directed paths between any ordered pair of vertices in a symmetric directed graph and assigning them minimum number of colours such that every two dipaths sharing an edge have distinct colour. We prove several exact results on the number of colours for some Cartesian product graphs, including 2-dimensional (toroidal) square meshes of odd side, which completes previous results for even sided square meshes.

## 1 Introduction

Optical processing of information is a new and fast developing area of informatics. Optical processing finds its applications in such important areas as: video conferencing, scientific visualisation, real-time medical imaging, high speed super-computing, distributed computing (covering local to wide area). Reasons are higher speed and accuracy of information processing. Optic technology gives possibility of creating high speed networks in the near future. The high speeds arise from the fact that the signals can be maintained in optical form in such networks and not to be converted during the transmission.

Multiple laser beams can be propagated over the same fiber on distinct optical wavelengts. A single optical link in an optical network can carry several logical signals if they are transmitted on different wavelengts. Multiple messages can be transmitted across the same channel simultaneously as long as they use distinct wavelengths - this technique is known as wavelength division multiplexing (WDM). We may consider that light rays have different colors. The corresponding input and output terminals are modulated to emit and receive the signal on the prescribed wavelength. For large scale networks switching is necessary. The switching is performed directly on the optical signal, without translation into

* This research was supported by grant No. 95/5305/277 of Slovak Grant Agency and by grant of British Council to the Project Loughborough Reconfigurable Array and Theoretical Aspects of Interconnection Networks. The research of the last two authors was supported also by the grant of EU INCO-COP 96-0195.

electronic form. Optical switches do not modulate the wavelengths of the signals passing through them but they direct the incoming waves to one or more of their outputs. As buffering is not generally available, optical communication requires design of new algorithms amenable to this routing environment.

WDM-technology estalishes connectivity by finding transmitter-receiver paths, and assigning a wavelength (color) to each path, so that no two paths going through the same link use the same wavelength. It is clear that the number of wavelengths (colors), so called optical bandwidth, is a limiting factor. Our problem is to find a solution for an instance of connectivity requests such that the number of wavelengths is minimized. There are considered some classes of instances of connectivity requests. The most interesting is the so called All-to-All instance where all possible connectivity requests are realized.

Optical network is modelled as a symmetric directed graph $G = (V(G), A(G))$, where $V(G)$ is a set of vertices, $A(G)$ is a set of arcs, such that if $\alpha = (u, v) \in A(G)$ then $\alpha' = (v, u) \in A(G)$.

A request is an ordered pair of nodes $(x, y)$ in $G$ corresponding to a message to be sent by $x$ to $y$ and an instance is a set of requests. In our paper we consider the so called *All-to-All* instance consisting of requests $\{(x, y) : x \in V(G), y \in V(G), x \neq y\}$.

Let a routing $R$ in $G$ be a set of $|V(G)(V(G) - 1)|$ dipaths specified for all ordered pairs $x, y$ of vertices in $G$.

Our problem consists of finding a routing $R$ for $G$ and assigning each path a colour, so that no two dipaths of $R$ sharing an arc have the same colour. Given routing $R$ for $G$ the smallest number of colours is denoted by $\vec{w}(G, R)$. Let $\vec{w}(G)$ denote the smallest $\vec{w}(G, R)$ over all routings $R$.

Given a network $G$ and a routing $R$ the arc congestion of an arc $\alpha \in A(G)$ in the routing $R$, denoted by $\vec{\pi}(G, R, \alpha)$, is the number of dipaths of $R$ containing $\alpha$. The maximum congestion of any arc of $G$ in the routing $R$ is called arc congestion of $G$ in the routing $R : \vec{\pi}(G, R) = \max_{\alpha \in A(G)} \vec{\pi}(G, R, \alpha)$. The minimum congestion of $G$ in any routing $R$ is called arc forwarding index of $G$ and denoted by $\vec{\pi}(G)$.

Let $P_n$ and $C_n$ be an $n$−vertex symmetric directed path and an $n$−vertex symmetric directed cycle.
For $k \geq 2$ and $3 \leq n_1 \leq n_2 \leq ... \leq n_k$ define $C_k = \prod_{i=1}^{k} C_{n_i}$ and $\mathcal{P}_k = \prod_{i=1}^{k} P_{n_i}$.
Let $Q_k = \prod_{i=1}^{k} P_{n_i}, n_i = 2, i = 1, 2, ..., k$, denote the $k$-dimensional symmetric directed hypercube graph.
Let $K_n$ denote the complete directed graph on $n$ vertices.
Let $G(cd^m, d), 1 \leq c \leq d$ be a symmetric directed recursive circulant graph i.e. there are $cd^m$ vertices and the arcs connect the pairs $(u, u + d^i \mod cd^m)$, where $0 \leq i \leq m$ if $c \neq 1$, or $0 \leq i \leq m - 1$ if $c = 1$.
For $G$ and $H$ symmetric directed graphs define $G[H]$ a composition of the graphs created from the graph $G$ by replacing of each vertex of the graph $G$ by a copy of the graph $H$ and the set of edges is composed of edges of the copies of the graph $H$ and if there was an edge between two vertices of the graph $G$, then the edge is replaced by $|V(H)|$ edges connecting by a perfect matching two corresponding

copies of $H$. Let us recall that the graph known as hypercube of cliques [8] of dimension $k$ is actually $K_2^{\lfloor \log k \rfloor}[Q_{k-\lfloor \log k \rfloor}]$.

A graph $G$ is $(k, w)$–colorable (see [1]), if there exists a routing $R$ s.t. $w$ is the number of colours, two dipaths from $R$ sharing an arc of $G$ do not have the same colour, for each vertex $v \in G$ and for each colour there are at most $k$ dipaths starting in $v$ and there exists a colour that for each vertex $v \in G$ there are at most $k - 1$ dipaths starting in $v$.

Trivially $\vec{w}(G) \geq \vec{\pi}(G)$. One of the main open problem in this area is the question whether or not the equality holds for all symmetric directed graphs and all-to-all instance, [6]. So far the equality was proved for the following graphs:

$\vec{w}(C_n) = \vec{\pi}(C_n) = \lceil \frac{1}{2} \lfloor \frac{n^2}{4} \rfloor \rceil$, for $n \geq 3$, [3],

$\vec{w}(\prod_{i=1}^{k} C_{n_i}) = \vec{\pi}(\prod_{i=1}^{k} C_{n_i}) = \frac{n^{k+1}}{8}$, for $n_1 = n_2 = ... = n_k = n, k \geq 2, n$ is even, [2],

$\vec{w}(\prod_{i=1}^{k} K_{n_i}) = \vec{\pi}(\prod_{i=1}^{k} K_{n_i}) = n_1 n_2 ... n_{k-1}$, if $n_1 \leq n_2 \leq ... \leq n_k$, [2],

$\vec{w}(T) = \vec{\pi}(T)$, for all trees $T$ and the colouring can be found in polynomial time [7].

In [1] there was shown that if $G$ is $(k, w)$–colorable, then $K_n[G]$ is $(n, \max\{k, n\}w)$–colorable. Using this result the authors of [1] independently showed the above equality for $\prod_{i=1}^{k} K_{n_i}$, $\prod_{i=1}^{k} C_3$, $\prod_{i=1}^{k} C_4$, and for some recursive circulant graphs and hypercube of cliques:

$\vec{w}(G(2^m, 2) = \vec{\pi}(G(2^m, 2)) = 2^{m-2}$,

$\vec{w}(G(2^m, 4) = \vec{\pi}(G(2^m, 4)) = 2^{m-1}$,

$\vec{w}(G(3^m, 3) = \vec{\pi}(G(3^m, 3)) = 3^{m-1}$,

$\vec{w}(G(4^m, 4) = \vec{\pi}(G(4^m, 4)) = 2^{2m-1}$,

$\vec{w}(K_2^{\lfloor \log k \rfloor}[Q_{k-\lfloor \log k \rfloor}]) = \vec{\pi}(K_2^{\lfloor \log k \rfloor}[Q_{k-\lfloor \log k \rfloor}]) = 2^{k-1}$.

See also the survey paper [3].

In our paper we prove

$\vec{w}(C_n \times C_n) = \vec{\pi}(C_n \times C_n) = \frac{n(n^2-1)}{8}$, for odd $n \geq 3$,

$\vec{w}(C_m \times C_n) = \vec{\pi}(C_m \times C_n) = \lceil \frac{m}{2} \lfloor \frac{n}{2} \rfloor \lceil \frac{n}{2} \rceil \rceil$, for $m = 3, 4$ and $n \geq m$,

$\vec{w}(C_m \times C_n) = \vec{\pi}(C_m \times C_n) = \lceil \frac{m}{2} \lfloor \frac{n}{2} \rfloor \lceil \frac{n}{2} \rceil \rceil$, for $n \geq 2m, m \geq 2$,

$\vec{w}(P_n \times P_n) = \vec{\pi}(P_n \times P_n) = n \lfloor \frac{n}{2} \rfloor \lceil \frac{n}{2} \rceil$, for arbitrary $n \geq 2$,

$\vec{w}(P_m \times P_n) = \vec{\pi}(P_m \times P_n) = m \lfloor \frac{n}{2} \rfloor \lceil \frac{n}{2} \rceil$, for $2 \leq m \leq 4$, and $n \geq m$,

$\vec{w}(P_m \times P_n) = \vec{\pi}(P_m \times P_n) = m \lfloor \frac{n}{2} \rfloor \lceil \frac{n}{2} \rceil$, for $n \geq 2m, m \geq 2$,

$\vec{w}(\prod_{i=1}^{k} P_{n_i}) = \vec{\pi}(\prod_{i=1}^{k} P_{n_i}) = \frac{n^{k+1}}{4}$, for $n_1 = n_2 = ... = n_k = n, k \geq 2, n$ is even.

## 2 Optimal Numbers of Wavelengths

Our main result completes the wavelength problem for 2-dimensional square toroidal meshes $C_n \times C_n$. The even case was proved by Beauquier [2]. The odd case is mentioned without proof in [6] but in the meantime an error has been found in the proof [5].

**Theorem 2.1** *For every odd $n \geq 3$*

$$\overrightarrow{w}(C_n \times C_n) = \overrightarrow{\pi}(C_n \times C_n) = \frac{n(n^2 - 1)}{8}.$$

**Proof.** Lower bound. It is mentioned in [2] that $\overrightarrow{\pi}(C_n \times C_n) = n(n^2 - 1)/8$, for $n$ odd.

Upper bound. The basic idea of the upper bound is a partition of all dipaths into $n(n^2 - 1)/8$ blocks in such a way that each block consists of a set of arc disjoint dipaths. Thus all members of a block can use the same colour.

Assume that the vertices of $C_n$ are labelled by $0, 1, 2, ..., n - 1$ in a natural way. This implies a labelling of $C_n \times C_n$. Let diagonal $D_k$, for $k = 0, 1, 2, ..., n - 1$, consists of all nodes $(i, j)$ in $C_n \times C_n$ with $i + j = k \mod n$. Similarly let $D_k^T$ be the diagonal consisting of all nodes $(i, j)$ with $i - j = k \mod n$. First we describe dipaths between every ordered pair of vertices. Labels of vertices are calculated mod $n$.

For a given $0 < v \leq \lfloor n/2 \rfloor$ and a given $0 \leq k < n$ we connect all nodes $(x, y) \in D_k$ bidirectionally to $(x - v, y)$ and $(x, y - v)$ using shortest dipaths. This set of dipaths forms the block $B_k(0, v)$.

For given $0 < u < v \leq \lfloor n/2 \rfloor$ and given $0 \leq k < n$ we connect all nodes $(x, y) \in D_k$ bidirectionally to $(x - u, y - v)$ through $(x - u, y)$ and to $(x - v, y - u)$ through $(x, y - u)$, respectively, using shortest dipaths. This set of dipaths forms the block $B_k(u, v)$.

For given $0 < v < u \leq \lceil n/2 \rceil$ and given $0 \leq k < n$ we connect all nodes $(x, y) \in D_k^T$ bidirectionally to $(x - u, y + v)$ through $(x, y + v)$ and to $(x - v, y + u)$ through $(x - v, y)$, respectively, using shortest dipaths. This set of dipaths forms the block $B_k^T(u, v)$.

For $u = v > \lfloor n/2 \rfloor/2$ and given $0 \leq k < n$ we associate for every node $(x, y) \in D_k$ four shortest dipaths as follows:
from $(x, y)$ to $(x - u, y - u)$ through $(x, y - u)$,
from $(x - u, y - u)$ to $(x, y)$ through $(x - u, y)$,
from $(x - u, y)$ to $(x, y - u)$ through $(x - u, y - u)$, and
from $(x, y - u)$ to $(x - u, y)$ through $(x, y)$.
This set of dipaths forms the block $B_k(u, u)$.

For $u = v \leq \lfloor n/2 \rfloor/2$ and given $0 \leq k < n$ we associate for every node $(x, y) \in D_k^T$ four shortest dipaths as follows:
from $(x, y)$ to $(x - u, y + u)$ through $(x, y + u)$,
from $(x - u, y + u)$ to $(x, y)$ through $(x - u, y)$,
from $(x - u, y)$ to $(x, y + u)$ through $(x - u, y + u)$, and
from $(x, y + u)$ to $(x - u, y)$ through $(x, y)$.
This set of dipaths forms the block $B_k^T(u, u)$.

It is easy to see that the dipaths of all blocks form all to all connections in $C_n \times C_n$. Moreover, directed paths belonging to the same block are arc disjoint. Note that the number of blocks is $n(n^2 - 1)/4$. Our final step is a pairing of arc disjoint blocks and colouring all paths in a pair of blocks with the same colour which will imply the desired result. The pairing rules are:

For every $k$, every $u, v$, $0 \le u < v \le \lfloor n/2 \rfloor$ such that $u + v \le \lfloor n/2 \rfloor$ pair the block $B_k(u, v)$ with the block $B_{k-u-v}(\lceil n/2 \rceil - v, \lfloor n/2 \rfloor - u)$.

For every $k$, every $u, v$, $0 < v \le u \le \lfloor n/2 \rfloor$ such that $u + v \le \lfloor n/2 \rfloor$ pair the block $B_k^T(u, v)$ with the block $B_{k-u-v}^T(\lceil n/2 \rceil - v, \lfloor n/2 \rfloor - u)$.

Observe that the above pairing rules assure that the paired blocks are arc disjoint. □

Before proving our next theorem we need an optimal colouring of dipaths in $C_n$, for $n = 2 \mod 4$, satisfying a special property. Note that the optimal colouring of dipaths in $C_n$ proposed by Bermond et al. [6] does not satisfy the property.

**Lemma 2.1** For $n = 2$ (mod 4) there exists a colouring of dipaths in $C_n$ such that $\overrightarrow{w}(C_n) = \lceil \lfloor n^2/4 \rfloor / 2 \rceil$ and there exists one colour that is only used to colour 2 clockwise oriented dipaths between 2 opposite vertices.

**Proof.** First we design a colouring of all dipaths of lengths smaller than $n/2$. Pick 4 vertices $u, v, u', v'$ on $C_n$ in clockwise order such that $dist(u, u') = dist(v, v') = n/2$ and $dist(u, v) < dist(v, u')$. These 4 vertices form 4 clockwise and 4 anticlockwise dipaths. We colour them with 1 colour. There are $n/2$ choices for $u$ and $u'$ and $\lfloor n/4 \rfloor$ choices for $v$ and $v'$. Thus we obtain a colouring of all dipaths in $C_n$ of lengths smaller than $n/2$ and use $n \lfloor n/4 \rfloor / 2$ colours.

Now we describe a colouring of dipaths of the length $n/2$. Pick 2 opposite vertices $u$ and $u'$ in $C_n$. There are $n/2$ choices for each pair. Connect $\lfloor n/4 \rfloor$ of pairs clockwise and the rest anticlockwise. Thus we need $\lfloor n/4 \rfloor$ colours (used in both directions) and 1 colour only used in one direction (in this case anticlockwise).

We have used altogether $n \lfloor n/4 \rfloor / 2 + \lfloor n/4 \rfloor + 1$ colours which matches $\lceil \lfloor n^2/4 \rfloor / 2 \rceil$ for $n = 2 \mod 4$. □

**Lemma 2.2** Assume that an $n$-vertex graph $G$ satisfies $2\overrightarrow{w}(G) \ge n$, and at most 2 dipath of the same colour end in a vertex. Then

$$\overrightarrow{w}(P_2 \times G) \le 2\overrightarrow{w}(G).$$

**Proof.** Consider an optimal colouring of dipaths in $G$ with $m = \overrightarrow{w}(G)$ colours $c_1, c_2, ..., c_m$. Take a new set of $m$ corresponding colours $c_1', c_2', ..., c_m'$. First we design a new colouring of $G$ using at most $2m$ colours. If two dipaths of the same colour $c_i$ ended in a vertex then we recolour one of them by the colour $c_i'$.

Let $V(P_2) = \{0, 1\}$ and $V(G) = \{v_1, v_2, ..., v_n\}$. Now we design dipath in $P_2 \times G$ in a greedy way, i.e. a dipath from $(u, v_i)$ to $(w, v_j)$ first "follows" the dipath from $v_i$ to $v_j$ in $G$ and then the dipath from $u$ to $w$ in $P_2$ (if such dipaths exist).

Finally, we design a colouring of dipaths in $G$ by means of the new colouring of $G$. The colouring of the dipath from $(0, v_i)$ to $(0, v_j)$ (from $(1, v_i)$ to $(1, v_j)$) is induced by the new colouring of dipaths in $G$. Consider a dipath from $(0, v_i)$ to $(1, v_j), i \ne j$. Let $c$ be a colour of the dipath from $v_i$ to $v_j$ in the new colouring of

$G$. Then colour the dipath from $(0, v_i)$ to $(1, v_j)$ by the colour $c'$ (we assume that $(c')' = c$). Similarly, we colour dipaths from $(1, v_i)$ to $(0, v_j), i \neq j$. It remains to colour the dipaths between $(0, v_i)$ and $(1, v_i)$. Wlog consider the dipath from $(0, v_i)$ to $(1, v_i)$. Observe that the dipaths, coloured so far, containing the arc $((0, v_i)(1, v_i))$ used $n - 1 < 2m$ colours, which finishes the proof. $\qquad \square$

**Theorem 2.2**

(i) For $n \geq 3$

$$\overrightarrow{w}(P_2 \times C_n) = \overrightarrow{\pi}(P_2 \times C_n) = \begin{cases} 3, & \text{for } n = 3, \\ \lfloor \frac{n}{2} \rfloor \lceil \frac{n}{2} \rceil, & \text{for } n > 3. \end{cases}$$

(ii) For $3 \leq m \leq 4, m \leq n$

$$\overrightarrow{w}(C_m \times C_n) = \overrightarrow{\pi}(C_m \times C_n) = \left\lceil \frac{m}{2} \left\lfloor \frac{n}{2} \right\rfloor \left\lceil \frac{n}{2} \right\rceil \right\rceil.$$

(iii) For $2 \leq m, 2m \leq n$

$$\overrightarrow{w}(C_m \times C_n) = \overrightarrow{\pi}(C_m \times C_n) = \left\lceil \frac{m}{2} \left\lfloor \frac{n}{2} \right\rfloor \left\lceil \frac{n}{2} \right\rceil \right\rceil.$$

**Proof.** (Sketch). The first result. The lower bound follows from a simple observation that $\overrightarrow{w}(P_2 \times C_n) \geq \max\{n, \lfloor n/2 \rfloor \lceil n/2 \rceil\}$. Case $n = 3$ is an easy exercise. For $n > 3$ and $n \neq 2 \bmod 4$ we simply combine Lemma 2.2 with the result of Bermond et al. [6] $\overrightarrow{w}(C_n) = \lceil \lfloor n/2 \rfloor \lceil n/2 \rceil / 2 \rceil$. For $n = 2 \pmod 4$ we use the colouring of $C_n$ from Lemma 2.1. By a similar arguments as in the proof of Lemma 2.2 we can prove

$$\overrightarrow{w}(P_2 \times C_n) \leq 2\overrightarrow{w}(C_n) - 1 = 2\lceil \lfloor n/2 \rfloor \lceil n/2 \rceil / 2 \rceil - 1 = \lfloor n/2 \rfloor \lceil n/2 \rceil.$$

The second result. In case $m = 3$ and $n \neq 2 \pmod 4$ we can prove using a similar argument as in the proof of Lemma 2.2 that

$$\overrightarrow{w}(C_3 \times C_n) \leq 3\overrightarrow{w}(C_n) = 3\lceil \lfloor n/2 \rfloor \lceil n/2 \rceil / 2 \rceil = \lceil 3\lfloor n/2 \rfloor \lceil n/2 \rceil / 2 \rceil.$$

In case $m = 3$ and $n = 2 \pmod 4$ we use the colouring of $C_n$ from Lemma 2.1 and by similar arguments as in the proof of Lemma 2.2 we show that

$$\overrightarrow{w}(C_3 \times C_n) \leq 3\overrightarrow{w}(C_n) - 1 = 3\lceil \lfloor n/2 \rfloor \lceil n/2 \rceil / 2 \rceil - 1 = \lceil 3\lfloor n/2 \rfloor \lceil n/2 \rceil / 2 \rceil.$$

In case $m = 4$ we combine the first result with Lemma 2.2 and the identity $C_4 \times C_n = P_2 \times (P_2 \times C_n)$.

The third result. Firstly we construct an optimal colouring for $n$ even. The colouring is based on a colouring of a ring of length $n$ similar to the one described in Lemma 2.1. Let the colouring is used for the first row of $C_m \times C_n$ which is a ring of length $n$. Then the same scheme of colouring shifted by $i - 1$ vertices to the right will be used for the $i$-th row. For each colour $c$ used in the ring by a dipath from a vertex $i$ to a vertex $j$ we create $m$ colours for $m$ dipaths from the vertex $i$ to $m$ vertices in the $j$-th column of $C_m \times C_n$. The details of the construction will appear in the journal version of the paper. For $n$ odd our optimal colouring is based on the colouring of a ring of length $n$ which we get from the optimal colouring of the ring of length $n + 1$ by removing of a vertex and by removing of the dipaths of length $(n + 1)/2$, shrinking the dipaths containing the vertex as an internal vertex and creating new dipaths (not longer than $\lfloor n/2 \rfloor$) from two dipaths of the same colour where one dipath ends in the vertex and the another starts in this vertex. □

**Lemma 2.3** *Let $R$ be a routing of shortest dipaths for $C_k$ such that any colour $c$ is used exclusively by either dipaths going in clockwise or counterclockwise direction. Then*

$$\overrightarrow{w}(\mathcal{P}_k) \le 2\overrightarrow{w}(C_k, R).$$

**Proof.** Consider $m = \overrightarrow{w}(C_k, R)$ colours $c_1, c_2, ..., c_m$ used to colour dipaths in $R$. Take $m$ new corresponding colours $c_1', c_2', ...c_m'$. Clearly $\mathcal{P}_k$ can be obtained from $C_k$ by deleting wrap-around arcs. Consider a dipath $p$ in $C_k$ coloured by a colour $c$ and starting in a vertex $u$ and ending in a vertex $v$. Assume that $p$ contains a wrap-around arc. Let $p'$ be the longest subdipath of $p$ containing the wrap-around arc such that it uses arcs of one dimension only. Let $p'$ starts in $u'$ and ends in $v'$. Delete the dipath $p'$ from $p$ and connect $u'$ to $v'$ by the shortest dipath in $\mathcal{P}_k$. We get a new dipath from $u$ to $v$. Colour this dipath by $c'$. If $p$ does not contain a wrap-around arc then we use it without any changes. One can see that by repeating of the above procedure for all dipaths we get a routing for $\mathcal{P}_k$ coloured by $2m$ colours. □

**Corollary 2.1**

(i) Let $k \ge 2$ and $n_1 = n_2 = ... = n_k = n \ge 4$ be an even number, then

$$\overrightarrow{w}(\mathcal{P}_k) = \overrightarrow{\pi}(\mathcal{P}_k) = \frac{n^{k+1}}{4}.$$

(ii) For arbitrary $n \ge 2$

$$\overrightarrow{w}(P_n \times P_n) = \overrightarrow{\pi}(P_n \times P_n) = n \left\lfloor \frac{n}{2} \right\rfloor \left\lceil \frac{n}{2} \right\rceil.$$

(iii) For $2 \le m \le 4, m \le n$

$$\overrightarrow{w}(P_m \times P_n) = \overrightarrow{\pi}(P_m \times P_n) = m \left\lfloor \frac{n}{2} \right\rfloor \left\lceil \frac{n}{2} \right\rceil$$

(iv) For $2 \le m, 2m \le n$

$$\overrightarrow{w}(P_m \times P_n) = \overrightarrow{\pi}(P_m \times P_n) = m \left\lfloor \frac{n}{2} \right\rfloor \left\lceil \frac{n}{2} \right\rceil$$

**Proof.** The upper bound for (i) follows from Lemma 2.3 and a result of Beauquier [2]. He proved that $\vec{w}(C_k) = n^{k+1}/8$ for even $n$. Moreover his routing uses shortest dipaths only. For (ii), (iii) and (iv) we combine Theorem 2.2 with Lemmas 2.2 and 2.3. □

**Note:** The same result as it is in Corollary 2.1 (i), was shown by Beauquier and Paulraja [4].

# References

1. Amar, D., Raspaud, A., Togni, O., Total exchange in the optical networks created by composition of cliques. Manuscript, September 1996.
2. Beauquier, B., All-to-all communication for some wavelength-routed all-optical networks, submitted to *Networks*.
3. Beauquier, B., Bermond, J-C., Gargano, L., Hell, P., Perennes, S., Vaccaro, U., Graph problems arising from wavelength routing in all optical networks, to appear in: *Proc. of WOCS'97*, 1997.
4. Beauquier, B., Paulraja, P. personal communication, 1997.
5. Bermond, J.-C., personal communication, 1997.
6. Bermond, J.-C., Gargano, L., Perennes, S., Rescigno, A., Vaccaro, U., Efficient collective communications in optical networks, in: *Proc. of 23rd Intl. Colloquium on Automata Languaages and Programming*, Lecture Notes in Computer Science 1099, Springer Verlag, Berlin, 1996.
7. Gargano, L., Hell, P., Perennes, S., Coloring all directed paths in symmetric tree, Manuscript, 1996.
8. Leighton, F. T., Introduction to Parallel Algorithms and Architectures. Morgan Kaufmann Publishers, Inc.,1992.

# Parallelizing Self-Organizing Maps

David Štrupl[1] and Roman Neruda[2] *

[1] Faculty of Mathematics and Physics
Malostranské nám. 25, Prague
email: strupl@kti.ms.mff.cuni.cz
[2] Institute of Computer Science, ASCR
P.O. Box 5, 182 07, Prague
email: roman@uivt.cas.cz

**Abstract.** Several ways of parallelizing the self-organizing network (Kohonen maps) are studied on the BSP-like parallel machine model. Optimal number of processors and criteria for choosing the right task decomposition are presented. Theoretical results are verified in the PVM environment on a cluster of workstations.

## 1 Introduction

Although neural network is considered to be a highly parallel device it is often simulated on sequential machines. Thus, quite naturally, there arises a question of whether it is efficient to make these simulations parallel. Neural networks consist of large number of relatively simple computational units and connections providing information exchange among these units. On the contrary, a parallel computer—as seen from todays point of view—is a device consisting of smaller number of quite powerful processors connected by a communication network. While the processors in a parallel machine are relatively powerful, their communication is usually slow.

Two different approaches to parallel implementation of a neural network learning are described in literature. The first one stays at some level of abstraction (as in [2]), assumes an arbitrary neural net (taken from a wide range of models) of a given size and estimates a hardware configuration required to run such a simulation. The second (more common) approach is limited to implementation of a particular algorithm (e.g. SOM) on a given hardware configuration ([6], [5]).

In our paper we discuss possible ways of distributing parts of the task considering a computing model that is general enough for the results to hold on different hardware architectures. Parallel run time is estimated for each task decomposition. Having the task size and parameters of the parallel machine model (especially the speed of interprocessor communication) we are able to supply two following results: the number of processors needed to compute the given task as

* This research was partially supported by GAASCR under grant no. A2030602, and by GACR under grant no. 201/95/0976.

fast as possible, and the way of dividing the load among these processors. Theoretical results are supported by a implementation of the parallel SOM algorithm in the PVM environment on the cluster of workstations.

The structure of the paper is as follows: in section 2 we briefly introduce Kohonen SOM model and BSP model. Main theoretical results are presented in section 3. Last section deals with implementation and experimental results.

## 2 Preliminaries

The *Self-Organizing Map (SOM)* was introduced by Kohonen in 1981 ([3]) — it is sometimes referred to as a Kohonen map. The structure of the Kohonen map can be illustrated on the figure 1.

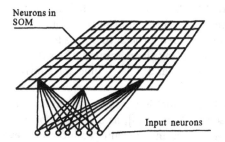

**Fig. 1.** Scheme of the Self-Organizing Map

Consider a layer of neurons (units) that are connected to input via set of synapses with real-valued parameters (weights) $w_i$. These neurons are ordered in some topological structure, such as two-dimensional mesh (or a ring, 3-D mesh, 2-D hexagonal mesh, etc). This structure defines a neighborhood $N(g)$ of neuron $g$, which is important during the learning phase.

Network learning is a typical unsupervised competition algorithm, in which we cycle through the training set of patterns and for each pattern $x_t$ do the following: a neuron $c$ closest to the $x_t$ is determined and it's weights are modified to better represent this pattern (i.e. the neuron is shifted in the input space towards the pattern). Moreover, all neurons belonging to $N(c)$ are shifted as well:

$$w_j = \begin{cases} w_j + \theta(x_t - w_j) & j \in N(c) \\ w_j & \text{otherwise,} \end{cases}$$

where $c = \arg\min\{||x_t - w_l||; l = 1, \ldots, n\}$. Parameter $0 \le \theta \le 1$ called *learning rate* is usually set large (say 0.9) at the beginning and decreases to zero during the learning. The size of the neighborhood also decreases from the initial quite large value (encompassing about one half of the network) to the smallest possible value, meaning that only the winner unit is shifted at the end.

As usual with the gradient-based learning algorithms there are two learning variants with respect to weight updates. If the updates are performed after each step we get the *on-line* learning, while with the *off-line* version we accumulate the updates and the changes are applied after the whole training set is exhausted.

Time complexity of the sequential algorithm is an important measure for comparison with the time complexity achieved by a parallel program. Assume that we charge zero time for initialization and also for control instructions of all the loops. This particular assumption is important to keep our estimates simple. This leads to the following formula describing the sequential time of one iteration of SOM:

$$T_S = ntdT^{SQ} + ntT^R + N_{BRS}tT^{SMA} \tag{1}$$

where $n$ is number of neurons, $t$ number of patterns in the training set, $d$ is dimension of the input space. The first term is time for computing the distances of the neurons from the patterns — $T^{SQ}$ being the time of subtracting and squaring. The second part represents the time for computing the distances by taking the square root ($T^R$ being the actual time for square root operation). The last term is time for computing the neurons shifts with $N_{BRS}$ being the number of neighbors of each neuron and $T^{SMA}$ time for subtracting, multiplying and adding numbers.

Our parallel machine model is based on the *Bulk-Synchronous Parallel Machine (BSP)* proposed by Valiant [4], as a possible bridging model between the programming language level and machine architecture level in parallel computation. It is an abstract model described by small number of parameters and it is to high extend "hardware independent".

A parallel machine (computer) consists of $p$ sequential processors (with local memory). These processors act independently and are mutually connected by communication network of unspecified type. The network is characterized by a parameter $g$ that stands for a ratio of the number of local computational operations performed per second by all the processors to the total number of data units (float numbers) delivered by the network per second. This is a measure of the interprocessor communication speed which is in most cases the crucial factor of parallel computation.

The network comprise a device that is capable of global synchronization. Thus, another parameter characterizing the network is the minimal time $\bar{L}$ (measured in local computational steps) between successive synchronization operations. If the synchronization facility is fast enough the parallel run time analysis can be done without considering this parameter. In our paper the synchronization analysis is omitted and the ratio between computation and communication is emphasized. A new parameter $c$ is considered in our estimates. It stands for the time (in local computational steps) needed to transfer a number from one processor to another. The parameter $c$ is not a basic parameter of the BSP model but it is an intuitive one and can be easily computed from basic BSP parameters as $c = g/p$. Additionally, we use $C$ to denote the time of transmitting one floating point number — e.g. $C = 4c$.

# 3 Parallel learning

The task of specifying a parallel algorithm consists in dividing the steps of the sequential algorithm among available processors in such a way that the overhead for communication will be as small as possible (to reach maximal speedup). We will concentrate on the off-line version of the SOM algorithm.

## 3.1 Partitioning the neural net

The first obvious approach to parallel implementation of SOM is to assign a subset of neurons to each processor. This approach introduces communication overhead due to the fact that for every input pattern we seek a neuron with minimal distance.

Our parallel algorithm works as follows: every processor computes distances and its own winning neurons. These neurons are sent to the master processor and the overall winners are computed on the master processor. The master process then broadcasts the winners to all others. After this, the appropriate neurons are shifted towards their patterns according to the SOM algorithm. Another problem arises here: the neighbors of the winning neuron have to be shifted as well. But those neighbors can dwell on another processor — more communication is required. Shift requests have to be sent potentially from every processor to every another processor. This can be time consuming especially in the case of more complicated neural net topology.

We state the discussed communication overhead more precisely in terms of our basic operations. The total time for one iteration ($T_N$) is a sum of three independent parts: time for initial data distribution among processors, computing time and time for gathering the resulting neurons locations. Since large number of iterations are typically performed the distribution and gathering times can be abandoned:

$$T_N = \frac{T_S}{p} + Cpt + ptT^C + Cpt + CpN_{REQ}, \tag{2}$$

where $T^C$ denotes time needed to compare two numbers.

The first term shows that we speed up the computation proportionally with the number of processors. The rest of the equation reflects the communication overhead. The time $Cpt$ is needed twice — both for getting the distances of local winning neurons to all patterns in the training set and for distributing back the information about global winners. Then there is the time $ptT^C$ spent on the master processor comparing the respective winners. And the last term represents the time taken by communicating shifting requests among processors. This time depends on number of shifting requests $N_{REQ}$. This number varies and depends heavily on the task computed.

Having defined the total time depending on the values $p, C, n, t, d$ our main goal is to determine the optimal number of processors, i. e. the number for which the total time is minimal.

**Theorem 1.** *Optimal number of processors $p$ for partitioning the network with the above described parameters $t, C, n, d$ is:*

$$p = \sqrt{\frac{n(dT^{SQ} + T^R) + N_{BRS}T^{SMA}}{2C + T^C + CN_{REQ}/t}} \tag{3}$$

*If the computed number of processors is less than 2 it makes no sense to use a parallel program of this kind.*

*Proof.* To get the minimal time needed for the computation we take the time as a continuous function of processors $T_N(p)$ and determine the derivative:

$$\frac{\partial T_N}{\partial p} = -\frac{T_S}{p^2} + 2Ct + tT^C + CN_{REQ}. \tag{4}$$

Making it equal to 0 we find a local extreme of $T_N(p)$. This leads to the following value of $p^2$:

$$p^2 = \frac{ndtT^{SQ} + tnT^R + tN_{BRS}T^{SMA}}{2Ct + tT^C + CN_{REQ}}.$$

As we are interested only in positive values of $p$ (it is still number of processors) we can take (3) as the only solution.

## 3.2 Distributing the training set

The other possible way of SOM parallelizing is splitting the training set. Every processor gets a part of it and an independent copy of the neural net, computes an iteration over its part of training set. At the end of the iteration the resulting neuron shifts are sent to the master processor. Here the new location of neurons is computed and then sent back to all worker processes. We decided to combine the resulting neurons shifts in terms of simply getting the arithmetic mean of the vectors obtained by individual worker processes.

In this approach we can similarly try to estimate the time needed for one iteration over the training set. The communication overhead in this case is much easier to compute because it does not depend on the neural net topology nor on any other hardly predictable parameter (e.g. $N_{REQ}$ from the previous section).

The time for one iteration also consist of three parts: time for distributing the input data, time for computing and time for gathering the results. Again, we will not count the initial distribution and final gathering and put immediately:

$$T_T = \frac{T_S}{p} + Cndp + ndT^A p + Cndp, \tag{5}$$

where the first term is the speedup obtained by $p$ processors and the rest is the communication and synchronization overhead. Time $Cndp$ is needed by master to collect the partial results, the next term is time for computing new positions of neurons by taking their arithmetic mean ($T^A$ being the addition time) and the last term is for distributing these global results.

**Theorem 2.** *Optimal number of processors p for partitioning the training set with the above described parameters $t, C, n, d$ is:*

$$p = \sqrt{\frac{t(dT^{SQ} + T^R + N_{BRS}T^{SMA}/n)}{d(2C + T^A)}} \tag{6}$$

*If $p < 2$, the optimal number of processors is 1 (i.e. the sequential algorithm is faster).*

*Proof.* Again, to get the minimal time needed for the computation we take the time as a continuous function of processors $T_T(p)$ and determine the derivative:

$$\frac{\partial T_T}{\partial p} = -\frac{T_S}{p^2} + 2Cnd + ndT^A \tag{7}$$

Making it equal to 0 we find a local extreme of $T_T(p)$:

$$p = \sqrt{\frac{T_S}{2Cnd + nT^A}}$$

### 3.3 Comparing the types of partitioning

An interesting question arises which of the above proposed ways of division performs better considering the parameters of the machine. The comparison leads to the following inequation: $T_T < T_N$. Solving this inequation for the parameter $p$ we obtain this result:

**Theorem 3.** *Using the same number of processors $p$ the algorithm partitioning the training set is faster if*

$$nd(2C + T^A) \le t(2C + T^C + N_{BRS}T^{SMA}/p + CN_{REQ}/t) \tag{8}$$

The inequation stated in the theorem usually holds in practical applications. This is due to the fact that in common tasks the size of the training set is usually much larger than the size of the neural net. Thus, we can conclude that is is often better to use the dividing training set approach.

## 4 Experiments

Experiments using PVM (Parallel Virtual Machine) have been carried out to verify our theoretical results. PVM is a software library and runtime support system for parallel programming on a cluster of workstations or a parallel machine ([1]). There is one master process coordinating the whole computation. This process reads the input data, creates other processes, coordinates the work and at the end gathers the results. A corresponding part of the neural net and the training set is assigned to every worker process running on different host machines.

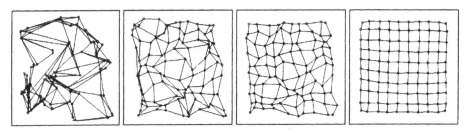

**Fig. 2.** Illustration of the convergence of the parallel SOM

The set of experiments was performed to compare the running times of the parallel programs with theoretical estimates. The main goal was to verify that the parallel solution leads to lower times of overall learning. The running times of 10 iterations of the Task 1 ($n = 100$, $t = 10000$, dimension of input space: 2,neural net topology: 2 dimensional mesh) and Task 2 ($n = 100$, $t = 10000$, dimension of input space: 15, topology: ring of neurons) are shown in table 1 and figure 3. The task sizes are chosen to reflect task sizes usually used in practical applications.

To easily verify the correct performance of learning algorithms we have chosen training patterns from the uniform distribution over a hypercube. Figure 4 illustrates the convergence of the network in four consecutive steps during the learning. These familiar pictures correspond to the classical experiments described in [3, pages 82, 140].

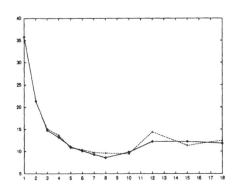

**Fig. 3.** Parallel run times depending on the number of processors

## 5 Conclusions

We have shown the conditions under which it makes sense to parallelize the SOM algorithm. Both theoretical and practical results show that for common task sizes

| Task \ processors | 1 | 2 | 3 | 4 | 5 | 6 |
|---|---|---|---|---|---|---|
| $T_{N,1}$ | 35.31 | 130.64 | 117.24 | 118.84 | 113.60 | 112.52 |
| $T_{N,1}$ | 37.51 | 127.64 | 108.71 | 122.46 | 117.30 | 119.02 |
| $T_{T,1}$ | 35.81 | 21.26 | 15.11 | 13.60 | 10.82 | 10.40 |
| $T_{T,1}$ | 35.78 | 21.40 | 14.75 | 13.13 | 11.06 | 10.11 |
| $T_{N,2}$ | 146.67 | 188.14 | 212.52 | 160.09 | 228.02 | 142.08 |
| $T_{T,2}$ | 145.71 | 77.37 | 115.72 | 88.02 | 69.30 | 60.04 |

| Task \ processors | 7 | 8 | 10 | 12 | 15 | 18 |
|---|---|---|---|---|---|---|
| $T_{N,1}$ | 127.38 | 125.58 | 135.70 | 131.88 | 157.29 | 160.01 |
| $T_{N,1}$ | 116.42 | 120.39 | 132.11 | 127.00 | 145.29 | 223.01 |
| $T_{T,1}$ | 9.72 | 9.61 | 9.46 | 14.37 | 11.31 | 12.43 |
| $T_{T,1}$ | 9.30 | 8.60 | 9.82 | 12.21 | 12.18 | 11.79 |
| $T_{N,2}$ | 197.25 | 152.20 | 180.28 | 178.11 | 228.34 | 371.18 |
| $T_{T,2}$ | 49.54 | 47.37 | 30.53 | 24.98 | 29.23 | 23.50 |

**Table 1.** Running times of parallel computations

it is better to partition the training set. Theorem 2 shows that optimal number of processors needed for parallel computation of SOM simulation depends on the square root of the size of the training set while using the training set partitioning approach. On the other hand the Theorem 1 indicates that in the case of splitting the neural net the optimal number of processors depends on the square root of the network size. From these facts we conclude the superiority of the former approach since the size of the training set in the common tasks is bigger than the number of neurons in the network.

# References

1. Geist A., Beguelin A., Dongarra J., Jiang W., Manchek R., Sunderam V.: PVM 3 User's Guide and Reference Manual, 1994.
2. Ghosh J., Hwang K.: Mapping Neural Networks onto Message-Passing Multicomputers. Journal of Parallel and Distributed Computing 6, pp. 291-330, 1989.
3. Kohonen T.: Self-Organizing Maps Springer-Verlag, Berlin, 1995.
4. Valiant L. G.: A Bridging Model for Parallel Computation. Communications of ACM, 33(8), 103–111, 1990.
5. N. Vassilas and P. Thiran and P. Ienne: How to modify Kohonen self-organizing feature maps for an efficient digital parallel implementation. Fourth International Conference on Artificial Neural Networks (Conf. Publ. No.409), 86–91, IEE, 1995.
6. V. Demian and J.-C. Mignot: Implementation of the self-organizing feature map on parallel computers. Computers and Artificial Intelligence, 14(1), 63–80, 1996.

# Author Index

* invited speaker

# Springer
# and the
# environment

At Springer we firmly believe that an international science publisher has a special obligation to the environment, and our corporate policies consistently reflect this conviction.

We also expect our business partners – paper mills, printers, packaging manufacturers, etc. – to commit themselves to using materials and production processes that do not harm the environment. The paper in this book is made from low- or no-chlorine pulp and is acid free, in conformance with international standards for paper permanency.

# Lecture Notes in Computer Science

For information about Vols. 1–1265

please contact your bookseller or Springer-Verlag